Quantitative Solutions in HYDROGEOLOGY and GROUNDWATER MODELING

Quantitative Solutions in

HYDROGEOLOGY and GROUNDWATER MODELING

Neven Kresic

LAW
ENGINEERING AND ENVIRONMENTAL SERVICES, INC.

CRC

LEWIS PUBLISHERS

Boca Raton New York

Acquiring Editor: Ken McCombs
Project Editor: Joan Moscrop
Direct Marketing Manager: Arline Massey
Cover design: Denise Craig
Manufacturing: Sheri Schwartz

Library of Congress Cataloging-in-Publication Data

Krešić, Neven.
Quantitative solutions in hydrogeology and groundwater modeling / Neven Kresic
p. cm.
Includes bibliographical references and index.
ISBN 0-56670-219-4
1. Hydrogeology—Mathematics. 2. Hydrogeology—Mathematical models. I. Title
GB1001.72.M7K74 1997
551.49′01′5118—dc20 96-54297
CIP

International Standard Book Number 0-56670-219-4
Library of Congress Card Number 96-54297
Printed in the United States of America 1 2 3 4 5 6 7 8 9 0
Printed on acid-free paper

To my wife Joanne
for all her love and help

and to the memory of
Senator Fulbright

PREFACE

Soon after submerging myself in the study of groundwater, I noticed a very definite lack of literature that could thoroughly explain how to apply all those strange and unfriendly equations that I was beginning to learn. This frustration continued as I started consulting for industry and various agencies while facing very diverse quantitative problems. Many of these problems seemed to contain all of the unanswered questions since my school days. Eventually, I even began to enjoy the application of all kinds of equations, including those I never had the chance to see during my undergraduate studies. As a teacher, however, I often heard from my students those same frustrations that I had years ago.

This book addresses many quantitative groundwater problems that both students and groundwater professionals will encounter at some point. It also solves them in a "user-friendly" way, partly the result of my own needs – I wanted to have one condensed source that would refresh my aging memory while quickly and conveniently finding answers to the most frequently asked questions.

I would like to thank all of my colleagues at the Water Resources Group, Law Engineering and Environmental Services, for their understanding and support during the writing of this book. I am also grateful to the following people for their friendship and advice: Dr. Gerald Feder, Dr. William Back and Dr. Leonard Konikow of the United States Geological Survey, Reston, Virginia, Dr. Kenneth Johnson of the Oklahoma Geological Survey, Roger Pennifill and Tom Fowler of Foster-Wheeler Environmental Corporation, Steven Ristorcelli of Mine Development Associates, Reno, Nevada, and Dr. John Sharp and Glenn Ouzts of Austin, Texas.

Atlanta, Georgia
October 1996

ABOUT THE AUTHOR

Neven Kresic, Ph.D., is a Senior Hydrogeologist at the National Center of Law Engineering and Environmental Services, Inc. in Kennesaw, Georgia where he heads the Center for Groundwater Modeling.

Dr. Kresic received his B.S. in Hydrogeologic Engineering from Belgrade University, Yugoslavia in 1981. He obtained his M.S. and Ph.D. from Belgrade University in 1984 and 1987, respectively. After graduation he worked as a Project Hydrogeologist with Energoprojekt Holding Corporation in Belgrade, and in 1982 he joined the Department of Hydrogeology at Belgrade University as a teaching assistant.

Dr. Kresic taught courses in hydrogeology and hydrogeologic mapping, and in 1988 was appointed Professor of Groundwater Dynamics at Belgrade University.

In 1991 Dr. Kresic came to the United States as a Senior Fulbright Scholar at The George Washington University, Washington, D.C. and the National Center of the United States Geological Survey in Reston, Virginia. In 1992 he was appointed Senior Hydrogeologist with Foster-Wheeler Environmental Corporation in its headquarters in Lyndhurst, New Jersey.

Before joining his present firm, Dr. Kresic was Assistant Professor at Texas Christian University in Fort Worth, Tcxas where he taught courses in hydrogeology, groundwater modeling, and groundwater remediation.

Dr. Kresic is a Professional Hydrogeologist, a Certified Ground Water Professional, and a member of the American Institute of Hydrology, the National Ground Water Association, the International Association of Hydrogeologists (IAH), and the Karst Commission of IAH.

He has been a consultant for the environmental industry and state and federal agencies in the United States and abroad. Before leaving Yugoslavia, Dr. Kresic served as Secretary of the Yugoslav Society for the Application of Mathematical Methods and Computers in Geology and Mining, and was Editor-in-Chief of its Bulletin.

The author of more than 40 papers, two books and one monograph,. in 1988 Dr. Kresic received the Milicevic Foundation Award for the best published geological work in Yugoslavia. His current research interests are quantitative hydrogeology and groundwater modeling, particularly in karst aquifers.

TABLE OF CONTENTS

Chapter 7. Water Wells, Steady State Flow 143

Chapter 8. Water Wells, Transient Flow 195

Chapter 1

POROSITY AND RELATED PARAMETERS

PROBLEM 1
POROSITY, EFFECTIVE POROSITY, DENSITY, SATURATION

The volume of a moist sand specimen is 72.5 cm^3 and its weight is 152 g. The volume is determined by submerging the sample in a container with mercury and measuring the weight of the spilled mercury. After drying in the oven for 24 hours at 105 °C, the weight of the specimen is 145 g and its volume is 71.2 cm^3 (a 1.8% decrease). Determine the following for the sand sample: porosity, specific weights and densities, void ratio, moisture content, degree of saturation, effective porosity.

Note: Confusion is often present when the mass and the weight of a sample are expressed with the same unit, such as gram (g) or kilogram (kg; 1 kg = 1000 g). The unit of mass in the SI system is the kilogram, while the unit of weight is the newton (N), which is also the unit of force (weight is a force). However, in laboratory and field engineering practices the weight is measured by scales that are usually calibrated in kilograms (grams). The obtained value for weight is then equaled to the value of mass which means, for example, that a sample that has a "weight" of 152 g has a mass of 152 g in the gravitational field of the earth (weight = mass x acceleration of gravity). This means that the mass of a body (sample) is a constant quantity, while its weight will change if the acceleration of gravity changes. For practical purposes the weight will here be expressed in grams.

The porosity is given as the ratio of volume of voids to the total initial volume of the specimen before drying, which includes both voids and solids (as can be seen, a sample can change the volume upon drying in the oven):

$$n = \frac{V_v}{V} \quad (\times\ 100\%) \tag{1.1}$$

The volume of voids can be determined using the three-phase diagram approach, or a water displacement test which is used to find the *effective porosity*.

1.1 THREE-PHASE DIAGRAM

Specific Weight and Density

Weight-volume relationships of a soil sample, *i.e.*, its specific weight, density, porosity, void ratio, and saturation, can be determined using three-phase diagrams as shown in Figure 1.1. If one of the volumes or weights is unknown, others should be indicated on the diagram in terms of that unknown. The weight (moist and dry) of the sample is easily measured in the laboratory. The total volume is determined by measuring the weight of the spilled mercury of known specific weight after the sample is submerged in a container with mercury. Specific weight and gravity of the sample, sample solids and water are calculated and/or found in tables. When the specific weight of water is assumed to be unity (1 g/cm^3), then knowing the weight (in grams) of water lost during oven-drying, which is the difference between the sample weights before and after drying, gives the same number for the volume of that water in cm^3 (see Figure 1.2).

The *specific weight* of the sample (sand specimen) is given as:

$$\gamma = \frac{W}{V} = \frac{152\ g}{72.5\ cm^3} = 2.09\ g/cm^3 \tag{1.2}$$

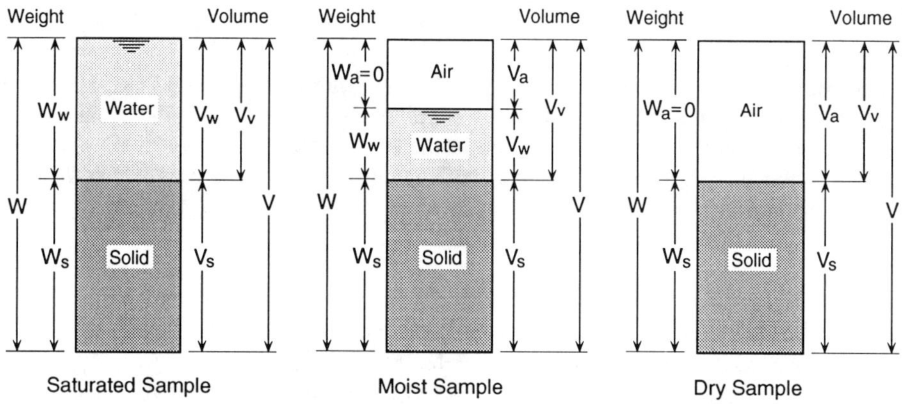

Figure 1.1 Three-phase diagrams for the samples with different degrees of saturation.

where W is the weight of the sample, *i.e.*, the weight of the specimen before oven-drying, and V is the volume of the sample. The specific weight is numerically equal to the sample *density* (ρ) if both weight and mass are expressed in the same units, for example, grams. Density is defined as the sample mass divided by the sample volume and its (physically correct) unit in the SI system is g/cm^3.

The *dry specific weight* is the weight of the sample after oven drying divided by the total initial volume of the sample before oven drying (a sample can change its volume during drying in the oven, as is the case with our sand specimen; these changes are usually small for sand and gravel, but may be significant for clayey sediments and clays). The weight of the sample after drying equals the weight of the solids or sample mineral matter (W_s).

$$\gamma_d = \frac{W_s}{V} \quad (g/cm^3) \tag{1.3}$$

The dry specific weight numerically equals the *bulk density* (ρ_b) which is defined as the mass of the oven-dried sample divided by the sample volume. The bulk density is always smaller than the overall density of a moist sample since the mass of water present in the voids is excluded from the calculation. The bulk density depends on the total pore space and varies accordingly -- the fine texture materials, such as silts and clays, have more pore space and lower bulk densities than sand and gravel.

The *specific weight of the solids* (mineral particles), which numerically equals the *particle density* (ρ_s), is the ratio of the weight of solids to the volume of the sample solids excluding pore space:

$$\gamma_s = \frac{W_s}{V_s} \quad (g/cm^3) \tag{1.4}$$

It is very difficult, if practically impossible, to measure directly the volume of solids in a soil or aquifer material sample. Rather, it is determined indirectly by assuming the particle density and using equation (3.4). Particle density of mineral soils and unconsolidated aquifer materials vary a great deal. They are not dependent on moisture content, particle size or void space, and are a material constant property. Sample particle density is an average of the densities of all present minerals. Most soils and unconsolidated aquifer materials reflect densities of the most abundant minerals -- quartz (2.65 g/cm^3) and feldspars (2.56 - 2.76 g/cm^3). Therefore, the particle density of 2.65 g/cm^3 is usually assumed unless very precise data is required.

The specific weight of the solids (particle density) can be expressed also as:

$$\gamma_s = G\gamma_w \tag{1.5}$$

where G is the *specific gravity* of the sample solids (dimensionless) and γ_w is the specific weight of water. If the specific weight of water is considered the unity (the average γ_w is 1 g/cm^3), the specific weight of solids (given in g/cm^3) equals the specific gravity the value of which can be found in tables (see Appendix A). The terms *specific weight of solids, specific gravity* and *particle density* are used as synonyms throughout literature and are numerically equal. For this problem it is assumed that the sand specimen has the specific gravity of 2.65 which gives the following value of the volume of solids using equation (4):

$$V_s = \frac{W_s}{\gamma_s} = \frac{145\ g}{2.65\ g/cm^3} = 54.7\ cm^3 \tag{1.6}$$

The volume of voids is then:

$$V_v = V - V_s = 72.5\ cm^3 - 54.7\ cm^3 = 17.8\ cm^3 \tag{1.7}$$

and the volume of voids filled with air is:

$$V_a = V_v - V_w = 17.8\ cm^3 - 7.0\ cm^3 = 10.8\ cm^3 \tag{1.8}$$

Again, note that the volume of water numerically equals the weight of water if they are expressed in cubic centimeters (cm^3) and grams (g), respectively, and if we assume that the specific weight of water is unity, *i.e.*, 1 g/cm^3.

The phase-diagram in Figure 1.2 shows all the measured and calculated (bold underlined) values of the weights (masses) and volumes for this problem.

Porosity

The porosity of the sample is calculated using equation (1.1):

$$n = \frac{V_v}{V} = \frac{17.8\ cm^3}{72.5\ cm^3} = 0.245 \quad or \quad 24.5\%$$

Void Ratio

The void ratio is given as:

$$e = \frac{V_v}{V_s} = \frac{17.8\ cm^3}{54.7\ cm^3} = 0.325 \tag{1.9}$$

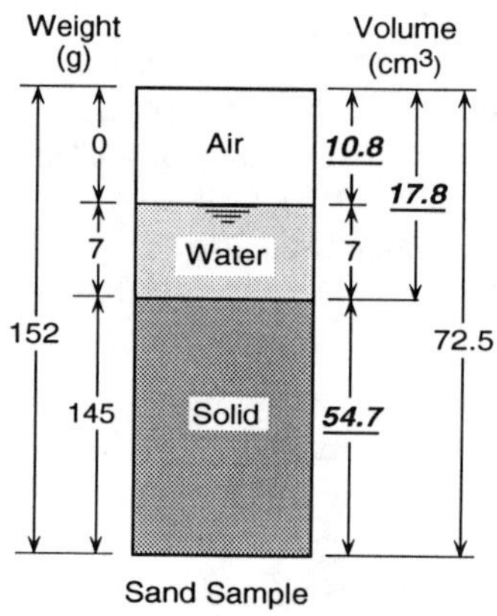

Figure 1.2 Weights and volumes of the three phases of the sand sample.

or

$$e = \frac{n}{1-n} = \frac{0.245}{1-0.245} = 0.324 \tag{1.9a}$$

where V_V is the volume of voids in the sample, V_S is the volume of solids in the sample, and n is the porosity of the sample. The void ratio is inversely related to the particle density (specific gravity): the greater the particle density, the smaller the void ratio, and vice versa.

Note: Unlike porosity, the void ratio can have values greater than 1, in which case it means that the percentage of voids in the unit volume of the sample is greater than the percentage of solids (mineral particles).

Moisture Content

The moisture content (w) is the ratio of the weight of water (W_W) in the sample to the weight of solids (W_S):

$$w = \frac{W_W}{W_S} = \frac{W - W_S}{W_S} = \frac{7\ g}{145\ g} = 0.0483 \quad or \quad 4.8\% \tag{1.10}$$

Degree of Saturation (Saturation Index)

The degree of saturation is usually expressed with the saturation index which is the percentage of sample voids filled with water:

$$S_I = \frac{V_W}{V_v} = \frac{7\ cm^3}{17.8\ cm^3} = 0.3933 \quad or \quad 39.3\% \tag{1.11}$$

1.2 WATER DISPLACEMENT TEST

Effective Porosity

A water displacement test is used to determine the effective porosity (n_{ef}) of the sample. Effective porosity is the volume of interconnected voids that allow free water (fluid) flow divided by the total sample volume. The oven-dried sample of the known volume (V = 72.5 cm^3) is immersed in the chamber containing 500 cm^3 of water, and left until it becomes saturated (the chamber is sealed to prevent water evaporation loss). It is assumed that during the oven-drying all the free water contained in the voids is expelled. The volume of water used to completely fill the voids, *i.e.*, saturate the sample, is therefore equal to the volume of voids. However, this volume will yield the value of effective porosity which is smaller than the overall porosity since it excludes micro pores which do not allow the free flow of water, and the pores that are not interconnected. The volume of the interconnected voids (V_{vi}) is determined as the difference between the initial water volume in the chamber (500 cm^3) and the water volume after the saturated sample is removed (483.5 cm^3):

$$V_{vi} = V_{w1} - V_{w2} = 500\ cm^3 - 483.5\ cm^3 = 16.5\ cm^3 \tag{1.12}$$

The effective porosity is:

$$n_{ef} = \frac{V_{vi}}{V} = \frac{16.5\ cm^3}{72.5\ cm^3} = 0.2276 \quad or \quad 22.8\% \tag{1.13}$$

As mentioned before, the effective porosity is less than the overall porosity. In the case of homogeneous medium-to-coarse sand and gravel, this difference is only a few percent. It dramatically increases, however, with an increase in the percentage of silt and clay particles in the rock. Some clays that have a total porosity of more than 60 % may have an effective porosity of less than 1-3 %. Values of the porosity of various rocks are given in Appendix A.

PROBLEM 2
VOLUME AND WEIGHT OF CONTAMINATED SOIL

Find the volume and the weight of the soil contaminated by a chemical spill. The affected area was sampled with a uniform network of soil borings as shown in Figure 2.1. Table 2.1 includes data on surface elevation, depth of contamination, and elevation of the bottom of the contaminated soil for each of the borings. The highly leached soil is weathered from the shaley sandstone bedrock and consists of silt and clay particles, mainly kaolinite (around 65%), K-feldspar (20%), and quartz (15%). The average soil porosity is 14% and its saturation index at the time of sampling was 22%.

2.1 VOLUME OF CONTAMINATED SOIL

The easiest way to find the volume of a large soil mass in the field is to use a software package that can produce contour maps of the surfaces of interest, and then calculate the volume in between them. One such package is SURFER for Windows by Golden Software,

Table 2.1 X,Y,Z data for the problem: X coordinate (column A), Y coordinate (column B), elevation of the land surface at each boring (column C), depth of the contamination (column D), and elevation of the bottom of the contamination (column E).

	A	B	C	D	E
Row 1	X	Y	Top	Depth	Bottom
2	10	10	541.1	0.0	541.1
3	10	40	541.4	0.0	541.4
4	10	70	541.9	0.0	541.9
5	10	100	542.5	0.0	542.5
6	10	130	542.9	0.0	542.9
7	10	160	543.2	0.0	543.2
8	40	10	541.0	0.0	541.0
9	40	40	541.4	0.9	540.5
10	40	70	541.9	0.0	541.9
11	40	100	542.7	1.1	541.6
12	40	130	543.4	0.8	542.6
13	40	160	543.4	0.0	543.4
14	70	10	541.4	0.0	541.4
15	70	40	541.7	1.7	540.0
16	70	70	542.0	1.2	540.8
17	70	100	543.0	3.2	539.8
18	70	130	543.8	2.3	541.5
19	70	160	543.6	0.0	543.6
20	100	10	541.9	0.0	541.9
21	100	40	542.4	0.7	541.7
22	100	70	542.2	2.9	539.3
23	100	100	543.2	4.2	539.0
24	100	130	543.9	4.8	539.1
25	100	160	544.0	0.0	544.0
26	100	190	544.1	0.0	544.1
27	130	10	542.0	0.0	542.0
28	130	40	542.5	0.0	542.5
29	130	70	543.0	1.8	541.2
30	130	100	543.4	3.6	539.8
31	130	130	544.0	3.1	540.9
32	130	160	544.3	2.7	541.6
33	130	190	544.4	0.0	544.4
34	160	40	542.4	0.0	542.4
35	160	70	542.7	1.2	541.5
36	160	100	543.4	2.0	541.4
37	160	130	543.9	0.0	543.9
38	160	160	543.3	0.0	543.3
39	160	190	544.4	0.0	544.4
40	190	40	541.8	0.0	541.8
41	190	70	542.2	0.0	542.2
42	190	100	543.2	0.0	543.2
43	190	130	543.7	0.0	543.7

2.3 Choose the **Gridding Method**. SURFER's default is **Kriging** which in most cases produces best results, especially if the right variogram model is specified. Several other methods are also available (more information about gridding methods can be found in the program's manual). The linear variogram model with a scale of 1 is chosen in our case (this is also a recommendation when in doubt about the right variogram model).

2.4 Name and save the **Output Grid File**. It is a good practice to give the same file name to all files, just with different (default) extensions. In our case the file to be created is named C:\PROJECT**TOP.GRD** (which includes the path). Press **OK**.

3. The next step is to create a *contour map* from the TOP.GRD file. This is done by choosing the **Contour** command from the **Map** menu (which is in the main menu). After specifying the grid file to be used, the **Contour Map** dialog box is displayed. The defaults usually result in a reasonable contour map. However, SURFER offers numerous options for editing the created contour map, including smoothing of contours, changing of contour levels, labels, axes attributes (scaling, display of grid lines, tick marks, font/title options, etc.), map orientation including 3D views, and much more. Figure 2.1 shows the contour lines of the top soil surface created with SURFER. The completed map should be saved in a .SRF file which can later be opened and edited if needed (e.g., **TOPMAP.SRF**).

3.1 SURFER accommodates the import of base maps in various bitmap and vector formats that can be overlaid over created contour maps. Base maps are used to show geographic information (roads, streams, lakes, county boundaries, etc.), geologic features, or, in our case, the limits of contamination. This is done with the **Load Base Map** command from the **Map** menu. The desired graphics file containing the base map is chosen in the **Import File** dialog box which displays all recognized (compatible) files and their formats. Base maps can also be created in the SURFER's **Plot** window using drawing tools, and then saved in various formats including AutoCAD [.DXF] files, Windows Metafile [.WMF] files, and Encapsulated PostScript [.EPS] format for possible editing in other graphic programs. This is done using the **Export** command from the **File** menu.

Boundary lines are placed on a contour map by importing the base map file into the **Plot** window that displays previously created contour map. The two maps are first both selected by pressing the **F2** key, and then overlaid with the **Overlay Maps** command from the **Map** menu. The two maps must use the same coordinate systems or the maps will not overlay correctly (more about overlaying maps and adjusting coordinate systems can be found in the program's manual).

3.2 The grid file, and consequently the contour map, can be modified to incorporate the interpreter's knowledge about the "real" field situation. Specific geologic and hydrogeologic conditions very often require the adjustment of contours "blindly" drawn by the program. The **Grid Editor** from the **Grid** menu allows changing the Z values at individual grid nodes in a grid [.GRD] file. The Grid Editor window displays a contour map and the position of the grid nodes which are indicated with small crosses ("+"). Clicking a node makes it the active node and displays X and Y coordinates at the top of the screen. The Z value is given in the edit box and can be changed as desired. The contour map is redrawn each time the change is implemented (e.g., by pressing ENTER key or by highlighting a new node). The edited grid file can be saved to the same name or to a different name using the **Save As** command.

4. Even though it is not necessary for our volume computation, it is useful to create a *3D surface plot* that, in the combination with the contour map, helps the understanding of spatial relationships in the field. This surface plot and the map, also in a 3D orientation, can be displayed at the same time.

4.1 Surface plot is created by choosing the **Surface** command from the **Map** menu, and by specifying the grid file in the **Open Grid** dialog box. Accepting defaults would usually produce an acceptable surface. Various **Map** menu commands can be used to modify the surface plot, including axis attributes and scaling (**Axis** command), plot scaling (**Scale** command; it also controls vertical exaggeration), and plot orientation (**3D View**).

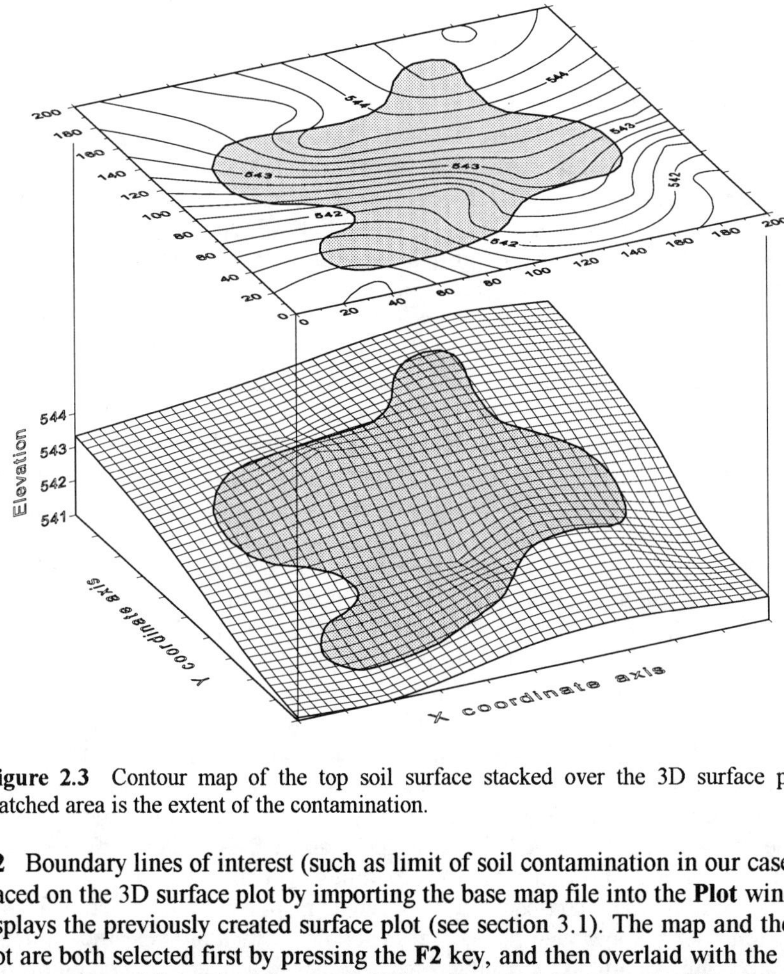

Figure 2.3 Contour map of the top soil surface stacked over the 3D surface plot. Hatched area is the extent of the contamination.

4.2 Boundary lines of interest (such as limit of soil contamination in our case) can be placed on the 3D surface plot by importing the base map file into the **Plot** window that displays the previously created surface plot (see section 3.1). The map and the surface plot are both selected first by pressing the **F2** key, and then overlaid with the **Overlay Maps** command from the **Map** menu. They must use the same coordinate systems or the base map will not overlay correctly (more about overlaying maps and adjusting coordinate systems can be found in the program's manual).

4.3 A plot of a contour map over a surface plot can be produced using the **Stack Maps** command from the **Map** menu. Both the map and the surface plot should be displayed in the same window. After selecting both, the **3D View** command should be used to set the rotation and tilt parameters to the desired values (use an *Orthographic* projection). Clicking OK will draw the 3D surface plot and the contour map at the same orientation. The next step is to select the surface plot and move (position) it to the bottom half of the page, and move the contour map to the top half of the page. Selecting both (**F2** key) and choosing the **Stack Maps** command will align the coordinate systems of the two maps. Figure 2.3 shows the final result for our example.

5. The entire procedure (sections 2 through 4.3) is repeated for the *bottom surface* of the contaminated soil. All the parameters for gridding, contouring and creating the 3D surface plot should be the same (except, of course, the Z values chosen in section 2.1, *i.e.*, **Scattered Data Interpolation** dialog box: it is column E in Table 2.1 which contains elevations of the bottom of the contaminated soil). The created file can be saved as BOTTOM.GRD.

6. SURFER can calculate *net volume* of material between two grid files that define two surfaces. The net volume is the difference between the *positive volumes* (cuts) and the *negative volumes* (fills). Positive volumes are the volumes of material in those places

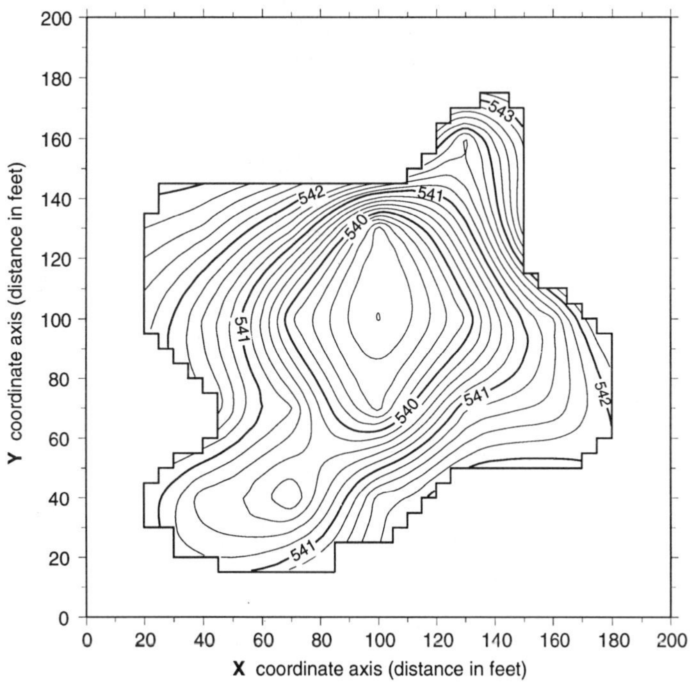

Figure 2.4 Contour map of the bottom surface of the contaminated soil. The area outside the contamination limits is blanked using the Grid Editor.

where the upper surface is above the lower surface. Negative volumes are the volumes of material in those places where the lower surface is above the upper surface. As it can be concluded, the upper surface does not have to be above the lower surface everywhere. *However, the grids for both surfaces must be the same size (use the same number of rows and columns), and use the same X,Y limits.*

6.1 In our case both grids are the same size and use same X,Y limits (from 0 to 200 feet). However, we are interested *only* in the volume of the contaminated soil, and not in the entire volume of material between two surfaces (grids). Non-contaminated area can be excluded from consideration during the volume calculations using the **Grid Editor** command from the **Grid** menu. After specifying *either* top or bottom surface grid file (e.g. TOP.GRD or BOTTOM.GRD) in the **Open Grid** dialog box, the grid file is represented by a contour map and the location of each of the grid nodes is marked with a small cross ("+"). The **Blank Node** command assigns the blanking value (1.70141e+038) to the selected node. This can also be done by typing **CTRL+B**.

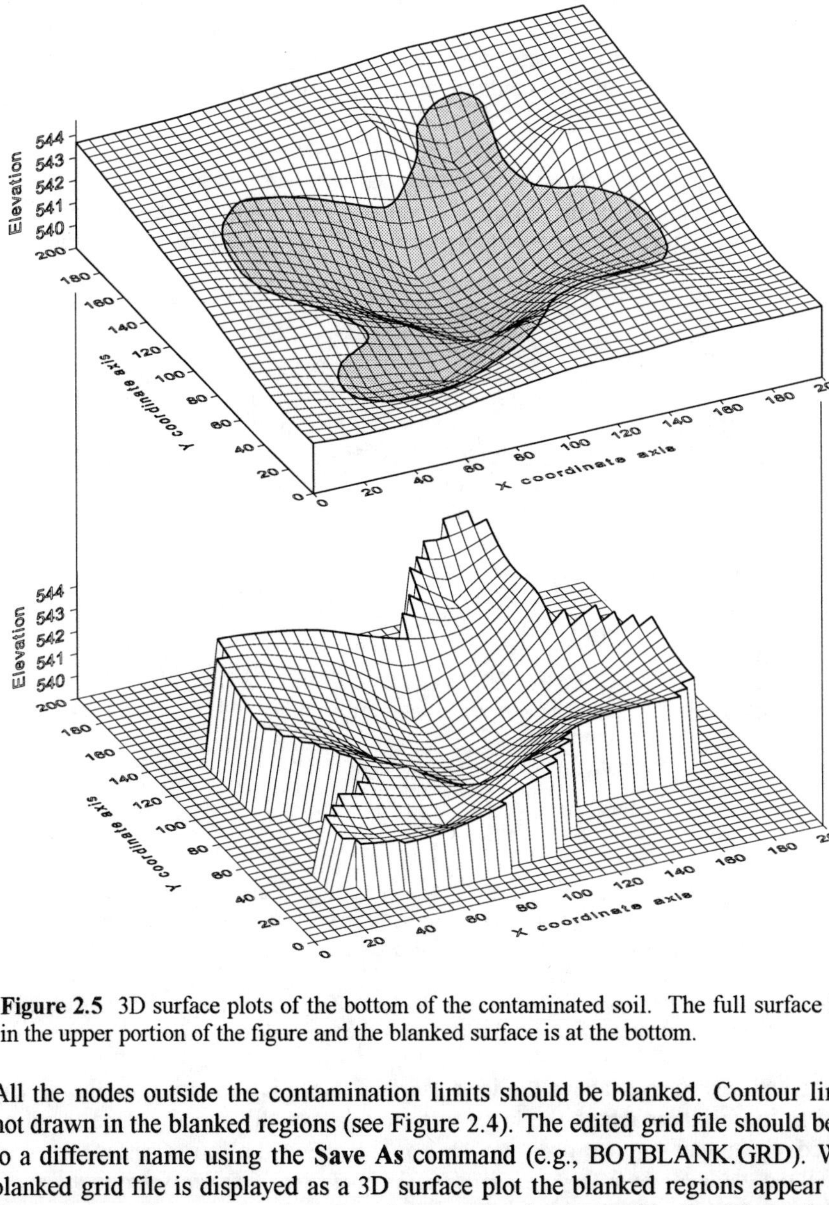

Figure 2.5 3D surface plots of the bottom of the contaminated soil. The full surface is in the upper portion of the figure and the blanked surface is at the bottom.

All the nodes outside the contamination limits should be blanked. Contour lines are not drawn in the blanked regions (see Figure 2.4). The edited grid file should be saved to a different name using the **Save As** command (e.g., BOTBLANK.GRD). When a blanked grid file is displayed as a 3D surface plot the blanked regions appear as low flat regions at the same level as the minimum Z value in the file (see Figure 2.5). The blanked regions on the lower surface should theoretically have the same elevations as corresponding areas on the upper surface (the thickness of the contamination is zero).Blanked regions on either the upper or lower surface are excluded from consideration during the volume calculations.

6.2 The volume between the upper and lower surfaces of the contaminated soil is calculated using the **Volume** command from the **Grid** menu. In the **Open Grid** dialog box specify the grid file for the upper surface (TOP.GRD) which will activate the **Grid Volume** dialog box. The box allows definition of the upper and lower surfaces by clicking the **Grid File** option button (the upper surface - TOP.GRD - is already displayed, and the file BOTBLANK.GRD should be selected for the lower surface).

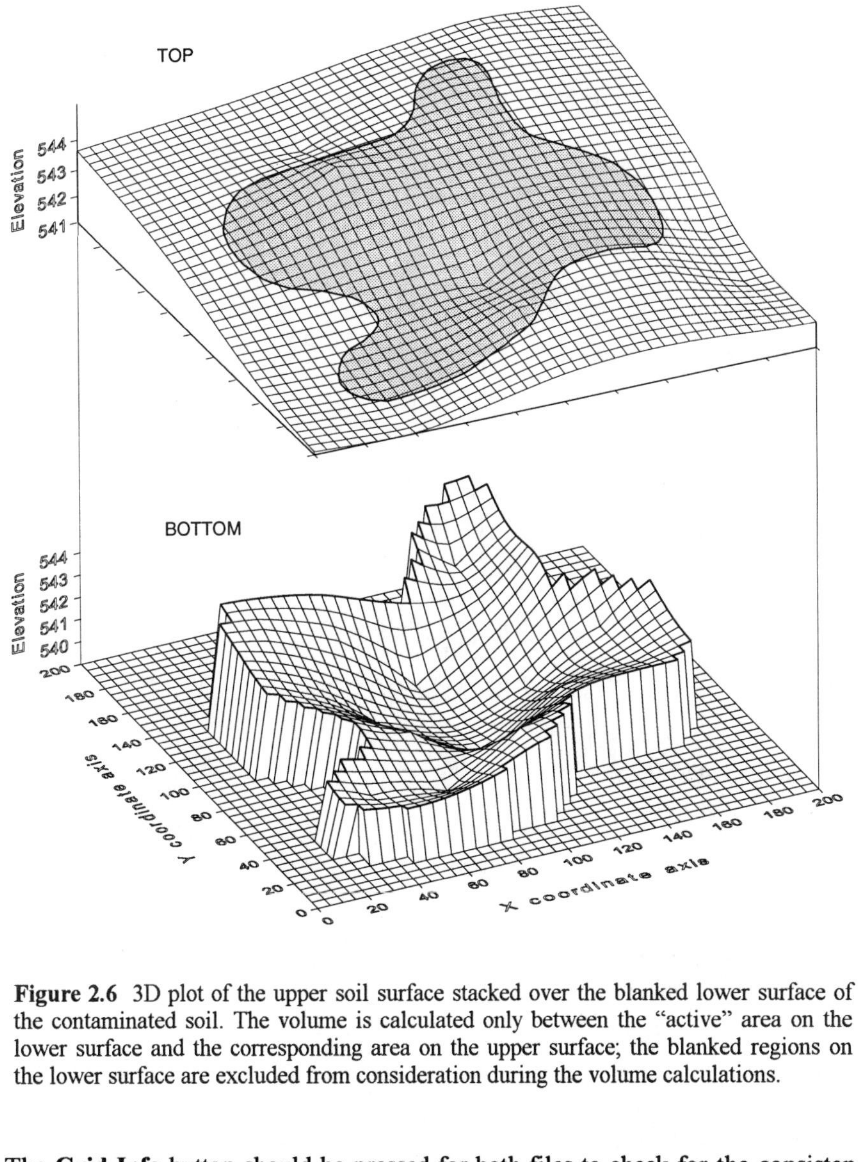

Figure 2.6 3D plot of the upper soil surface stacked over the blanked lower surface of the contaminated soil. The volume is calculated only between the "active" area on the lower surface and the corresponding area on the upper surface; the blanked regions on the lower surface are excluded from consideration during the volume calculations.

The **Grid Info** button should be pressed for both files to check for the consistency of grid parameters such as grid limits and number of grid lines (they must be the same for both files). After clicking the OK button, an **Editor** window is automatically opened and the volume and area calculations are reported. To save the information in a text file, choose the **Save As** command from the Editor window **File** menu and specify the name for the file (e.g., VOLUME.TXT). Figure 2.6 shows a 3D presentation of the space between the upper and the lower surfaces of the contaminated soil for which volume was calculated. Table 2.2 is the text file generated by the SURFER's volume calculation. The volumes are given in cubic feet, and the areas in square feet. These dimensions are based on the dimensions used for the initial X,Y and Z data (feet in our case).

Table 2.2 Text file generated by the SURFER's Volume command from the Grid menu.

VOLUME COMPUTATIONS

UPPER SURFACE

Grid File:	C:/PROJECT/TOP.GRD
Rows:	0 to 40
Cols:	0 to 40
Grid size as read:	41 cols by 41 rows
Delta X:	5
Delta Y:	5
X-Range:	0 to 200
Y-Range:	0 to 200
Z-Range:	540.976 to 544.639

LOWER SURFACE

Grid File:	C:/PROJECT/BOTBLANK.GRD
Rows:	0 to 40
Cols:	0 to 40
Grid size as read:	41 cols by 41 rows
Delta X:	5
Delta Y:	5
X-Range:	0 to 200
Y-Range:	0 to 200
Z-Range:	538.992 to 543.719

VOLUMES

Approximated Volume by	
Trapezoidal Rule:	33434
Simpson's Rule:	33463
Simpson's 3/8 Rule:	33439.1

CUT & FILL VOLUMES

Positive Volume [Cuts]:	33434
Negative Volume [Fills]:	0
Cuts minus Fills:	33434

AREAS

Positive Planar Area	
(Upper above Lower):	16175
Negative Planar Area	
(Lower above Upper):	0
Blanked Planar Area:	23825
Total Planar Area:	40000
Positive Surface Area	
(Upper above Lower):	16209.8
Negative Surface Area	
(Lower above Upper):	0

The volume of the contaminated soil is the average of three results obtained by the three different calculation methods: 33,445 cubic feet or 3,110 cubic meters (1 foot = 0.305 meters; 1 cubic foot = 0.093 cubic meters).

Volume of excavated soil

In most cases, the volume of excavated soil will be larger than the volume of contaminated soil because of the requirements of equipment for excavation, slope stability, etc. Soil is usually excavated down to a planar horizontal surface slightly below the bottom of contamination. The slope of the excavation will depend on the cohesive and other properties of the soil. The volume of soil to be excavated can also be determined with SURFER:

- The **Grid Editor** command from the **Grid** menu can be used to change Z values of nodes and to redraw the initial contour map of the upper surface "forcing" it to display uniform slopes and a horizontal plane at the desired constant Z value. The volume between the upper irregular soil surface and the new edited surface is calculated using the **Volume** command from the **Grid** menu as explained in section 6.2. Note that the blanking of the area outside excavation limits should be done for the new edited file representing the lower

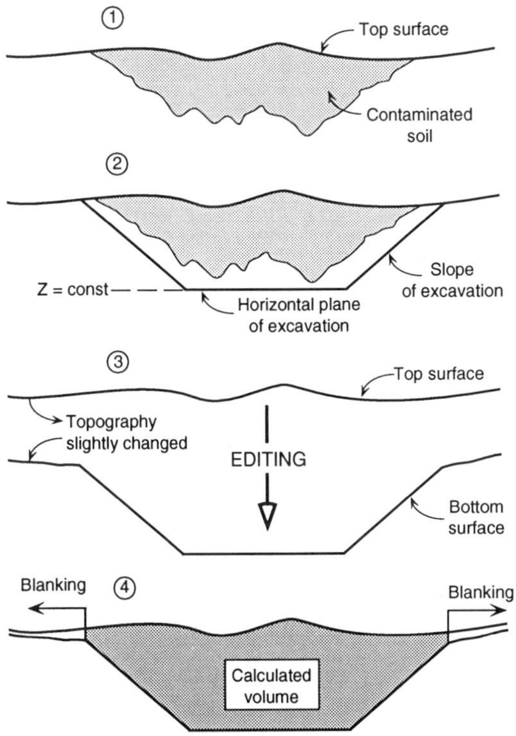

Figure 2.7 Schematic presentation of the steps to calculate volume of excavated soil. 1: Define depth of contamination. 2: Choose slope and horizontal plane of excavation. 3: Edit top soil contour map using the Grid Editor to reflect the shape of excavation (this new edited file will be the bottom surface). 4: Blank areas outside the limits of contamination and use the Volume command.

surface. This will ensure that only the volume of soil within the excavation limits is calculated. (Editing of the grid nodes and the consequent redrawing of the original map always slightly changes the contours outside the edited area.) Figure 2.7 shows the phases of volume calculation in this case.

2.2 WEIGHT OF CONTAMINATED SOIL

The overall weight of contaminated soil is an important parameter when designing its possible removal, transportation and subsequent cleanup or disposal. It depends on the following main factors:

- density of soil particles (*i.e.*, specific gravity of the constituting minerals),
- amount of water (or liquid) in the soil,
- porosity of soil.

In general, the weight of soil increases with the increasing density of its solids, and decreases with increasing porosity. Also, the weight of the same soil will be greater if the soil is more saturated. In our case, the summary report on laboratory testing of soil samples states that the average soil porosity is 14% and its saturation index at the time of sampling was 22%. Knowing the total volume of the contaminated soil and percentages of minerals present, the weight of the soil is calculated as follows:

Find the volume of solids

The volume of solids is easily determined using equation (1.1) from Problem 1 and knowing that volume of solids equals total volume minus volume of voids

$$n = \frac{V_v}{V} \quad (\times\ 100\%) \tag{2.1}$$

or

$$V_v = V_{total} \times n \tag{2.1a}$$

and

$$V_s = V_{total} - V_v \tag{2.2}$$

Combining (2.1a) and (2.2) gives

$$V_s = V_{total} - V_{total} \times n \tag{2.3}$$

or

$$V_s = V_{total}(1 - n) \tag{2.3a}$$

Since the total volume of contaminated soil is 3,110 m^3 and the porosity is 14%, the volume of solids is:

$$V_s = 3110\ m^3 (1 - 0.14) \ = \ 2675\ m^3$$

Find specific weight of soil particles

Specific weight of soil particles is the ratio of the weight of solids to the volume of solids:

$$\gamma_s = \frac{W_s}{V_s} \quad (g/cm^3) \tag{2.4}$$

It is numerically equal to particle density (ρ_s) in the constant gravitational field of the earth and, for practical purposes, can be expressed in the units of mass, *i.e.*, grams (g) or kilograms (kg). (Note that weight = mass times acceleration of gravity.) The specific weight of solids in soil is also given as:

$$\gamma_s = G\gamma_w \tag{2.5}$$

where G is the specific gravity of the mineral matter (solids) and γ_w is the specific weight of water. The average specific gravity of solids is calculated as a weighted average of the specific gravity of the three constituting minerals: kaolinite, K-feldspar, and quartz (specific gravity of the major rock-building minerals is given in Appendix A):

$$G_{av} = \frac{G_{kaolinite} \times 65\% + G_{K-feldspar} \times 20\% + G_{quartz} \times 15\%}{100\%} \tag{2.6}$$

$$G_{av} = \frac{2.60 \times 65\% + 2.58 \times 20\% + 2.65 \times 15\%}{100\%} = 2.61$$

Assuming that the specific weight of water is unity (γ_w=1g/cm^3), the specific weight of soil particles is then 2.61 g/cm^3.

Find weight of solids

From equation (2.4) the weight of the solid matter in soil is

$$W_s = \gamma_s V_s \tag{2.7}$$

$$W_s = 2.61\ g/cm^3 \cdot 2675 \cdot\ 1 \times 10^6\ cm^3 = 6.98 \times 10^9\ g$$

(Note that 1 m^3 equals 1x10^6 cm^3.)

$$W_s = 6.98 \times 10^6\, kg = 6980\ tons$$

Find weight of water (liquid) in soil

The volume of water in soil is the product of saturation index (S_I) and the volume of voids (V_v):

$$V_w = S_I \times V_v \tag{2.8}$$

or

$$V_w = S_I(V_{total} - V_{solids}) \tag{2.8a}$$

$$V_w = 0.22 \cdot (3110\ m^3 - 2675\ m^3) = 95.7\ m^3 = 96\ m^3$$

The weight of water is found from the specific weight of water (which can be considered unity for most practical purposes, *i.e.*, γ_w=1g/cm^3) and the volume of water in the soil:

$$W_w = \gamma_w \times V_v \tag{2.9}$$

$$W_w = 1\ g/cm^3 \cdot 96 \cdot 1 \times 10^6\, cm^3 = 96\ tons$$

Find the overall weight of soil

The overall weight of contaminated soil is the sum of two weights: that of solid matter and that of water present in the voids:

$$W_{total} = W_{solids} + W_{water} \tag{2.10}$$

$$W_{total} = 6980\ tons + 96\ tons = 7076\ tons$$

Chapter 2

LABORATORY PERMEAMETERS

PROBLEM 3
CONSTANT HEAD PERMEAMETER

Determine the hydraulic conductivity and permeability of a coarse sand sample that was tested in the laboratory with a constant head permeameter. The results of the test are given in Table 3.1. The fluid used for testing was de-aired water with a temperature of 23°C. Figure 3.1 is a schematic drawing of the laboratory permeameter. The inside diameter of the sample tubing is $1^{7/8}$ inches (4.8 cm) and the length of the sand sample is 6 inches (15.2 cm). The experiment consisted of ten individual tests each with a different position of the constant head (the head difference between the constant head and the head of the permeameter outlet is given in Table 3.1). The flow rate for each test was determined by measuring the volume of water collected at the outlet and the corresponding time. After determining the hydraulic conductivity for water as the test fluid, calculate what the hydraulic conductivity of the sample would be if, instead of water, the test fluid was gasoline at 18 °C (kinematic viscosity of gasoline at 18 °C is 6.5×10^{-7} m^2/s).

Table 3.1 Results of the constant head permeameter test of a sand sample.

Test #	1	2	3	4	5	6	7	8	9	10
Head differ. (cm)	5	6	7	8	9	10	12	14	16	18
Volume (cm^3)	4	3.5	5	5.5	3	6	3.5	3	6	4.5
Time (sec)	125	92	106	112	46	92	48	34	61	45

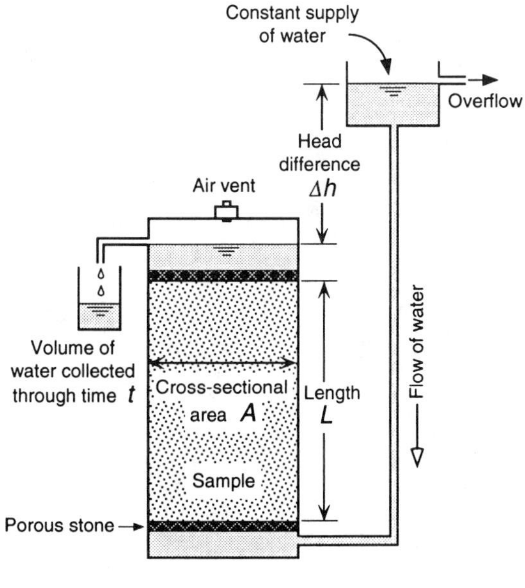

Figure 3.1 Schematic presentation of the constant head permeameter used for testing the sand sample.

3.1 HYDRAULIC CONDUCTIVITY

The permeameter shown in Figure 3.1 is based on Darcy's apparatus named after Henry Darcy, a French civil engineer who was using it "to determine the laws of flow of water through sand" (Domenico and Schwartz, 1990). The famous Darcy's law (1856), which describes the flow of fluids through (intergranular) porous media, was derived experimentally after hundreds of laboratory tests with the apparatus. In its basic form, this linear law states that the rate of a fluid flow (Q) through a sand sample is directly proportional to the cross-sectional area of the flow (A) and the loss of the hydraulic head between two points of measurements (Δh), and it is inversely proportional to the length of the sample (l):

$$Q = KA\frac{\Delta h}{l} \qquad [\mathrm{m^3/s}] \qquad (3.1)$$

K is the proportionality constant of the law called *hydraulic conductivity* and has units of velocity.

Note: It is not uncommon to find a minus sign in front of equation (3.1) in the literature with the following explanation: it indicates that the flow is in the direction of a decrease in the hydraulic head (i.e., from a higher toward a lower hydraulic head). However, although the explanation is correct, the minus sign should not be placed in front of equation (3.1) which is the solution of a differential form of Darcy's equation. The minus sign, which is correctly present in the differential equation, "disappears" during its integration and it is mathematically incorrect to show it as a part of equation (3.1).

The general hydraulic equation of the continuity of flow, which results from the principle of conservation of mass, is (for incompressible fluids):

$$Q = v_1 A_1 = v_2 A_2 = constant \qquad (3.2)$$

which means that the volumetric flow rate (Q) through the cross-sectional areas A_1 and A_2 in a stream of fluid is the same. Average flow velocities at cross-sections 1 and 2 are v_1 and v_2 respectively. From equation (3.2) the velocity of flow can be generally expressed as:

$$v = \frac{Q}{A} \qquad [\mathrm{m/s}] \qquad (3.3)$$

Relating equations (3.1) and (3.3) gives another form of Darcy's equation:

$$v = K\frac{\Delta h}{l} \qquad (3.4)$$

Expressed in words, equation (3.4) states that the velocity of fluid flow is proportional to the hydraulic gradient. The hydraulic gradient is the loss of the hydraulic head (Δh) along the distance l. It is usually designated as i :

$$i = \frac{\Delta h}{l} \qquad [\text{dimensionless}] \qquad (3.5)$$

which brings another common form of Darcy's equation:

$$v = Ki \tag{3.6}$$

Finally, equation (3.6) is used to determine the hydraulic conductivity of the sample:

$$K = \frac{v}{i} \qquad \text{[m/s]} \tag{3.7}$$

The sand sample inserted into the permeameter is a cylindrical specimen of the cross-sectional area A and length l (Figure 3.1). It is maintained in place by two porous stones. A steady-state flow of water through the sample is established by maintaining the upgradient and downgradient hydraulic heads. The upgradient hydraulic head is adjustable (it can be changed) while the downgradient hydraulic head is fixed by the position of the outlet. For each position of the upgradient hydraulic head the flow rate (Q) is determined by measuring the volume of water (V) collected at the outlet through time t ($Q=V/t$). Measurements for at least ten head differences should be performed in order to plot a graph *velocity* versus *hydraulic gradient*. The graph (see Figure 3.2) is used to check the validity of the experiment and to calculate a representative value of the hydraulic conductivity. Measurements and calculations needed for plotting the graph are shown in Table 3.2.

Table 3.2 Elements for plotting the graph *velocity* versus *hydraulic gradient* shown in Figure 3.2.

Test #	1	2	3	4	5	6	7	8	9	10
V (cm^3)	4	3.5	5	5.5	3	6	3.5	5.5	7	5
t (sec)	125	92	106	112	46	92	48	74	87	54
Q = V/t	0.032	0.038	0.047	0.049	0.065	0.065	0.072	0.074	0.080	0.092
● ***v = Q/A***	***0.0018***	***0.0021***	***0.0026***	***0.0027***	***0.0036***	***0.0036***	***0.0040***	***0.0041***	***0.0044***	***0.0051***
Δh (cm)	5	6	7	8	9	10	11	12	13	14
● ***i = Δh/l***	***0.329***	***0.395***	***0.460***	***0.526***	***0.592***	***0.658***	***0.724***	***0.789***	***0.855***	***0.921***

The cross-sectional area of the sample, *i.e.*, the fluid flow, is determined using the following equation:

$$A = r^2 \pi = \left(\frac{d}{2}\right)^2 \pi \tag{3.8}$$

where r is the radius of the sample, and d is the diameter of the sample.

$$A = (2.4\ cm)^2 \cdot 3.14 = 18.1\ cm^2$$

If the experiment is correctly set, the individual tests for various head differences should group along a straight line on the graph velocity versus hydraulic gradient as shown in Figure 3.2. In addition, it should be possible to start the best-fit straight line drawn through the experimental data at the graphs origin. This would indicate that the assumptions of Darcy's law, *i.e.*, that the flow is laminar (linear) and in steady state, are satisfied and that equation (3.7) can be applied.

The representative *hydraulic conductivity* of the sample is determined by choosing a point anywhere on the straight line and by substituting its coordinates into equation (3.7):

$$K = \frac{0.0032\ cm/s}{0.6} = 5.3 \times 10^{-3}\ cm/s$$

$$K = 5.3 \times 10^{-5}\ m/s$$

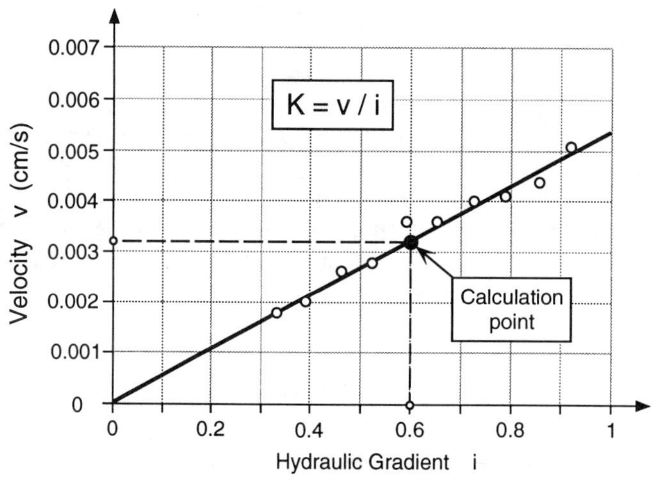

Figure 3.2 Graph *velocity* (*v*) versus *hydraulic gradient* (*i*) for the constant head permeameter test which is used to determine the hydraulic conductivity of the sand sample.

3.2 PERMEABILITY

The permeability of a porous medium is the ease with which a fluid can flow through that medium. In other words, permeability characterizes the ability of a porous medium to transmit a fluid. In hydrogeology the permeability is also referred to as the *intrinsic permeability*. It is dependent only on the physical properties of the porous medium: grain size, grain shape and arrangement, pore interconnections, etc. On the other hand, hydraulic conductivity is dependent on properties of both the porous medium and the fluid.

The relationship between the permeability (K_i) and the hydraulic conductivity (K) is expressed through the following formula:

$$K_i = K\frac{\mu}{\rho g} \qquad [\mathrm{m}^2] \qquad (3.9)$$

where μ is the *absolute viscosity* of the fluid (also called *dynamic viscosity* or simply *viscosity*), ρ is the *density* of the fluid, and g is the *acceleration of gravity*. The viscosity and the density of the fluid are related through the property called *kinematic viscosity* (ν):

$$\nu = \frac{\mu}{\rho} \qquad [\mathrm{m}^2/\mathrm{s}] \qquad (3.10)$$

Inserting the kinematic viscosity into equation (3.9) somewhat simplifies the calculation of the permeability since only one value (that of ν) has to be obtained from tables or graphs (for most practical purposes the value of the acceleration of gravity (g) is 9.806 m/s^2):

$$K_i = K\frac{\nu}{g} \qquad (3.11)$$

The kinematic viscosity of water used in the constant head permeameter test is found from the graph in Figure 3.3: the water temperature is 23 °C and the kinematic viscosity is 9.4x10^{-7} m^2/s. Substituting these values and the value of the hydraulic conductivity into equation (3.11) gives:

$$K_i = 5.3 \times 10^{-5}\ m/s \cdot \frac{9.4 \times 10^{-7}\ m^2/s}{9.81\ m/s^2}$$

$$K_i = 5.08 \times 10^{-12}\ m^2$$

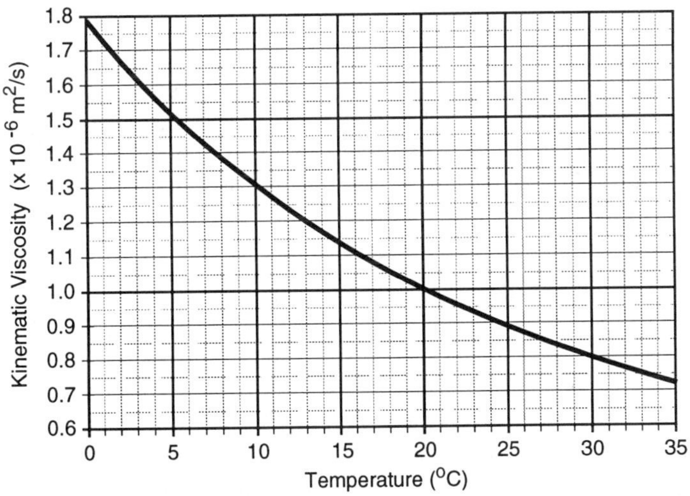

Figure 3.3 Kinematic viscosity of water versus temperature (data from Maidment, 1993).

Although it is much better to express the permeability in the units of area (m^2 or cm^2), for reasons of consistency and easier use in other formulas, in practice it is more commonly given in darcys (which is a "tribute" to oil industry):

$$1\ darcy = 9.87 \times 10^{-9}\ cm^2 \qquad \text{(or approximately } 1x10^{-8}\ cm^2\text{)}$$

Knowing that 1 m^2 = 10,000 cm^2, the permeability of our sand sample is

$$K_i = \frac{5.08 \times 10^{-12} \cdot 10{,}000\ cm^2}{9.87 \times 10^{-9}\ cm^2} = \frac{5.08 \times 10^{-8}\ cm^2}{9.87 \times 10^{-9}\ cm^2}$$

$$K_i = 5.15\ darcys$$

Hydraulic conductivity with gasoline as a test fluid

As already mentioned, the permeability of a porous medium is independent of the characteristics of the fluid flowing through that medium. This means that, once determined, the permeability can be used to estimate the hydraulic conductivity of the medium for various fluids or temperatures using equation (3.11) in the following form:

$$K = K_i \frac{g}{\nu} \tag{3.12}$$

Knowing that the kinematic viscosity of gasoline at 18 °C is 6.5x10^{-7} m^2/s and that the permeability of our sand sample is 5.08x10^{-12} m^2, the hydraulic conductivity is easily calculated:

$$K = 5.08 \times 10^{-12}\ m^2 \cdot \frac{9.81\ m/s^2}{6.5 \times 10^{-7}\ m^2/s}$$

$$K = 7.67 \times 10^{-5}\ m/s$$

The result shows that the hydraulic conductivity of the sand sample is higher for gasoline than for water: 7.7x10^{-5} m/s versus 5.3x10^{-5} m/s. It also means that, for example, contamination of groundwater with gasoline would spread much faster in that medium than contamination with oil which is a much more viscous fluid.

PROBLEM 4
FALLING HEAD PERMEAMETER

Determine the hydraulic conductivity and permeability of a silty sand sample that was tested in the laboratory with a falling head permeameter. The results of the test are given in Table 4.1. The fluid used for testing was de-aired water with a temperature of 26 °C. Figure 4.1 is a schematic drawing of the laboratory permeameter. The inside diameter of the sample tubing is 1$^{7/8}$ inches (4.8 cm) and the length of the silty sand sample is 6 inches (15.2 cm). The experiment consisted of ten individual tests each with a different time period and position of the initial head (h_O). After determining the hydraulic conductivity for water as the test fluid, calculate what the hydraulic conductivity of the sample would be if, instead of water, the test fluids were (separately) heavy oil at 70 °F and carbon tetrachloride at 84 °F.

The sand sample inserted into the permeameter is a cylindrical specimen of the diameter $2r_s$ and length L (Figure 4.1). It is maintained in place by two porous stones. The flow of water through the sample is established by adding water to the tall, narrow burette. In time t the head falls from the position h_O to the position h. Measurements for at least ten changes of head and corresponding times should be performed in order to plot a graph *time* versus *ratio between the initial and final heads*. The graph (see Figure 4.2) is used to check the validity of the experiment and to calculate a representative value of the hydraulic conductivity. Measurements and calculations needed for plotting the graph are shown in Table 4.1.

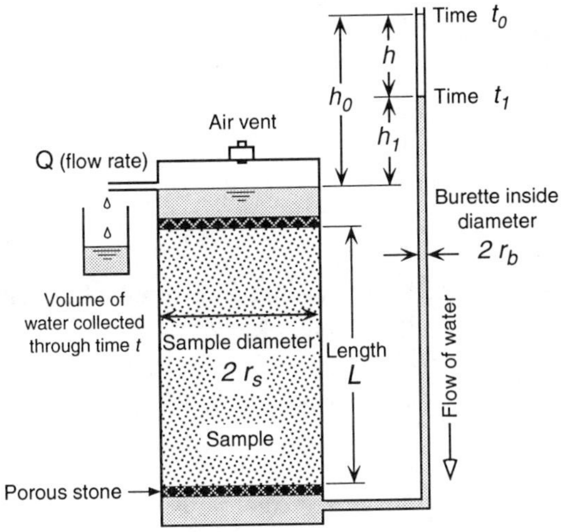

Figure 4.1 Schematic presentation of the falling head permeameter used for testing silty sand sample.

4.1 HYDRAULIC CONDUCTIVITY

The hydraulic conductivity of the sample is found starting with the general hydraulic equation of the continuity of flow (see Problem 3 and Equation 3.2): fluid flow through the burette equals flow through the sample:

$$Q_{burette} = Q_{sample} \tag{4.1}$$

In general, the flow rate is the product of the cross-sectional area of flow (A) and the velocity of flow (v):

$$Q_{burette} = A_{burette} \times v_{burette} \tag{4.2}$$

Knowing the radius of burette (r_b) and understanding that the flow velocity equals the change of head in time gives

$$Q_{burette} = r_b^2 \pi \times -\frac{dh}{dt} \tag{4.3}$$

The minus sign indicates that the hydraulic head, which causes the flow in the burette, changes from the higher one to the lower one.

Table 4.1 Elements for plotting the graph time versus ratio between the initial (h_0) and the final (h_1) heads shown in Figure 4.2.

h_0 (cm)	48.0	50.5	59.0	60.0	50.0	49.0	55.0	52.0	60.0	65.0
h_1 (cm)	42.5	44.0	50.5	47.0	35.0	32.0	37.0	45.0	49.0	47.0
• ***h_0/h_1***	***1.13***	***1.16***	***1.17***	***1.28***	***1.42***	***1.52***	***1.48***	***1.15***	***1.22***	***1.38***
• ***Time (sec)***	***422***	***461***	***505***	***781***	***1162***	***1423***	***1270***	***421***	***644***	***1000***

The fluid flow through the sand sample is:

$$Q_{sample} = A_{sample} \times v_{sample} \tag{4.4}$$

or

$$Q_{sample} = r_s^2 \pi \times K \frac{h}{L} \tag{4.5}$$

where r_s is the sample radius, L is the sample length, h is the change in hydraulic head and K is the hydraulic conductivity of the sample. Equating (4.3) and (4.5) gives

$$r_b^2 \pi \times (-\frac{dh}{dt}) = r_s^2 \pi \times K \frac{h}{L} \tag{4.6}$$

This equation is then integrated with the following boundary conditions: at time t_0 the hydraulic head is h_0, and at time t_1 the hydraulic head is h_1 (see also Figure 4.1):

$$r_b^2 L \times (-\int_{h_0}^{h_1} \frac{dh}{h}) = r_s^2 K \times \int_{t_0}^{t_1} dt \tag{4.7}$$

Since both integrals are elementary, they are easily solved (see Appendix G):

$$r_b^2 L \times -(\ln h_1 - \ln h_0) = r_s^2 K \times (t_1 - t_0) \tag{4.8}$$

$$r_b^2 L \times \ln \frac{h_0}{h_1} = r_s^2 K \times t \tag{4.9}$$

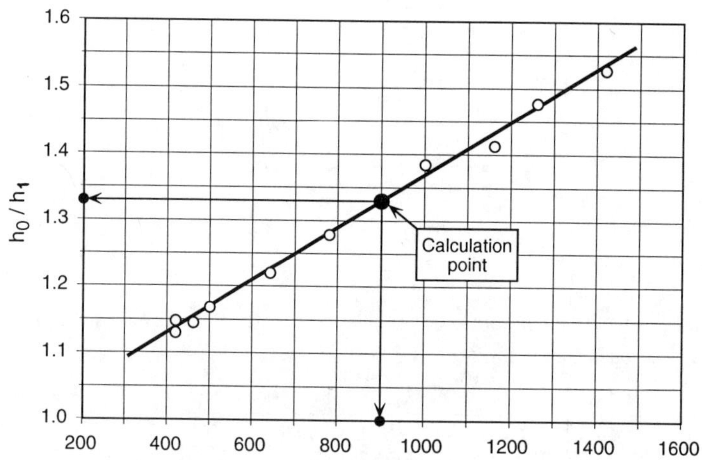

Figure 4.2 Graph time versus ratio between the initial and the final heads for calculating hydraulic conductivity with the falling head permeameter.

Finally, the hydraulic conductivity is:

$$K = \frac{r_b^2 L}{r_s^2 t} \ln \frac{h_0}{h_1} \tag{4.10}$$

The value of the hydraulic conductivity is found by inserting known values of the burette's radius (r_b), length (L) and radius (r_s) of the sample, and the coordinates of a point on the best-fit line drawn through the experimental data as shown in Figure 4.2:

$$K = \frac{(1cm)^2 \times 15.2cm}{(2.4cm)^2 \times 900\sec} \ln 1.338$$

$$K = 0.00085\ cm/s\ =\ 8.5 \times 10^{-4}\ cm/s\ =\ 8.5 \times 10^{-6}\ m/s$$

4.2 PERMEABILITY

As already mentioned in Problem 3, the permeability of a porous medium is the ease with which a fluid can flow through that medium. Permeability characterizes the ability of a porous medium to transmit a fluid and its value is independent of the fluid characteristics. Permeability is dependent only on the physical properties of the porous medium such as grain size, grain shape and arrangement, pore interconnections, etc. On the other hand, hydraulic conductivity is dependent on properties of both the porous medium and the fluid flowing through it.

The relationship between the permeability (K_i) and the hydraulic conductivity (K) is expressed through the following formula:

$$K_i = K \frac{\mu}{\rho g} \tag{4.11}$$

where μ is the *absolute viscosity* of the fluid (or *dynamic viscosity* or simply *viscosity*), ρ is the *density* of the fluid, and g is the *acceleration of gravity*. Remember that the viscosity and the density of the fluid are related through the property called *kinematic viscosity* (ν):

$$\nu = \frac{\mu}{\rho} \tag{4.12}$$

Inserting the kinematic viscosity into equation (4.11) simplifies the calculation of the permeability since only one value (that of ν) has to be obtained from tables or graphs (for most practical purposes the value of the acceleration of gravity is 9.806 m/s^2):

$$K_i = K \frac{\nu}{g} \tag{4.13}$$

The dynamic viscosity of water used in the falling head permeameter test is found from the graph in Figure 4.3: the water temperature is 26 °C and the dynamic viscosity is 8.8×10^{-4} Ns/m^2. Density of water at 26 °C is found from the graph in Figure 4.4: 996.56 kg/m^3 (or 0.99656 g/cm^3). The values of dynamic viscosity and density can be inserted into equation (4.12) to find kinematic viscosity (if a graph like the one shown in Problem 3 and Figure 3.3 is not readily available). Together with the acceleration of gravity, they can also be inserted into equation (4.11) to give the permeability:

$$K_i = 8.5 \times 10^{-6}\, m/s \frac{8.8 \times 10^{-4}\, \frac{Ns}{m^2}}{996.56 \frac{kg}{m^3} \cdot 9.806 \frac{m}{s^2}}$$

$$K_i = 7.65 \times 10^{-13}\, m^2$$

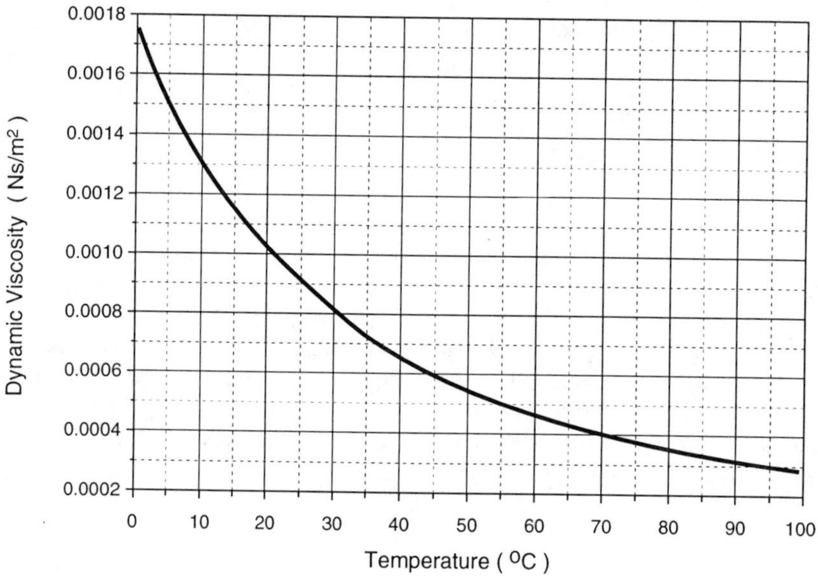

Figure 4.3 Dynamic viscosity of water as a function of temperature (data from Giles et al., 1994).

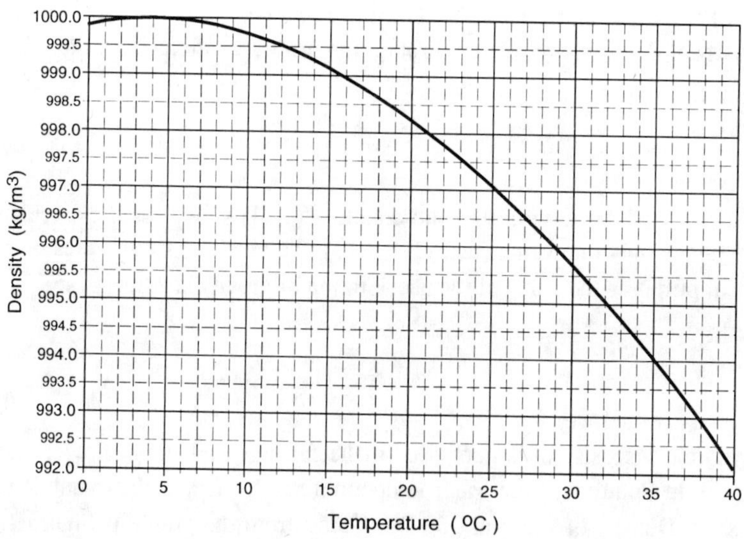

Figure 4.4 Density of water as a function of temperature (data from Maidment, 1993).

Note: Equations with physical properties that have complex derived units can often lead to dimensionally inconsistent (erroneous) results if conversions between various units are not properly applied. In our case the key for obtaining permeability in square meters is knowing that

$$1\ Newton\ (N) = 1\ kgm/s^2$$

Although it is much better to express the permeability in the units of area (m^2 or cm^2) for reasons of consistency and easier use in other formulas, it is more commonly given in darcys:

$$1\ darcy = 9.87 \times 10^{-9}\ cm^2 \qquad \text{(or approximately } 1 \times 10^{-8}\ cm^2\text{)}$$

Knowing that 1 m^2 = 10,000 cm^2, the permeability of our sand sample is

$$K_i = \frac{7.65 \times 10^{-13} \cdot 10{,}000\ cm^2}{9.87 \times 10^{-9}\ cm^2} = \frac{7.65 \times 10^{-9}\ cm^2}{9.87 \times 10^{-9}\ cm^2}$$

$$K_i = 0.775\ darcys$$

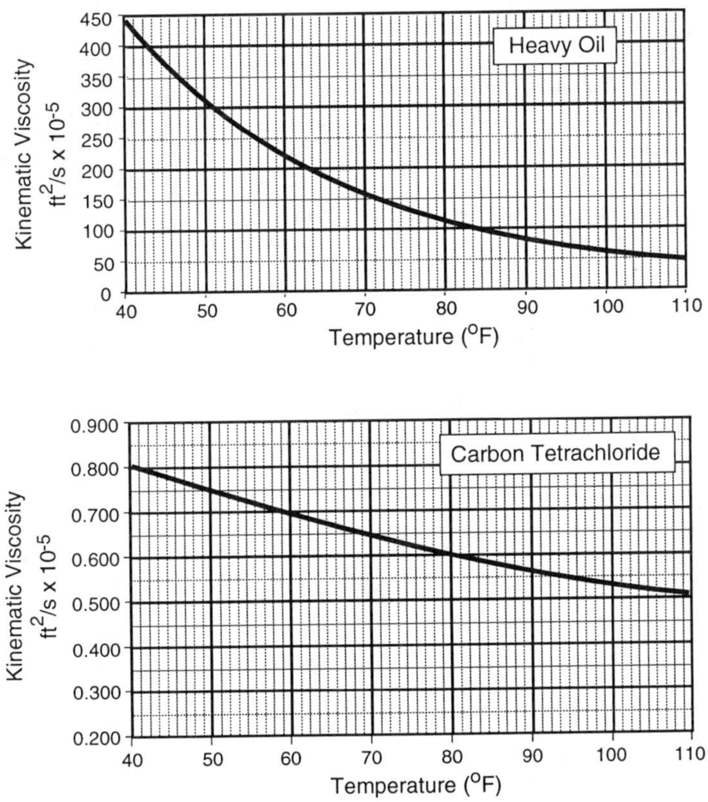

Figure 4.5 Dependence of kinematic viscosity of two common groundwater contaminants, heavy fuel oil and carbon tetrachloride, on temperature (data from Giles et al., 1994).

Hydraulic conductivity with heavy fuel oil and carbon tetrachloride as test fluids

Once the permeability of a porous medium as its constant property is found, it is easy to calculate hydraulic conductivities for various fluids if their kinematic viscosity is known. Figure 4.5 shows dependence of the kinematic viscosity of two common groundwater contaminants, heavy fuel oil and carbon tetrachloride, on temperature. Kinematic viscosity of heavy oil at 70 °F is 160x10^{-5} ft^2/s and kinematic viscosity of carbon tetrachloride at 84 °F is 0.558x10^{-5} ft^2/s. Knowing that 1 ft^2/s equals 0.09290 m^2/s, the hydraulic conductivities are found rearranging equation (4.13):

$$K = K_i \frac{g}{\nu} \tag{4.14}$$

Heavy fuel oil: $$K = 7.65 \times 10^{-13} m^2 \frac{9.806 \frac{m}{s^2}}{1.4864 \times 10^{-4}\ m^2/s} = 4.68 \times 10^{-8}\ m/s$$

C. tetrachloride: $$K = 7.65 \times 10^{-13} m^2 \frac{9.806 \frac{m}{s^2}}{5.1838 \times 10^{-7}\ m^2/s} = 1.45 \times 10^{-5}\ m/s$$

As can be seen, the hydraulic conductivity of the same porous material can vary several orders of magnitude depending upon the fluid that flows through it. Fluids with very high kinematic viscosity, such as heavy fuel oil, travel much more slowly through porous media than water. In addition, their viscosity, and consequently the hydraulic conductivity of a porous medium, change quite significantly with temperature: kinematic viscosity of heavy fuel oil decreases more than 4 times when temperature changes from 40 to 80 °F while this change for water is 1.8 times, and for carbon tetrachloride 1.3 times.

Chapter 3

GRAIN SIZE ANALYSIS

PROBLEM 5
GRAIN SIZE DISTRIBUTION CURVES, HYDRAULIC CONDUCTIVITY

Three sets of standard ASTM (American Society for Testing Materials) sieves were used to determine grain size distribution of three aquifer samples. Using the test results from Table 5.1 do the following for the each sample:

1. Draw the grain size distribution curve.
2. Determine the uniformity coefficient and the effective grain size.
3. Calculate hydraulic conductivity of the aquifer material using appropriate empirical formulas.

Table 5.1 Results of laboratory sieve tests for three aquifer samples.

Sample 1 (500 g)

Retained at Sieve #	6	10	12	16	20	30	40	60	100	Bottom Pan
Diameter of Sieve (mm)	***3.350***	***2.000***	***1.700***	***1.180***	***0.850***	***0.600***	***0.425***	***0.250***	***0.150***	-
Weight of Fraction (mg)	45.2	95.1	204.0	97.8	39.5	13.6	2.2	1.9	1.0	0.1
Content of Fraction (%)	9.04	19.02	40.80	19.56	7.90	2.72	0.44	0.38	0.19	0.28
Percent Coarser than Sieve Diameter	***9.04***	***28.06***	***68.86***	***88.42***	***96.32***	***99.04***	***99.48***	***99.86***	***100.05***	***100.33***

Sample 2 (500 g)

Retained at Sieve #	6	10	16	20	30	40	60	100	140	Bottom Pan
Diameter of Sieve (mm)	***3.350***	***2.000***	***1.180***	***0.850***	***0.600***	***0.425***	***0.250***	***0.150***	***0.106***	-
Weight of Fraction (mg)	25.1	52.4	58.3	72.9	124.7	89.3	52.2	18.8	4.6	2.8
Content of Fraction (%)	5.02	10.48	11.66	14.58	24.94	17.86	10.44	3.76	0.92	0.56
Percent Coarser than Sieve Diameter	**5.02**	**15.50**	**27.16**	**41.74**	**66.68**	**84.54**	**94.98**	**98.74**	**99.66**	**100.22**

Sample 3 (500 g)

Retained at Sieve #	10	16	20	30	40	60	100	140	200	Bottom Pan
Diameter of Sieve (mm)	***3.350***	***1.180***	***0.850***	***0.600***	***0.425***	***0.250***	***0.150***	***0.106***	***0.075***	-
Weight of Fraction (mg)	18.7	69.3	47.9	45.4	34.2	72.5	39.9	52.6	26.2	91.7
Content of Fraction (%)	3.74	13.86	9.58	9.08	6.84	14.50	7.98	10.52	5.24	18.34
Percent Coarser than Sieve Diameter	***3.74***	***17.60***	***27.18***	***36.26***	***43.10***	***57.60***	***65.58***	***76.10***	***81.34***	***99.68***

5.1 GRAIN SIZE DISTRIBUTION CURVE

Porosity, hydraulic conductivity and permeability are hydrogeological parameters that all greatly depend on the size of sediment grains and the percentage of various sediment fractions. The most widely used method to determine grain size distribution is by *laboratory sieve analysis*. As shown in Figure 5.1, a sediment sample of known weight is placed in the top sieve of a set of sieves with decreasing mesh openings/diameters. The set is then placed in a mechanical shaker for up to 10 minutes to separate the sample into fractions which are retained at individual sieves. The weight of each fraction is measured and expressed as percent of the initial (total) weight of the sample. The grain size distribution curve is plotted on a semilogarithmic paper: the cumulative percent coarser by

Sieve #	**Sieve Diam. (mm)**	Weight of the Fraction (mg)	Content of the Fraction (%)	**Percent Coarser than the Sieve Diameter**
6	**3.35**	45.2	9.04	**9.04**
10	**2.00**	95.1	19.02	**28.06**
12	**1.70**	204.0	40.80	**68.86**
16	**1.18**	97.8	19.56	**88.42**
20	**0.850**	39.5	7.90	**96.32**
30	**0.600**	13.6	2.72	**99.04**
40	**0.425**	2.2	0.44	**99.48**
60	**0.250**	1.9	0.38	**99.86**
100	**0.150**	1.0	0.19	**100.05**
Bottom Pan	-	0.1	0.38	**100.33**

Figure 5.1 Schematic presentation of the calculations for plotting the grain size distribution curve of Sample 1. Note that , due to weighing imprecision, the final cumulative percent may be different from 100.

weight (as shown in the last column in Figure 5.1) is plotted on the vertical arithmetic scale, while sieve openings diameter is plotted on the horizontal logarithmic scale (Figure 5.2). Alternatively, the cumulative percent finer by weight can be used to plot the curve as well. As shown in Figure 5.2, it is common practice to have two vertical arithmetic scales: one (usually on the right) with the cumulative percent coarser than the sieve diameter, and one (usually on the left) with the cumulative percent finer than the sieve diameter. Logarithmic scale of the grain size decreases from left to right, but the opposite direction is not uncommon, especially in foreign literature.

The lower limit of the applicability of sieve analysis is for grains retained at ASTM sieve #200 with an opening size (diameter) of 0.075 mm. This is also the boundary between sand and "fines" (silt and clay) according to *engineering classification* of sediment grain sizes adopted by ASTM. Grains smaller than 0.075 mm are separated and analyzed by several methods, the most common being the *hydrometer method*. After removal of the coarser fraction by sieving, remaining silt and clay particles are suspended in a vertical cylinder of water and left to settle. A hydrometer is submerged in the cylinder at various time intervals to measure the density of the suspension at given elevations. The diameter of the grains is determined from the suspension density which progressively decreases over time at a given elevation. The relationship is explained by Stoke's law which, simply put, states that grains of different size have different fall velocities in a suspension which causes them to separate (for a detailed explanation of the hydrometer method see Lambe, 1951).

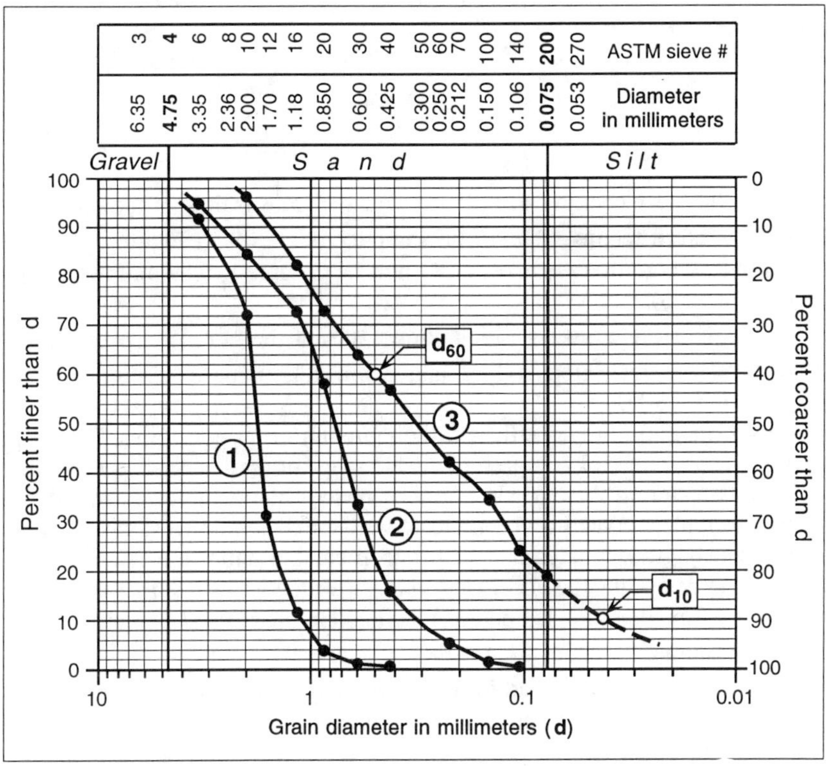

Curve No.	Coefficient of Uniformity $U = d_{60}/d_{10}$	Effective Grain Size $d_e = d_{10}$ (mm)
1	1.67	1.12
2	2.78	0.32
3	11.63	0.043

Figure 5.2 Grain size distribution curves for three aquifer samples.

If the percentage of the fines in the sample is smaller than 10%, the hydrometer method is usually not performed in common hydrogeological studies (see sections 5.2 and 5.3). When the fines constitute a significant portion of a sample, as is the case in Sample 3, and the hydrometer analysis is not readily available, the grain size curve may be extrapolated following the shape established by the sieve method. For more precise studies, (especially for groundwater protection and remediation) and when the hydraulic conductivity of low permeable porous media is important to determine, both sieve and hydrometer methods should be performed.

5.2 THE UNIFORMITY COEFFICIENT, EFFECTIVE GRAIN SIZE

Grain size distribution curves serve four main purposes in hydrogeological practice:

First is to determine the range of grain size present in the sample, *i.e.*, if the sample is more or less *uniform*. This could be done just visually, or using some numerical parameters from the curve which also allow classification and comparison of different materials. The slope of the curve, or range of the grains present, is most commonly defined by the *uniformity coefficient* (U):

$$U = \frac{d_{60}}{d_{10}} \tag{5.1}$$

where:

d_{60} is the sieve opening size (diameter) which allows 60% of the sample by weight to pass and therefore retains 40% of the sample. As shown in Figure 5.2, d_{60} is easily determined from the grain size curves (1.87 mm for Sample 1, 0.89 mm for Sample 2, and 0.5 mm for Sample 3).

d_{10} is the sieve diameter which allows 10% of the sample to pass, thus retaining 90%. Values of d_{10} for Samples 1, 2 and 3 are 1.12 mm, 0.32 mm and 0.043 mm respectively (Figure 5.2).

The lower the uniformity coefficient, the more uniform ("well graded", "well sorted") the material. Large values of U represent a wide range of grain sizes and generally "poorly graded" porous material. A somewhat arbitrarily set criterion for well-sorted material is $U<5$. In our example, curves 1 and 2 indicate well-sorted (uniform) samples while curve 3 represents a poorly-graded material since the coefficient of uniformity is more than 5.

Second is to determine the *effective grain size,* or the grain size that is most important for groundwater flow in a sampled porous medium. On the basis of numerous experimental data, it is now common practice to use d_{10} as the effective grain size ($d_{ef} = d_{10}$) although some other values can still be found in the literature (e.g. d_{17} and d_{20}). The effective grain size expresses the equivalent permeability of a porous material: the smallest 10% of grains, which also fill the pore spaces between larger grains, is what actually determines the material's permeability. In our example, values of $d_{ef} = d_{10}$ as determined from the grain size distribution curves, are given in Figure 5.2. The value for Sample 3 is taken from the extrapolated part of the curve since the hydrometer method was not available for determining the grain size of the fraction passing through #200 sieve (grains smaller than 0.075 mm). Such extrapolation, however, is not recommended since almost 20% of the sample are fines (silt and clay) and the slope of the curve may vary considerably.

Third is to design a *gravel pack* around well screens and drains. A detailed explanation of gravel pack calculations is given in Problem 6.

Fourth is to determine the *hydraulic conductivity* and/or *permeability* of the sampled porous medium. Calculations of hydraulic conductivity using the grain size curves are explained in the next section. Once the hydraulic conductivity is determined, it is easy to find permeability following the example in Problem 3 (and vice versa).

5.3 HYDRAULIC CONDUCTIVITY

Quite a few empirical formulas for calculating hydraulic conductivity using grain size analysis can be found in today's literature. The vast majority, however, are not dimensionally homogeneous and represent various modifications of the few initial works such as those of Hazen or Kozeny (see Freeze and Cherry, 1979; Fetter, 1994). It should be clearly understood that all these empirical formulas have various limits of application and give just approximate values of hydraulic conductivity for small point samples. Since they are derived for different experimental materials and conditions, it is very common that several formulas applied to the same sample will yield several very different values of hydraulic conductivity.

The following *dimensionally correct* general formula of hydraulic conductivity can be used for expressing most empirical equations currently in use (Vukovic and Soro, 1992):

$$K = \frac{g}{\nu} C f(n)\, d_e^2 \tag{5.2}$$

where g is gravitational acceleration, ν is kinematic viscosity of the fluid, C is a dimensionless coefficient which depends on various parameters of the porous medium such as grain shape, structure, heterogeneity, etc., $f(n)$ is a function of porosity n, and d_e is the effective grain size (usually d_{10}).

Three empirical formulas, all having specific application limits, are expressed in the form of equation (5.2) and used to calculate the hydraulic conductivity of the three samples (adapted from Vukovic and Soro, 1992):

The Hazen equation is:

$$K = \frac{g}{\nu} C_h f(n) d_{10}^2 \tag{5.3}$$

where

$$C_h = 6 \times 10^{-4} \tag{5.3a}$$

$$f(n) = [1 + 10(n - 0.26)] \tag{5.3b}$$

The Hazen equation is applicable for sediments with the coefficient of uniformity less than 5 (U<5) and the effective grain size between 0.1 and 3 mm (0.1 mm < d_{10} < 3 mm). Equation (5.3) is a more complex form of the initial Hazen formulas of 1892 and 1901 that can be found elsewhere in the literature (e.g., Fetter, 1994) as:

$$K = C d_{10}^2$$

where values of C are arbitrarily taken from the range of 1200 for clean uniform sand to 400 for silty/clayey heterogeneous sand. This formula was later modified by Lange (1958) to account for the sample porosity (equation 5.3).

The Kozeny equation is:

$$K = \frac{g}{\nu} C_k f(n) d_{10}^2 \tag{5.4}$$

where

$$C_k = 8.3 \times 10^{-3} \tag{5.4a}$$

$$f(n) = \frac{n^3}{(1 - n)^2} \tag{5.4b}$$

The formula is applicable for coarse sand.

The Breyer equation is:

$$K = \frac{g}{\nu} C_b\, d_e^2 \tag{5.5}$$

where

$$C_b = 6 \times 10^{-4} \log \frac{500}{U} \tag{5.5a}$$

As can be seen, the formula does not express the hydraulic conductivity as a function of porosity. It is applicable for $1 < U < 20$, and $0.06 \text{ mm} < d_{10} < 0.6 \text{ mm}$ which makes it useful for analyzing heterogeneous porous media with the poorly sorted grains, as is the case with Sample 3 in our example.

For comparison, the hydraulic conductivities of all three samples are calculated using all three equations and the results are shown in Table 5.2. Gravity acceleration (g) is assumed to be 9.807 m/s^2, and the kinematic viscosity (ν) is $1.14\text{x}10^{-6}$ m^2/s for the groundwater having temperature of 15 $^\circ$C. Porosity of samples 1, 2 and 3, as determined in the laboratory, is 0.35, 0.27, and 0.21, respectively.

Table 5.2 Hydraulic conductivity (in m/s) of the three samples calculated using three empirical formulas.

Formula	Sample 1	Sample 2	Sample 3
Hazen	$1.23\text{x}10^{-2}$	$5.81\text{x}10^{-4}$	$4.77\text{x}10^{-6}$
Kozeny	$9.09\text{x}10^{-3}$	$2.70\text{x}10^{-4}$	$1.96\text{x}10^{-6}$
Breyer	$1.60\text{x}10^{-3}$	$1.19\text{x}10^{-3}$	$1.56\text{x}10^{-5}$

As can be seen from Table 5.2, the difference between calculated hydraulic conductivities increases with the increasing uniformity coefficient. When choosing hydraulic conductivity that will be used for further calculations, one should always have in mind specific limitations of these formulas. For example, a simple averaging of several results (three in this case), which is a surprisingly common practice in the industry, would probably not lead to a big mistake if applied to Sample 1. If applied to Sample 3, however, it would probably cause some erroneous assessments. Having in mind the applicability of each equation, the following values should be chosen:

Sample 1 (coarse sand): $9.09\text{x}10^{-3}$ m/s (Kozeny formula).

Sample 2 (medium sand): $5.81\text{x}10^{-4}$ m/s (Hazen formula) or
$8.85\text{x}10^{-4}$ m/s (average of Hazen and Breyer formulas).

Sample 3 (silty sand): $1.56\text{x}10^{-5}$ m/s (Breyer formula).

PROBLEM 6
GRAVEL PACK SIZE

Determine the well gravel pack sizes for the two aquifers represented with the grain size curves shown in Figure 6.1. The wells tapping the aquifers have the same outside screen diameter of 273 mm ($10^{3/4}$ inches). The diameter of the boreholes in their screen sections is also the same -- 445 mm ($17^{1/2}$ inches). Note that the diameter of the borehole corresponds to the diameter of the drilling bit ("drilling diameter") which is one of the main criteria for the well design.

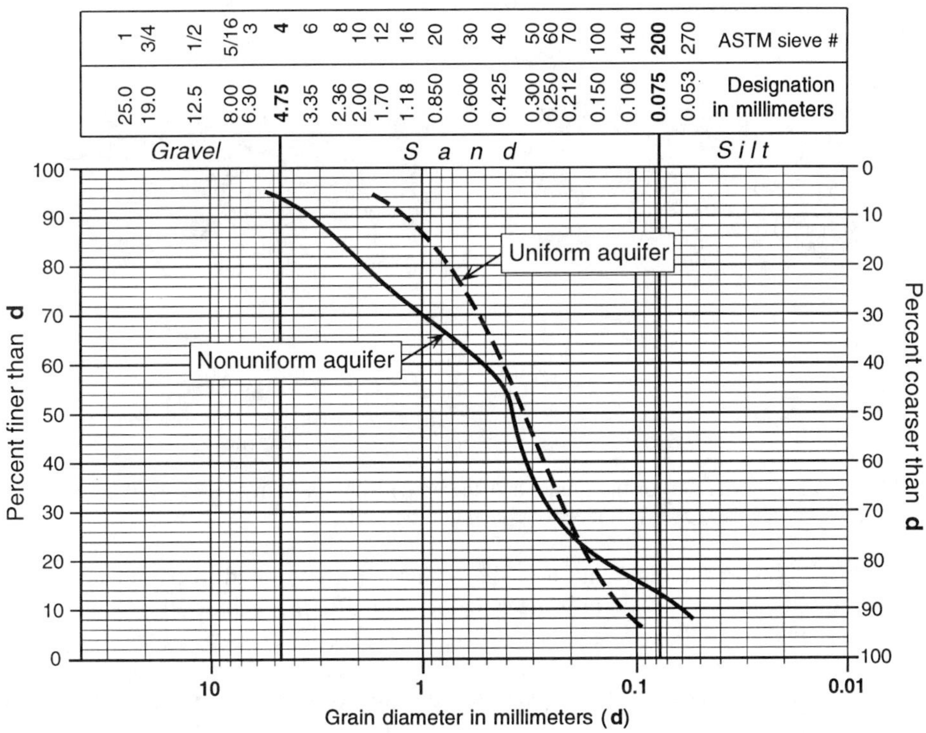

Figure 6.1 Grain size distribution curves for the analyzed aquifers.

6.1 PURPOSE OF GRAVEL (FILTER) PACK

Following are the main purposes of a well gravel pack:

- to stabilize the formation,
- to prevent or reduce sand pumping,
- to enable larger screen openings,
- to establish transitional velocity and pressure fields between the formation and the well screen.

The placement of a filter pack makes the zone around the well screen more permeable and increases the effective hydraulic diameter of the well. The gravel pack allows the removal of some formation material during well development, and it retains most of the aquifer fine material during the well exploitation.

A gravel pack is particularly useful in fine-grained, uniformly graded formations, and in extensively laminated aquifers (e.g., alternating layers of silt, sand and gravel) where it eliminates/reduces sand pumping and enables larger screen openings thus increasing otherwise small well yields.

In addition to sand pumping, which can destroy the pump and the screen, another very important problem in well maintenance can be avoided or deferred by a properly installed gravel pack -- chemical and biological incrustation of the well screen and the material adjacent to the well screen. Carbonate, iron and manganese incrustation are the most common problems. They are related to the velocity-induced pressure changes that disturb the chemical equilibrium of the groundwater in the well screen zone.

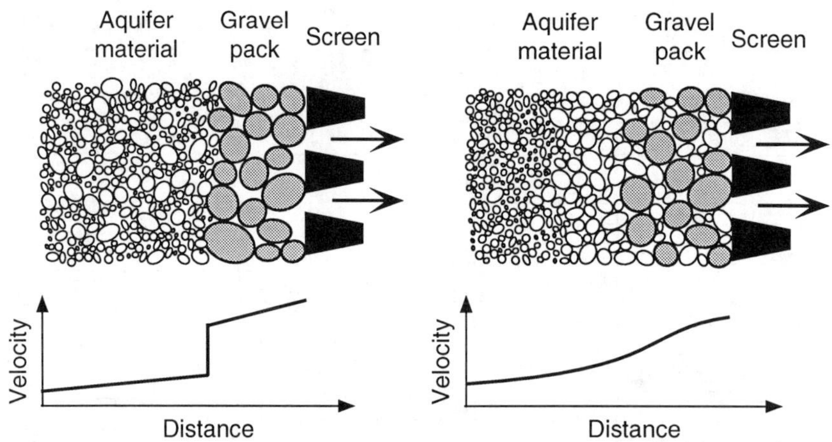

Figure 6.2 Distribution of the grains and the groundwater velocity in the well screen zone before (left) and after (right) development of the well.

Figure 6.2. shows the distribution of the grains and the groundwater velocity in the screen section of a well with the gravel pack, before and after the well development. Due to the permeability change, there is an abrupt increase of groundwater velocity at the contact between the aquifer material and the gravel pack before the well development. This increase corresponds to an equally abrupt decrease of pressure which causes the precipitation of calcium carbonate, iron and manganese. This precipitation may occur either in an improperly developed well with a gravel pack, or in a well without a gravel pack (here the pressure drop occurs at the contact between the aquifer material and the screen).

There is a significant change in the velocity/pressure field and the permeability in the screen zone after well development. Most of the fine grains are removed from the aquifer material resulting in an increased permeability. At the same time the permeability of the gravel pack portion is somewhat smaller due to the filling of the pore space between the pack grains with the aquifer material. The result is a gradually increasing groundwater velocity and a gradually decreasing pressure in the well screen zone which allows groundwater to carry its chemical load into the well rather than to precipitate it onto the screen and the aquifer/pack material.

6.2 THICKNESS OF GRAVEL PACK

In order to successfully retain the formation particles, the thickness of the gravel (filter) pack in ideal conditions does not have to be more than 0.5 inches according to laboratory tests made by Johnson Division : "Filter-pack thickness does little to reduce the possibility of sand pumping, because the controlling factor is the ratio of the grain size of the pack material in relation to the formation material" (Driscoll, 1986). However, for practical purposes, the thickness of the gravel pack should be at least 3 inches to ensure its accurate placement and complete surrounding of the screen. On the other hand, a filter pack that is more than 8 in. (203 mm) thick can make final development of the well more difficult "...because the energy created by the development procedure must be able to penetrate the pack to repair the damage done by drilling, break down any residual drilling fluid on the borehole wall, and remove fine particles (from the formation) near the borehole" (Driscoll, 1986).

Well design depends on many factors, one of which is the grain size distribution of the aquifer and, consequently, the size and thickness of the gravel pack. The drilling diameter (diameter of the bit), as the main element of a well design, should be chosen with the need for a gravel pack in mind, among other things.

The gravel pack thickness (GPT) corresponds to the thickness of the annulus between the well screen and the borehole wall :

$$GPT = \frac{D_b - D_s}{2} \tag{6.1}$$

where D_b is the diameter of the borehole in the screen section, and D_s is the outside diameter of the screen. In our case this is

$$GPT = \frac{445mm - 273mm}{2} = 86mm$$

The gravel pack thickness of 86 mm (~3.4 in.) satisfies the two criteria: it is more than 3 in. and less than 8 in. It is generally better to choose gravel packs of smaller thickness since drilling and design costs significantly increase with increasing well diameter. However, in some cases it may be necessary to place a thick gravel pack. This includes reverse rotary drilling which is most efficient for large-diameter boreholes (>406 mm). The completed borehole may be significantly larger than an optimum screen diameter thus requiring stabilization of the aquifer material (the cost of a thicker gravel pack is incomparably lower than the cost of a larger screen).

6.3 SIZE OF GRAVEL PACK

The choice of the gravel pack size presented here is based on the recommendations of The Institution of Water Engineers and Scientists, London, England (Brandon, 1986). These recommendations are a compromise between numerous procedures that can roughly be divided into two groups:

1) A more pragmatic approach followed mainly in the United States, which emphasizes simplicity of design (e.g., it is hard to find recommendations for design of a graded filter pack with the uniformity coefficient grater than 2.5), and

2) A strict approach that emphasizes the problems of choosing gravel packs for heterogeneous aquifers (the practice in most European countries) including design of filters with two or more layers.

If the aquifer material is uniform and well sorted, a *"uniform"* gravel pack is needed. The grading of the filter pack is based on the grain size distribution curve of the finest aquifer material within the well screen section. A *"graded"* gravel pack should be considered for aquifers with a wide range of particle sizes and large uniformity coefficients. In this case the grading of the pack depends on both the coarsest and the finest aquifer materials.

Uniform Gravel Pack

Uniform gravel packs are designed for aquifers with uniformity coefficients less than 3. The first criterion for choosing the gravel pack is:

$$D_{50} = 4 \text{ to } 6 \text{ times } d_{50} \tag{6.3}$$

where D_{50} is the sieve opening size which would allow 50 % of the gravel pack material to pass, and d_{50} is the sieve opening size which would allow 50 % of the aquifer material to pass (see Problem 5). A greater multiplier is used for aquifers that have fines such as silt and clay.

The second criterion is:

$$U_{pack} < 2.5 \tag{6.4}$$

where U_{pack} is the uniformity coefficient of the gravel pack.

Figure 6.3. illustrates the procedure for choosing gravel pack characteristics for the aquifer material whose uniformity coefficient is less or close to 3:

$$U_{aquifer} = \frac{d_{60}}{d_{10}} = \frac{0.42\ mm}{0.18\ mm} = 2.3$$

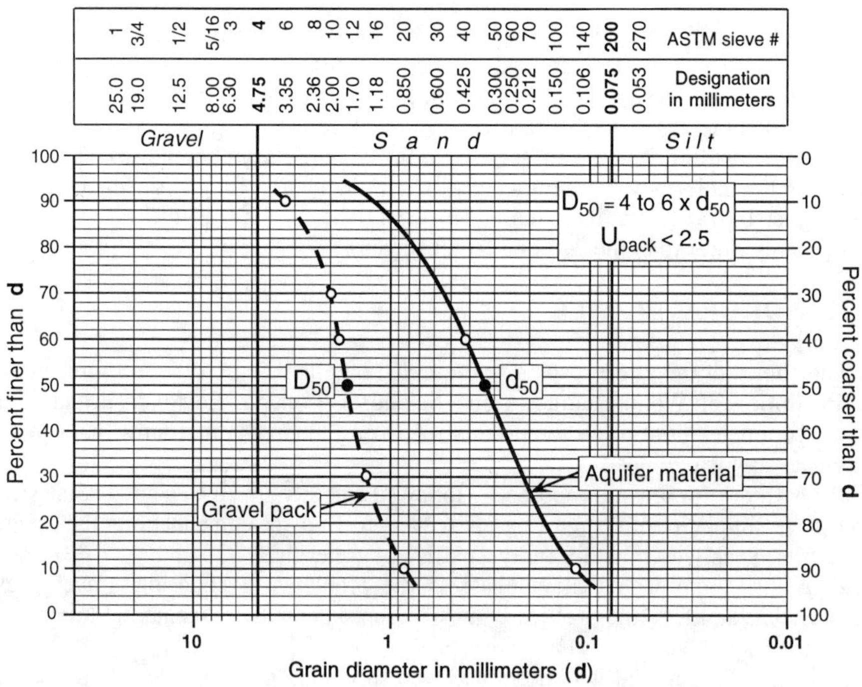

Figure 6.3 Grain size distribution curves for the uniform aquifer material and the corresponding gravel (filter) pack.

Table 6.1 Characteristic grain sizes of the gravel pack curve in Figure 6.3 and the corresponding standard ASTM sieves.

	D_{10}	D_{30}	D_{50}	D_{70}	D_{90}
Grain Size (mm)	0.850	1.40	1.70	2.00	3.35
ASTM Sieve #	20	14	12	10	6

First, an *approximate* ("temporary") D_{50} is determined knowing that the d_{50} is 0.34 mm and applying equation (6.3):

$$D^{*}_{50} = 4.5 \times 0.34\ mm = 1.53\ mm$$

Second, the grain diameter closest to a standard ASTM sieve is chosen for the *final* D_{50} (#12 or 1.70 mm in our case). A "temporary" gravel pack curve is then drawn through the D_{50} to roughly follow the shape of the aquifer curve. This "temporary" curve would usually have to be adjusted further since it lies in the coarser grain zone and, because of the logarithmic scale, the uniformity coefficient becomes larger than that of the aquifer material.

Third, the *final* gravel pack curve and its characteristic grain sizes (such as D_{10}, D_{30}, D_{50}, D_{70}, D_{90}) are chosen to match the closest standard ASTM sieves, and at the same time satisfy the requirement regarding the uniformity coefficient (U<2.5). In our case this requirement is satisfied since U=2.1 (D_{60}=1.80 mm, and D_{10}=0.85 mm). Choosing characteristic grain sizes that correspond to the standard sieves simplifies the process of constructing a suitable gravel pack if a proper commercial pack is not readily available. Table 6.1. shows the set of standard sieves that can be used to produce the gravel pack curve in Figure 6.3 (dimensions of the standard ASTM sieves are given in Appendix B).

Graded Gravel Pack

Graded gravel pack should satisfy the following criteria:

- D_{15} of the pack is at least 4 times greater than d_{15} of the aquifer sample (if there is more than one aquifer sample, this criterion should be applied for the coarsest sample):

$$D_{15} \geq 4 \text{ to } 6 \text{ times } d_{15} \tag{6.5}$$

A greater multiplier is used for more heterogeneous aquifers that include silt and clay.

- D_{85} of the pack should be slightly less than 4 times the d_{85} (if there is more than 1 aquifer sample, this criterion should be applied for the finest sample):

$$D_{85} \leq 4 \text{ times } d_{85} \tag{6.6}$$

- The curve of the gravel pack should match the shape of the aquifer sample (if there is more than one aquifer sample, the gravel pack curve should match the shape of the average sample):

$$U_{pack} \approx U_{aquifer} \tag{6.7}$$

Figure 6.4 shows a gravel pack curve that satisfies the above criteria. Two starting points for drawing an *approximate* ("temporary") pack curve are determined knowing that d_{15} is 0.095 mm and d_{85} is 2.50 mm:

$$D_{15}^* = 6 \times 0.095\,mm = 0.57\,mm$$

$$D_{85}^* = 3.5 \times 2.50\,mm = 8.75\,mm$$

Then the *final* points are chosen to correspond to the closest standard sieves:

$$D_{15} = 0.60\,mm \text{ or } \#30 \text{ sieve}$$

$$D_{85} = 8.00\,mm \text{ or } 5/16 \text{ sieve}$$

and the curve is drawn to follow the shape of the aquifer curve, without trying to match minor "irregularities". Characteristic grain sizes on the curve (such as D_{10}, D_{30}, D_{50}, D_{70}, D_{90}) are also chosen to match the closest standard ASTM sieves, and at the same time satisfy the requirement that the uniformity coefficient of the gravel pack is similar to that of the aquifer. Table 6.2 shows the set of standard sieves that can be used to produce the gravel pack curve in Figure 6.4.

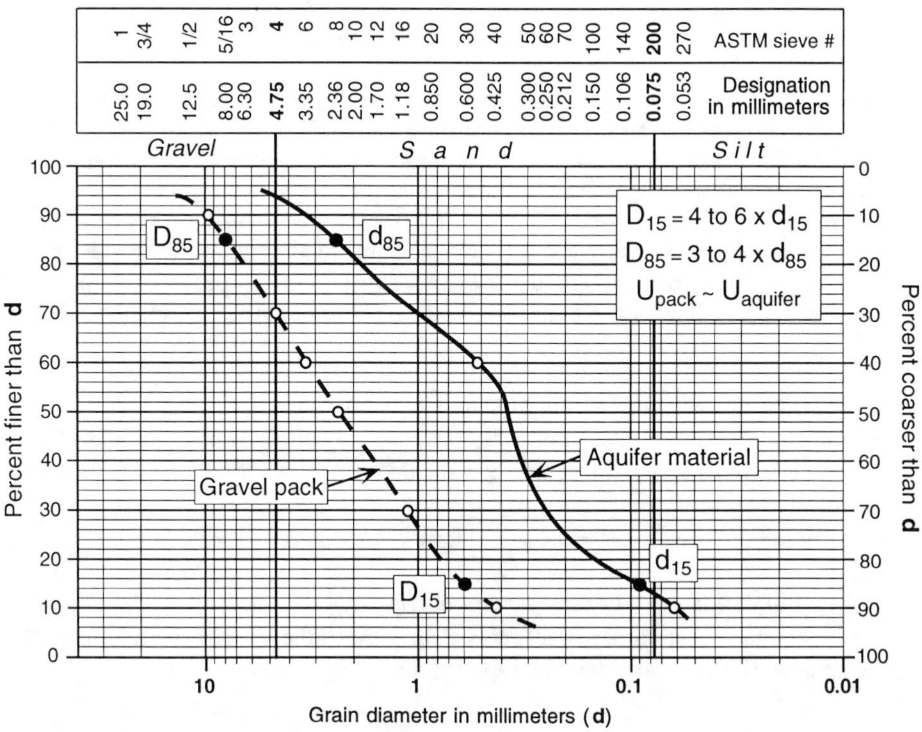

Figure 6.4 Grain size distribution curves for the non-uniform aquifer material and the corresponding gravel (filter) pack.

Table 6.2 Characteristic grain sizes of the gravel pack curve in Figure 6.4 and the corresponding standard ASTM sieves.

	D_{10}	D_{15}	D_{30}	D_{50}	D_{70}	D_{85}	D_{90}
Grain Size (mm)	0.425	0.60	1.18	2.36	4.75	8.00	9.5
ASTM Sieve #	40	30	16	8	4	5/16	3/8

Chapter 4

ONE-DIMENSIONAL STEADY STATE FLOW

PROBLEM 7
CONFINED HOMOGENEOUS AQUIFER

a) Calculate groundwater flow rate per unit width between the river and the drainage channel shown in Figure 7.1. The average elevation of the water surface in the river is 144.60 meters, and in the channel 142.20 meters. The hydraulic conductivity of the confined intergranular aquifer developed in medium alluvial sand is 3.5×10^{-4} m/s. The hydrogeologic cross-section in Figure 7.2 shows the relationship between the aquifer, the overlying silty clay (aquitard) and the underlying dense (impermeable) clay.

b) Calculate the position of the potentiometric surface at a midpoint between the river and the channel.

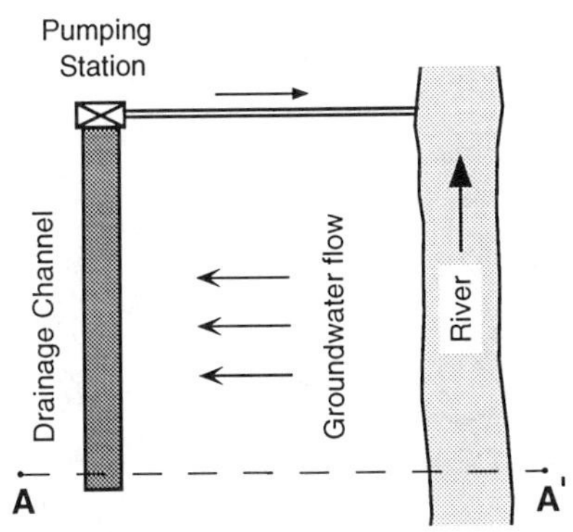

Figure 7.1 Plan view of the river and the drainage channel.

7.1 GROUNDWATER FLOW PER UNIT WIDTH

Based on the hydrogeologic cross-section A-A' in Figure 7.2 and the available information, a scheme for the required calculation is developed as shown in Figure 7.3. The equation of the continuity of groundwater flow is:

$$Q = Av \tag{7.1}$$

where Q is the flow rate in cubic meters per second, A is the cross-sectional area of flow in square meters, and v is Darcy's velocity of groundwater in meters per second.

The cross-sectional area of flow is:

$$A = ab \tag{7.2}$$

where a is the width of the flow in meters, and b is the aquifer thickness in meters.

Darcy's velocity, also called the *specific discharge*, is given as (see also Problem 3):

$$v = -Ki \tag{7.3}$$

where K is the hydraulic conductivity of the aquifer material in meters per second, and i is the hydraulic gradient. The minus sign denotes that Darcy's velocity is oriented in the direction of the hydraulic head decrease, *i.e.*, down the hydraulic gradient. Hydraulic gradient is an infinitesimally small change of hydraulic head (dh) along an infinitesimally small length of the flow path (dl):

$$i = \frac{dh}{dl} \tag{7.4}$$

In the case of one-dimensional groundwater flow in an isotropic, homogeneous, confined aquifer, the flow rate per unit width (also called the *unit flow*) is given as:

$$q = -bK\frac{dh}{dl} \tag{7.5}$$

or

$$q = -T\frac{dh}{dl} \tag{7.5a}$$

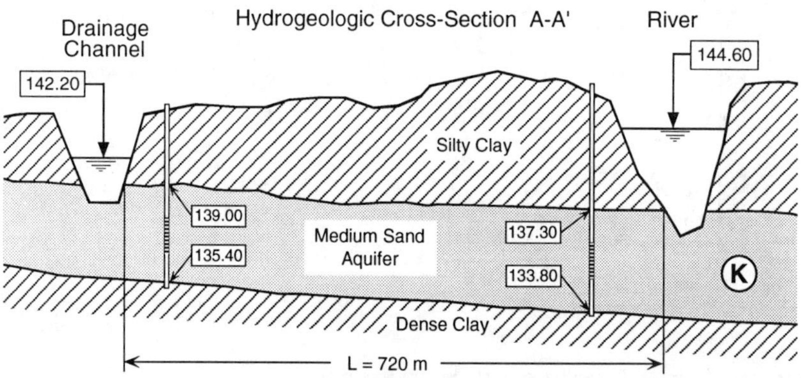

Figure 7.2 Hydrogeologic cross-section between the river and the drainage channel shown in Figure 7.1 as determined by field investigations.

where T is the aquifer's transmissivity (T=Kb). The linear differential equation (7.5a) can be written in the following form:

$$dh = -\frac{q}{T}dl \tag{7.6}$$

Its integration with the boundaries $l=0$, $h=h_1$ (River), and $l=L$, $h=h_2$ (Drainage Channel):

$$\int_{h_1}^{h_2} dh = -\frac{q}{T}\int_0^L dl \tag{7.7}$$

gives the solution:

$$h_2 - h_1 = -\frac{q}{T}L \quad \text{or} \quad \frac{q}{T}L = h_1 - h_2 \tag{7.8}$$

Finally, the discharge rate per unit width of the aquifer is:

$$q = T\frac{h_1 - h_2}{L} \quad (m^2 / s) \tag{7.9}$$

From Figure 7.2 it can be seen that the aquifer thickness remains fairly constant and that an average value for b of 3.5 meters can be adopted. If, however, aquifer thickness varies significantly, it should be integrated (see Problem 10). Also, if the slope of the aquifer is significant (e.g., more than 10 degrees), the real thickness should be determined (see Problem 9). Substituting the values from the calculation scheme (Figure 7.3) gives the

$$q = 0.00122 \; \frac{m^2}{s} \; \frac{144.60\ m \; - \; 142.20\ m}{720\ m} \; = \; 0.000004067 \;\; m^2 / s$$

$$q = \; 4.07 \times 10^{-6} \;\; m^2 / s$$

i.e., groundwater flow rate is 4.07×10^{-6} m^3/s per one meter of aquifer width.

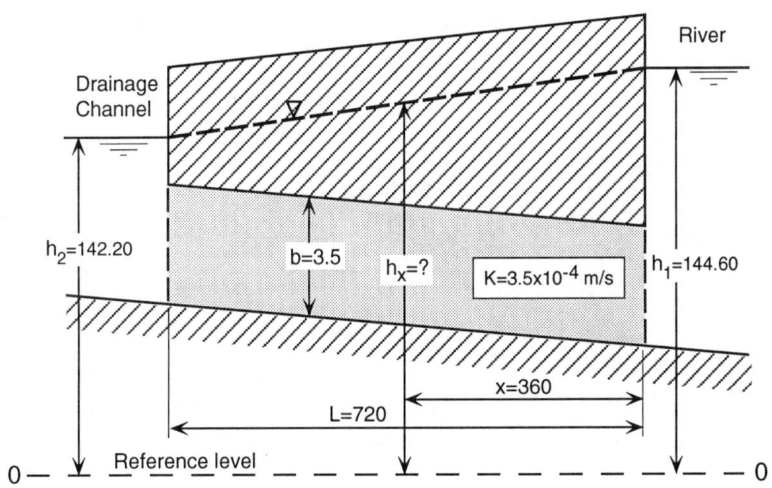

Figure 7.3 Calculation scheme for Problem 7.

7.2 POSITION OF THE HYDRAULIC HEAD

The position of the potentiometric surface at distance x from the higher hydraulic head (River) using equation (7.9) is given as:

$$h_x = h_l - \frac{q}{T}x) \tag{7.10}$$

Substituting the values from the calculation scheme (Figure 3) into equation (7.10) gives the final solution:

$$h_x = 144.60\ m \quad - \quad \frac{4.07 \times 10^{-6}\ m^2 / s}{1.22 \times 10^{-3}\ m^2 / s} \times 360\ m = 143.40\ m$$

Note: It can be seen that a drop of the potentiometric surface at midpoint is exactly one half the difference between the water surface elevations in the river and the channel. Decrease of the hydraulic gradient in a confined, homogeneous, isotropic aquifer is linear, i.e., the potentiometric surface shown in the cross-section A-A' is a straight line.

PROBLEM 8
CONFINED HETEROGENEOUS AQUIFER

Calculate the groundwater flow per unit width of a confined aquifer in the section between the two parallel rivers shown in Figure 8.1. The aquifer can be divided into three homogeneous areas with different hydraulic conductivities: $K_1 = 6.2 \times 10^{-5}$ m/s, $K_2 = 7.9 \times 10^{-4}$ m/s, $K_3 = 4.1 \times 10^{-4}$ m/s. Draw a calculation scheme showing all parameters necessary for the solution.

Determine the position of the piezometric surface on the cross-section.

Determine the "equivalent hydraulic conductivity" and compare it with the average hydraulic conductivity. Also compare the results of groundwater flow calculation using these two conductivities with the result obtained using the original calculation scheme.

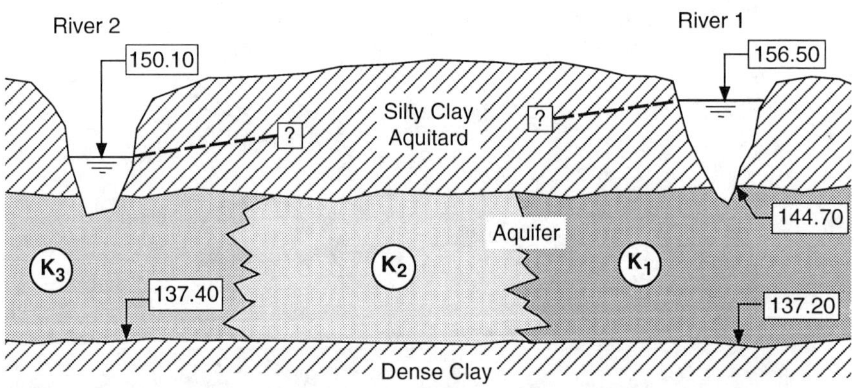

Figure 8.1 Hydrogeologic cross-section of a confined aquifer with three homogeneous areas of different hydraulic conductivity.

8.1 UNIT GROUNDWATER FLOW

The groundwater flow rate per unit width of a confined, homogeneous and isotropic aquifer is given as:

$$q = bK\frac{h_1 - h_2}{L} \tag{8.1a}$$

$$q = T\frac{h_1 - h_2}{L} \tag{8.1b}$$

The total loss of groundwater potential along a flow path is:

$$h_1 - h_2 = \Delta h = \frac{qL}{T} \tag{8.2}$$

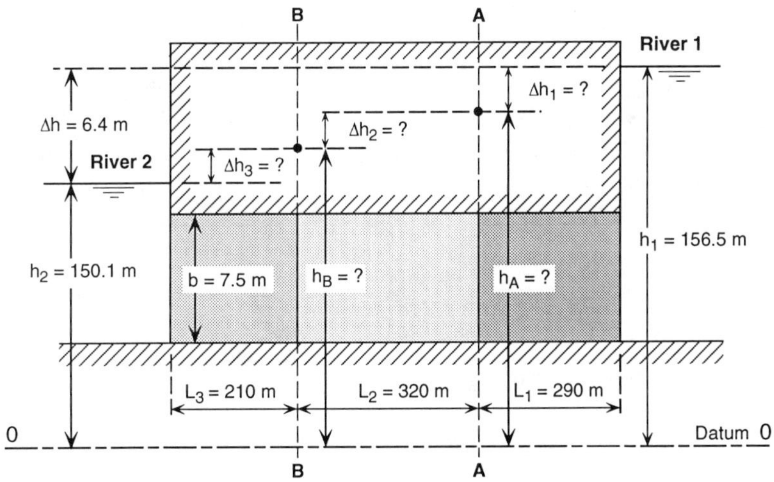

Figure 8.2 Calculation scheme with information necessary for the solution.

The loss of potential (hydraulic head) in each section is:

$$\Delta h_1 = \frac{q_1 L_1}{T_1} \tag{8.3a}$$

$$\Delta h_2 = \frac{q_2 L_2}{T_2} \tag{8.3b}$$

$$\Delta h_3 = \frac{q_3 L_3}{T_3} \tag{8.3c}$$

The total loss is the sum of the three (all) partial losses:

$$\Delta h = \Delta h_1 + \Delta h_2 + \Delta h_3 \tag{8.4}$$

$$\Delta h = \sum_{i=1}^{i=3} \Delta h_i \tag{8.5}$$

The flow rate through each section is the same because the cross-sectional area of the flow is the same: $q_1 = q_2 = q_3$, which gives the following form for the equation (8.5):

$$\Delta h = q \sum_{i=1}^{i=3} \frac{L_i}{T_i} \tag{8.6}$$

The unit flow rate (flow rate per unit width of the aquifer) is then:

$$q = \frac{\Delta h}{\sum_{i=1}^{i=3} \frac{L_i}{T_i}} \tag{8.7}$$

$$q = \frac{b\Delta h}{\sum_{i=1}^{i=3} \frac{L_i}{K_i}} \tag{8.8}$$

where b is the thickness of the aquifer assumed to be constant since the elevations of the aquifer top and bottom do not change significantly, *i.e.*, they remain roughly parallel to each other and horizontal. If this was not the case, the aquifer thickness would have to be integrated (see Problem 10).

Substituting information from the calculation scheme into equation (8.8) gives the final solution:

$$q = \frac{7.50m \times 6.40m}{\frac{290m}{6.2 \times 10^{-5} m/s} + \frac{320m}{7.9 \times 10^{-4} m/s} + \frac{210m}{4.1 \times 10^{-4} m/s}}$$

$$q = 8.58 \times 10^{-6} m^2/s$$

Note: In equations (8.1) and (8.2) one can use either actual values for the hydraulic heads in the rivers (h_1=56.5 m; h_2=50.1 m) or some "new" values if a changing of the reference level is preferred. For example, the reference level can be placed at the base of the aquifer below the River 1 since it is approximately horizontal. This would give the following values for the hydraulic heads: h_1=9.3 m; h_2=12.9 m. In both cases the head difference is the same: Δh=6.4 m. In the case of unconfined aquifers one should be very careful in choosing a reference level because the related equation for calculating flow rate is not linear (see Problem 11). In other words, the hydraulic head used for the calculation should correspond to the actual saturated thickness of the aquifer; using water table altitudes (height above the sea level) for the hydraulic heads will yield incorrect results.

8.2 POSITION OF THE PIEZOMETRIC SURFACE

The position (elevation) of the piezometric surface at verticals A and B is calculated using equations 8.3a, 8.3b and 8.3c:

$$\Delta h_1 = \frac{q_1 L_1}{T_1} = 5.35m$$

$$\Delta h_2 = \frac{q_2 L_2}{T_2} = 0.46m$$

$$\Delta h_3 = \frac{q_3 L_3}{T_3} = 0.59m$$

$$h_A = h_1 - \Delta h_1 = 151.15m$$

$$h_B = h_A - \Delta h_2 = 150.69m$$

Figure 8.3 is a schematic presentation of the solution. The largest hydraulic loss of groundwater flow is in the least permeable section of the aquifer (the slope of the piezometric surface is the steepest in the section with the hydraulic conductivity K_1). Accordingly, this loss is the smallest in the central, most permeable section (K_2).

Note that, in general, the position (slope) of the piezometric surface is dependent on the aquifer's transmissivity. It means that, in addition to hydraulic conductivity, the thickness of the aquifer is also important. For the same aquifer material, the slope will be more steep where the aquifer is thin and less steep where the aquifer is thicker. This is true for both confined and unconfined aquifers.

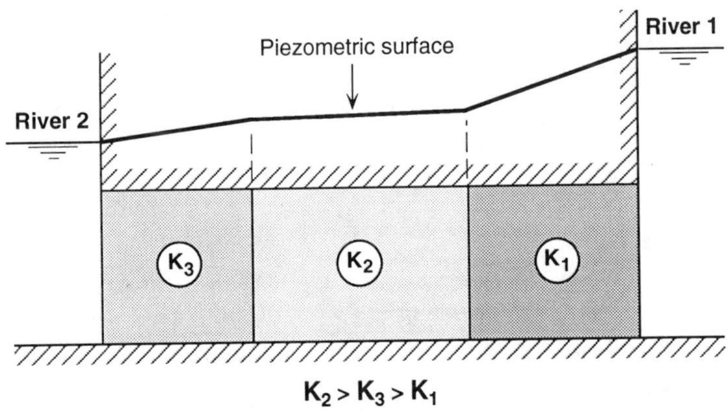

Figure 8.3 Schematic presentation of the position of the piezometric surface in the aquifer. Note that the linear loss of the hydraulic head is the largest in the least permeable section of the aquifer (K_1).

8.3 EQUIVALENT AND AVERAGE HYDRAULIC CONDUCTIVITIES

The "*equivalent hydraulic conductivity*" represents the aquifer as homogeneous throughout the calculation domain. It gives the same value for the rate of groundwater flow as varying conductivities. However, the distribution (slope) of the hydraulic head in such a case is uniform and does not reflect the reality.

Introducing the equivalent hydraulic conductivity (K_{eqv}) into the equation for the unit discharge rate (8.1a) gives:

$$q = bK_{eqv} \frac{h_1 - h_2}{L} \tag{8.9}$$

The unit discharge rate (discharge rate per unit width) is also given by equation (8.8) as:

$$q = \frac{b\Delta h}{\sum_{i=1}^{i=3} \frac{L_i}{K_i}} \tag{8.10}$$

Relating equations (8.9) and (8.10) gives the final expression for the equivalent hydraulic conductivity:

$$K_{eqv} = \frac{L}{\sum_{i=1}^{i=3} \frac{L_i}{K_i}} \tag{8.11}$$

Substitution of the values from the calculation scheme into equation (8.11) leads to the final solution for the equivalent hydraulic conductivity:

$$K_{eqv} = 1.47 \times 10^{-4}\, m/s$$

The *average (mean) hydraulic conductivity* is given as:

$$K^* = \frac{\sum_{i=1}^{i=3} K_i}{n} \tag{8.12}$$

or, by using actual values:

$$K^* = \frac{6.2 \times 10^{-5}\, m/s + 7.9 \times 10^{-4}\, m/s + 4.1 \times 10^{-4}\, m/s}{3}$$

$$K^* = 4.2 \times 10^{-4}\, m/s$$

The unit groundwater flow rate, if calculated with the mean hydraulic conductivity, is:

$$q^* = bK^* \frac{\Delta h}{L} \tag{8.13}$$

or, after substituting numeric values:

$$q^* = 2.46 \times 10^{-5}\, m^2/s$$

This result is considerably higher than the real value of the groundwater flow rate per unit width calculated using all three hydraulic conductivities (equation 8.8).

PROBLEM 9
CONFINED ONE-DIMENSIONAL STEADY STATE FLOW AQUIFER WITH FORKS

Calculate groundwater flow per unit width downgradient from the monitoring well 3 (MW3) shown on the hydrogeological cross-section in Figure 9.1. Hydraulic conductivity of the confined sand aquifer is 3.7×10^{-5} m/s and it may be considered homogeneous throughout the study area. Water levels registered in the monitoring wells MW1, MW2, and MW3 are 144.6, 145.0, and 142.4 meters above sea level respectively. The general slope of the "upper fork" of the aquifer is 20 degrees, and the "lower fork" slopes at approximately 15 degrees. Monitoring wells MW1 and MW2 are clustered, *i.e.*, very close to each other (about 1 meter apart). The distance between the MW3 and the MW1/MW2 cluster is 170 meters. Draw a calculation scheme for solving the problem.

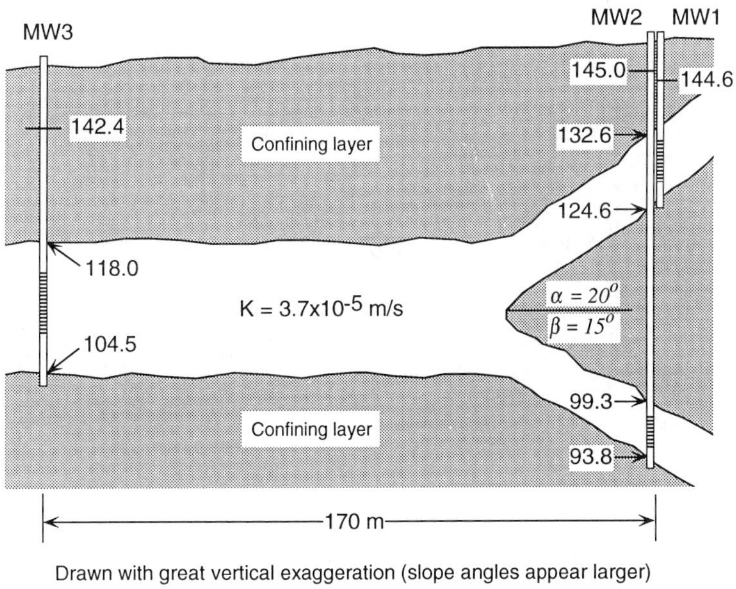

Figure 9.1 Simplified hydrogeologic cross-section of the confined aquifer with forks. Note that, because of the vertical exaggeration, sloping angles a and b appear bigger than their true values of 20 and 30 degrees respectively.

9.1 UNIT GROUNDWATER FLOW

Groundwater flow per unit width in a confined aquifer of uniform thickness is a product of the aquifer transmissivity and the hydraulic gradient (see also Problem 7 and equation 7.9):

$$q = T\frac{h_1 - h_2}{L} \tag{9.1}$$

where h_1 is a higher hydraulic head and h_2 is a lower hydraulic head measured in a homogeneous part of the aquifer along the same flow line, and L is the distance between these two heads.

The aquifer shown in Figure 9.1 has three sections -- the horizontal portion, "the upper fork", and "the lower fork" -- all with different uniform thickness, and with only one known hydraulic head per section. Calculation scheme in Figure 9.2 shows all the elements needed to determine the flow downgradient from MW3. As explained by the general hydraulic equation of the continuity of flow (see Problem 3 and equation 3.2), the unit groundwater flow through the horizontal section of the confined aquifer remains constant since groundwater neither enters nor leaves the aquifer. The same is true for the flows in two forks which combine to generate the flow in the horizontal section:

$$q_1 + q_2 = q_3 \tag{9.2}$$

In order to calculate all three unit flows, the following elements should be determined for each section:
- one additional hydraulic head,
- distance between this new calculated head and the known head,
- thickness of the aquifer.

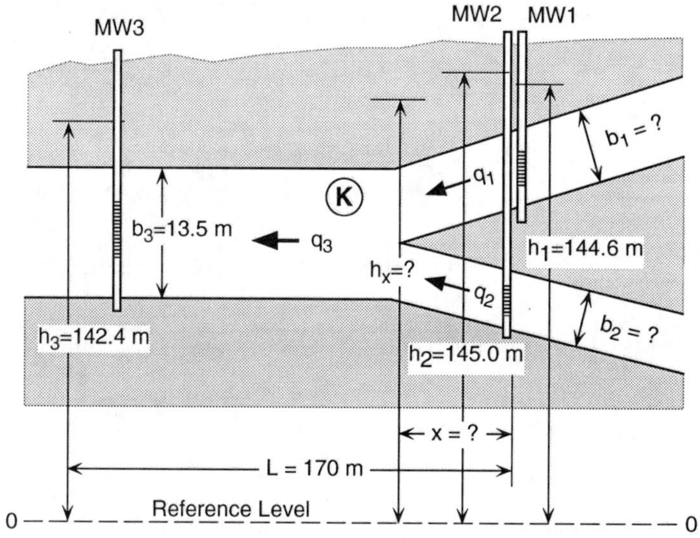

Figure 9.2 Calculation scheme for the problem.

As shown in Figure 9.2, the hydraulic head h_x is common for all three sections of the aquifer: it is a lower (downgradient) head for both forks, and at the same time it is a higher (upgradient) head for the horizontal section. Its position is determined by the distance x, *i.e.*, the distance between the MW1/MW2 cluster and the beginning of the horizontal section where two forks meet. The unit groundwater flows for the three sections of the aquifer are given with the following equations:

Upper fork:
$$q_1 = T_1 \frac{h_1 - h_x}{x} \tag{9.3}$$

Lower fork:
$$q_2 = T_2 \frac{h_2 - h_x}{x} \tag{9.4}$$

Horizontal section:
$$q_3 = T_3 \frac{h_x - h_3}{L} \tag{9.5}$$

where T_1, T_2, and T_3 are transmissivities determined from the aquifer hydraulic conductivity and the aquifer thickness ($T_1=Kb_1$; $T_2=Kb_2$; $T_3=Kb_3$).

The thicknesses of the three sections, and the position of the unknown hydraulic head h_x (distance x), can be easily found by performing simple trigonometric calculations. Once these values are found, equations (9.2) through (9.5) can be solved as a system of 4 equations with 4 unknowns (q_1, q_2, q_3, *and* h_x).

9.2 AQUIFER THICKNESS

In calculations of groundwater flow rate and aquifer transmissivity it is very important to use aquifer true thickness and not its apparent thickness (which is often the case in practice). Aquifer true thickness is the perpendicular distance between its top and bottom, whereas aquifer apparent thickness is the vertical distance between its top and bottom as determined by exploratory drilling. If the aquifer sloping angle is equal or less than 10 degrees, it is a common practice to use the apparent thickness since the error introduced is only 1.5% or less. However, for steeper slopes the errors can be significant. For example, the angles of 20, 30, and 40 degrees would yield errors of 6.0, 13.4 and 23.4 % respectively if the apparent thickness is used instead of the aquifer true thickness. Figure 9.3 shows the trigonometric relationship between the angle of slope (α), the aquifer true thickness (b), and the aquifer apparent thickness (Δh).

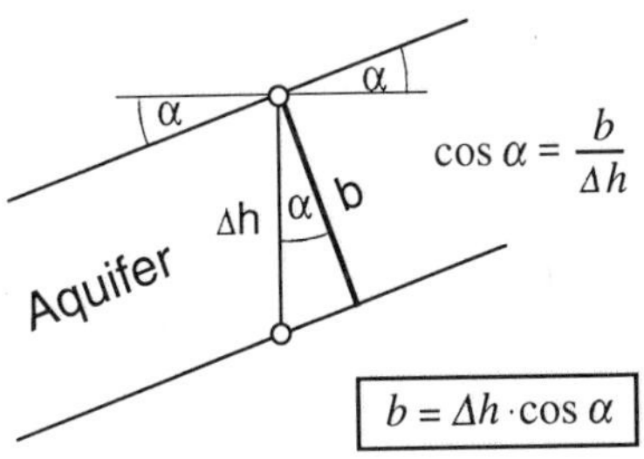

Figure 9.3 Determination of the aquifer true thickness (b) from the angle of slope (α) and the apparent thickness (Δh).

True thickness of the "upper fork" is:

$$b_1 = \Delta h_1 \cos\alpha \tag{9.6}$$

$$b_1 = (132.6m - 124.6m) \times \cos 20^o = 7.5m$$

True thickness of the "lower fork" is:

$$b_2 = \Delta h_2 \cos\beta \tag{9.7}$$

$$b_1 = (99.3m - 93.8m) \times \cos 15^o = 5.3m$$

The thickness of the horizontal section, as determined from the MW3 log, is:

$$118.0m - 104.5m = 13.5m$$

9.3 POSITION OF THE HYDRAULIC HEAD h_x

Figure 9.4 shows elements needed for the trigonometric determination of the distance x. Combining basic equations for tangents of angles α and β leads to:

$$x \tan\alpha + x \tan\beta = y + z \tag{9.8}$$

and

$$x = \frac{y+z}{\tan\alpha + \tan\beta} \tag{9.9}$$

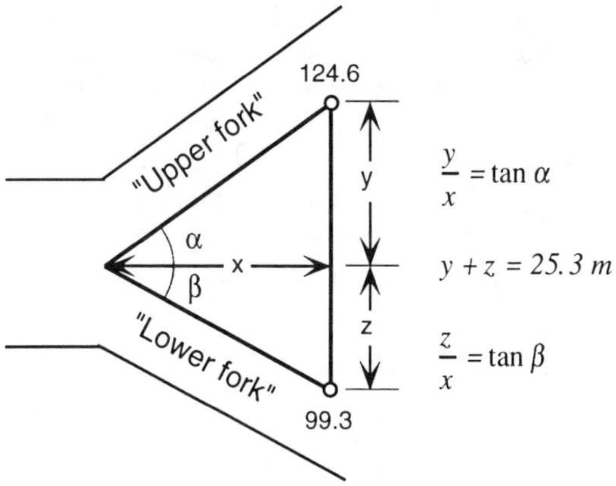

Figure 9.4 Elements for the trigonometric determination of the distance x (position of the hydraulic head h_x).

Substitution of known values into equation (9.9) gives the distance x:

$$x = \frac{25.3\ m}{\tan 20^o + \tan 15^o} = \frac{25.3\ m}{0.27 + 0.36} = 40\ m$$

9.4 HYDRAULIC HEAD h_x

After finding the thickness of each of the three sections, and the position of the hydraulic head h_x, its value is easily determined by combining equations (9.2) through (9.5):

$$Kb_1 \frac{h_1 - h_x}{x} + Kb_2 \frac{h_2 - h_x}{x} = Kb_3 \frac{h_x - h_3}{L} \tag{9.10}$$

Taking out the hydraulic conductivity (K) and solving for h_x as the only unknown gives:

$$h_x = \frac{b_1Lh_1 + b_2Lh_2 + b_3xh_3}{b_1L + b_2L + b_3x} \tag{9.11}$$

$$h_x = 144.3\ m$$

Figure 9.5 is a schematic presentation of the position of the hydraulic head in all three sections of the aquifer (note that all lines are straight indicating the linear decrease of the hydraulic head characteristic for confined aquifers).

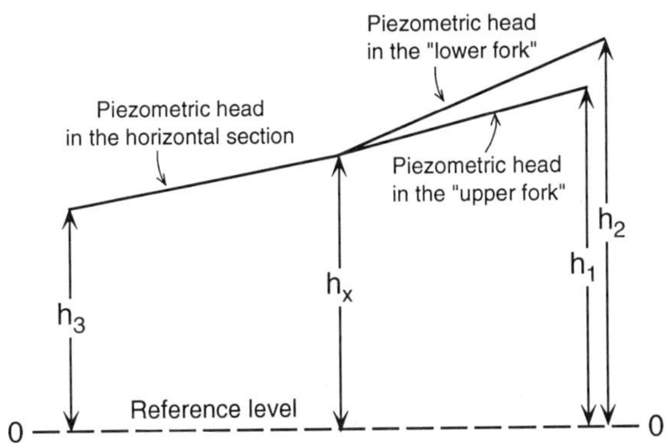

Figure 9.5 Schematic presentation of the position of the hydraulic head in all three sections of the aquifer.

9.5 UNIT GROUNDWATER FLOWS

Unit groundwater flow downgradient from MW3, or flow in the aquifer's horizontal section (which is the same), is determined by equation (9.5):

$$q_3 = T_3 \frac{h_x - h_3}{L} = Kb_3 \frac{h_x - h_3}{L}$$

$$q_3 = 3.7 \times 10^{-5}\, m/s \cdot 13.5\, m \cdot \frac{144.3\, m - 142.4\, m}{170\, m}$$

$$q_3 = 5.58 \times 10^{-6}\, m^2/s$$

Similarly, unit groundwater flows (*i.e.*, groundwater flow rates per 1 meter of aquifer width) in the upper and lower forks are calculated using equations (9.3) and (9.4) respectively:

$$q_1 = 3.7 \times 10^{-5}\, m/s \cdot 7.5\, m \cdot \frac{144.6\, m - 144.3\, m}{40\, m}$$

$$q_1 = 2.08 \times 10^{-6}\, m^2/s$$

and

$$q_2 = 3.7 \times 10^{-5}\, m/s \cdot 5.3\, m \cdot \frac{145.0\, m - 144.3\, m}{40\, m}$$

$$q_2 = 3.43 \times 10^{-6}\, m^2/s$$

As mentioned earlier, the unit groundwater flow downgradient from MW3 is the sum of unit groundwater flows in two forks:

$$q_3 = q_1 + q_2$$

$$q_3 = 2.08 \times 10^{-6} m^2/s \,+\, 3.43 \times 10^{-6} m^2/s \,=\, 5.51 \times 10^{-6} m^2/s$$

This result is very close to one obtained using equation (9.5) and hydraulic heads h_x and h_3.

PROBLEM 10
CONFINED ONE-DIMENSIONAL FLOW AQUIFER WITH VARYING THICKNESS

Calculate unit groundwater flow in a confined aquifer shown in Figure 10.1. Also find the piezometric surface in the monitoring well MW2. The hydraulic conductivity of the aquifer, which can be considered as homogeneous and isotropic, is 5.6×10^{-5} m/s.

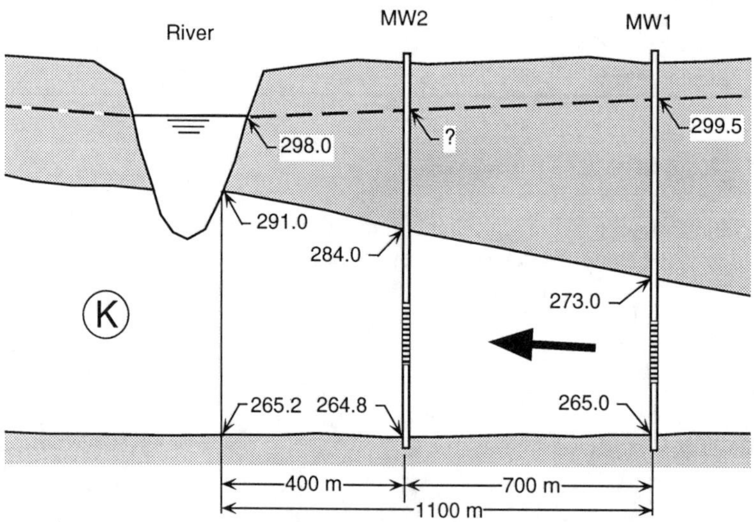

Figure 10.1 Schematic hydrogeological cross section of a confined aquifer with varying thickness.

10.1 UNIT GROUNDWATER FLOW

As can be seen in Figure 10.1, the aquifer has a varying thickness and therefore a varying transmissivity. This means that a common equation describing one-dimensional groundwater flow in a confined aquifer (equation 7.9), derived in Problem 7, cannot be applied directly since the aquifer thickness has to be integrated. In many cases the varying thickness can be described with a simple linear function or a combination of several linear functions with different slopes. For the situation shown in Figure 10.1 the linear function is easily derived by setting the coordinate system X-Y so that it corresponds to the length-aquifer thickness system in the field (L-b) as shown in Figure 10.2.

The equation of the straight line in the X-Y coordinate system is

$$y = ax + b \tag{10.1}$$

where a is the slope of the line ($a = \Delta y/\Delta x$) and b is the y-axis intercept. The corresponding equation of the top of the confined aquifer is

$$b = \frac{b_3 - b_1}{L} x + b_1 \tag{10.2}$$

where $(b_3-b_1)/L$ is the slope and b_1 is the intercept. The bottom of the aquifer can be considered horizontal and the reference level is placed at the elevation 265.0 meters (which is the elevation of the contact between the sand aquifer and dense clay at monitoring well MW1).

The general equation of the unit groundwater flow (groundwater flow per unit width of the aquifer) is given as the product of the aquifer thickness (b) and the flow velocity (v):

$$q = b \cdot v \tag{10.3}$$

Groundwater velocity (or Darcy's velocity - see Problem 3) is:

$$v = -Ki \tag{10.4}$$

where K is the hydraulic conductivity of the aquifer material in meters per second, and i is the hydraulic gradient. The minus sign denotes that Darcy's velocity is oriented in the direction of the hydraulic head decrease, *i.e.*, down the hydraulic gradient. The hydraulic gradient is an infinitesimally small change of hydraulic head (dh) along an infinitesimally small length of the flow path (dx):

$$i = \frac{dh}{dx} \tag{10.4a}$$

In the case of one-dimensional groundwater flow in an isotropic, homogeneous, confined aquifer, the flow rate per unit width is then:

$$q = -bK\frac{dh}{dx} \tag{10.5}$$

The unit flow in the aquifer with the changing thickness, as shown in Figures 10.1 and 10.2, is expressed by combining equations (10.2) and (10.5):

$$q = -(\frac{b_3 - b_1}{L}x + b_1)\cdot K\frac{dh}{dx} \tag{10.6}$$

In order to solve the differential equation (10.6), the variables x and h first have to be separated:

$$-dh = \frac{q}{K}\cdot\frac{dx}{\frac{b_3 - b_1}{L}x + b_1} \tag{10.7}$$

Equation (10.7) is then solved for the known boundary conditions: for $x=0$ the hydraulic head is h_1, and for $x=L$ the hydraulic head is h_2:

$$-\int_{h_1}^{h_3} dh = \frac{q}{K}\cdot\int_0^L \frac{dx}{\frac{b_3 - b_1}{L}x + b_1} \tag{10.8}$$

The integral on the left is an elementary integral, and the integral on the right is solved knowing that, in general notification (see also Appendix G):

$$\int \frac{dx}{ax+b} = \frac{1}{a}\ln(ax+b)+C$$

The solution for equation (10.8) is

$$h_1 - h_3 = \frac{q}{K}\cdot\frac{1}{\frac{b_3-b_1}{L}}\cdot\left[\ln\left(\frac{b_3-b_1}{L}L+b_1\right)-\ln\left(\frac{b_3-b_1}{0}0+b_1\right)\right] \tag{10.9}$$

$$h_1 - h_3 = \frac{q}{K}\cdot\frac{L}{b_3-b_1}\cdot\ln\frac{b_3}{b_1} \tag{10.10}$$

and the unit flow rate is then:

$$q = K\frac{h_1-h_3}{L}\cdot\frac{b_3-b_1}{\ln\frac{b_3}{b_1}} \tag{10.11}$$

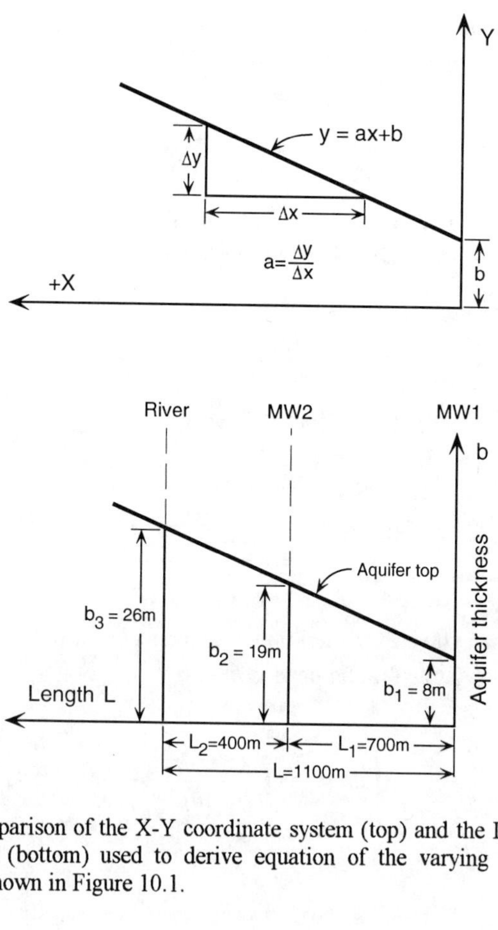

Figure 10.2 Comparison of the X-Y coordinate system (top) and the Length-Thickness coordinate system (bottom) used to derive equation of the varying thickness of the confined aquifer shown in Figure 10.1.

The numeric value of the unit flow is found by inserting numbers from the calculation scheme shown in Figure 10.2 into equation (10.11):

$$q = 5.6 \times 10^{-5}\, m/s \cdot \frac{299.5m - 298.0m}{1100m} \cdot \frac{26m - 8m_1}{\ln \frac{26m}{8m}}$$

$$q = 1.16 \times 10^{-6}\ m^2/s$$

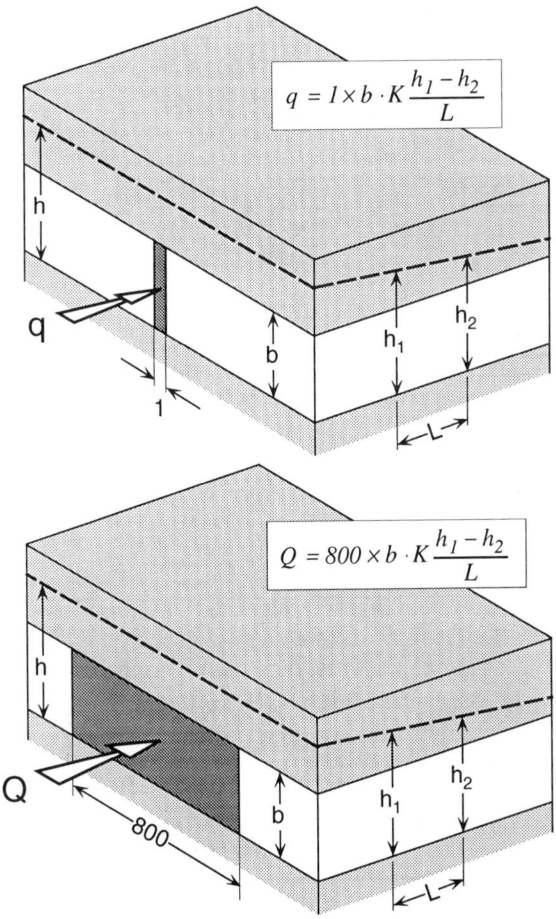

Figure 10.3 Schematic presentation of the unit flow rate (top) and the volumetric flow rate (bottom) in a confined aquifer.

Note that the volumetric flow rate is obtained when the unit flow is multiplied by aquifer width of interest. For example (see Figure 10.3), the total flow rate through an 800 meter wide portion of the aquifer would be

$$Q = 1.16 \times 10^{-6} m^2/s \cdot 800m \ = \ 9.28 \times 10^{-4} m^3/s$$

10.2 POSITION OF THE PIEZOMETRIC SURFACE

Piezometric surface at MW2 is found by solving equation (10.7) with the following boundary conditions (see Figure 10.2): for $x=0$ the hydraulic head is h_1, and for $x=L_1$ the hydraulic head is h_2:

$$-\int_{h_1}^{h_2} dh = \frac{q}{K}\cdot\int_0^{L_1}\frac{dx}{\frac{b_2-b_1}{L_1}x+b_1} \tag{10.12}$$

$$h_2 = h_1 - \frac{q}{K}\cdot\frac{L_1}{b_2-b_1}\cdot\ln\frac{b_2}{b_1} \tag{10.13}$$

Inserting values from the calculation scheme shown in Figure 10.2 gives:

$$h_2 = 299.5m - \frac{1.16\times10^{-6}m^2/s}{5.6\times10^{-5}m/s}\cdot\frac{700m}{19m-8m}\cdot\ln\frac{19m}{8m}$$

$$h_2 = 298.36m$$

Position of the piezometric surface at various distances from MW1 (h_x), and for known aquifer thickness at those distances (b_x), can be calculated using the following general equation based on equation (10.13):

$$h_x = h_1 - \frac{q}{K}\cdot\frac{L_1}{b_x-b_1}\cdot\ln\frac{b_x}{b_1} \tag{10.14}$$

Table 10.1 shows the hydraulic heads calculated for several verticals between MW1 and the River which are used to draw the piezometric surface in Figure 10.4.

If two hydraulic heads (h_1 and h_2) are known in a confined aquifer that has a linearly varying thickness, the position of the piezometric surface at any point between these two heads can be found using the following equation:

$$h_x = h_1 - (h_1-h_2)\cdot\frac{L_x}{L}\cdot\frac{b_2-b_1}{b_x-b_1}\cdot\frac{\ln\frac{b_x}{b_1}}{\ln\frac{b_2}{b_1}} \tag{10.15}$$

Table 10.1 Hydraulic heads between MW1 and the River calculated using equation 10.14

Distance from MW1 in meters	0	200	400	600	700	800	1000	1100
Aquifer thickness in meters	8	11.3	14.5	17.9	19.0	21.0	24.3	26.0
Hydraulic head in meters above mean sea level	299.5	299.06	298.74	298.49	298.36	298.26	298.09	298.0

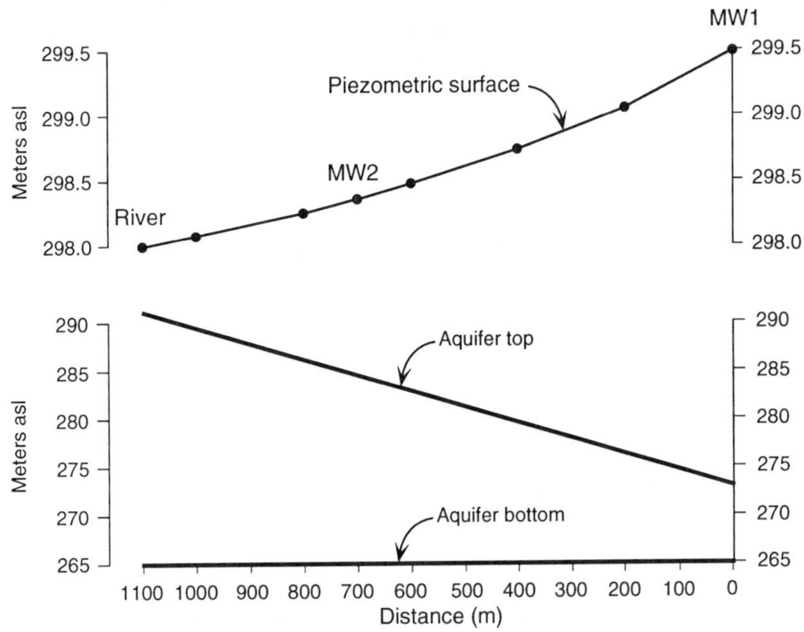

Figure 10.4 Position of the piezometric surface between MW1 and the River. Note a decrease of the piezometric surface gradient with the increasing thickness of the aquifer.

Equation (10.15) is obtained by combining equation (10.14) with a general form of equation (10.11) in which both heads are known. It is useful and simple to use since one need not to know the hydraulic conductivity and the unit flow in the aquifer. As shown in Figure 10.5, equation (10.15) is valid for confined aquifers where the top changes linearly

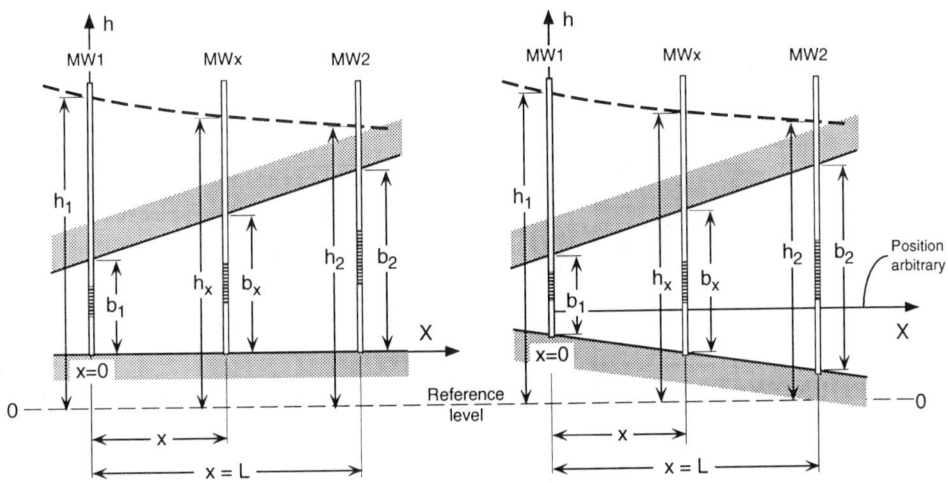

Figure 10.5 Schematic presentation of two confined aquifers with linearly varying thickness. In both cases the hydraulic head h_x can be calculated using equation (10.15).

By applying Darcy's law, and assuming that the velocity vector is horizontal (Dupuit's hypothesis), groundwater flow through the elementary prism of unit width is given as:

$$q = -Kh\frac{dh}{dx} \tag{11.2}$$

Differentiating equation (11.2) and combining it with equation (11.1) gives the expression

$$\frac{d^2(h^2)}{dx^2} = -\frac{2w}{K} \tag{11.3}$$

Differential equation (11.3) can be integrated to give the following general solution with the constants of integration C_1 and C_2:

$$h^2 = -\frac{w}{K}x^2 + C_1 x + C_2 \tag{11.4}$$

The constants of integration, *i.e.*, the particular solution, can be found if the boundary conditions are known. From the calculation scheme in Figure 11.3, the boundary conditions for the flow between the Infiltration Channel and the River are: at $x=0$ (Infiltration Channel) the hydraulic head is $h=h_1$; at $x=L$ (River) the hydraulic head is $h=h_2$. With these boundary conditions equation (11.4) becomes

$$h^2 = h_1^2 - (h_1^2 - h_2^2)\frac{x}{L} + \frac{w}{K}(L-x)x \tag{11.5}$$

Equation (11.5) enables calculation of the hydraulic head (water table elevation) for any given position between two points with known saturated thickness. Water table elevations at several points between the Infiltration Channel and the River, when the proposed drainage gallery is not operating, are calculated using equation (11.5) and the results are shown in Table 11.1.

Table 11.1 Water table elevations between the Infiltration Channel and the River when the aquifer is under the influence of infiltration.

Distance from the Infiltration Channel (x)	100	200	300	450	600	700	800
Saturated thickness above the impermeable horizontal base (h)	10.21	10.31	10.29	9.95	9.57	9.07	8.40
Water table elevation in meters	283.21	283.31	283.29	282.95	282.57	282.07	281.40

The infiltration rate, which has dimension of velocity, was calculated from the effective precipitation data for May:

$$w = \frac{P_e}{t} = \frac{2.5\ inches}{31\ days} = \frac{0.0635\ meters}{2.6784 \times 10^6\ seconds} = 2.371 \times 10^{-8}\ m/s$$

The saturated thickness (h) was calculated from the base of the aquifer which is assumed to be horizontal and chosen for the new reference level (see Figures 11.2 and 11.3). The water table elevation (or altitude above sea level) was obtained by adding calculated saturated thickness to the elevation of the new reference level (273.0 m asl) placed at the bottom of the Drainage Gallery as shown in Figures 11.3 and 11.4.

If there is no gain of water due to infiltration of precipitation, or loss of water due to evapotranspiration, equation (11.5) reduces to

$$h^2 = h_1^2 - (h_1^2 - h_2^2)\frac{x}{L} \tag{11.6}$$

Table 11.2 shows water table elevations at the same points as in Table 11.1 when there is no infiltration of precipitation, and the proposed drainage gallery is not operating (equation (11.6) is used in this case).

Table 11.2 Water table elevations between the Infiltration Channel and the River when the aquifer is not under the influence of infiltration.

Distance from the Infiltration Channel (x)	100	200	300	450	600	700	800
Saturated thickness above the impermeable horizontal base (h)	9.78	9.57	9.34	9.00	8.64	8.39	8.13
Water table elevation (in meters)	282.78	282.57	282.34	282.00	281.64	281.39	281.13

Groundwater flow per unit width at any distance x from the coordinate beginning (*i.e.*, the origin, where $x=0$) is given as

$$q = -Kh\frac{dh}{dx} \tag{11.7}$$

Differentiation of equation (11.5) leads to

$$h\frac{dh}{dx} = -\frac{h_1^2 - h_2^2}{2L} - \frac{w}{K}(x - \frac{L}{2}) \tag{11.8}$$

The equation for the unit groundwater flow in an unconfined aquifer under the influence of precipitation is obtained by combining equations (11.7) and (11.8):

$$q = K\frac{h_1^2 - h_2^2}{2L} + w(x - \frac{L}{2}) \tag{11.9}$$

For the case shown in Figure 11.3, unit groundwater flow at the point of contact between the Infiltration Channel and the aquifer is calculated for the vertical where $x=0$:

$$q_0 = K\frac{h_1^2 - h_2^2}{2L} - w\frac{L}{2} \tag{11.10}$$

$$q_0 = 2.1 \times 10^{-4}\, m/s \frac{10^2 m - 8^2 m}{2(450m + 400m)} - 2.371 \times 10^{-8}\, m/s \frac{850m}{2}$$

$$q_0 = -5.64 \times 10^{-6}\ m^2/s$$

The negative sign indicates that groundwater flow is in the negative direction of the X axis, *i.e.*, toward the Infiltration Channel. Unit groundwater flow entering the River is calculated for the vertical where $x=L$ (or 850 m):

$$q_L = K\frac{h_1^2 - h_2^2}{2L} + w\frac{L}{2} \tag{11.11}$$

$$q_L = 1.453 \times 10^{-5}\ m^2/s$$

From the above analysis it can be seen that both the Infiltration Channel and the River are receiving groundwater flow generated by the infiltration of precipitation due to formation of a local groundwater divide ("mound"). The total groundwater flow per unit width in the section between the Channel and the River is therefore the sum of the absolute values of two flows, and equals the rate of infiltration multiplied by the distance between the Channel and the River:

$$q_{total} = q_0 + q_L = \left| -5.64 \times 10^{-6}\ m^2/s \right| + \left| 1.453 \times 10^{-5}\ m^2/s \right| \tag{11.12}$$

$$q_{total} = 2.017 \times 10^{-5}\ m^2/s$$

or

$$q_{total} = wL = 2.371 \times 10^{-8}\ m/s\ \times 850\ m = 2.015 \times 10^{-5}\ m^2/s \tag{11.13}$$

11.2 GROUNDWATER FLOW TOWARD A DRAINAGE GALLERY IN AN UNCONFINED AQUIFER

The unit groundwater inflow into the Drainage Gallery, as shown in Figure 11.4, is the sum of flows from two directions. From the Infiltration Channel direction, the flow is generated by the infiltration of precipitation and by the induced recharge from the Channel due to the reversed hydraulic gradient:

$$q_1 = K\frac{h_1^2 - h_G^2}{2L_1} + w\frac{L_1}{2} \tag{11.14}$$

$$q_1 = 2.773 \times 10^{-5}\ m^2/s$$

Similarly, the unit inflow from the River direction is generated by the infiltration of precipitation and by the induced recharge from the River:

$$q_2 = K\frac{h_2^2 - h_G^2}{2L_2} + w\frac{L_2}{2} \tag{11.15}$$

$$q_2 = 2.049 \times 10^{-5}\ m^2 / s$$

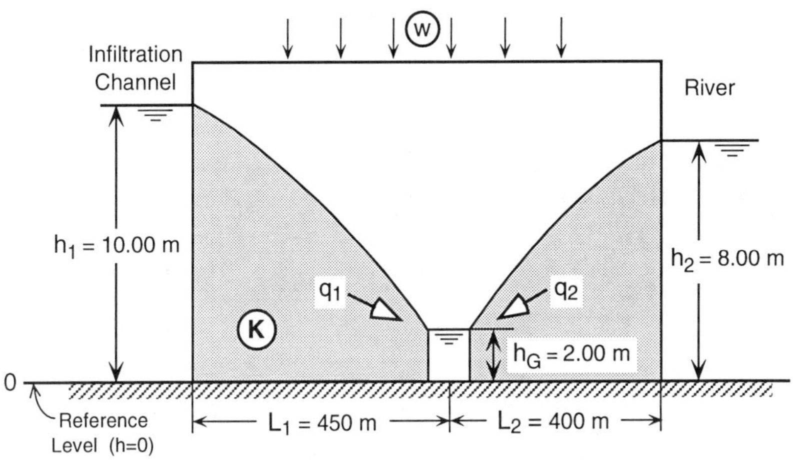

Figure 11.4 Calculation scheme for Problem 11.

The total unit inflow into the Drainage Gallery, from both directions, is:

$$q_{total} = q_1 + q_2 = 4.822 \times 10^{-5}\ m^2 / s \tag{11.16}$$

The calculated groundwater inflow per unit width of the Drainage Gallery can be used to estimate total inflow for a desired length. For example, for a 200 meters long gallery, the total inflow would be:

$$Q = q_{total} \times 200\ m = 9.644 \times 10^{-3}\ m^3 / s \tag{11.17}$$

PROBLEM 12
UNCONFINED AQUIFER WITH SLOPING BASE

Find the hydraulic conductivity in the section of the unconfined aquifer between the river and the vertical at point A as shown in Figure 12.1. The base of the aquifer can be divided into two sections: the first one slopes with a gradient $i_o>0$, while the second one (closer to the channel) is horizontal. The hydraulic conductivity of the second section, as determined by testing a sample from the piezometer, is $K_2 = 3.8 \times 10^{-3}$ cm/s. Also find the hydraulic head at the point where the base changes the slope (point A on the cross-section in Figure 12.1) and the unit groundwater flow between the river and the drainage channel.

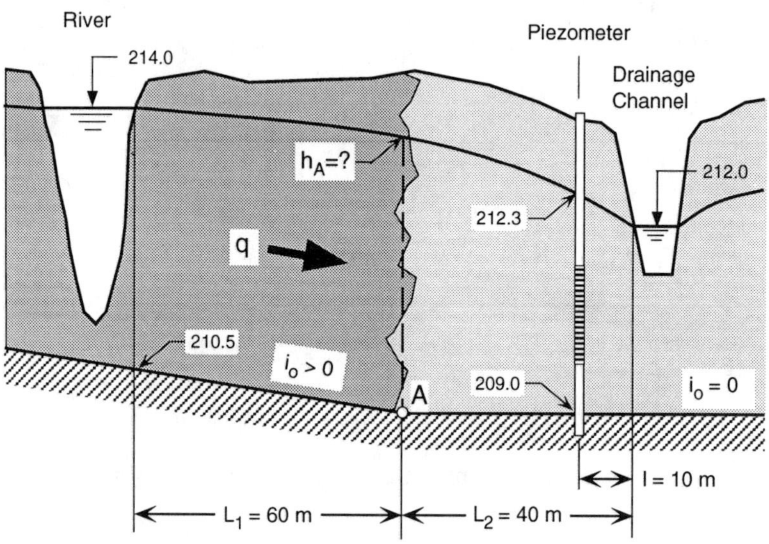

Figure 12.1 Hydrogeological cross-section of an unconfined aquifer with a sloping and a horizontal impermeable base.

12.1 GENERAL EQUATION OF STEADY STATE GROUNDWATER FLOW IN AN UNCONFINED AQUIFER

Figure 12.1 shows possible cases of steady state flow in an unconfined aquifer. The uniform flow (case a), in which the stream lines are parallel and the slope of the hydraulic head is equal to the slope of the aquifer base, is far less common then non-uniform flow in which the stream lines are not parallel and the hydraulic head gradient is different from the aquifer base gradient (cases b through e).

An important element to be considered in deriving the equation of flow for both uniform and non-uniform conditions is the *normal aquifer depth*, denoted h_0 in Figure 12.1a. It is the depth perpendicular to both the hydraulic head line and the aquifer base line which are parallel in uniform flow conditions. Having in mind the principle of the conservation of mass (see Problem 3 and equation 3.2), the flow rate in uniform conditions is the product of the velocity of flow and the cross-sectional area of flow:

$$Q = A \cdot v \tag{12.1}$$

The unit flow, or the flow rate per unit width of the aquifer, is a product of the aquifer thickness (*i.e.*, the *normal depth* -- h_o) and the flow velocity:

$$q = h_o \cdot v \tag{12.2}$$

The flow velocity is in a discrete, non-differential form given as:

$$v = K \cdot i_o \tag{12.3}$$

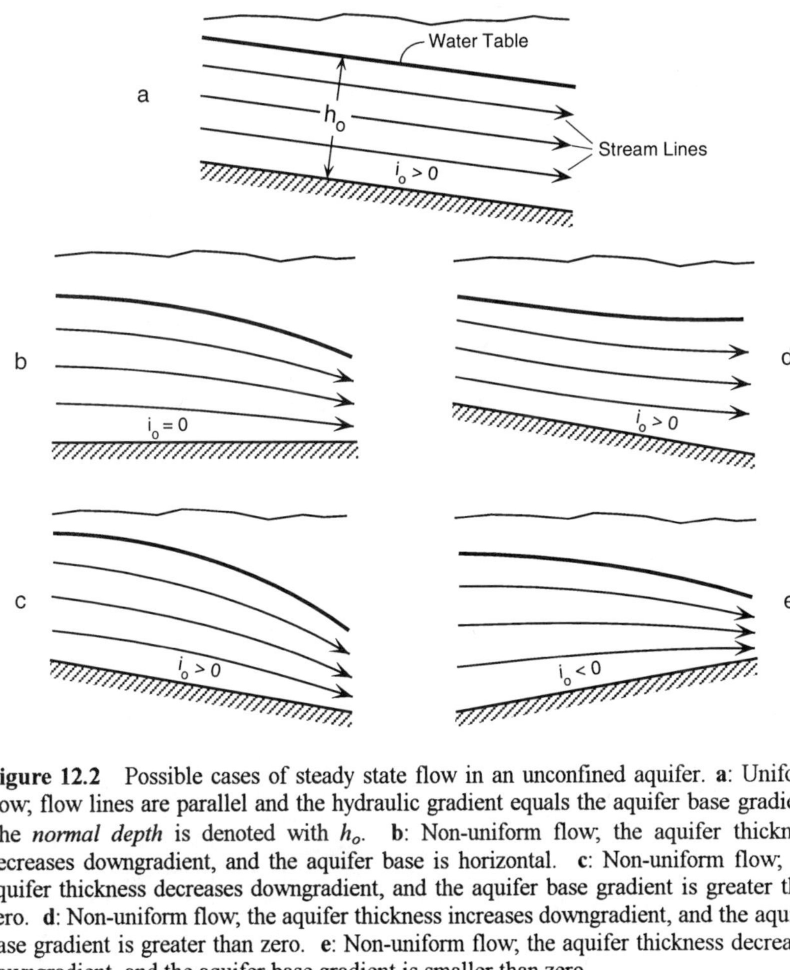

Figure 12.2 Possible cases of steady state flow in an unconfined aquifer. **a**: Uniform flow; flow lines are parallel and the hydraulic gradient equals the aquifer base gradient. The *normal depth* is denoted with h_o. **b**: Non-uniform flow; the aquifer thickness decreases downgradient, and the aquifer base is horizontal. **c**: Non-uniform flow; the aquifer thickness decreases downgradient, and the aquifer base gradient is greater than zero. **d**: Non-uniform flow; the aquifer thickness increases downgradient, and the aquifer base gradient is greater than zero. **e**: Non-uniform flow; the aquifer thickness decreases downgradient, and the aquifer base gradient is smaller than zero.

where i_o is the hydraulic gradient equal to the aquifer base gradient. The unit flow is then:

$$q = h_o \cdot K \cdot i_o \tag{12.4}$$

As already mentioned, the uniform flow is rare in natural conditions and a more general and complex equation has to be derived to describe non-uniform flow in which aquifer thickness changes.

Dupuit's hypothesis allows the derivation of an equation which gives an accurate calculation of the flow rate, but is less accurate for calculating the position of the hydraulic head, especially near the points of aquifer discharge. Figure 12.3 illustrates the hypothesis which states that:

- Equipotential lines are vertical.
- Velocity is constant along the vertical, *i.e.*, along the equipotential line.
- The velocity vector has only the horizontal component.

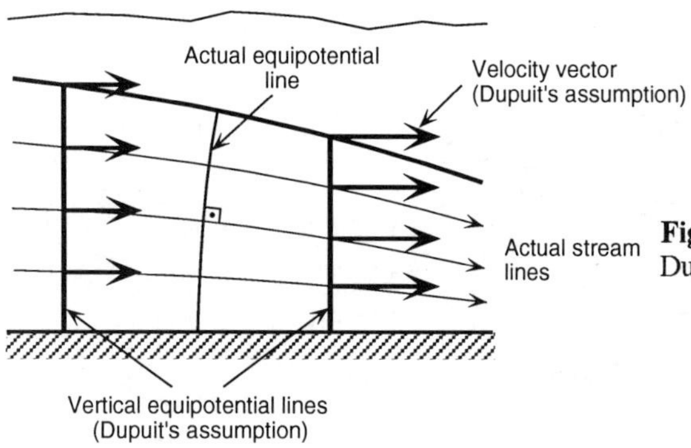

Figure 12.3 Schematic presentation of Dupuit's hypothesis.

The elements for deriving the general differential equation of non-uniform flow are shown in Figure 12.4. Note that the hydraulic head (H) is the sum of the pressure head (h) and the elevation head (z). The equation is derived for an infinitesimally small decrease of the hydraulic head (dH) along an infinitesimally small distance (dx). The difference in hydraulic heads between verticals 2 and 1 is:

$$dH = dz + dh \tag{12.5}$$

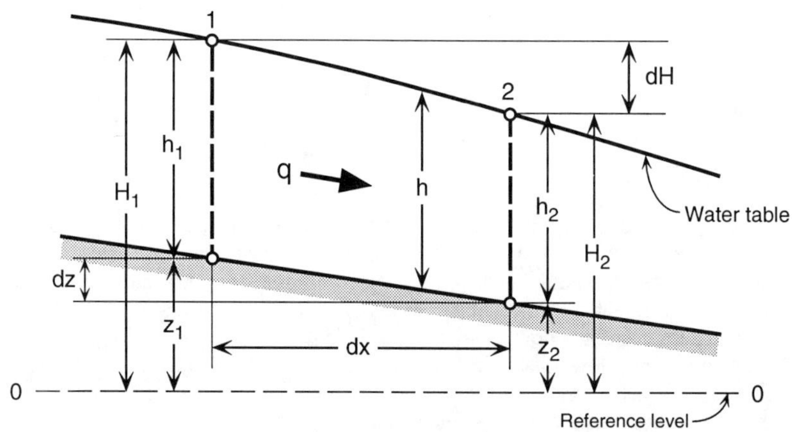

Figure 12.4 Elements for deriving the general differential equation of non-uniform flow in an unconfined aquifer with a sloping base.

The slope of the impermeable base is:

$$i_o = -\frac{dz}{dx} \tag{12.6}$$

Note that the minus sign appears because i_o decreases in the direction of flow. From equation (12.6) it follows that

$$dz = -i_o \cdot dx \tag{12.7}$$

Inserting equation (12.7) into equation (12.5) gives:

$$dH = -i_o dx + dh \tag{12.8}$$

Equation (12.8) can be divided by *dx* and multiplied by the hydraulic conductivity (*K*):

$$K\frac{dH}{dx} = K(\frac{dh}{dx} - i_o) \tag{12.9}$$

If equation (12.9) is multiplied by (−1), the left-hand side is equal to Darcy's velocity which gives:

$$v = -K(\frac{dh}{dx} - i_o) \tag{12.10}$$

Finally, the general differential equation of unit groundwater flow in an unconfined aquifer is given as the product of the flow velocity and the saturated thickness:

$$q = -h \cdot K(\frac{dh}{dx} - i_o) \tag{12.11}$$

Solution for The Horizontal Aquifer Base

If the aquifer base is horizontal then $i_o=0$ and equation (12.11) becomes

$$q = -h \cdot K\frac{dh}{dx} \tag{12.12}$$

To solve this equation the variables *x* and *h* first have to be separated:

$$\frac{q}{K} \cdot dx = -h \cdot dh \tag{12.13}$$

and then integrated for known boundary conditions. Figure 12.5 shows a general case with the following boundary conditions: when $x=0$ the hydraulic head is h_1, and when $x=L$ the hydraulic head is h_2. For these boundaries equation (12.13) is:

$$\frac{q}{K} \cdot \int_0^L dx = -\int_{h_1}^{h_2} h \cdot dh \tag{12.14}$$

and its solution is:

$$q = K\frac{h_1^2 - h_2^2}{2L} \tag{12.15}$$

Equation (12.15) is valid for conditions where there are no gains of water due to aquifer recharge by precipitation or seepage through underlying aquitards, and where there are no losses such as evapotranspiration. The equation that includes the influence of external (positive or negative) fluxes is (see Problem 11, equation 11.10):

$$q = K\frac{h_1^2 - h_2^2}{2L} \pm w\frac{L}{2} \tag{12.16}$$

where w is the vertical gain or loss of water with the velocity dimension. As can be seen, equation (12.16) reduces to equation (12.15) for $w=0$.

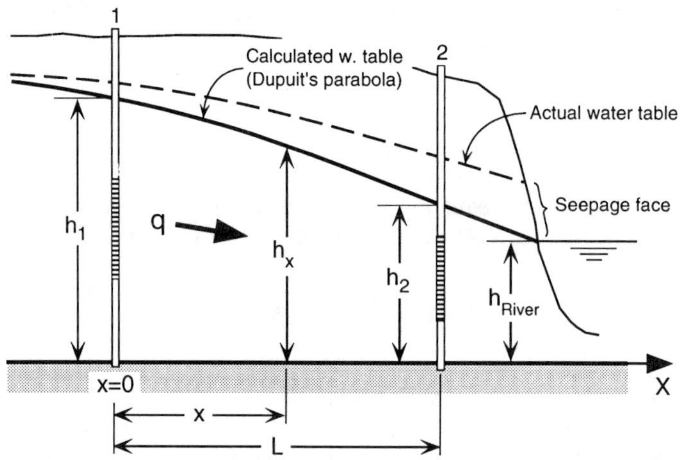

Figure 12.5 Boundary conditions for solving the general differential equation of non-uniform flow in an unconfined aquifer with the horizontal base. The position of the hydraulic head h_x can be found if the heads h_1 and h_2 are known. Note that the calculated water table is below the real water table because of the Dupuit's assumption. The difference between the two is the biggest in the aquifer discharge zone where a seepage surface (face) commonly form.

If two hydraulic heads are known, the position of the hydraulic head at any distance x from the origin (see Figure 12.5) is found using the following equation (see also equation 11.6 in Problem 11):

$$h_x = \sqrt{h_1^2 - (h_1^2 - h_2^2)\frac{x}{L}} \tag{12.17}$$

As mentioned earlier, the water table line calculated with equations (12.15) and (12.17) is always below the actual water table because of Dupuit's assumption. This difference increases downgradient from the higher hydraulic head used in the calculation. In the zone of groundwater discharge, such as the one shown in Figure 12.5, this difference is the biggest. It is partly a consequence of the formation of a seepage face at aquifer contacts with surface water bodies, levees, etc. The position of the actual water table is difficult to formulate mathematically although various graphoanalytical methods for its determination have been suggested (e.g., see Kashef, 1987).

Solution for The Sloping Aquifer Base

The velocity of non-uniform flow in an unconfined aquifer with a gradient of its impermeable base greater than 0 ($i_o > 0$) is (see equation 12.10):

$$v = K(i_o - \frac{dh}{dx}) \tag{12.18}$$

The unit flow, in general, is:

$$q = v \cdot h \tag{12.19}$$

where h is the saturated aquifer thickness. The equivalent unit flow in uniform conditions would be:

$$q = v \cdot h_o \tag{12.20}$$

or, as expressed with equation (12.4):

$$q = K \cdot h_o \cdot i_o \tag{12.21}$$

Combining equations (12.19) through (12.21) gives another equation for the flow velocity in non-uniform conditions:

$$v = \frac{K \cdot h_o \cdot i_o}{h} \tag{12.22}$$

As can be seen, the velocity is expressed in terms of the aquifer base gradient (i_o), saturated thickness (h) and the equivalent normal aquifer depth (h_o) that would produce the same unit flow if the flow conditions were uniform.

Equations (12.18) and (12.22) give:

$$K \cdot \frac{h_o \cdot i_o}{h} = K \cdot (i_o - \frac{dh}{dx}) \tag{12.23}$$

or, after simplifying and rearranging:

$$\frac{dh}{dx} = i_o \cdot (1 - \frac{dh}{dx}) \tag{12.24}$$

To solve this basic differential equation of non-uniform flow when $i_o > 0$, a new variable, called the *relative aquifer depth* (η), is introduced so that

$$\eta = \frac{h}{h_o} \tag{12.25}$$

This transforms equation (12.24) into

$$\frac{dh}{dx} = i_o \cdot (\frac{\eta - 1}{\eta}) \tag{12.26}$$

Differentiation of equation (12.25) gives

$$dh = d\eta \cdot h_o \tag{12.27}$$

which transforms equation (12.26) into

$$\frac{d\eta \cdot h_o}{dx} = i_o \cdot (\frac{\eta - 1}{\eta}) \tag{12.28}$$

To solve this equation variables η and x have to be separated:

$$\frac{\eta}{\eta - 1} \cdot d\eta = \frac{i_o}{h_o} \cdot dx \tag{12.29}$$

Further preparation of equation (12.29) is needed in order to use elementary integrals for solving it:

$$d\eta + \frac{d\eta}{\eta - 1} = \frac{i_o}{h_o} \cdot dx \tag{12.30}$$

Note that:

$$\frac{\eta}{\eta - 1} = \frac{\eta - 1 + 1}{\eta - 1} = 1 + \frac{1}{\eta - 1}$$

Equation (12.30) is solved for the following general boundary conditions: when $x = x_1$ the relative aquifer depth is $\eta = \eta_1$, and when $x = x_2$ the relative depth is $\eta = \eta_2$:

$$\frac{i_o}{h_o} \cdot \int_{x_1}^{x_2} dx = \int_{\eta_1}^{\eta_2} d\eta + \int_{\eta_1}^{\eta_2} \frac{d\eta}{\eta - 1} \tag{12.31}$$

All three integrals in equation (12.31) are elementary (see Appendix G) and the solution is easy:

$$\frac{i_o}{h_o} \cdot (x_2 - x_1) = \eta_2 - \eta_1 + \ln \frac{\eta_2 - 1}{\eta_1 - 1} \tag{12.32}$$

where

$$\eta_2 = \frac{h_2}{h_o} \qquad \text{and} \qquad \eta_1 = \frac{h_1}{h_o}$$

and (x_2-x_1) is the distance between heads h_2 and h_1.

Equation (12.32) is the solution of the differential equation of non-uniform flow in an unconfined aquifer with the sloping base. Since h_o is not known, the equation is solved implicitly, *i.e.*, the normal aquifer depth (h_o) is found by trial. All constant, known terms in equation (12.32) can be grouped as follows:

$$i_o(x_2 - x_1) = h_o\left(\frac{h_2}{h_o} - \frac{h_1}{h_o}\right) + h_o \ln\frac{h_2 - h_o}{h_1 - h_o} \tag{12.33}$$

Note that:

$$\ln\frac{\eta_2 - 1}{\eta_1 - 1} = \ln\frac{\dfrac{h_2}{h_o} - 1}{\dfrac{h_1}{h_o} - 1} = \ln\frac{\dfrac{h_2 - h_o}{h_o}}{\dfrac{h_1 - h_o}{h_o}} = \ln\frac{h_2 - h_o}{h_1 - h_o}$$

Equation (12.33) is further simplified:

$$i_o(x_2 - x_1) + h_1 - h_2 = h_o \ln\frac{h_2 - h_o}{h_1 - h_o} \tag{12.34}$$

or, since all the values on the left-hand side are known:

$$Constant = h_o \ln\frac{h_2 - h_o}{h_1 - h_o} \tag{12.34a}$$

The normal depth is found by trial so that the constant equals the value on the right-hand side of equation (12.34a).

12.2 POSITION OF THE HYDRAULIC HEAD AT POINT A

Three equations can describe the unit flow in Section 2 of the aquifer where the impermeable base is horizontal (see equation 12.15 and Figure 12.6). The boundaries for the first equation are the piezometer and the drainage channel:

$$q = K \cdot \frac{h_P^2 - h_2^2}{2l} \tag{12.35}$$

The boundaries for the second one are the vertical at point A and the drainage channel:

$$q = K \cdot \frac{h_A^2 - h_2^2}{2L_2} \tag{12.36}$$

The boundaries for the third one are the vertical at point A and the piezometer:

$$q = K \cdot \frac{h_A^2 - h_P^2}{2 \cdot (L_2 - l)} \tag{12.37}$$

The hydraulic head at point A can be found by combining any two of the above equations. For example, equations (12.35) and (12.36) give

$$\frac{h_P^2 - h_2^2}{l} = \frac{h_A^2 - h_2^2}{L_2} \tag{12.38}$$

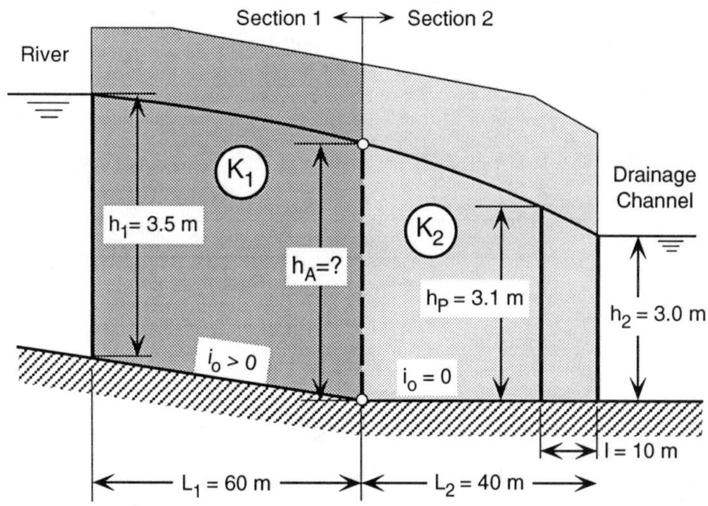

Figure 12.6 Calculation scheme for solving the problem.

and h_A, as the only unknown, is found as follows:

$$L_2 \cdot (h_P^2 - h_2^2) = l \cdot h_A^2 - l \cdot h_2^2$$

$$h_A = \sqrt{\frac{L_2 \cdot (h_P^2 - h_2^2) + l \cdot h_2^2}{l}}$$

$$h_A = \sqrt{\frac{40m \cdot (3.1^2 m - 3.0^2 m) + 10m \cdot 3.0^2 m}{10m}}$$

$$h_A = 3.38m$$

12.3 UNIT GROUNDWATER FLOW

Knowing that the unit groundwater flow has to be the same in both sections of the aquifer (the principle of the conservation of mass), its numeric value is found using any one of equations (12.35) through (12.37):

$$q = K \cdot \frac{h_P^2 - h_2^2}{2l}$$

$$q = 3.8 \times 10^{-5}\, m/s \cdot \frac{3.1^2\, m - 3.0^2\, m}{2 \cdot 10m}$$

$$q = 1.16 \times 10^{-6}\, m^2/s$$

The obtained value for q can be used to calculate unknown elements of the flow in Section 1 of the aquifer where the base gradient is greater than zero.

12.4 HYDRAULIC CONDUCTIVITY IN SECTION 1

The unit non-uniform groundwater flow in an unconfined aquifer with a sloping base is described with the following equation (see also equation 12.21):

$$q = K \cdot h_o \cdot i_o \tag{12.39}$$

where K is the hydraulic conductivity, h_o is the normal aquifer depth for the equivalent uniform flow, and i_o is the gradient of the impermeable aquifer base. The unit flow is the same as in the horizontal section, and the base gradient can be easily determined from the available elevation data which leaves two unknowns in equation (12.39).

The flow conditions in Section 1 are described with equation (12.34) for the known distance between the river and the vertical at point A, for the known hydraulic heads h_1 and h_A, and for the known gradient of the aquifer base (i_o):

$$i_o \cdot L_1 + h_1 - h_A = h_o \ln \frac{h_1 - h_o}{h_A - h_o} \tag{12.40}$$

The base gradient is found as (see Figure 12.1):

$$i_o = \frac{210.5m - 209.0m}{60m} = 0.025$$

After inserting known numeric values into equation (12.40):

$$0.025 \cdot 60m + 3.5m - 3.38m = h_o \ln \frac{3.5m - h_o}{3.38m - h_o}$$

$$1.62m = h_o \ln \frac{3.5m - h_o}{3.38m - h_o}$$

the normal depth (h_o) can be found by trial as shown in Table 12.1. Its value, with an error of 0.6%, is 3.20m (actual constant value is 1.62m, and for h_o=3.20m the constant is 1.63m).

Table 12.1 Implicit solution for the normal aquifer depth (h_o).

H_0 (m)	3.00	3.30	3.10	3.15	***3.20***	3.19
Constant (m)	0.82	3.02	1.11	1.32	***1.63***	1.56

Finally, the hydraulic conductivity in Section 2 is found using equation (12.39) and by inserting all known numeric values:

$$K = \frac{q}{h_o \cdot i_o} \tag{12.41}$$

$$K = \frac{1.16 \times 10^{-6} m^2 / s}{3.20m \cdot 0.025}$$

$$K = 1.45 \times 10^{-5} \ m / s$$

Chapter 5

ONE-DIMENSIONAL TRANSIENT FLOW

PROBLEM 13
SUDDEN CHANGE AT THE BOUNDARY
HYDROGEOLOGIC PARAMETERS

Find the hydrogeologic parameters of the unconfined aquifer shown in Figure 13.1. The aquifer is in a direct hydraulic connection with a large perennial river. The fluctuations of the water table are recorded in a monitoring well 30 meters from the river and are shown in Figure 13.2. The hydrograph shows a sudden rise in the river stage and a consequent rise in the aquifer water table. From the available data find the hydraulic conductivity of the aquifer assuming its homogeneity and isotropy. Also find what would be the rise in the water table 8 days after the flood at 100 meters from the river.

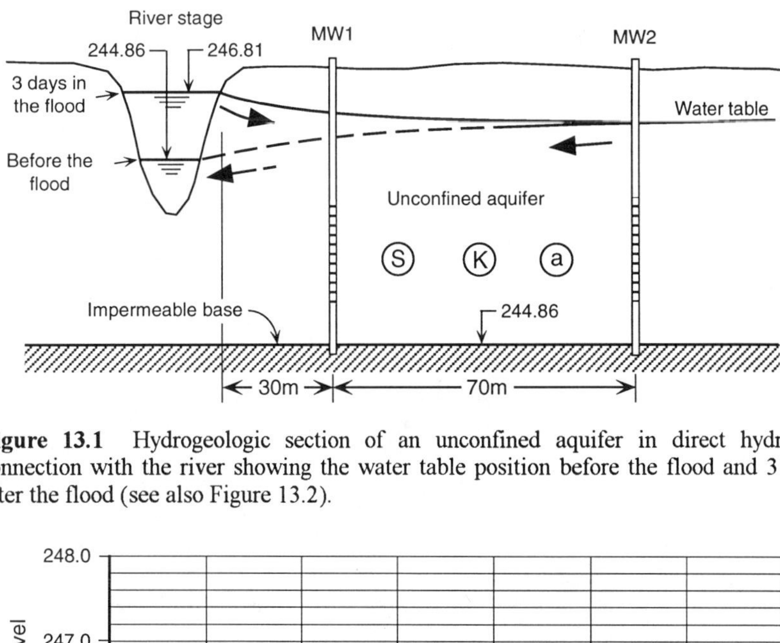

Figure 13.1 Hydrogeologic section of an unconfined aquifer in direct hydraulic connection with the river showing the water table position before the flood and 3 days after the flood (see also Figure 13.2).

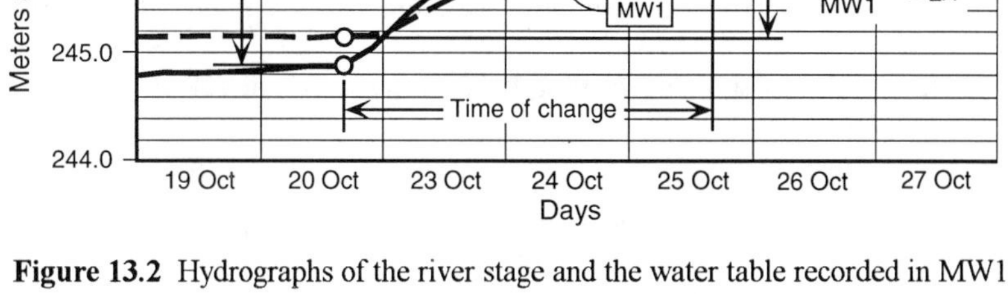

Figure 13.2 Hydrographs of the river stage and the water table recorded in MW1.

13.1 GENERAL EQUATION OF TRANSIENT GROUNDWATER FLOW

The general equation of two-dimensional transient groundwater flow (*i.e.,* flow that changes in time) is derived by examining the water budget for a Representative Elementary Volume (REV) of the aquifer. REV is a volume that has all the characteristics of the aquifer, including its homogeneity/heterogeneity and isotropy/anisotropy. In addition, its dimensions are, for the purposes of mathematical derivation of the differential equation, infinitesimally small. This equation, known as Boussinesq equation, will later be reduced to one-dimensional conditions for solving this problem. Figure 13.3 shows REV for a homogeneous, isotropic, unconfined aquifer resting on a horizontal, impermeable base. The aquifer is subject to a vertical gain or loss of water due to infiltration and/or evapotranspiration. Elements of the water budget shown in Figure 13.3 are:

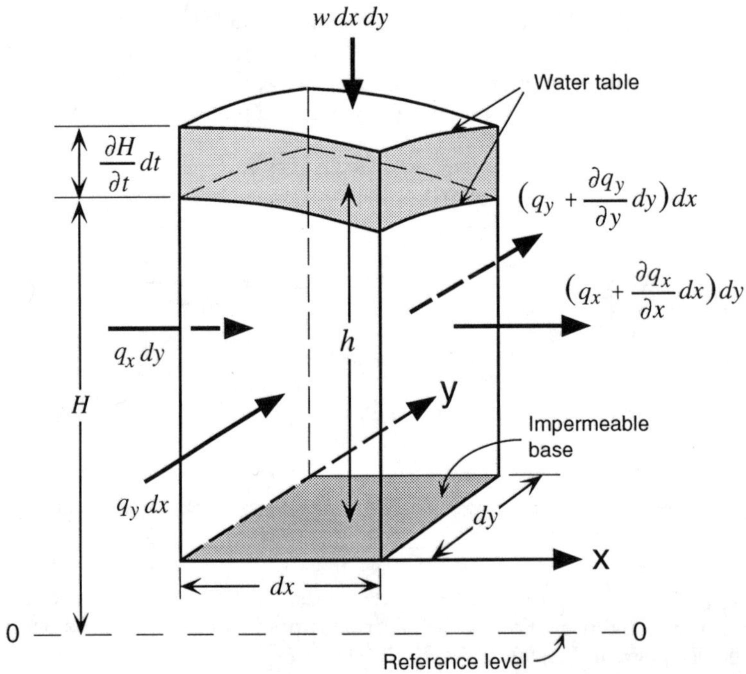

Figure 13.3 Representative Elementary Volume (REV) of an isotropic, homogeneous, unconfined aquifer resting on a horizontal, impermeable base, with elements for deriving Boussinesq's differential equation of groundwater flow (adapted from Sestakov, 1979).

q_x Unit (specific) flow per elementary aquifer thickness in the X axis direction.

q_y Unit (specific) flow per elementary aquifer thickness in the Y axis direction.

$q_x dy$ Total flow entering REV in the X axis direction (*dy* is the width of the flow).

$q_y dx$ Total flow entering REV in the Y axis direction (*dx* is the width of the flow).

$(q_x + \frac{\partial q_x}{\partial x}dx)dy$ Total flow leaving REV in the X axis direction.

$(q_y + \frac{\partial q_y}{\partial y}dy)dx$ Total flow leaving REV in the Y axis direction.

$\frac{\partial q_x}{\partial x}dx$	Change of the unit flow in the X axis direction.
$\frac{\partial q_y}{\partial y}dy$	Change of the unit flow in the Y axis direction.
$wdxdy$	Vertical element of the water budget for REV (w is the intensity of the net gain or loss).

The specific flow in the X axis direction can be expressed as the product of Darcy's velocity in the X axis direction and the aquifer thickness (note that the hydraulic gradient in Darcy's velocity is a change of the hydraulic head (H) along the X axis direction):

$$q_x = v_x h \tag{13.1}$$

$$v_x = -K\frac{\partial H}{\partial x} \tag{13.2}$$

$$q_x = -Kh\frac{\partial H}{\partial x} \tag{13.3}$$

Similarly, the specific flow in the Y axis direction is:

$$q_y = -Kh\frac{\partial H}{\partial y} \tag{13.4}$$

The volume of water stored in (or released from) REV as the water table changes its position in time is:

$$V = \frac{\partial H}{\partial t}dt \cdot dx \cdot dy \cdot S \tag{13.5}$$

where S is the aquifer storativity which, for an unconfined aquifer, equals specific yield. The specific yield is roughly equal to the effective porosity which is a percentage of voids within porous medium that store or release water without retention (see Problem 1).

Note that the partial differentials in all of the above equations are used to describe changes in time of the hydraulic head and the specific flow along an individual coordinate axis while change in an other direction is not considered. This is because both the hydraulic head and the specific flow are functions of both space (represented by two coordinate axes x and y) and time:

$$q = f(x, y, t)$$

$$H = f(x, y, t)$$

The water balance equation for the REV states that the difference between the flows entering REV (inputs) and leaving REV (outputs) in time must be equal to the volume of water stored in (or released from) REV:

$$q_x dydt - (q_x + \frac{\partial q_x}{\partial x} dx)dydt + q_y dxdt - (q_y + \frac{\partial q_y}{\partial y} dy)dxdt +$$

$$+ wdxdydt = \frac{\partial H}{\partial t} dtdxdyS \tag{13.6}$$

or, after simplifying:

$$-\frac{\partial q_x}{\partial x} - \frac{\partial q_y}{\partial y} + w = S\frac{\partial H}{\partial t} \tag{13.7}$$

By expressing unit flows q_x and q_y in terms of Darcy's law, equation (13.7) becomes (see equations 13.1 and 13.4):

$$\frac{\partial (hK\frac{\partial H}{\partial x})}{\partial x} + \frac{\partial (hK\frac{\partial H}{\partial y})}{\partial y} + w = S\frac{\partial H}{\partial t} \tag{13.8}$$

This nonlinear partial differential equation of two-dimensional groundwater flow in an unconfined aquifer is known in literature as the Boussinesq equation after a French engineer who first suggested its linearization. He introduced the following assumptions:

- the aquifer impermeable base is horizontal and the reference level can be placed at the base so that $h=H$ (see Figure 13.3);
- the change of the hydraulic head along flow lines is very slow so that the average thickness of the aquifer can be considered equal to the hydraulic head: $h_{av}=h=H$.
- the aquifer is homogeneous and isotropic so that $K=const.$

These assumptions give the following linearized form of the Boussinesq equation:

$$\frac{\partial^2 h}{\partial x^2} + \frac{\partial^2 h}{\partial y^2} + \frac{w}{Kh_{av}} = \frac{S}{Kh_{av}}\frac{\partial h}{\partial t} \tag{13.9}$$

By introducing a new parameter called *hydraulic diffusivity* (a) which is given as:

$$a = \frac{Kh_{av}}{S} = \frac{T}{S} \tag{13.10}$$

equation (13.9) can be written as:

$$\frac{\partial^2 h}{\partial x^2} + \frac{\partial^2 h}{\partial y^2} + \frac{w}{T} = \frac{1}{a}\frac{\partial h}{\partial t} \tag{13.11}$$

or

$$a\frac{\partial^2 h}{\partial x^2} + a\frac{\partial^2 h}{\partial y^2} + \frac{w}{S} = \frac{\partial h}{\partial t} \tag{13.12}$$

It has been proven that, even under the most unfavorable circumstances, the above described linearization yields an error of less than 20% (Lebedev, 1968). For most practical purposes, the linear Boussinesq equation will give an acceptable result.

The linear Boussinesq equation for one-dimensional groundwater flow is:

$$\frac{\partial h}{\partial t} = a\frac{\partial^2 h}{\partial x^2} + \frac{w}{S} \tag{13.13}$$

where

$a = \frac{Kh_{av}}{S}$	is the hydraulic diffusivity,
h_{av}	is the average aquifer thickness (the thickness of the saturated zone),
K	is the hydraulic conductivity,
S	is the specific yield of the unconfined aquifer (its storativity),
h	is the hydraulic head (equal to H, see Figure 13.3),
x	is the distance along X axis,
w	is the intensity of aquifer recharge at the water table (e.g., recharge from precipitation).

If the unconfined aquifer of interest is receiving water from (or losing water to) an underlying aquifer, either directly or by leakage through an aquitard, equation (13.13) becomes

$$\frac{\partial h}{\partial t} = a\frac{\partial^2 h}{\partial x^2} + \frac{w}{S} \pm \frac{\varepsilon}{S} \tag{13.14}$$

where ε is the intensity of leakage.

Partial differential equation (13.13) can be solved for known boundary and initial conditions of the transient groundwater flow.

13.2 SOLUTION FOR ONE-DIMENSIONAL FLOW WITH A SUDDEN CHANGE AT A BOUNDARY

The following assumptions are made in order to solve equation (13.13) for an unconfined aquifer that has a strong equipotential boundary (e.g., river, channel, lake) on one side and has no boundary on the other side (note that, even though there must be a second boundary, for practical purposes it is placed far enough so that it does not influence the flow field):

- the change at the boundary (where $x=0$) is sudden, and is given as (see Figure 13.4):

$$\Delta H(0,t) = \Delta H_0 \qquad (for\ x = 0,\ and\ \ t = t) \tag{13.15}$$

- the flow field is not disturbed on the other side (where there is no boundary) and the change in water table is 0:

$$\frac{\partial(\Delta H(x,t))}{\partial x} = 0 \qquad (for\ \ x \to \infty,\ \ and\ \ t = t) \tag{13.16}$$

- there is no aquifer recharge either from the top or from the bottom:

$$w = 0 \qquad and \qquad \varepsilon = 0 \tag{13.17}$$

For these boundary and initial conditions, the solution is "borrowed" from the theory of heat transfer and has the following form (modified from Lebedev, 1968):

$$\Delta H(x,t) = \Delta H_0 \cdot erfc(\lambda) \tag{13.18}$$

where

$\Delta H(x,t)$	is a function of space and time,
ΔH_0	is the change at the boundary,
$erfc(\lambda)$	is the complementary error function given as $erfc(\lambda) = 1 - erf(\lambda)$.

Error function (*erf*) is an integral of probability and is readily found in tables (see Appendix D). For our solution of transient 1D flow with a sudden change at a boundary, it has the following form:

$$erf(\lambda) = \frac{2}{\sqrt{\pi}} \int_0^{\lambda} e^{-\lambda^2} d\lambda \tag{13.19}$$

where λ is a parameter given as:

$$\lambda = \frac{x}{2\sqrt{at}} \tag{13.20}$$

In equation (13.20), *x* is the distance from the boundary (river) at which the change in groundwater table (ΔH) is recorded, *a* is the hydraulic diffusivity ($a = Kh_{av}/S$, see also equation 13.10), and *t* is the time of change.

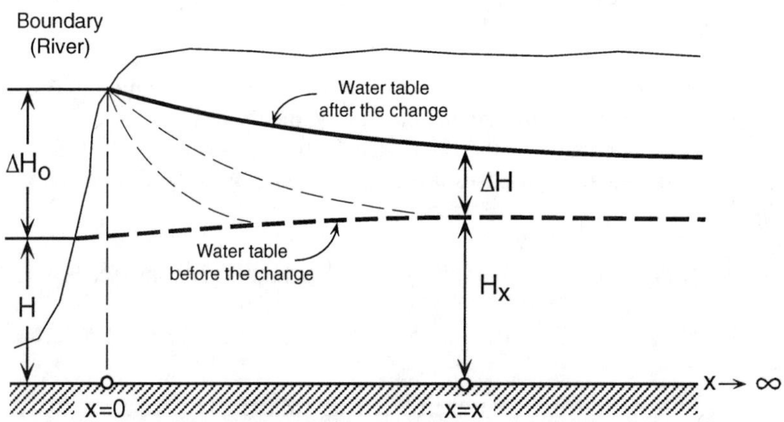

Figure 13.4 Boundary conditions for a transient 1D flow with a sudden change at a boundary. The flow field has one strong equipotential boundary (river) while the boundary on the other side is placed in the infinity.

The average aquifer thickness needed to find the hydraulic diffusivity is calculated as follows (see Figure 13.4):

$$h_{av} = \frac{(H + \Delta H_0) + (H_x + \Delta H)}{2} \tag{13.21}$$

Equation (13.18) can be used to predict a rise in water table at some distance x from the boundary after a sudden change of river stage such as during floods, or it can be used to estimate aquifer parameters if water levels in the river and the aquifer are recorded simultaneously. If two monitoring wells are placed along a stream line, the one closer to the boundary and having a bigger change can act as a boundary (ΔH_0) thus avoiding measurements of the river stage.

13.3 HYDROGEOLOGIC PARAMETERS

The basic hydrogeologic parameter of transient groundwater flow described with equation (13.11) is hydraulic diffusivity (a). It is expressed in terms of hydraulic conductivity (K), average aquifer thickness (h_{av}) and aquifer storativity (S, which is termed *specific yield* for unconfined aquifers):

$$a = \frac{Kh_{av}}{S} = \frac{T}{S} \tag{13.22}$$

Figure 13.5 shows a calculation scheme with the numeric values necessary to find the hydraulic diffusivity in our case.

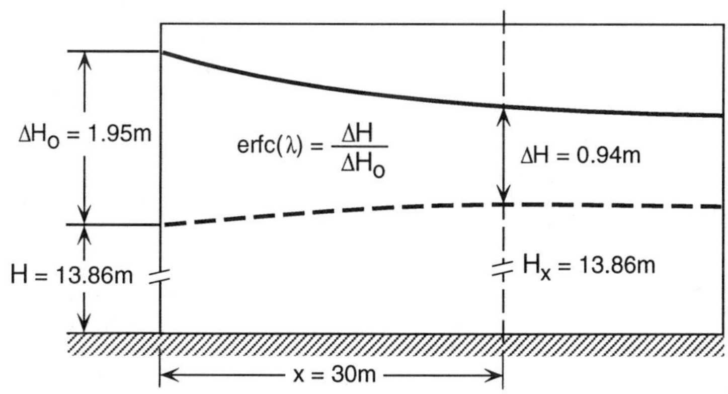

Figure 13.5 Calculation scheme for solving the problem.

The complementary error function is found from equation (13.18) and knowing the values of ΔH and ΔH_0:

$$erfc(\lambda) = \frac{\Delta H(at\ 30m,\ after\ 3\ days)}{\Delta H_0} \tag{13.23}$$

$$erfc(\lambda) = \frac{0.94m}{1.95m} = 0.48$$

Parameter λ is found from Table 13.1 for the known value of *erfc(λ): λ=0.50*. The hydraulic diffusivity is then calculated from equation (13.20):

$$\lambda = \frac{x}{2\sqrt{at}}$$

$$a = \frac{x^2}{4\lambda^2 t} \tag{13.24}$$

$$a = \frac{(30m)^2}{4 \cdot (0.50)^2 \cdot 3 \cdot 86400 \sec} = \frac{900m^2}{259200 \sec} = 3.47 \times 10^{-3} m^2 / s$$

Table 13.1 Values of the complementary error function (adapted from Lebedev, 1968)

(λ)	erfc (λ)	(λ)	erfc (λ)	(λ)	erfc (λ)
0.00	1.000	0.42	0.552	0.84	0.235
0.01	0.989	0.43	0.543	0.85	0.229
0.02	0.977	0.44	0.534	0.86	0.224
0.03	0.966	0.45	0.524	0.87	0.219
0.04	0.955	0.46	0.515	0.88	0.213
0.05	0.944	0.47	0.506	0.89	0.208
0.06	0.933	0.48	0.497	0.90	0.203
0.07	0.921	0.49	0.488	0.91	0.198
0.08	0.910	0.50	0.480	0.92	0.193
0.09	0.899	0.51	0.471	0.93	0.188
0.10	0.887	0.52	0.462	0.94	0.184
0.11	0.876	0.53	0.453	0.95	0.179
0.12	0.865	0.54	0.445	0.96	0.175
0.13	0.854	0.55	0.437	0.97	0.170
0.14	0.843	0.56	0.428	0.98	0.166
0.15	0.832	0.57	0.420	0.99	0.161
0.16	0.829	0.58	0.412	1.00	0.157
0.17	0.810	0.59	0.404	1.05	0.138
0.18	0.799	0.60	0.396	1.10	0.120
0.19	0.789	0.61	0.388	1.15	0.103
0.20	0.777	0.62	0.381	1.20	0.090
0.21	0.766	0.63	0.373	1.25	0.072
0.22	0.756	0.64	0.365	1.30	0.066
0.23	0.745	0.65	0.358	1.35	0.056
0.24	0.734	0.66	0.351	1.40	0.048
0.25	0.723	0.67	0.343	1.45	0.040
0.26	0.713	0.68	0.336	1.50	0.034
0.27	0.703	0.69	0.329	1.60	0.024
0.28	0.692	0.70	0.322	1.70	0.016
0.29	0.682	0.71	0.315	1.80	0.011
0.30	0.671	0.72	0.309	1.90	0.007
0.31	0.661	0.73	0.302	2.00	0.0048
0.32	0.651	0.74	0.295	2.10	0.0030
0.33	0.641	0.75	0.289	2.20	0.0019
0.34	0.631	0.76	0.282	2.30	0.00117
0.35	0.621	0.77	0.276	2.40	0.00070
0.36	0.611	0.78	0.270	2.50	0.00042
0.37	0.601	0.79	0.264	2.60	0.00024
0.38	0.591	0.80	0.258	2.70	0.00013
0.39	0.581	0.81	0.252	2.80	0.00008
0.40	0.572	0.82	0.252	2.90	0.00004
0.41	0.562	0.83	0.240	3.00	0.00002

If the specific yield of the aquifer is known (e.g., based on laboratory analyses of the samples' effective porosity), the aquifer transmissivity can be found using equation (13.22) and vice-versa. If both parameters are unknown, it is better to assume a value for the specific yield and find the transmissivity since the former has a smaller range and is easier to estimate. For example, if our aquifer consists of a uniform medium-size sand it is reasonable to assume a specific yield of 0.25. The transmissivity is calculated from equation (13.22):

$$a = \frac{T}{S}$$

$$T = a \cdot S = 3.47 \times 10^{-3} m^2 / s \cdot 0.25 = 8.675 \times 10^{-4} m^2 / s$$

The hydraulic conductivity is then

$$K = \frac{T}{h_{av}} = \frac{8.675 \times 10^{-4} m^2 / s}{15.46m} = 5.61 \times 10^{-5} m / s$$

where aquifer average thickness is estimated using equation (13.21) (see also Figure 13.5):

$$h_{av} = \frac{(H + \Delta H_0) + (H_x + \Delta H)}{2} = \frac{(13.86m + 1.95m) + (14.18m + 0.94m)}{2} = 15.46m$$

13.4 CHANGE IN WATER TABLE

Once the hydraulic diffusivity is found, the change in water table at any distance from the boundary and at any time can be predicted using equation (13.18). The rise in water table at 100 meters from the river 8 days after the flood is found as follows:

$$\Delta H(x,t) = \Delta H_0 \cdot erfc(\lambda)$$

$$\Delta H(x,t) = 1.95m \cdot erfc(\lambda)$$

$$\lambda = \frac{x}{2\sqrt{at}} = \frac{100m}{2\sqrt{3.47 \times 10^{-3} m^2 / s \cdot 8 \cdot 86400 \sec}} = 1.02$$

For λ=1.02 the complementary error function is approximately 0.148 (see Table 13.1) and the change in water table is then:

$$\Delta H(100m,\ 8\ days) = 1.95m \cdot 0.148 = 0.29m$$

Figure 13.6 shows the change in water table at 100 meters from the river for 15 days after the sudden change at the boundary (see also Table 13.2). The values are obtained using equations (13.18), (13.20), determined hydraulic diffusivity (parameter a), and complementary error function (Table 13.1).

Table 13.2 Calculation of the change in water table at 100 meters from the river using equations (13.18), (13.20), determined parameter a, and complementary error function given in Table 13.1.

Day	1	2	3	4	5	6	7	8
λ	2.888	2.042	1.667	1.444	1.291	1.179	1.091	1.021
erfc(λ)	0.000	0.004	0.017	0.042	0.067	0.098	0.122	0.148
ΔH (m)	0.000	0.007	0.033	0.082	0.131	0.191	0.238	0.289

Day	9	10	11	12	13	14	15
λ	0.963	0.913	0.871	0.834	0.801	0.772	0.746
erfc(λ)	0.174	0.197	0.219	0.238	0.258	0.275	0.291
ΔH (m)	0.339	0.384	0.427	0.464	0.503	0.536	0.567

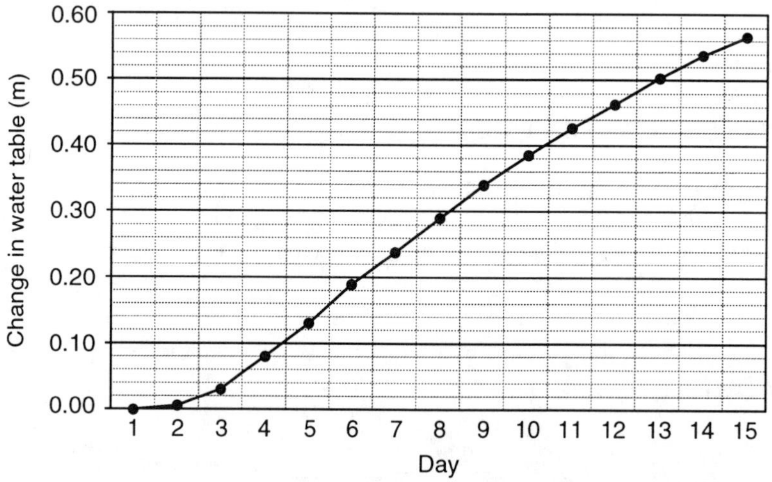

Figure 13.6 Change in water table at 100 meters from the river calculated using equations (13.18), (13.20), determined parameter a, and complementary error function (Table 13.1).

Note: it is somewhat arbitrary to choose between "a sudden" and "a gradual" change at the boundary. Theoretically, when solving equation (13.13), a sudden change is considered to be instantaneous. In real field conditions, this should correspond to not more than 2–3 days of significant change in the water level at the boundary. Longer lasting change is better to simulate with a gradual change at the boundary, especially if there is a significant precipitation in the area.

PROBLEM 14
GRADUAL LINEAR CHANGE AT THE BOUNDARY INFILTRATION RATE

Figure 14.1 shows a hydrogeologic cross-section of an unconfined aquifer under direct influence of a large perennial river. Water table fluctuations are monitored by two wells placed along an average streamline of groundwater flow (which is roughly perpendicular to the river). The hydrographs in Figures 14.2 and 14.3 show two episodes of a gradual increase in the water table which are consequences of the following:

1. a change in the river stage during a period without precipitation in the study area (first episode), and
2. a change in the river stage accompanied with infiltration from precipitation during a two-week period with frequent rains (second episode).

From the available data, find the hydrogeologic parameters of the sand aquifer whose effective porosity, determined from laboratory tests of core samples, is 0.20. Also find the precipitation infiltration rate during the second episode.

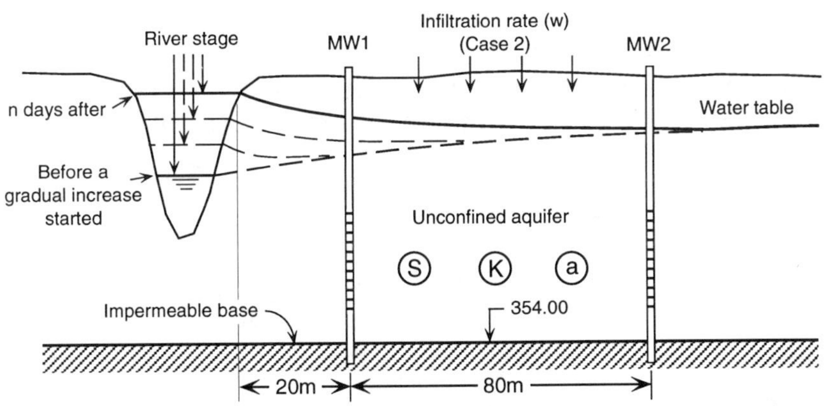

Figure 14.1 Hydrogeologic section of an unconfined aquifer in direct hydraulic connection with the river showing a general position of the water table before and at the end of a gradual increase (see Figure 14.2 for the exact elevations of water table for the two cases discussed).

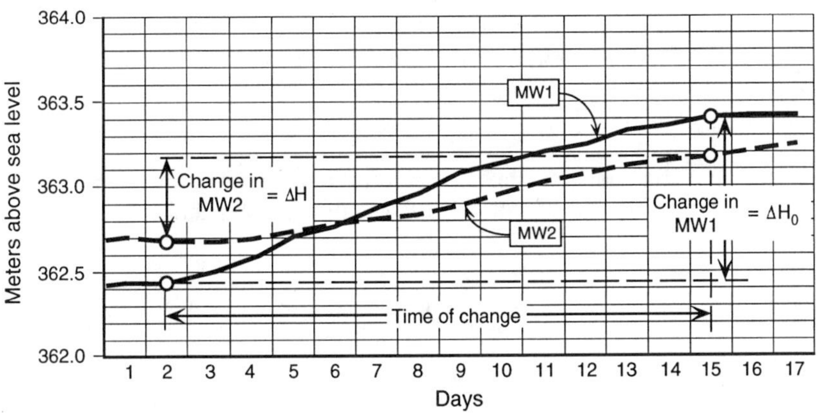

Figure 14.2 Hydrograph of the water table elevation recorded in two monitoring wells during an episode of a gradual linear change in river stage, and without significant precipitation in the study area.

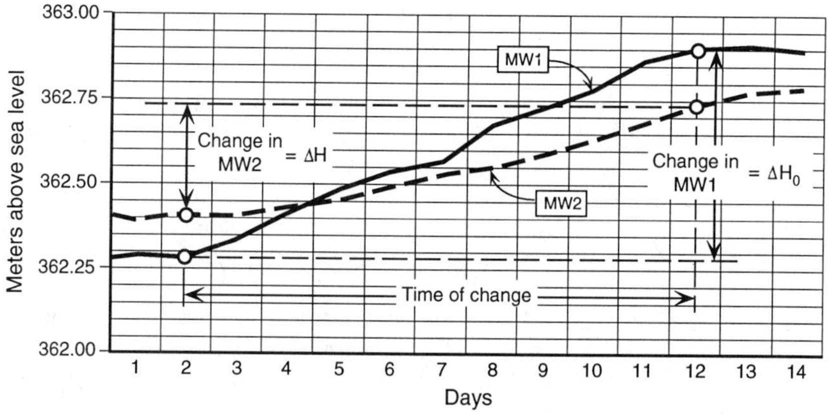

Figure 14.3 Hydrograph of the water table elevation recorded in two monitoring wells during an episode of gradual linear change in the river stage with significant precipitation in the study area.

14.1 SOLUTION OF THE TRANSIENT ONE-DIMENSIONAL FLOW EQUATION FOR A GRADUAL CHANGE AT THE BOUNDARY

The general linear differential equation of a transient one-dimensional groundwater flow is (see also Problem 13, equation 13.13):

$$\frac{\partial h}{\partial t} = a\frac{\partial^2 h}{\partial x^2} + \frac{w}{S} \tag{14.1}$$

where

$a = \dfrac{Kh_{av}}{S}$ is the hydraulic diffusivity,

h_{av} is the average aquifer thickness (*i.e.*, the thickness of the saturated zone),
K is the hydraulic conductivity,
S is the specific yield of the unconfined aquifer (*i.e.*, its storativity),
h is the hydraulic head (equal to H, see Figure 13.3),
x is the distance along the X axis,
w is the intensity of aquifer recharge at the water table (e.g., recharge from precipitation),
t is the time.

If the unconfined aquifer of interest is receiving water from (or losing water to) an underlying aquifer, either directly or by leakage through an aquitard, equation (14.1) becomes

$$\frac{\partial h}{\partial t} = a\frac{\partial^2 h}{\partial x^2} + \frac{w}{S} \pm \frac{\varepsilon}{S} \tag{14.2}$$

where ε is the intensity of leakage.

Partial differential equation (14.2) can be solved for known boundary and initial conditions of the transient groundwater flow. Figure 14.1 shows a situation in the vicinity of a strong equipotential boundary (river) developed after a gradual increase in the river stage.

If data on the river stage changes are not readily available or recorded, measurements in two monitoring wells shown can be used to solve equation (14.2). In this case, the monitoring well closer to the river (MW1) becomes the boundary and the coordinate beginning of the X axis is placed at its location at the impermeable base of the aquifer (Figure 14.4). Boundary and initial conditions of the flow field are then:

1. $\Delta H(0,t) = b \cdot t$ where b is the velocity of change in the water table at $x=0$, and t is the time of change;

2. $\dfrac{\partial[\Delta H(x,t)]}{\partial x} = 0 \quad for\ x \to \infty$

   i.e., the flow field is not disturbed (there is no change in the water table) at distance $x \to \infty$ from the boundary;

3. There is a net infiltration across the water table: $w \neq 0$ (note that, in general, this net infiltration is the difference between the effective infiltration and the evaporation from the water table). The aquifer recharge by infiltration starts at time $t=0$ and it has constant intensity during time t;

4. There is no leakage into the aquifer from below: $\varepsilon = 0$.

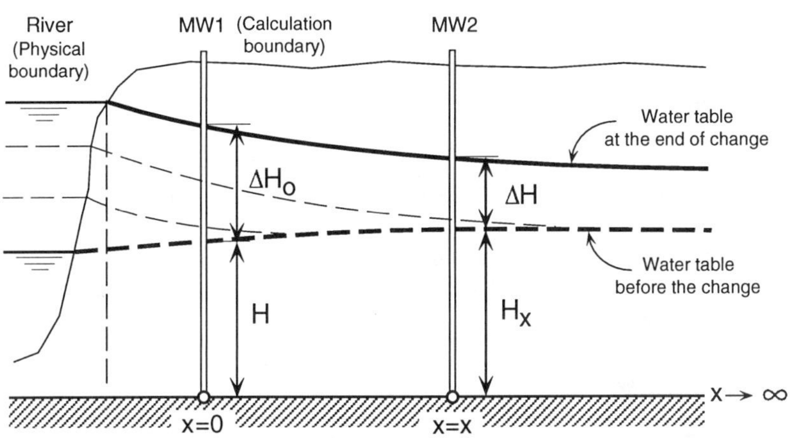

Figure 14.4 Boundary conditions for a transient 1D flow with a gradual change at a boundary. The flow field has one strong equipotential boundary (river) while the boundary on the other side is placed in infinity.

Solution of equation (14.2) for these initial and boundary conditions is (Lebedev, 1968):

$$\Delta H(x,t) = \Delta H_0 \cdot R(\lambda) + \frac{wt}{S}[1 - R(\lambda)] \tag{14.3}$$

where

$\Delta H(x,t)$ is a function of space and time,

ΔH_0 is the change at the boundary,

$R(\lambda)$	is a special function, values of which are found in tables (see Table 14.1 and Appendix D),
w	is the intensity of infiltration,
t	is the time,
S	is the aquifer storativity (specific yield).

Function R(λ) is given as:

$$R(\lambda) = 4i^2 erfc(\lambda) \tag{14.4}$$

where $erfc(\lambda)$ is the complementary error function $(erfc(\lambda) = 1 - erf(\lambda))$. The error function (*erf*) is an integral of probability and is also readily found in tables (see Appendix D). For our solution of transient 1D flow with a gradual change at a boundary, it has the following form:

$$erf(\lambda) = \frac{2}{\sqrt{\pi}} \int_0^{\lambda} e^{-\lambda^2} d\lambda \tag{14.5}$$

where λ is a parameter given as:

$$\lambda = \frac{x}{2\sqrt{at}} \tag{14.6}$$

In equation (14.6), x is the distance between the boundary (MW1) and MW2 where the change in groundwater table (ΔH) is recorded, a is the hydraulic diffusivity ($a = Kh_{av}/S$, see also equation 13.10), and t is the time of change.

14.2 HYDROGEOLOGIC PARAMETERS

If a gradual increase in the water table is not accompanied by significant precipitation and a consequent recharge of the aquifer by infiltration, and if there is no leakage into the aquifer from other sources, *i.e.*,:

$w = 0$ and $\varepsilon = 0$

equation (14.3) reduces to

$$\Delta H(x,t) = \Delta H_0 \cdot R(\lambda) \tag{14.7}$$

Figure 14.2 shows hydrographs of water table recorded in two monitoring wells (MW1 and MW2) during such an episode. The value of water table change in both wells is easily determined from the hydrographs as the difference between the highest level recorded at the end of a noticeable increase, and the level recorded just before the change started. The time period between these two recordings is the time base for the calculations. For MW1 the change is:

$$\Delta H_0 = 363.40m - 362.44m = 0.96m$$

and for MW2 the change is:

$$\Delta H = 363.18m - 362.68m = 0.50m$$

As can be seen from the hydrographs in Figure 14.2, the water table in MW2 lags behind MW1 for about 2 days. The time base for the calculation is 13 days.

Once the change in water table is determined for both monitoring wells, the function $R(\lambda)$ is found using equation (14.7):

$$R(\lambda) = \frac{\Delta H}{\Delta H_0} = \frac{0.50m}{0.96m} = 0.52$$

Parameter λ is then found from tables (such as Table 14.1) knowing the value of $R(\lambda)$:

$$\lambda = 0.27$$

The relationship between hydraulic diffusivity (a), which is the basic hydrogeologic parameter of transient flow described with equation (14.1), and parameter λ is given with equation (14.6) and can be rewritten as:

$$a = \frac{x^2}{4\lambda^2 t} \tag{14.8}$$

where x is the distance between MW1 and MW2 (or, in general, the distance between the boundary and any point of water table measurement along the same streamline), and t is the time of change. Substituting known values for λ, x, and t into equation (14.8) determines the hydraulic diffusivity:

$$a = \frac{(80m)^2}{4 \cdot 0.52^2 \cdot 86400 \sec \cdot 13} = 5.27 \times 10^{-3} m^2 / s$$

Once the hydraulic diffusivity of the aquifer is determined, it is possible to easily calculate infiltration rates for the periods when the rise in the water table can partially be attributed to aquifer recharge from precipitation.

It should be noted that the hydraulic diffusivity relates to the ability of the aquifer to transmit change imposed at the boundary to some distant location within the aquifer (or to transmit change imposed by a certain stress such as a pumping well) . The time of response to this change is inversely proportional to the hydraulic diffusivity (Bureau of Reclamation, 1985).

14.3 INFILTRATION RATE

Figure 14.3 shows hydrographs of the water table elevations recorded in two monitoring wells during an episode of aquifer recharge due to infiltration of precipitation as well as change in the river stage. The infiltration intensity (w) can be calculated using equation (14.3) and knowing the changes in the water table recorded in MW1 and MW2, the time of change, the aquifer's hydraulic diffusivity, and the specific yield. As shown in the previous chapter, the hydraulic diffusivity can be determined by analyzing water table measurements in monitoring wells during a gradual change at the boundary. This procedure does not require knowledge on aquifer hydraulic conductivity, transmissivity, or specific yield. Note, however, that if the value of hydraulic diffusivity determined this way

If data on total, or gross precipitation in the area are available, it is possible to determine the percentage of net infiltration using the following formula:

$$I = \frac{P_{ef}}{P_{total}} \times 100\% \tag{14.12}$$

where

I is the infiltration in percent,

P_{ef} is the effective precipitation recharging the aquifer,

P_{total} is the total, or gross precipitation recorded over the study area.

In our case, the gross precipitation recorded during the 10-day period is 121.4 millimeters which gives the following value of infiltration:

$$I = \frac{34mm}{121.4mm} \times 100\% = 28\%$$

Chapter 6

FLOW NETS

PROBLEM 15
CONSTRUCTION AND CHARACTERISTICS OF FLOW NETS
GROUNDWATER FLOW RATE

Find the groundwater flow rate between the two lakes shown in Figure 15.1 using a flow net. The hydraulic conductivity of the confined aquifer is 7.3×10^{-6} m/s. The aquifer thickness at borehole B1 near the Upper Lake is 14.6 meters.

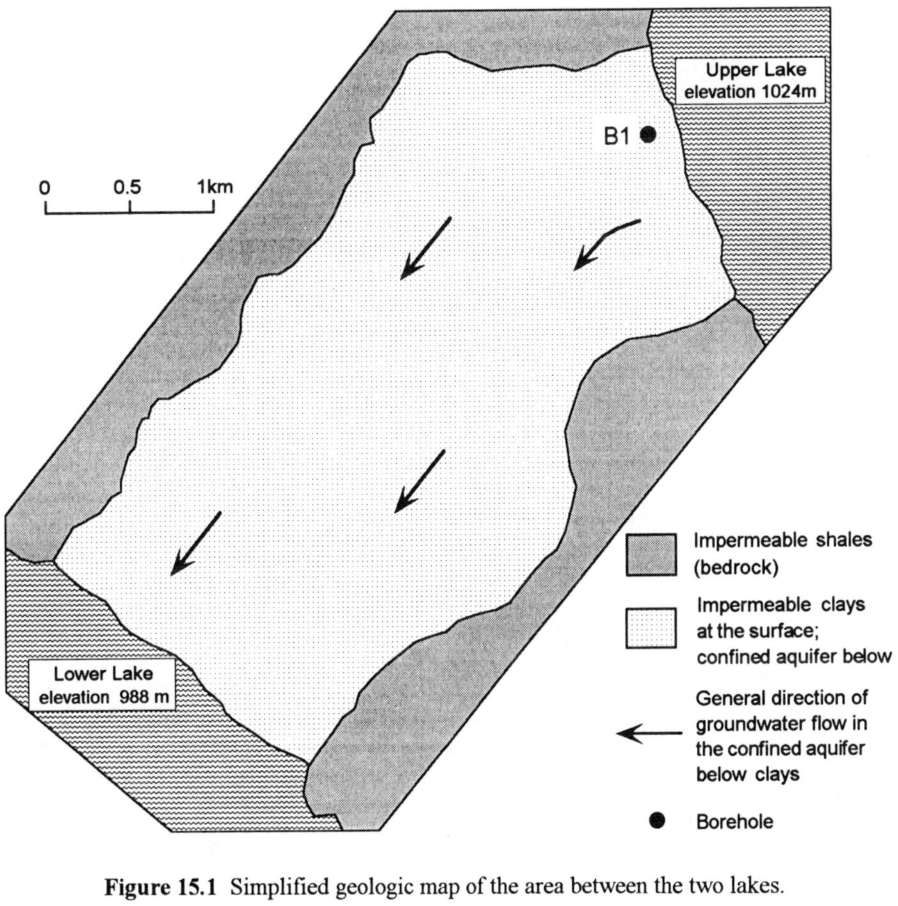

Figure 15.1 Simplified geologic map of the area between the two lakes.

15.1 CHARACTERISTICS OF FLOW NETS

A flow net is a set of streamlines and equipotential lines which are perpendicular to each other (see Figure 15.2). The streamline (or flow line) is an imaginary line representing the path of a groundwater particle as it flows (or would flow) through an aquifer. Two streamlines bound a flow segment of the flow field and never intersect, *i.e.*, they are roughly parallel when observed in a relatively small portion of the aquifer. Similarly, two adjacent equipotential lines never intersect and can also be considered parallel within a small aquifer portion. These characteristics are the main reason why a flow net in a homogeneous, isotropic aquifer is sometimes called the net of small (curvilinear) squares.

Flow nets describe two-dimensional flow fields and are a simple, yet powerful tool for analyzing groundwater flow rates, velocities and distribution of transmissivity (hydraulic conductivity) within the aquifer. Practically every groundwater map that shows the contours of the water table or the piezometric surface represents a flow net without streamlines which can be easily added following simple rules. Groundwater contour maps and flow nets can be constructed:

1. manually, from data recorded in the field (monitoring wells);
2. using a computer program for contouring (data are also recorded in the field), and adding the streamlines manually;
3. using a numerical or analytical groundwater flow model whose final result is the distribution of hydraulic heads within the aquifer, *i.e.*, an XYZ computer file that can be graphically contoured during model post-processing.

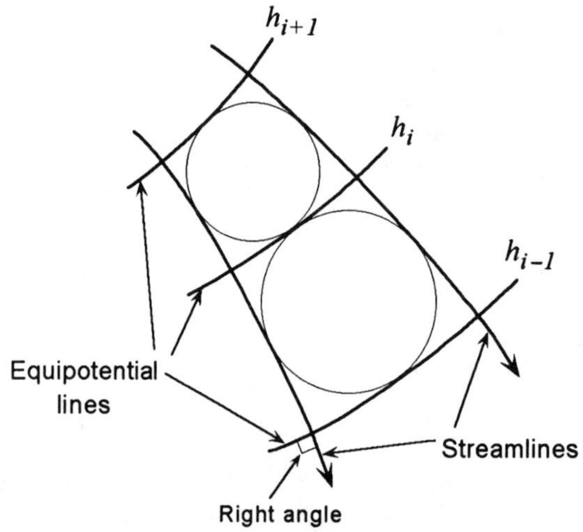

$$h_{i+1} - h_i = \Delta h = h_i - h_{i-1}$$

Figure 15.2 A flow net is a set of equipotential lines and streamlines which are perpendicular to each other. The equipotential line connects points with the same groundwater potential, *i.e.*, hydraulic head -- *h*. The streamline is an imaginary line representing the path of a groundwater particle as it flows (or would flow) through an aquifer.

When constructing a flow net, and later using it for the flow rate calculations, it is very important to remember the following hydrodynamic characteristics:

- The velocity vector (of a moving groundwater particle) is a tangent to the streamline (see Figure 15.3). Therefore, a water particle cannot cross a streamline, *i.e.*, there is no groundwater flow across a streamline. It follows that an impermeable boundary is also a streamline. This is the reason why streamlines are sometimes used as hydraulic ("artificial") no-flow boundaries in groundwater modeling.

- The velocity vector (of a moving groundwater particle) is perpendicular to an equipotential line and is oriented toward the decreasing hydraulic head (see Figure 15.3). It follows, as previously mentioned, that in homogeneous, isotropic aquifers the streamlines and equipotential lines are perpendicular to each other. Although not theoretically true, for practical purposes this can be applied to most intergranular aquifers if they are not highly anisotropic. In the case of a pronounced anisotropy, flow nets can not be constructed without complex analytical solutions which defeats their purpose (for a discussion on flow nets in anisotropic conditions see Fetter, 1994, p. 153–155).
- A drop in the hydraulic head between successive equipotential lines is the same (Δh=const). The value of this decrease corresponds to the contour interval on groundwater contour maps.

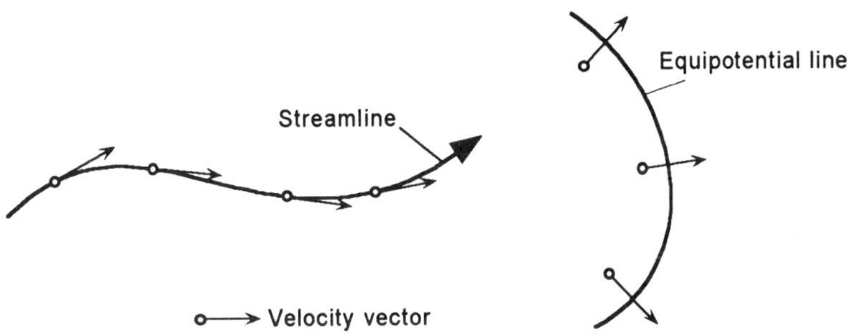

Figure 15.3 The groundwater flow velocity vector is a tangent to the streamline (left), and is perpendicular to the equipotential line (right).

Note that in highly fissured and karst aquifers, where groundwater flow is discontinuous, (i.e., it takes place mainly along preferential flow paths), and Darcy's law does not apply, flow nets are not an appropriate method for the flow characterization.

A flow net does not change over time for steady state flow. In transient conditions a flow net represents the instantaneous flow field at a particular time, and it changes for any other time. In practice, it can be viewed as a continuous succession of steady states, each of which is "steady" for short periods of time (Domenico and Schwartz, 1990).

The principles of the flow net construction and groundwater flow calculations are the same for confined and unconfined aquifers (see Figure 15.4). In both cases the flow field is defined by the geometric characteristics of the net, *i.e.*, spacing between the successive streamlines (ΔW), spacing between the successive equipotential lines (ΔL), aquifer thickness, and the numeric value of the equipotential lines (*i.e.*, the hydraulic head -- h).

15.2 GROUNDWATER FLOW RATE

Flow net is a graphical solution for the two-dimensional flow field and in order to obtain the volumetric flow rate this solution has to be extended throughout the aquifer. If the flow net represents a map view, the solution is extended through the overall aquifer thickness. If the net is constructed for a cross-sectional view, then the solution is extended through the aquifer width. Figure 15.5 shows elements for the flow calculations for both map and cross-sectional views.

The base for the flow determination is the unit flow rate per flow net segment - Δq. The flow net segment is a portion of the aquifer bound by two streamlines and it extends

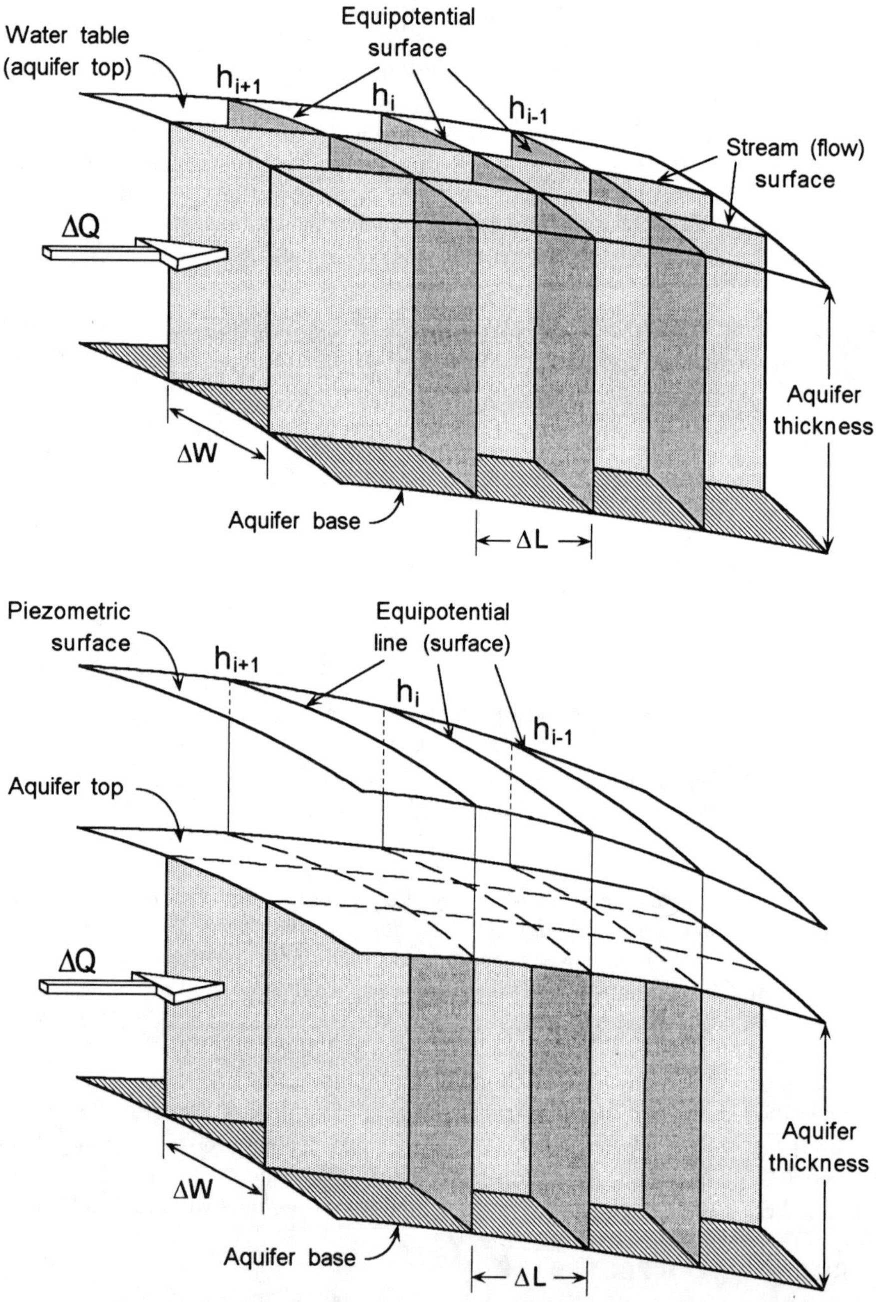

Figure 15.4 **Top**: A flow net for an unconfined aquifer. Streamlines and equipotential lines are intersections of the aquifer water table with the stream surfaces and equipotential surfaces respectively. **Bottom**: A flow net for a confined aquifer. Streamlines are intersections of the stream surfaces and the aquifer top, while equipotential lines are the lines of equal hydraulic head at the imaginary piezometric surface.

throughout the aquifer. For a map view, when visualized in 3 dimensions, these bounding streamlines correspond to intersections of vertical stream surfaces and the aquifer top (see Figure 15.5A). The thickness of the segment equals the aquifer thickness (*b*), and the width of the segment is the distance between two bounding streamlines (ΔW). For a cross-sectional view (Figure 15.5B), the streamlines are intersections of the stream surfaces with the plane of the drawing. The thickness of the flow net segment is the spacing between the two bounding streamlines (ΔW), while the width of the segment equals the width of the aquifer.

Note that, since the flow net method is a two-dimensional graphical solution of a three-dimensional flow field, it is very important to understand basic differences between the map view and the cross-sectional view approaches.

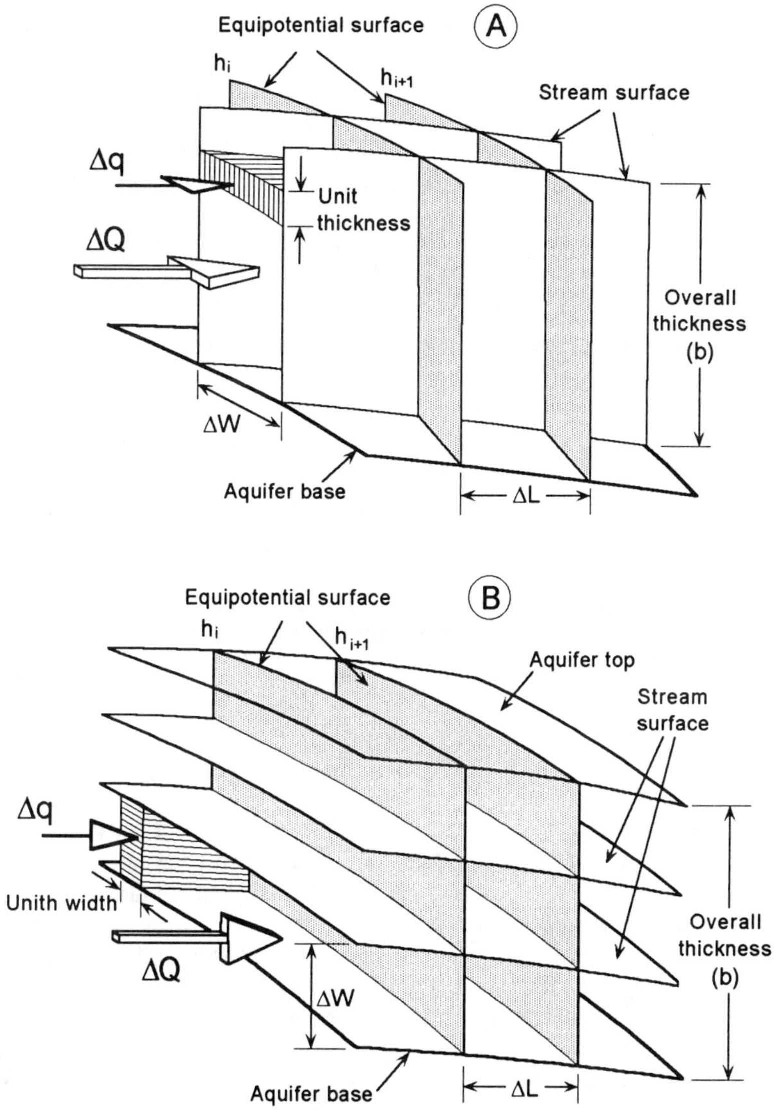

Figure 15.5 Elements for the calculations of flow rates using flow nets. A: map view approach; B: cross-sectional view approach.

Map View Approach

The unit flow rate per flow net segment (Δq) is a product of the flow velocity (v) and the flow area (A):

$$\Delta q = v \cdot A \tag{15.1}$$

where the velocity is given as the product of the hydraulic conductivity and the hydraulic gradient:

$$v = K \cdot \frac{h_i - h_{i-1}}{\Delta L} = K \cdot \frac{\Delta h}{\Delta L} \tag{15.2}$$

and the flow area is the spacing between two bounding streamlines multiplied by one (unity):

$$A = \Delta W \times 1 \tag{15.3}$$

When combining equations (15.2) and (15.3) the unit flow rate per net segment is:

$$\Delta q = K \Delta W \frac{\Delta h}{\Delta L} \tag{15.4}$$

The volumetric flow rate per net segment (ΔQ) is obtained by multiplying the unit flow rate with the aquifer thickness (b):

$$\Delta Q = \Delta q \cdot b \tag{15.5}$$

Finally, the total flow rate in the aquifer (Q), or in its part that is of interest, is found by summing the flow rates in the corresponding number of segments (ΔQ_1, $\Delta Q_2, \ldots, \Delta Q_N$).

Cross-sectional View Approach

As seen from Figure 15.5, the above equation of the unit flow per net segment derived for the map view approach remains the same:

$$\Delta q = K \Delta W \frac{\Delta h}{\Delta L} \tag{15.6}$$

where K is the aquifer hydraulic conductivity, ΔW is the spacing between two adjacent streamlines, Δh is the hydraulic head difference between two adjacent equipotential lines (this corresponds to the contour interval on groundwater contour maps), and ΔL is the spacing between two adjacent equipotential lines. The volumetric flow rate per net segment is obtained by multiplying the unit flow with the aquifer width, and the total flow is the sum of flows in the corresponding number of segments.

The main difficulties in applying analytical solutions to groundwater flow, such as those explained in Chapter 4, occur when the aquifer studied has a varying hydraulic conductivity, and varying thickness and slope. A properly constructed flow net "takes care" of these irregularities and greatly simplifies calculations of groundwater flow (note, however, that design of a flow net is often not easy and requires significant time and skill). Figure 15.6 illustrates the principle of mass (flow) conservation for a net segment or a streamtube as is often called: the flow rate in the segment remains the same even though the aquifer thickness (and therefore the area of flow - A) changes. This means that, in order to calculate the volumetric flow rate in one segment, it is enough to know the aquifer thickness at only one location within the segment. Also, because the flow net in a homogeneous, isotropic aquifer is a set of curvilinear squares, the unit flow rate in adjacent net segments (streamtubes) is the same. This further simplifies the flow calculations (see Figure 15.7).

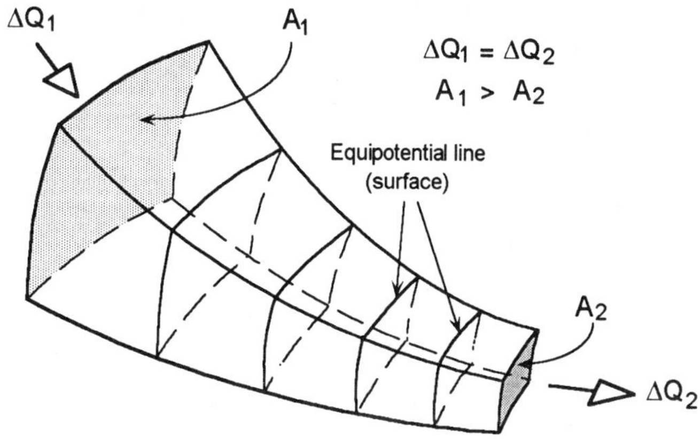

Figure 15.6 Conservation of flow rate within a streamtube. There is no gain or loss of water in the portion of the aquifer bound by two streamlines.

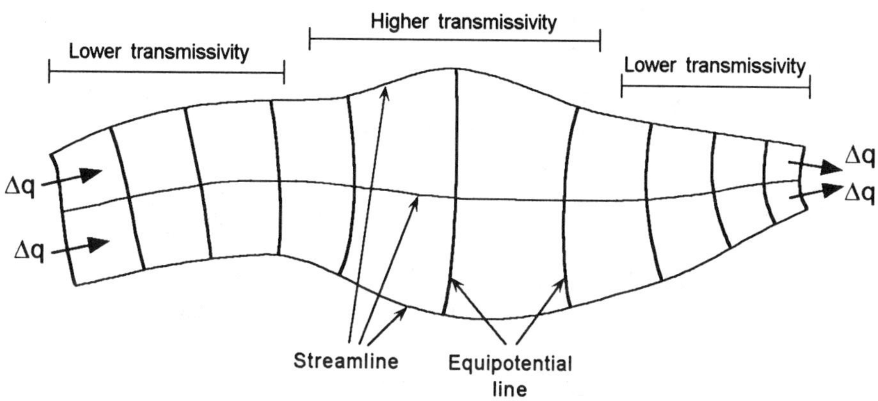

Figure 15.7 Flow rates through segments of the flow net are equal. The varying size of the curvilinear squares (*i.e.*, spacing of the equipotential and streamlines) is a consequence of varying transmissivity.

15.3 FLOW NET AND FLOW RATE FOR THIS PROBLEM

The groundwater flow field in the confined aquifer shown in Figure 15.1 has two types of hydrodynamic boundaries whose recognition is necessary for the flow net construction. These are:

1. Equipotential boundaries at the contact between the Upper and Lower lakes and the aquifer;
2. Impermeable or no-flow boundaries along the contact between the bedrock shales and the aquifer.

It is best to start the flow net by drawing several streamlines perpendicular to one of the equipotential boundaries. The lines should be roughly equally-spaced and parallel to each other (Figure 15.8). The next step is to draw an equipotential line parallel to the lake-aquifer contact (which is an equipotential boundary), and perpendicular to all the streamlines and the impermeable boundaries (remember that an impermeable boundary is also a streamline). The equipotential line and the streamlines should form "curvilinear squares", preferably of similar size.

After completing the first set of quadrangles, all streamlines are extended a bit further (keeping the outer ones parallel to the impermeable boundaries) and another equipotential line is drawn perpendicular to them. The process is repeated until the last set of squares is drawn at the lower equipotential boundary (Lower Lake).

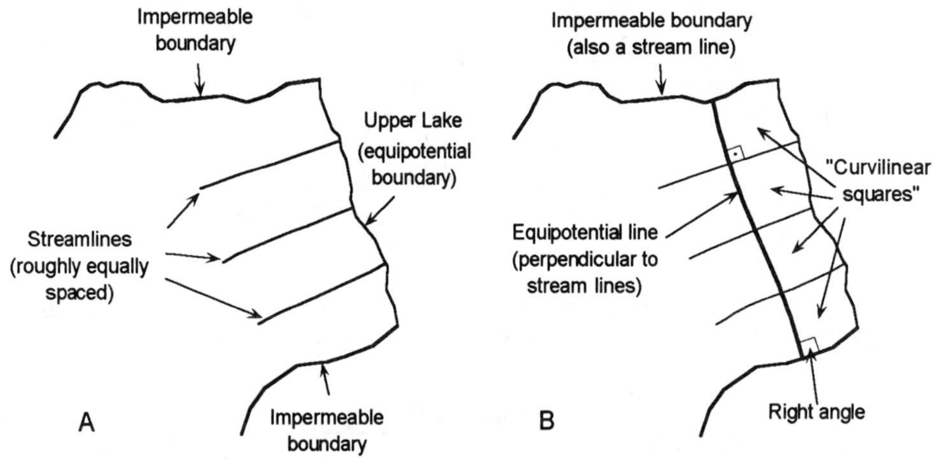

Figure 15.8 Beginning of the flow net construction by A: drawing a trial set of streamlines and B: continuing with the first equipotential line.

Alternatively, a trial set of streamlines can be drawn through the entire flow field followed by the equipotential lines which should be spaced to form curvilinear squares. In any case, it will often be necessary to adjust the initial shape of both streamlines and equipotential lines, and to change their number.

The flow net is correct if the following tests are satisfied (see Figure 15.9):

- a circle touching all four sides can be drawn into each curvilinear square, and
- diagonals of the squares can be drawn to form continuous lines, *i.e.,* another set of curvilinear squares.

Once the flow net is finished, it can be used for the calculations of groundwater flow rates in the aquifer and the analysis of transmissivity distribution. The flow net can easily be made denser if a more precise analysis is required for certain portions of the aquifer (Figure 15.10).

Groundwater Flow Rate

Elements needed for the calculation of the total groundwater flow rate through the confined aquifer, from the Upper Lake to the Lower Lake, are shown in Figure 15.11. As usual, the flow rate per unit thickness of the aquifer (q) is first given as the product of the flow velocity (v) and the cross-sectional area of flow (A):

$$q = v \cdot A \tag{15.7}$$

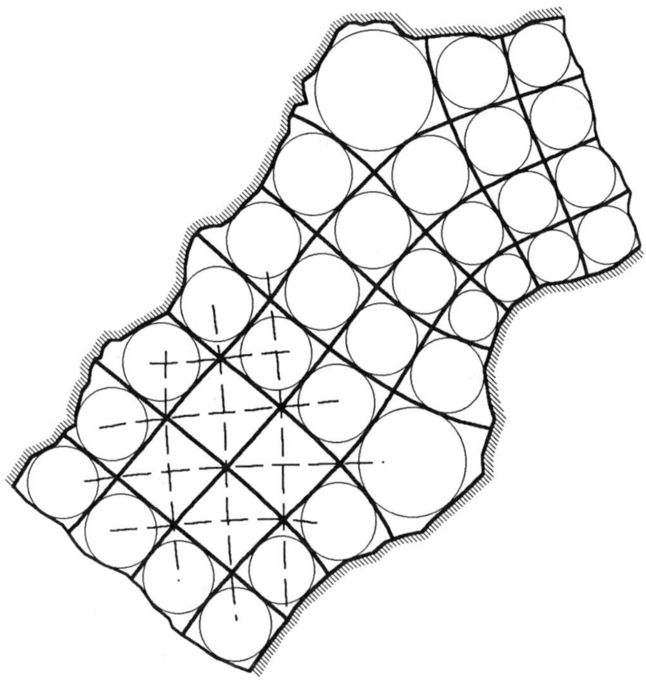

Figure 15.9 Tests of the flow net validity: circles touching all four sides can be drawn into each square; diagonals of the squares can be drawn to form continuous lines and another set of squares.

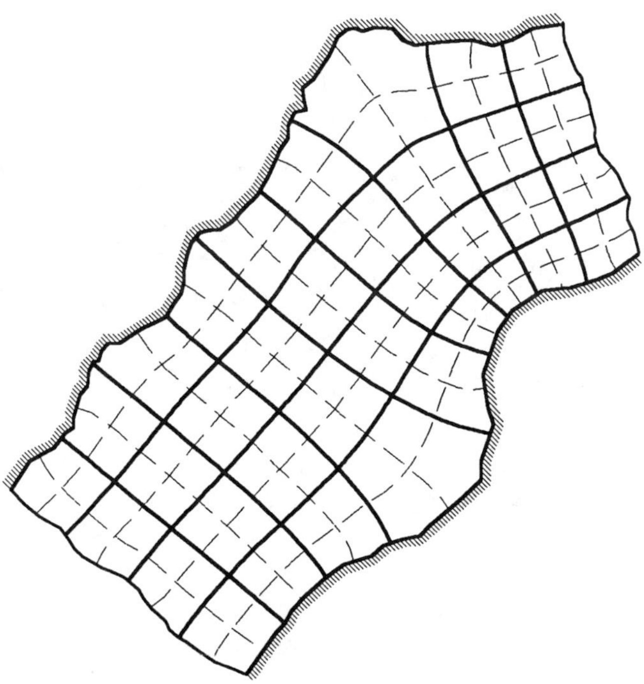

Figure 15.10 Once finished and tested, the flow net can be made denser for more precise calculations in the aquifer portions of interest.

The unit cross-sectional area of flow for the entire aquifer is:

$$A = 1 \cdot \Delta W \cdot N \tag{15.8}$$

where 1 is the unit aquifer thickness, ΔW is the distance between two streamlines forming a flow segment ("streamtube"), and N is the number of segments.

The average groundwater velocity (v) through the entire aquifer is:

$$v = K \cdot \frac{\Delta H}{M \cdot \Delta L} \tag{15.9}$$

where K is the hydraulic conductivity, ΔH is the head difference between the Upper and Lower Lakes, ΔL is the distance between two adjacent equipotential lines, and M is the number of squares along the flow path, *i.e.*, between two lakes.

Substituting equations (15.8) and (15.9) into equation (15.7) gives:

$$q = K\Delta H \cdot \frac{N}{M} \cdot \frac{\Delta W}{\Delta L} \tag{15.10}$$

Since the flow net in our case is a net of curvilinear squares (tests for the net validity are satisfied - see Figure 15.9), the distance between two equipotential lines can be considered equal to the distance between two streamlines: $\Delta L = \Delta W$, which reduces equation (15.10) to:

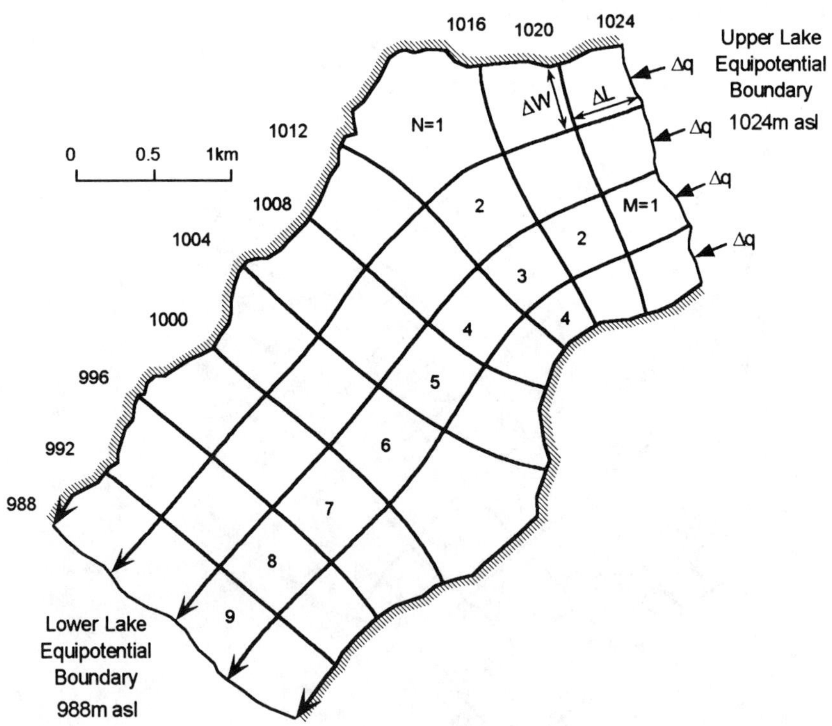

Figure 15.11 The finished flow net and the elements for the flow rate calculation.

$$q = K\Delta H \cdot \frac{N}{M} \tag{15.11}$$

As can be seen, the flow rate per unit thickness of the aquifer depends on the geometric characteristics of the flow net (*i.e.*, number of squares along the flow direction and perpendicular to it), hydraulic conductivity, and the total head loss between the two equipotential boundaries. Inserting numeric values into equation (15.11) gives the flow rate:

$$q = 7.3 \times 10^{-6} m/s \cdot (1024m - 988m) \cdot \frac{4}{9} = 1.168 \times 10^{-4} m^2/s$$

To obtain the total flow rate through the aquifer, the flow rate per unit thickness has to be multiplied by the overall aquifer thickness. From the flow net in Figure 15.11, it can be seen that the squares first receiving the flow, *i.e.*, those adjacent to the upper equipotential boundary (Upper Lake), are approximately of equal size. If the aquifer is homogeneous, *i.e.*, if the hydraulic conductivity of the porous medium remains constant throughout (which is our assumption), this means that the aquifer thickness is roughly the same in the first four squares. The data from borehole B1 (see Figure 15.1) indicate an aquifer thickness of 14.6m which gives the total flow of:

$$Q = q \cdot b = 1.168 \times 10^{-4} m^2/s \cdot 14.6m = 1.705 \times 10^{-3} m^3/s$$

15.4 TRANSMISSIVITY ANALYSIS

Groundwater contour maps and flow nets are useful tools for analyzing distribution of aquifer transmissivity and planning further investigations (e.g., location of monitoring and test wells). A portion of an aquifer with a wider spacing of equipotential lines (such as area A in Figure 15.12 for our case), indicates a higher transmissivity, while closely spaced contour lines mean a lower transmissivity (area B in Figure 15.12). From the definition of transmissivity:

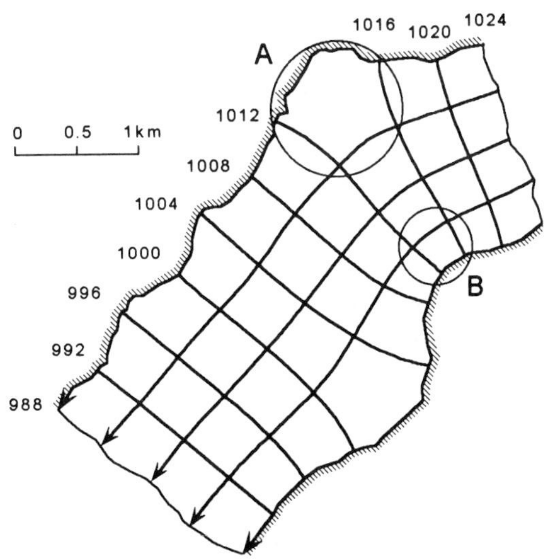

Figure 15.12 Varying spacing between equipotential lines indicates varying aquifer transmissivity (area A has a larger transmissivity than area B).

$$T = K \cdot b$$

it follows that the higher transmissivity can be a consequence of higher hydraulic conductivity, greater aquifer thickness, or both. If, based on geologic, sedimentologic, and other studies it is concluded that the aquifer hydraulic conductivity is uniform (as in our case), than the variation in the spacing of equipotential (contour) lines means that the aquifer thickness varies as well. Generally speaking, aquifer portions with widely spaced groundwater contours are better candidates for exploitation wells or other means of groundwater withdrawal. Note, however, that other hydrodynamic criteria must be considered as well. For example, even though area A in our case (see Figure 15.12) is the thickest portion of the aquifer, placing an exploitation well there would not be appropriate because of the closeness of the impermeable boundary. A well located near the aquifer center and equipotential boundaries would yield more.

PROBLEM 16

GROUNDWATER CONTOUR MAPS
INFLUENCE OF SURFACE STREAMS

The position of the water table in the unconfined alluvial aquifer of The Blue River is regularly recorded in 14 monitoring wells and 9 domestic wells as shown in Figure 16.1 (the alluvial plane and the aquifer extend beyond the map). Two sets of measurements representing the summer recession period of 1994 and the wet spring period of 1995 are given in Table 16.1. Construct and analyze the water table contour maps for both sets. A short 8-hour pumping test was conducted in the domestic well DW7 and the hydraulic conductivity of the aquifer was determined to be 3.47×10^{-5} m/s. The aquifer's impermeable base has an average elevation 605 ft above sea level in the area shown in Figure 16.1 (the base can be considered horizontal). Also analyze the potential for the possible spread of groundwater contamination around DW7 and present it graphically on a water table contour map.

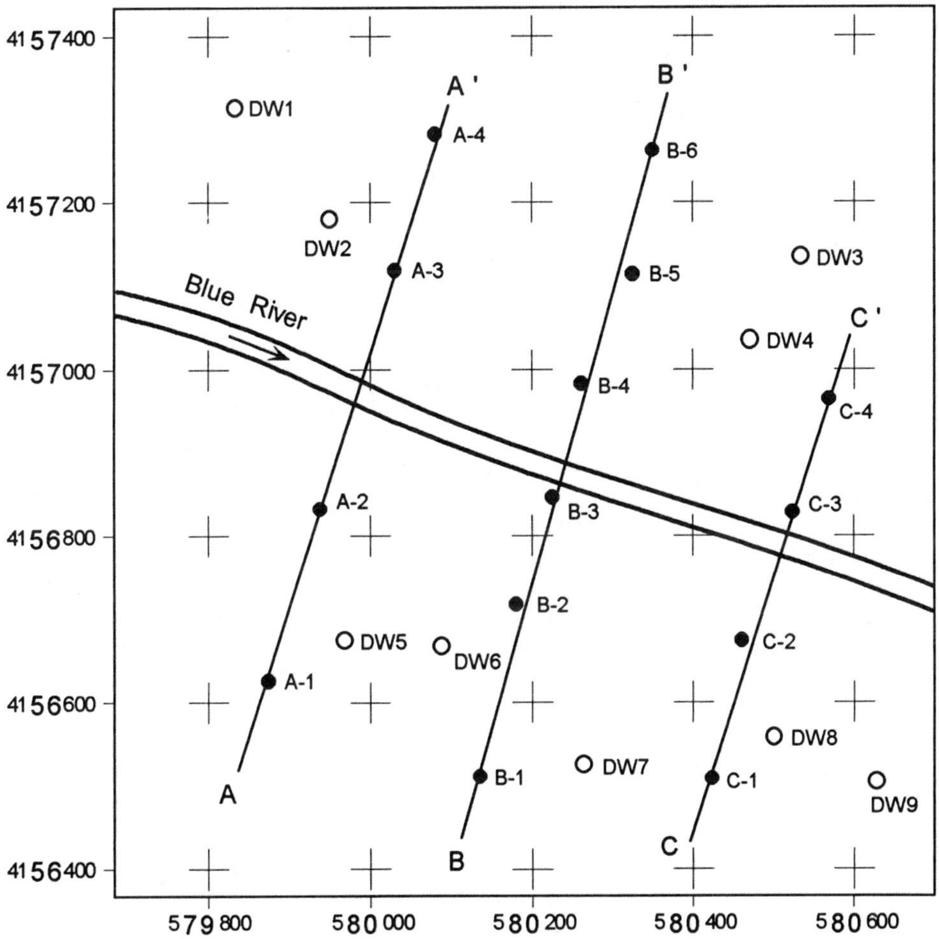

Figure 16.1 Monitoring wells (filled circles) and domestic wells (empty circles) used for measuring water table elevations in The Blue River alluvial aquifer.

Table 16.1 Alluvium aquifer water table elevations, in feet above sea level, measured in monitoring and domestic wells on July 15, 1994 and May 24, 1995. UTM coordinates are in meters.

Well	X coordinate	Y coordinate	July 15, 1994	May 24, 1995
A-1	579872.34	4156625.87	673.90	675.60
A-2	579938.25	4156833.42	672.60	675.55
A-3	580029.12	4157117.56	672.75	675.60
A-4	580080.49	4157282.19	674.65	675.65
B-1	580134.86	4156512.96	673.40	674.70
B-2	580180.35	4156717.51	672.70	674.40
B-3	580223.30	4156846.25	671.60	673.60
B-4	580259.27	4156984.09	672.35	674.15
B-5	580323.65	4157113.12	673.50	674.10
B-6	580350.45	4157268.09	674.70	675.00
C-1	580422.10	4156675.15	672.85	673.65
C-2	580463.54	4156675.23	672.10	673.05
C-3	580524.05	4156827.63	671.50	672.55
C-4	580570.72	4156966.07	672.70	672.70
DW1	579832.14	4257313.89	674.75	676.45
DW2	579949.16	4157180.20	673.45	675.90
DW3	580533.25	4157134.51	673.75	673.80
DW4	580470.81	4157036.78	672.90	673.35
DW5	579968.06	4156674.83	673.45	675.30
DW6	580087.19	4156669.12	673.20	674.85
DW7	580270.34	4156527.45	673.10	674.20
DW8	580500.18	4156558.78	672.65	673.10
DW9	580629.02	4156506.19	672.80	672.85

16.1 CONTOURING

Contour maps of the water table (unconfined aquifers) or the piezometric surface (confined aquifers) are made in the majority of hydrogeologic investigations and, when properly drawn, represent a very powerful tool in aquifer studies. Although commonly used for determination of groundwater flow direction only, contour maps, when accompanied with other data, also allow for the analyses and calculations of the flow velocity, particle travel time, hydraulic conductivity and transmissivity.

At least several data sets collected in different hydrologic seasons should be used to draw groundwater contour maps for the area of interest. In addition to recordings from piezometers, monitoring wells and other water wells, every effort should be made to record elevations of water surface in the nearby surface streams, lakes, ponds and other surface water bodies including cases when these bodies seem "too far" to influence groundwater flow pattern. In addition, one should gather information about hydrometeorologic conditions in the area for preceding months paying special attention to the presence of extended wet or dry periods. All of this information is essential for making a correct contour map.

16.1.1 Groundwater Flow Direction

The direction of groundwater flow in a localized area of an aquifer can be determined if at least three recordings of water table (or piezometric surface) elevations are available. Figure 16.2 illustrates the principle of finding the position of a water table in three dimensions using data from three monitoring wells. In a map, or two-dimensional view, the water table is represented by the contour lines which connect points with the same hydraulic head. The fastest way to construct contours is by linear triangulation as shown in Figure 16.2. The direction of groundwater flow is indicated by the arrow drawn perpendicular to and "down" the contour lines. This is also the direction of the dip of the water table approximated by the plate. It is very important to understand that the flow direction determined this way is representative only of the local area covered by the three monitoring wells. Depending on hydrogeologic conditions this direction may change in a nearby aquifer portion. For that reason more observation points are usually established during a hydrogeologic investigation in order to construct a reliable contour map(s).

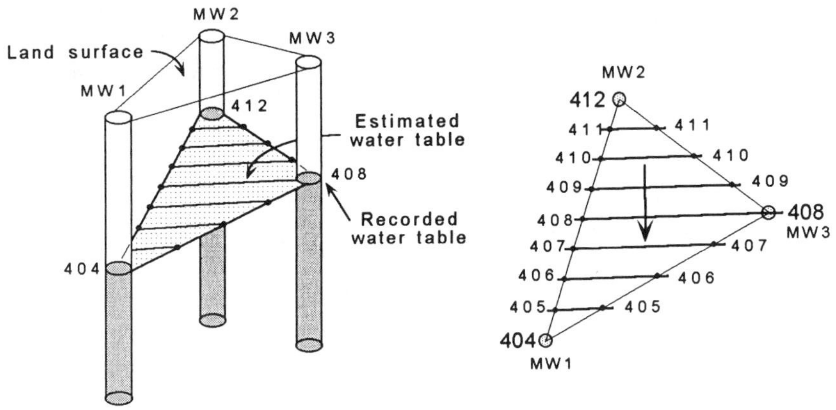

Figure 16.2 **Left**: Finding the position of the water table in three dimensions using data from three monitoring wells (numbers are water levels in meters or feet above seal level). **Right:** Construction of water table contour lines by triangulation with linear interpolation. The arrow, perpendicular to the contours, indicates the direction of groundwater flow.

16.1.2 Contouring Methods

Manual Contouring

Manual contouring is practically always utilized in groundwater studies, either as the only method or in conjunction with computer based methods. A complete reliance on software contouring could lead to erroneous conclusions since computer programs are unable to recognize interpretations apparent to a groundwater professional such as presence of geologic boundaries, varying porous media, influence of surface water bodies, or principles of groundwater flow. Thus manual contouring and/or manual reinterpretation of a computer generated maps are essential and integral parts of hydrogeologic studies.

Manual contouring is essentially based on the triangular linear interpolation (the principle of which is shown in Figure 16.2) combined with the hydrogeologic experience of the interpreter. The first draft map is not necessarily an exact linear interpolation between data points. Rather, it is an interpretation of the hydrogeologic and hydrologic conditions with contours that roughly follow numeric data on water table (or piezometric surface) elevations. Whenever possible the contours should be drawn to satisfy principles described in flow net analysis (see Problem 15). This means that almost inevitable local "irregularities" in water table elevations should not blur the overall tendency of groundwater flow (remember that a groundwater contour map is also a flow net without shown streamlines and it represents a graphical solution of the two-dimensional flow field). A novice is often caught in drawing various depressions in water table from which there is no "escape" of groundwater. Unless there is a valid hydrogeologic explanation (e.g., presence of a pumping well) these depressions are probably the result of erroneous contouring or data. Similarly "mysterious" local mounds in water table should be carefully examined.

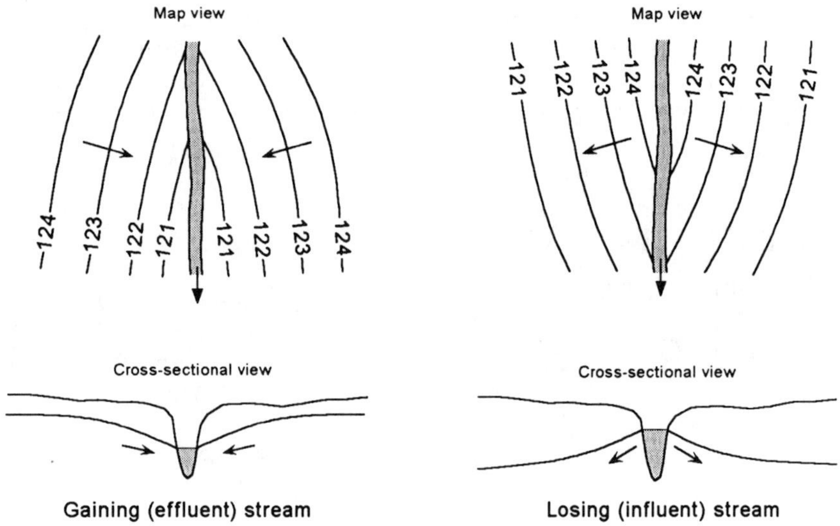

Figure 16.3 Two basic relationships between surface streams and aquifers. **Left**: Surface stream drains the aquifer, *i.e.,* it gains water (effluent stream). **Right**: Surface stream loses water to the aquifer (influent stream). Various combinations of these two relationships are also possible (see Section 16.3).

One of the most important aspects of constructing contour maps in alluvial aquifers (as in our case) is to determine the relationship between ground and surface waters. Figure 16.3 shows two basic cases which can coexist to form more complicated situations (see section 16.3). In hydraulic terms, the contact between the aquifer and the surface stream is an equipotential boundary. This does not mean that everywhere along the contact the

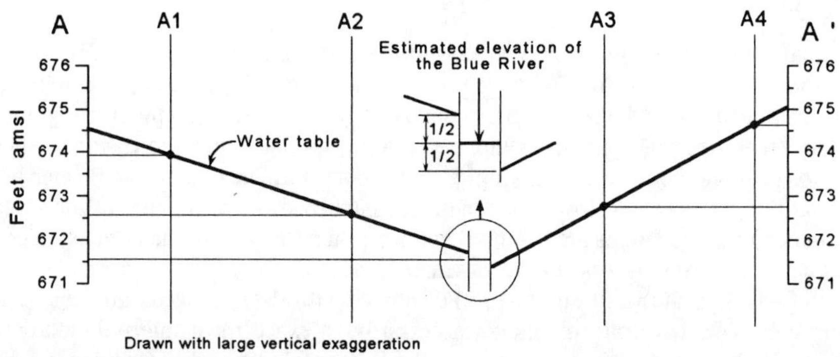

Figure 16.4 Estimating elevation of the water surface in the river channel using data from the groundwater monitoring wells. The estimated value is the average of the water table line intercepts at two banks.

hydraulic head is the same. Since both the river water and the groundwater are flowing, there is a hydraulic gradient along the contact. If enough measurements of a river stage are available, it is relatively easy to draw the water table contours in the vicinity of the river and to finish them along the river-aquifer contact. However, often little or no data is available and the river elevation, at the expense of precision, has to be estimated from a topographic map or from the monitoring well data. Figure 16.4 shows how the elevation of The Blue River on July 15, 1994 was estimated from the water table recordings in monitoring wells. Two water table lines are drawn connecting readings in two wells on each side of the river. If the intercepts of these two lines at river banks are not approximately at the same level (which is the consequence of different hydraulic gradients, *i.e.*, hydraulic conductivities in our case), then the average should be used. Elevations estimated this way for all three lines of monitoring wells (A, B, and C) are shown in Table 16.2. In addition, five more points with estimated elevation along the river are included based on a presumed linear change of the river channel hydraulic gradient.

Table 16.2 Estimated elevation of the water surface in The Blue River channel, in feet above mean sea level, July 15, 1994. UTM coordinates of the estimation points are in meters.

Est. point	X coordinate	Y coordinate	Elevation (ft)
A	579980.45	4156975.30	671.70
B	580235.80	4156878.25	671.25
C	580510.65	4156784.75	670.85
1	579711.25	4157075.45	672.35
2	579836.40	4157040.00	672.00
3	580075.23	4156938.75	671.50
4	580374.86	4156831.08	671.10
5	580700.00	4156725.00	670.50

Figure 16.5 shows manually drawn contour map of the water table in The Blue River alluvial aquifer for July 15, 1994.

Contouring with Computer Programs

Even though the early phase computer-generated contour maps are likely to be inaccurate because hydrogeologic variabilities may not be considered, it is desirable to have the final map as a (digital) XYZ computer file. This will allow its use within a geographic information system (GIS) for further computerized analysis which is the emerging standard in geosciences. Having an XYZ file of the water table (piezometric surface) will also significantly simplify data preparation for a possible future groundwater modeling.

If the available computer program cannot produce a satisfactory contour map (for example, there is a complicated mixture of impermeable and equipotential boundaries), and it cannot be forced to do so by the interpreter, the solution is to digitize a manually drawn map for controlling the interpolation. This, however, may be a lengthy process and it is better to acquire an appropriate software package for contouring. Quite a few computer programs today offer a wide range of contouring methods, allow the interpreter to interactively change generated grids, and display maps in three dimensions (e.g., SURFER by Golden Software, GRIDZO by Rockworks, Transform by Spyglass, QUICKSURF for AutoCAD by Schreiber Instruments).

Contouring programs require that the groundwater level data be organized as XYZ files where X and Y are plane coordinates of the measuring points (monitoring wells) and Z is the elevation of the water table or piezometric surface above a chosen reference level, usually above mean sea level (amsl). The most widely used plane coordinate system in geosciences and engineering is The Universal Transverse Mercator system (UTM) which is based on a transverse Mercator projection. In our example the lower left corner of the map shown in Figure 16.1 has the following UTM coordinates:

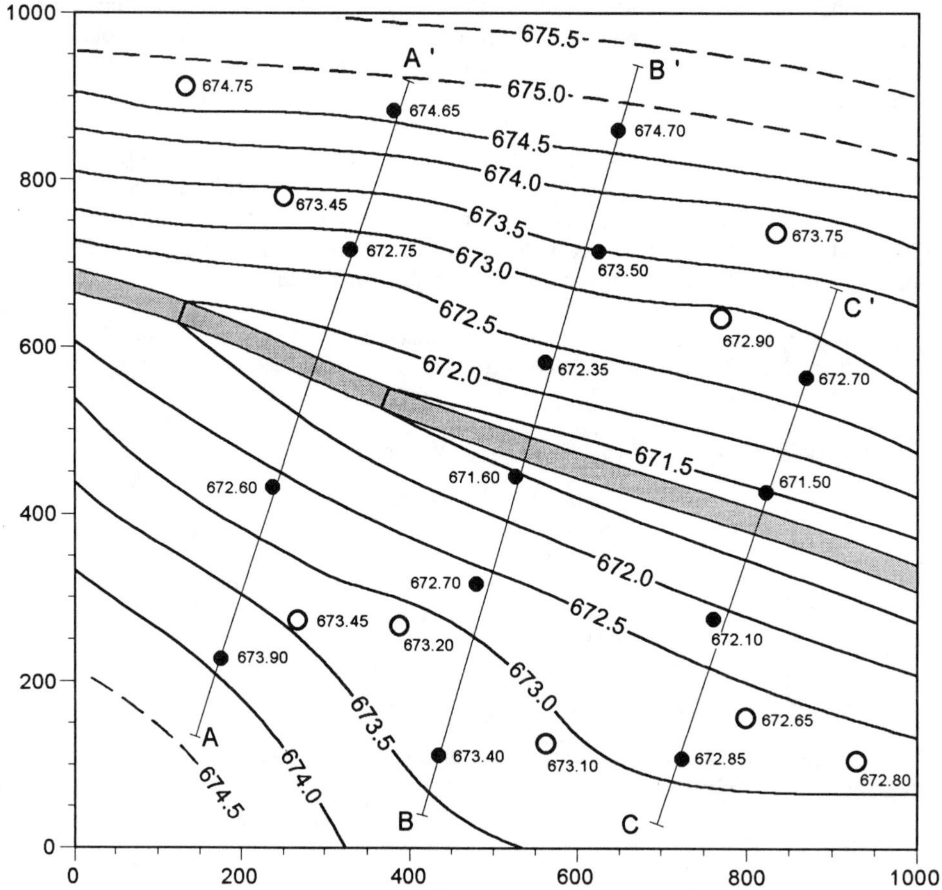

Figure 16.5 A manually drawn contour map of the water table in the Blue River alluvial aquifer on July 15, 1994. Numbers next to wells are recorded water table elevations in feet above sea level. Contour interval is 0.5 feet. (Note that, for comparison, the UTM coordinates are converted into local XY coordinates, also in meters.)

X: 579,700 meters EAST (the letters are omitted)
Y: 4,156,400 meters NORTH (the letters are also omitted).

This means that the corner of the map is 79,700 meters (79.7 km) east of the origin of the fifth UTM zone (number 5, the first number in the X coordinate, designates the UTM zone to which the map area belongs), and that it is 4,156,400 meters (4,156.4 km) north of the equator. UTM coordinates are shown on standard USGS topographic maps usually with blue numbers and ticks along the map boundaries.

Individual states also have state coordinate systems with unique origins and distances (coordinates) expressed in feet. Well and hydrologic data bases maintained by state agencies are often based on such a system and can be easily converted into the UTM coordinate system. If these official systems are not, for various reasons, yet applied to the data collected in the field, a simple local (temporary) system may be required. In our case, the lower left corner of the map would get an XY coordinate of 0,0 with distances along the X (horizontal) and Y (vertical) axes measured in meters as shown in Figure 16.5. Once the UTM coordinates of the origin (lower left corner) are determined, the local coordinates can be easily converted.

Although the latitude/longitude system is the most common way to specify geographic locations on the Earth, it is not used in hydrogeologic and engineering practice since it does not allow precise measurements of distances and areas.

Kriging

Kriging is one of the most robust and widely used methods for interpolation and contouring in different fields. It is a geostatistical method that takes into consideration the spatial variance, location, and sample distribution in data. It is particularly useful in heterogeneous porous media when used to contour hydrogeologic parameters such as hydraulic conductivity and storage. There are various interpolation (gridding) models within the kriging method and the most appropriate one for the existing data set can be chosen after determining the *semi-variogram* with *type* (spherical or exponential, for example), the *sill*, and the *nugget effect*. Although all major contouring programs on the market include kriging, few allow for the generation of a semi-variogram from the recorded data. Therefore, in order to select an appropriate kriging method, it will often be necessary to generate the semi-variogram using some other program.

Semi-Variogram

The semi-variogram, γ, is a graph (and/or formula) describing the expected difference in value between pairs of samples with a given relative orientation (Clark, 1979). A theoretical function (model) that best fits the recorded data is found after calculating the experimental semi-variogram. This procedure consists of choosing an oriented interval (say, 40 ft, west-east) and finding the variance of differences between all possible pairs of data that are 40 (± tolerance) ft apart. This variance is called the experimental *variogram* (it varies with distance and direction) and is given as:

$$2\gamma^*(h) = \frac{1}{n} \cdot \sum_{i=1}^{i=n} [g(x)_i - g(x+h)_i]^2 \tag{16.1}$$

where

$\gamma^*(h)$ is the experimental *semi-variogram*,
h is the oriented distance between two data points, (say, 40 ft, west-east)
g is the value of the sample (data point),
x is the position of one sample in the pair,
$x+h$ is the position of the other sample in the pair,
n is the number of pairs in calculation.

The calculated value (divided by two) is the measure of the difference between the two data points which are a distance h (40 ft) apart in the west-east direction. This measure is in squared units of data (in our case it is squared hydraulic head, or ft^2).

The process in now repeated for as many new distances as possible. Each new distance is increased by the basic interval ($h=40ft+40ft=80ft$; $h=120ft$; $h=160ft$, etc.), and the results are plotted in a graph called *semi-variogram* as shown in Figure 16.6. The distance between the data pairs is plotted along the horizontal axis, and the value of the semi-variogram (γ) along the vertical. Variograms can be plotted for various directions, which is recommended, especially if a certain degree of anisotropy in data is expected. Directions that produce noticeable different (while still meaningful) semi-variograms may indicate an actual anisotropy in the data. Note, however, that there is a substantial degree of subjectivity in interpreting semi-variograms.

For a statistically valid semi-variogram, there should be at least 30 pairs of data which is often not the case in groundwater studies (note that 50 pairs is already considered as a good sample set). When the data set is small, the directional calculation (e.g., west-east) will often not produce a meaningful semi-variogram. In such cases, the calculation should be performed in all possible directions and for all sample pairs that fall within the calculation interval. This will produce a *global semi-variogram.*

For irregularly spaced data points, the basic distance (h) has a tolerance interval which is larger for smaller data sets. This tolerance should not exceed one-half of the basic distance. For example, if the basic distance h is 40 ft, the tolerance should be maximum ±20 ft. This means that all the pairs that fall between 20 and 60 ft (in the given direction) will be included in the calculation of the semi-variogram. The next value of γ will then be calculated for 80±20 ft, and so on.

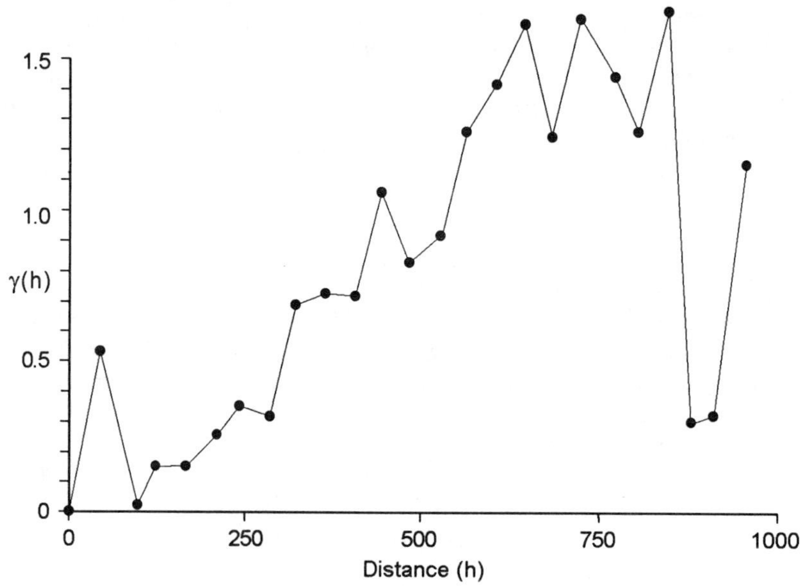

Figure 16.6 Experimental global semi-variogram for the water table measurements in the Blue River alluvial aquifer, July 15, 1994. The lag is 40 ft.

The shape of the experimental variogram should indicate which theoretical model is appropriate for the given data, and what are the values of the three important parameters of the semi-variogram:

- a range of influence,
- C sill,
- C_0 nugget effect.

The so-called "ideal shape" for the semi-variogram, which is derived theoretically and is known as the *spherical* or *Matheron* model, is shown in Figure 16.7a. The range of influence (a) is the distance at which data become independent of one another. After this point the graph is horizontal and the corresponding value of γ is called the *sill* of the semi-variogram (C). The theoretical spherical model is given as:

$$\gamma(h) = C \cdot \left(1.5\frac{h}{a} - 0.5\frac{h^3}{a^3}\right) \tag{16.2}$$

The sill value is equal to the sample variance which is given as:

$$s^2 = \frac{1}{n-1} \cdot \sum_{i=1}^{i=n} (g_i^2 - \bar{g}^2) \tag{16.3}$$

where ($\bar{g}$) is the sample mean:

$$\bar{g} = \frac{1}{n} \cdot \sum_{i=1}^{i=n} g_i \tag{16.4}$$

Figure 16.7b shows a semi-variogram with the *nugget effect* -- the graph's intercept at the vertical axis. Its presence may indicate a potential error in data collection, or any other unexplainable (random) component in the data. The nugget effect is a constant that raises a theoretical semi-variogram C_0 units along the vertical:

$$\gamma(h) = C_0 + \gamma'(h) \tag{16.5}$$

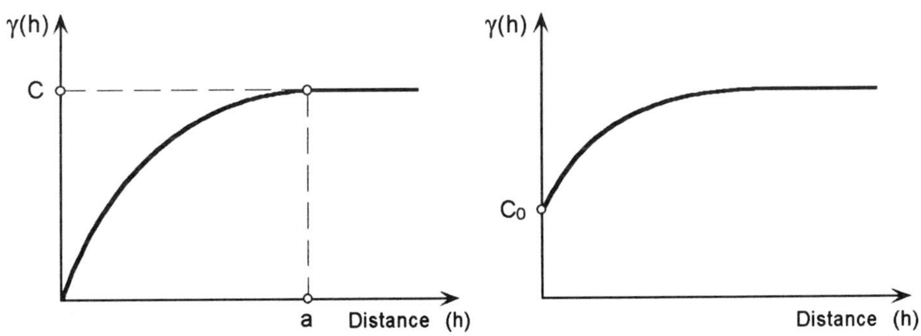

Figure 16.7 Left: The theoretical spherical semi-variogram with the sill -- *C*, and the range of influence – *a*. **Right**: The spherical semi-variogram with the nugget-effect -- C_0.

where $\gamma'(h)$ is one of the common theoretical semi-variograms. Figure 16.8 shows the three most often used theoretical models that are computer-fitted for the May 1995 water table in the Blue River aquifer: a spherical model (see equation 16.2), an exponential model, which in generally given with the following equation:

$$\gamma(h) = C \cdot \left[1 - e^{-\frac{h}{a}}\right] \tag{16.6}$$

and a linear model whose general equation is:

$$\gamma(h) = bh \tag{16.7}$$

where b is the slope of the straight line.

Once the "right" semi-variogram model and its parameters are selected, the kriging contouring method will use them to interpolate a map. Again, note that the selection of a model is rather subjective and requires experience, including the possibility of combining several models. For more information on semi-variograms and kriging see, for example, Clark (1979), and Isaaks and Srivastava (1989).

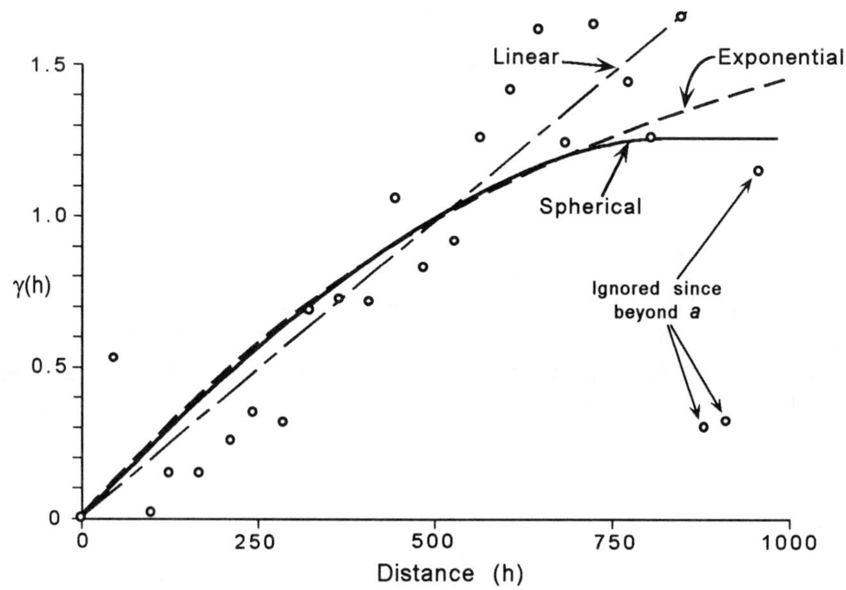

Figure 16.8 Computer-fitted theoretical semi-variogram models for the May 1995 water table recorded in the Blue River alluvial aquifer.

Figure 16.9 shows two maps of the water table in the Blue River alluvial aquifer generated with the Linear Kriging model within SURFER contouring software (published by Golden Software, Golden, Colorado). Neither map accurately describes the role of the Blue River as a gaining stream - map A because there is no data on the stream elevations, and map B because there is not enough data. In addition, contours along the maps' edges where there is less data, especially in northeast and southwest, tend to bend away from the river which does not seem "natural" for the base flow period. *(Note that during base flow conditions, i.e., during extended periods without significant precipitation, the river stage is at its lowest and the supply of water comes from the aquifer only. This means that groundwater is finding the shortest way to reach the river, i.e., the streamlines are more perpendicular to the river, and the contour lines are then more parallel to the river.)*

Figure 16.10 shows a much more acceptable map after five more data points are estimated along the river (see Table 16.2). However, the contours remain bent and questionable in the corners of the map indicating that more adjustments have to be done to generate the final XYZ computer file.

Even though kriging is widely considered as a reliable contouring method, it is recommended to try all the methods available within the chosen software package since some may work better with a particular set of data. Interpolation is not an exact process and can yield often very different result. It is iterative, with the final accepted result assessed qualitatively.

Triangulation with Linear Interpolation

This method is based on exact linear interpolation between three neighboring points as shown in Figure 16.2. Because the original data points are used to define the triangles, they are preserved (honored) very closely on the generated map (note that this is not always the case with kriging). In general, when there are few data points (e.g., less than 20), and/or data are not evenly distributed, triangulation is not effective since it creates triangular facets and "holes" in the map. In our case the method closely represents the expected water

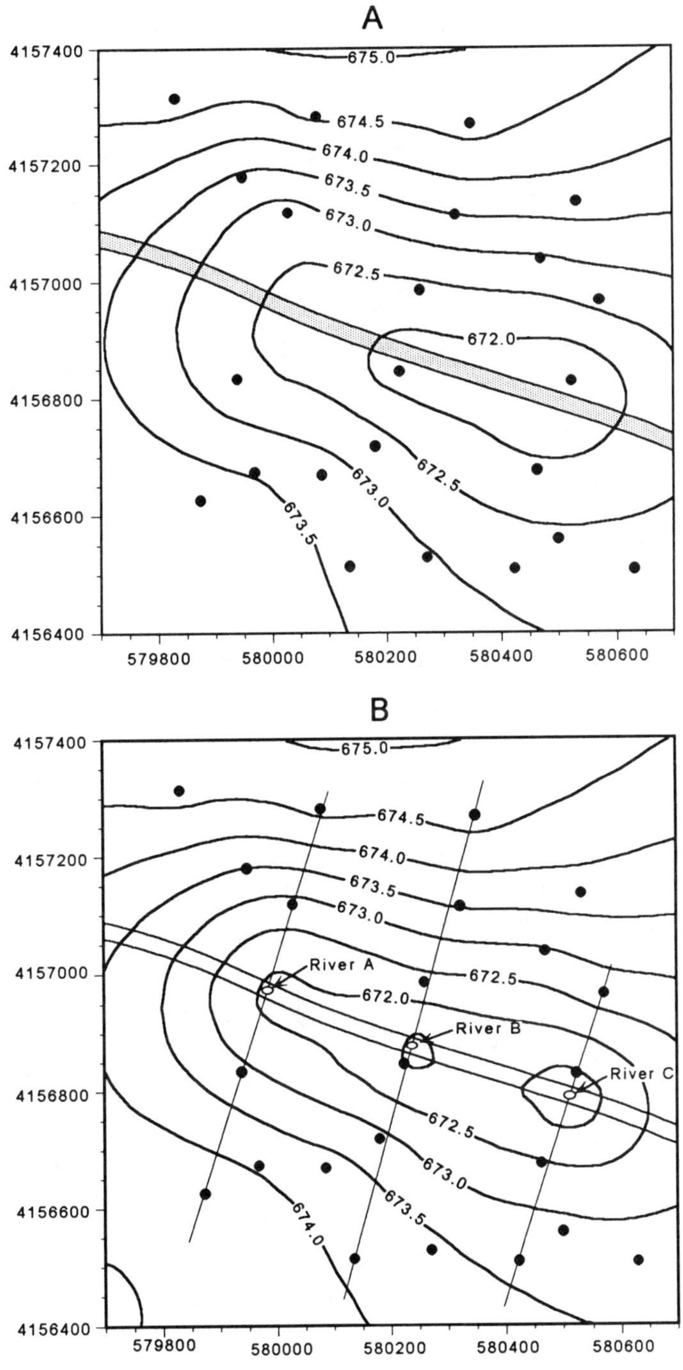

Figure 16.9 July 15, 1994 water table contour maps of the Blue River alluvial aquifer generated using a linear kriging model within SURFER contouring software (SURFER is published by Golden Software, Golden, Colorado). A: Map without water elevation data along the river. B: Map with three river elevations estimated as shown in Figure 16.4.

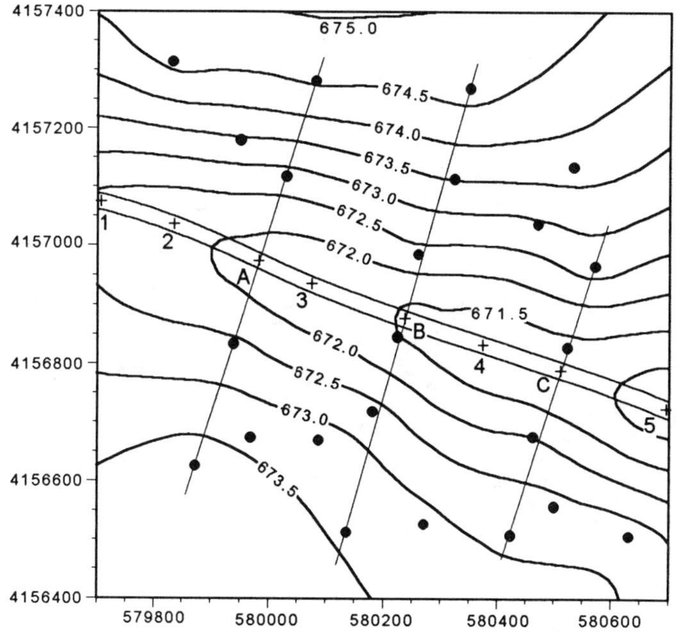

Figure 16.10 Water table contour map with additional estimated data points along the river, generated with the linear kriging.

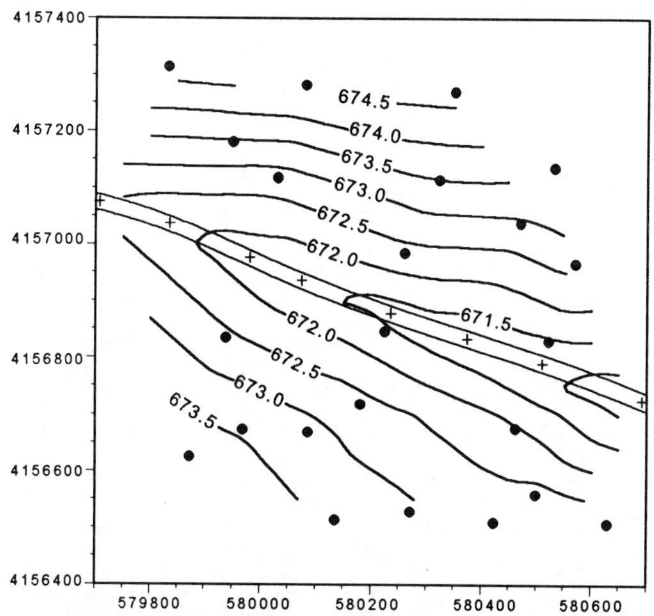

Figure 16.11 July 15, 1994 water table map generated using Triangulation with Linear Interpolation method within SURFER.

table in most of the map where the data are distributed evenly (Figure 16.11). The trend indicated by the map allows easy manual filling of the holes and completion of the contours. This also means that, as in case of the kriging method, further adjustments are needed to obtain the final XYZ computer file.

Triangulation is fast and accurate with 200 to 1000 data points evenly distributed over the map area. Note, however, that this number of original data is rarely available in groundwater investigations and one would have to estimate additional "auxiliary" data to satisfy the requirement. If enough data is available, the advantage of triangulation is that it preserves breaks in contours such as faults, geologic boundaries and streams indicated in a data file.

Inverse Distance to a Power

This interpolation method is usually present in various contouring packages because it is very fast and can handle large data sets. However, it does not produce good results when used to contour groundwater levels often represented with small data sets. Inverse distance

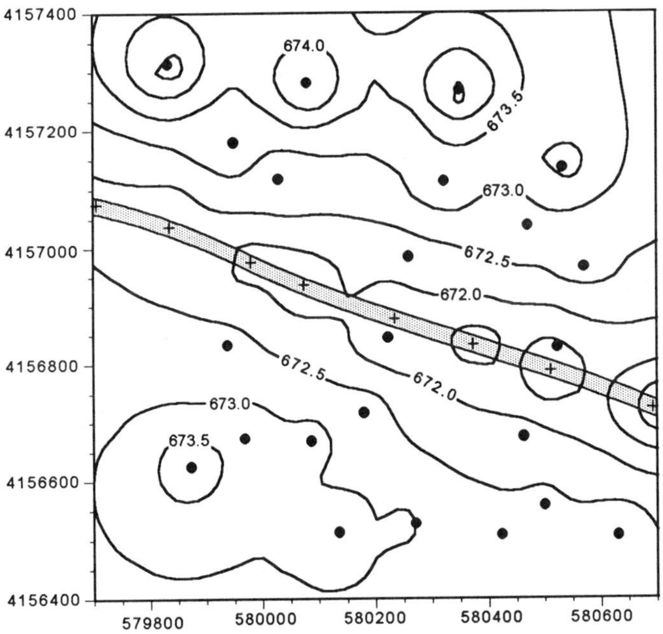

Figure 16.12 July 15, 1994 water table map generated using Inverse Distance to Power method within SURFER. The power is 2 and the smoothing factor is 1. Note a distinct "bull's eyes" pattern around the data points.

to a power weights data points during interpolation so that the influence of one data point relative to another decreases with the power of distance. The result is the generation of "bull's eyes" pattern around the positions of data points as shown in Figure 16.12. Software smoothing of contours can only slightly reduce this effect but does not eliminate unnatural depressions and mounds in water table.

Minimum Curvature

Contouring with minimum curvature is widely used in the earth sciences since this method generates the smoothest possible surface while still closely honoring the original data points. However, minimum curvature is not always accurate in preserving breaks in contours which is often most important when contouring piezometric surface near streams, faults and geologic boundaries. Figure 16.13 shows some unnatural "solutions" by the method in our case (note a depression in the water table in the river at intersection C-C').

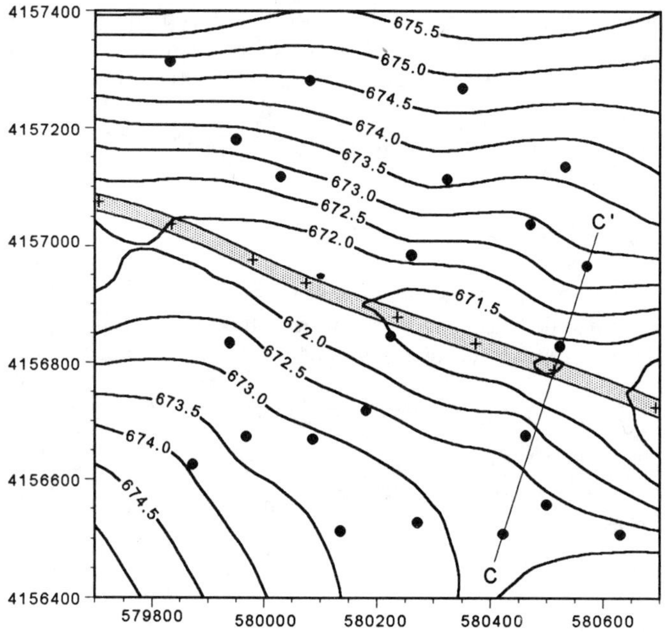

Figure 16.13 July 15, 1994 water table map generated using the Minimum Curvature method within SURFER.

Radial Basis Functions (Splines)

A diverse group of radial basis functions is analogous to variograms in kriging and produces similar results (see Figure 16.14). They are all exact interpolators and closely preserve original data. Splines are frequently used radial basis functions in geosciences. The natural cubic spline function is (Keckler, 1994):

$$B(h) = \left(h^2 + R^2\right)^{3/2} \tag{16.8}$$

and the thin plate spline function is:

$$B(h) = (h^2 + R^2) \cdot \log(h^2 + R^2) \tag{16.9}$$

where

h is the anisotropically rescaled, relative distance from the data point to the grid node, and

R^2 is the smoothing parameter specified by the user.

More detail on contouring with radial basis functions can be found in the work of Carlson and Foley, 1991: *Radial Basis Interpolation on Track Data*, Lawrence Livermore National Laboratory, UCRL-JC-1074238.

Comparing all the maps shown in Figures 16.9 through 16.14, it seems that the thin plate spline function most closely resembles the manually drawn map (Figure 16.5) and the expected natural look of the water table during the July 15, 1994 recession period. Since the computer-generated map (Figure 16.14B) still shows some "irregularities" near the central-eastern edge (the contours bend away from the river) and in the southwestern section, further adjustments are needed to create the final XYZ file. This is achieved by introducing new auxiliary data points throughout the map, and particularly in the parts with few data points such as along the map edges. Table 16.3 shows coordinates and

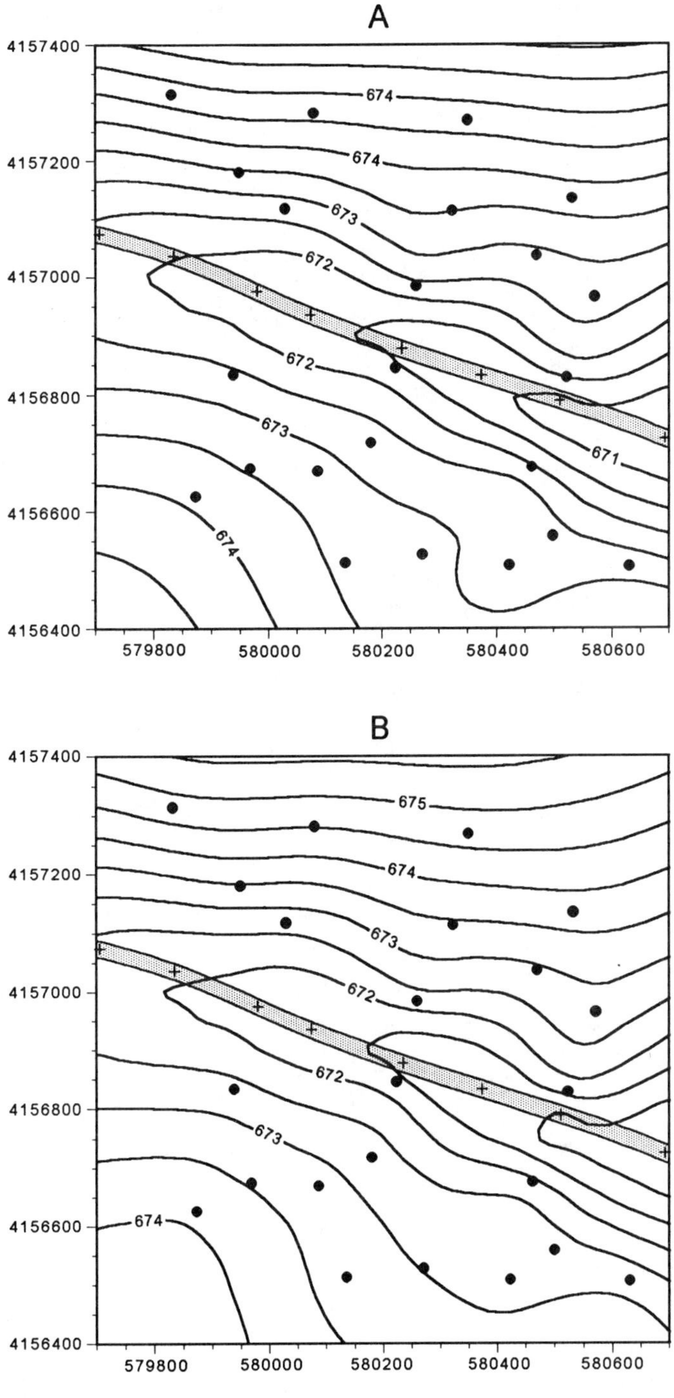

Figure 16.14 July 15, 1994 water table contour maps of The Blue River alluvial aquifer generated using Radial Basis Functions method within SURFER. A: natural cubic spline function B: thin plate spline function.

estimated elevations at 25 new locations. The elevations of the water table are estimated from the manually drawn map shown in Figure 16.5. The resulting final map (Figure 16.15) very closely resembles the expected groundwater flow conditions in the alluvial aquifer.

Table 16.3 Estimated elevation of the water table at auxiliary points, in feet above sea level, July 15, 1994. UTM coordinates of the estimation points are in meters.

No.	X	Y	Z
New1	579750.00	4156700.00	674.00
New2	579750.00	4156800.00	673.50
New3	579700.00	4157000.00	672.50
New4	579700.00	4157125.00	672.50
New5	579750.00	4157150.00	673.00
New6	579750.00	4157250.00	674.00
New7	579850.00	4157100.00	672.50
New8	579850.00	4156900.00	672.50
New9	579800.00	4156850.00	673.00
New10	580000.00	4156550.00	673.70
New11	580000.00	4156450.00	674.00
New12	580100.00	4157000.00	672.05
New13	580300.00	4156750.00	672.00
New14	580600.00	4156550.00	672.50
New15	580600.00	4156850.00	672.00
New16	580600.00	4157200.00	674.50
New17	580600.00	4157250.00	675.00
New18	580700.00	4156600.00	672.05
New19	580700.00	4156775.00	671.50
New20	580700.00	4156950.00	673.00
New21	580700.00	4157050.00	673.50
New22	580700.00	4156475.00	673.00
New23	580000.00	4157400.00	675.50
New24	580250.00	4156400.00	673.50
New25	580700.00	4156825.00	672.00

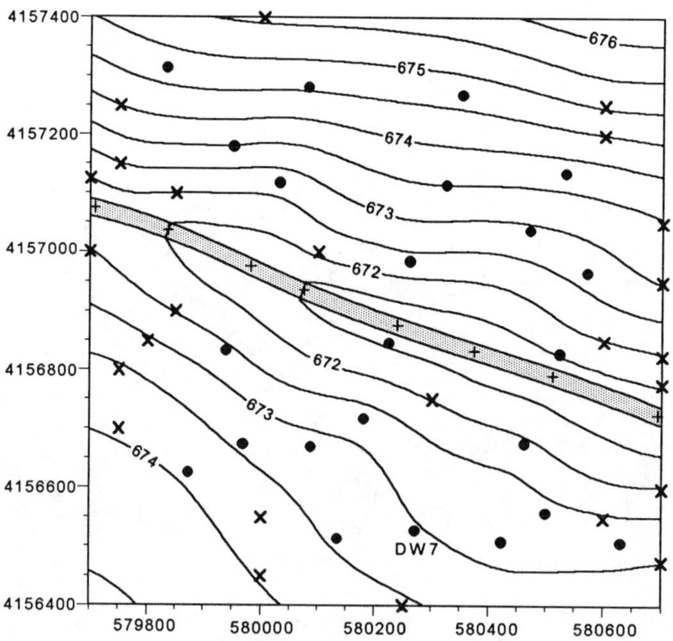

Figure 16.15 The final map of the water table on July 15, 1994 generated using the Thin Plate Spline contouring method within SURFER. New auxiliary data points with estimated water table elevations are shown with x marks while the original recorded data points (monitoring and domestic wells) are shown with circles.

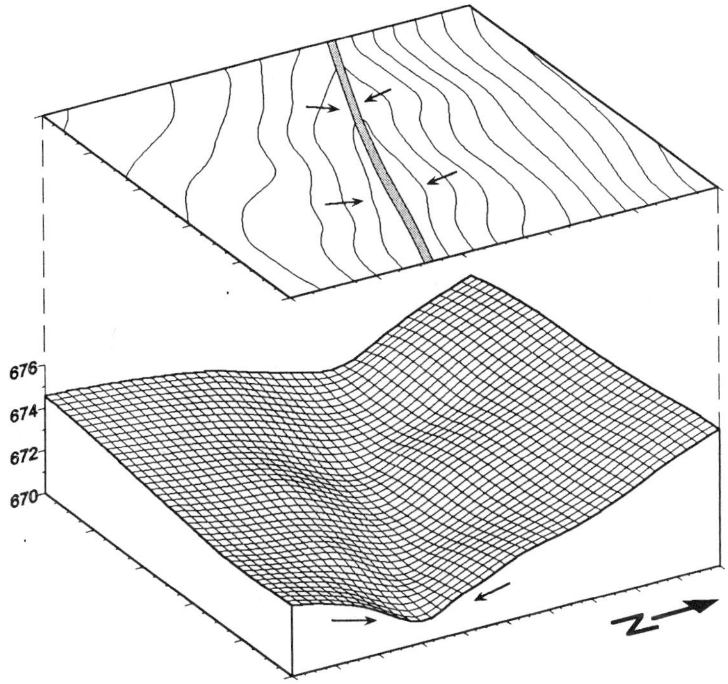

Figure 16.16 Map and block-diagram of the water table in the Blue River alluvial aquifer for July 15, 1994. The generally steeper hydraulic gradient north of the river which indicates a lower hydraulic conductivity. (Note that local aquifer heterogeneity is more apparent when the contour map is displayed in both map and 3D views.)

16.2 HYDRAULIC CONDUCTIVITY ANALYSIS

A contour map of an aquifer, accompanied by data on aquifer thickness, is an excellent tool for the analysis of transmissivity and hydraulic conductivity distributions. Wider spacing of water table contours indicates higher transmissivity of porous media, while closely spaced contours show parts of the aquifer with lower transmissivity. Since the transmissivity is the product of the hydraulic conductivity and the aquifer thickness ($T=Kb$), the following is true:

- In homogeneous aquifers with uniform hydraulic conductivity, widely spaced contours indicate larger aquifer thickness, while steeper hydraulic gradients (e.g., closely spaced contours) indicate smaller aquifer thickness;
- When an aquifer, or its portion of interest, has a uniform thickness, steeper hydraulic gradients indicate lower hydraulic conductivity of the porous media, while more widely spaced contours mean higher conductivity.

Based on the borehole data, it is concluded that the Blue River alluvial aquifer is (sub)horizontal within the map area with an average elevation of its impermeable base at 605 feet above sea level. The water table elevation varies between 675.0 and 670.6 feet which is an amplitude of 4.4 ft. This amplitude is only about 6.5% of the average aquifer thickness (*i.e.*, thickness of the saturated zone) which means that the variable spacing of the water table contours (see Figures 16.15 and 16.16) is the result of the varying hydraulic conductivity rather than the (slight) change in thickness.

The portion south of the Blue River, excluding a strip along the river, is generally more permeable than the portion north of the river where the hydraulic gradient is steeper. The highest hydraulic conductivity is expected around domestic well DW7 which is also the only well with the pumping test data available. However, keeping in mind the above-mentioned relationship between contour spacing and aquifer transmissivity, and knowing the numeric value of the hydraulic conductivity at one location, allows one to calculate the hydraulic conductivity for other portions of the aquifer. Figure 16.17 illustrates this procedure which consists of the following steps:

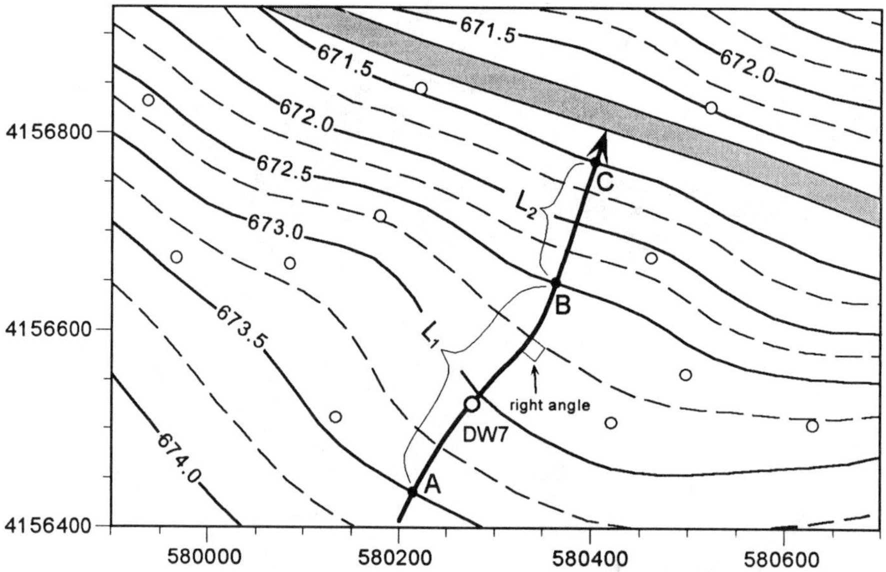

Figure 16.17 The streamline drawn through domestic well (DW7), a suspected source of groundwater contamination. The streamline is perpendicular to the water table contours. Note the auxiliary contours (dashed lines) which make drawing the streamline easier.

1. Draw a streamline through the aquifer portion of interest (in our case the streamline is drawn through DW7);
2. Identify sections along the streamline with similar hydraulic gradients, *i.e.*, with similar contour spacing (in our case there are two such sections: A-B, and B-C);
3. Calculate groundwater flow rate per unit aquifer width along the streamline in the section with the known hydraulic conductivity (A-B in our case) and then use this calculated value to find the hydraulic conductivity in the other section (B-C).

Groundwater flow rate per unit width in an unconfined aquifer without infiltration and with the horizontal base is described with the following equation (see also Problem 12 and equation 12.15):

$$q = K\frac{h_1^2 - h_2^2}{2L} \tag{16.10}$$

where

q is the unit flow,
K is the hydraulic conductivity,
h_1 is the higher hydraulic head,
h_2 is the lower hydraulic head,
L is the distance between the two heads measured along the streamline.

The unit flow in section A-B is then

$$q_1 = K_1 \frac{h_A^2 - h_B^2}{2L_1} \tag{16.11}$$

Hydraulic head h_A is the height of the water table above the reference level which is set at the horizontal base (see also Figure 12.5):

$$h_A = 673.5\,ft - 605\,ft = 68.5\,ft = 20.88m$$

Hydraulic head h_B is:

$$h_B = 672.5\,ft - 605\,ft = 67.5\,ft = 20.57m$$

Inserting all numeric values into equation (16.11) gives the unit flow along the streamline in section A-B:

$$q_1 = 3.47 \times 10^{-5}\,m/s \cdot \frac{(20.88m)^2 - (20.57m)^2}{2 \cdot 281m} = 7.93 \times 10^{-7}\,m^2/s$$

From the principle of the continuity of flow along the streamline, it follows that the unit flow in section A-B equals the unit flow in section B-C:

$$q_1 = q_2 = K_2 \frac{h_B^2 - h_C^2}{2L_2} \tag{16.12}$$

which gives the following expression for the hydraulic conductivity in section B-C:

$$K_2 = \frac{q_1 2L_2}{h_B^2 - h_C^2} \tag{16.13}$$

or

$$K_2 = \frac{7.93 \times 10^{-7}\,m^2/s \cdot 2 \cdot 130m}{(20.57m)^2 - (20.27m)^2} = 1.68 \times 10^{-5}\,m/s$$

The hydraulic conductivity in section B-C can also be found by inserting equation (16.11) into equation (16.13):

$$K_2 = K_1 \cdot \frac{L_2}{L_1} \cdot \frac{(h_A^2 - h_B^2)}{(h_B^2 - h_C^2)} \tag{16.14}$$

16.3 INFLUENCE OF SURFACE STREAMS

The importance of understanding the possible influences of a surface stream on a water table is illustrated with the situation recorded in the Blue River alluvial aquifer on May 24, 1995 following a five-week period of intensive precipitation. Figure 16.18 shows a computer-generated contour map of the water table that looks quite different from the map shown in Figure 16.15 which is representative of a recession period recorded in July 1994. Although the spring map needs further adjustments (it is based on the recorded data only), it is clear that the changes in river stage cause significant changes in the water table far into the aquifer. This is explained by a general example shown in Figure 16.19.

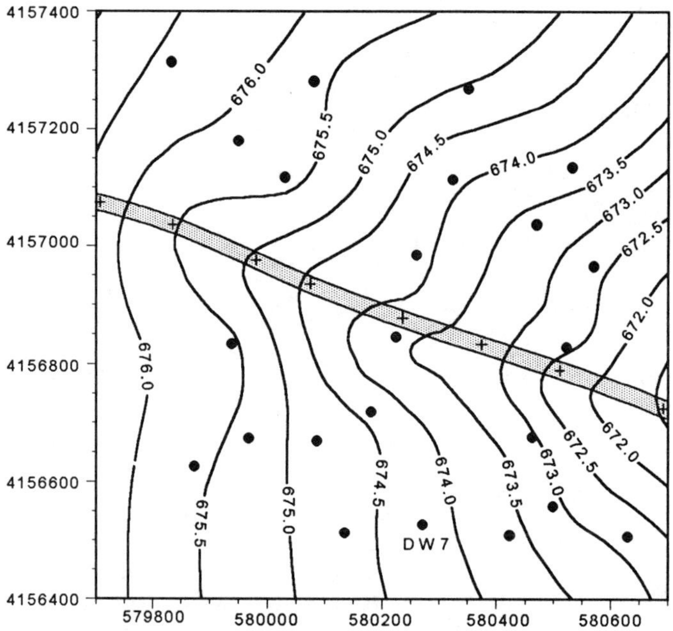

Figure 16.18 Computer-generated contour map of the water table in the Blue River alluvial aquifer for May 24, 1995. Contouring software is SURFER and the gridding method is the thin plate spline.

Cross-section 1 is characteristic for a gaining stream and groundwater flow directed toward the stream in the whole aquifer (see also Figure 16.3). In alluvial aquifers of large perennial streams this situation is present most of the time. Case 2 shows a reverse direction of groundwater flow close to the stream following a rapid rise in the river stage during and after major wet periods. Groundwater flow in most of the aquifer is still oriented toward the stream. After the river stage returns to normal, a rather complex groundwater pattern may develop as case 3 shows. The aquifer area affected by these changes depends on their duration and the hydrogeologic characteristics of the porous media.

The most pronounced influence is illustrated with the map view for case 3. Water table contours show two "ridges" parallel to the river that cause groundwater to flow in two opposite directions (small arrows). There are also two major (new) directions of groundwater flow parallel to the stream as indicated with large arrows. This picture may become even more complex if there are several close periods of significant changes in river stage. If the hydrometeorologic and hydrologic conditions in the area for the preceding few months are not known, it may, at times, be quite difficult to correctly interpret the water table recordings, especially if there is little data.

The computer-generated map shown in Figure 16.18 is adjusted manually so that the contours along northern and southern edges band more toward the river (Figure 16.20). Although there is no supporting data, this interpretation is made based on experience with software contouring during the recession period of July 1994 and the example shown in Figure 16.19.

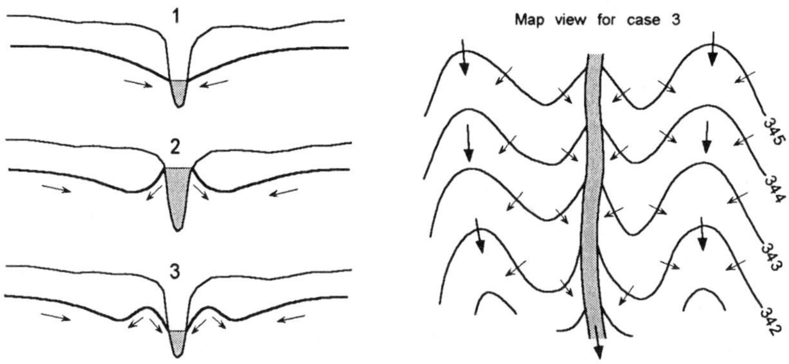

Figure 16.19 **Left**: Cross-sectional views of some common cases of the influence of surface streams on the water table in adjacent alluvial aquifers. 1) Basic case of a gaining (effluent) stream; 2) Position of water table after a significant rise in river stage; 3) Water table soon after the river stage returns to normal. **Right**: Map view for case 3. Small arrows indicate local directions of groundwater flow while large arrows show two major directions parallel to the river.

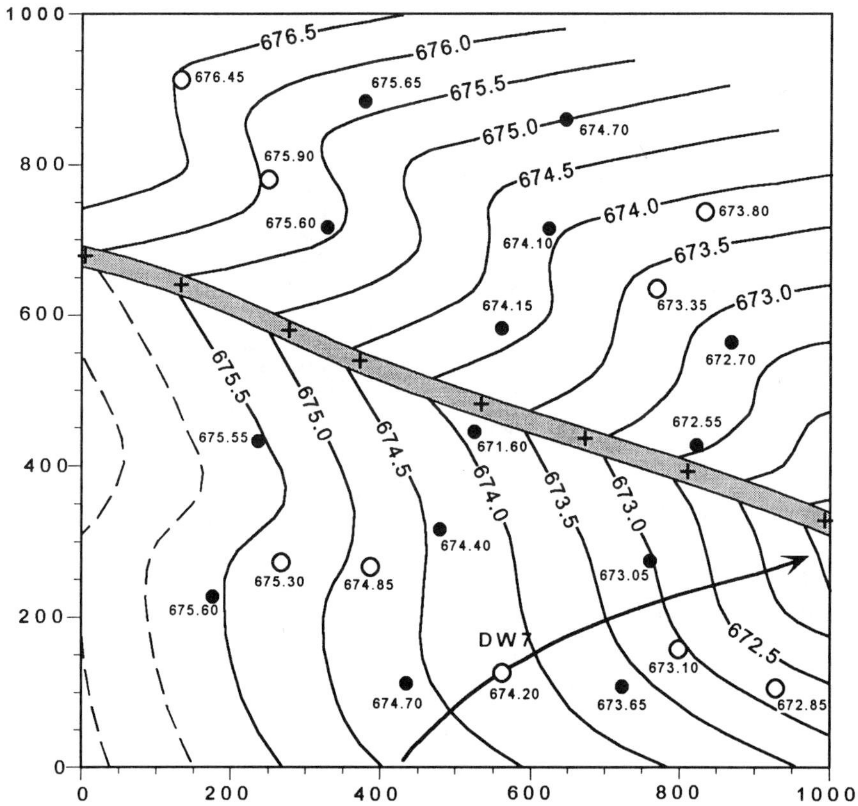

Figure 16.20 Manually adjusted water table contour map for May 24, 1995 with the streamline through domestic well DW7. Numbers next to the wells are water table elevations in feet above sea level. Contour interval is 0.5 feet and the local coordinate system is in meters.

16.4 SPREAD OF GROUNDWATER POLLUTION

In order to correctly assess spread of possible groundwater pollution, it is best to analyze water table contour maps for all distinct hydrologic seasons in the area of interest. Why one, or even several maps for similar hydrologic conditions, are not enough is illustrated with the situation shown in Figure 16.21. The streamline drawn through DW7 (the area of suspected pollution) is perpendicular to the river during the recession period of July 15, 1994. If this was the only data set available, and an assumption was made that, more or less, the groundwater flow direction is same throughout the year, an area affected by the pollution would be a relatively narrow strip around the streamline (assuming that the pollutant is transported primarily by groundwater, *i.e.*, due to advection). A similar mistake of underestimating the spread of possible pollution would be made if the situation recorded on May 24, 1995 (Figure 16.20) was considered representative for the aquifer.

Knowing that both data sets are recorded during two distinct and relatively extreme hydrologic periods (summer recession and spring high water stage), the estimated spread of possible pollution shown in Figure 16.21 seems accurate. It includes anticipated common lateral dispersion so that the plume limits are somewhat further than the two controlling streamlines. Of course, depending on "how wet was the wet period of May 24, 1995", the corresponding streamline may approach the river at a sharper angle (after higher floods) causing even wider spread of pollution.

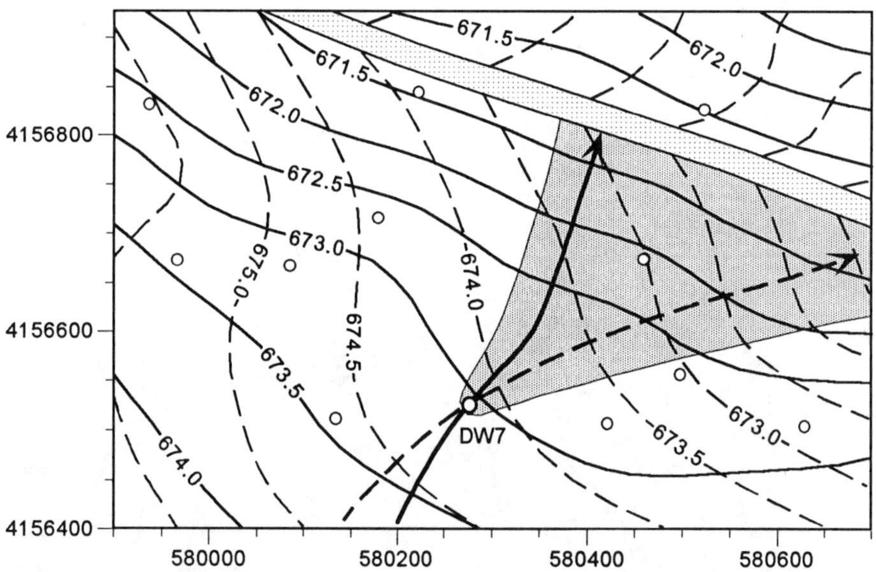

Figure 16.21 Estimated spread of a possible groundwater pollution (hatched area) around domestic well DW7 if no prevention or remediation were conducted. Solid contours are for July 1994 and dashed contours are for May 24, 1995. UTM coordinates are in meters and contour interval is 0.5 feet.

PROBLEM 17

FLOW NETS IN HETEROGENEOUS AQUIFERS REFRACTION OF STREAMLINES, FLOW VELOCITY

Estimate the path of groundwater contamination that would develop from the spill site shown in Figure 17.1 if no remedial action was taken. Three monitoring wells, MW1, MW2 and MW3, were drilled around the spill to determine the immediate direction of groundwater flow. Water table elevations recorded in the unconfined silty sand aquifer were 467.98, 467.82, and 467.53 meters above sea level respectively. The hydraulic conductivity of the porous medium was estimated to be 4.3×10^{-6} m/s based on laboratory analyses of core samples. Two short pumping tests were performed in the nearby domestic wells DW1 and DW2 and yielded hydraulic conductivity of 1.45×10^{-5} m/s and 6.11×10^{-6} m/s respectively. The water table elevations recorded in DW1 and DW2 before the test were 467.32 and 467.23 meters. The hydraulic conductivity of the main sandy gravel aquifer in the area, that has been alternately and discontinuously pumped by two municipal wells, is 2.81×10^{-5} m/s. The water table recorded in W1 during a long pumping break was 466.91 meters.

If it is shown that the main aquifer is threatened by the spill, estimate the time needed for the pollution to reach the aquifer assuming that there is no reaction between the pollutant and the porous media/groundwater.

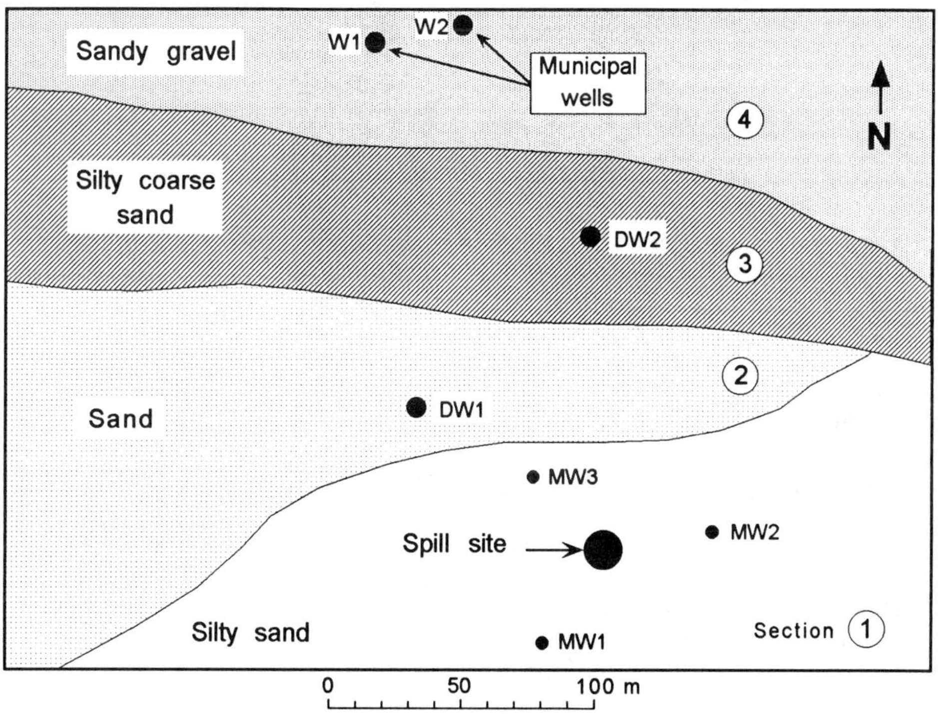

Figure 17.1 Geologic map of the spill area with the location of new monitoring wells (MW), domestic wells (DW) and municipal wells pumping the main unconfined aquifer (W).

17.1 GROUNDWATER FLOW DIRECTION

The first step in dealing with an accidental spill of hazardous materials that has the potential for groundwater pollution is to determine the direction of groundwater flow in the immediate vicinity of the spill. This is done by recording the water table elevations in the nearby existing water wells or, if they are not available, in newly drilled monitoring wells as in our case. A minimum of three wells placed around the spill is needed to determine the direction of groundwater flow as shown in Figure 17.2 (see also Problem 16 and Figure 16.2). The water table contour lines are drawn using a linear (exact) interpolation between the three points. The streamline (*i.e.*, the direction of groundwater flow) through the center of the spill is drawn perpendicular to the contours and is directed downgradient, toward north-northeast. This first analysis shows that, in general, the flow is indeed oriented toward the municipal wells.

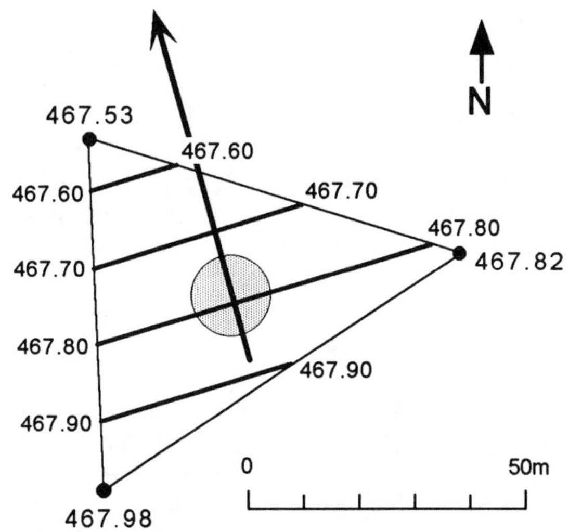

Figure 17.2 Water table contours and direction of groundwater flow determined by the triangular linear interpolation of data recorded in monitoring wells MW1, MW2 and MW3.

The flow direction determined in this way is valid only for the silty sand hydrogeologic unit. It will change at the boundary between two different porous media having different hydraulic conductivities -- silty sand and sand (see Figure 17.1). In other words, the streamline representing the flow in the spill area will *refract* at the boundary and a new direction of flow will have to be determined for the sand unit (see Figure 17.1). Similarly, the flow will change direction at each subsequent boundary between two porous media with different hydraulic conductivity.

17.2 REFRACTION OF STREAMLINES

Streamlines representing the direction of groundwater flow refract at the boundary between porous media with different hydraulic conductivity. (An analogy would be refraction of light rays when they enter a medium with different density, e.g., from air to water.) The refraction causes that the incoming angle, or *angle of incidence* and the *angle of refraction* are different (angle of incidence is the angle between the normal at the boundary and the incoming streamline; angle of refraction is the angle between the normal at the boundary and the outgoing streamline). The only exception is when the streamline is perpendicular to the boundary in which case both angles are the same -- 90°.

The mathematical relationship between the angle of incidence (α_1), angle of refraction (α_2), and the hydraulic conductivities of two porous media -- K_1 and K_2, is explained by the elements shown in Figure 17.3. The picture applies to both map and cross-sectional views as long as there is a clearly defined boundary between two porous media.

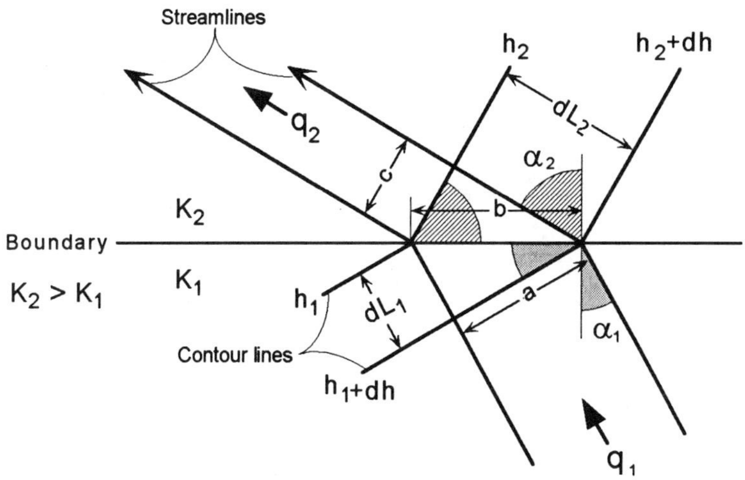

Figure 17.3 Refraction of two streamlines at the boundary between two porous media with different hydraulic conductivities (K_1 and K_2). The angle of incidence is α_1 and the angle of refraction is α_2. Other elements are explained in the text. (*Note that the same angles are shown elsewhere in the picture. This follows from the geometric rule that two angles with sides perpendicular to each other are equal.)*

The unit flow bound by two streamlines remains constant through the two media with different hydraulic conductivity since there is no gain or loss of water across streamlines:

$$q_1 = q_2$$

From the definition of the equipotential line (see Problem 15.1), it follows that the value of the hydraulic head along the same equipotential line remains constant regardless of the changes in hydraulic conductivity of the porous media:

$$h_1 = h_2$$

Also, the increment in hydraulic head between two successive contour lines is the same

$$h_1 + dh = h_2 + dh$$

By applying Darcy's law, the unit flows in two porous media are expressed as (see Figure 17.3):

$$q_1 = K_1 \cdot a \cdot \frac{dh}{dL_1} \tag{17.1}$$

$$q_2 = K_2 \cdot c \cdot \frac{dh}{dL_2} \tag{17.2}$$

As already mentioned, these two unit flows are equal:

$$K_1 \cdot a \cdot \frac{dh}{dL_1} = K_2 \cdot c \cdot \frac{dh}{dL_2} \tag{17.3}$$

From the rules of trigonometry (see Appendix G) it follows that:

$$a = b \cos \alpha_1$$

$$c = b \cos \alpha_2$$

$$\frac{dL_1}{b} = \sin \alpha_1$$

$$\frac{b}{dL_1} = \frac{1}{\sin \alpha_1}$$

$$\frac{dL_2}{b} = \sin \alpha_2$$

$$\frac{b}{dL_2} = \frac{1}{\sin \alpha_2}$$

When these transformations are included into equation (17.3), it becomes:

$$K_1 \cdot \frac{\cos \alpha_1}{\sin \alpha_1} = K_2 \cdot \frac{\cos \alpha_2}{\sin \alpha_2} \tag{17.4}$$

Knowing that:

$$\frac{\cos \alpha}{\sin \alpha} = \frac{1}{\tan \alpha}$$

equation (17.4) can also be written as:

$$\frac{K_1}{K_2} = \frac{\tan \alpha_1}{\tan \alpha_2} \tag{17.5}$$

which is the well known equation describing the relationship between the angles of incidence and refraction, and the hydraulic conductivities of two porous media.

Figure 17.4 shows the refraction of streamlines when they reach a boundary of higher hydraulic conductivity (top) and a boundary of lower hydraulic conductivity (bottom). The most noticeable effect is the change in spacing between both streamlines and contour lines. When interpreting or constructing a groundwater contour map, it is very important to understand the influence of geologic boundaries and a varying hydraulic conductivity on the overall picture. This knowledge is of particular importance during the calibration of groundwater models since the final result of modeling groundwater flow is a distribution of hydraulic heads in the flow field. This distribution, or the contour map of the piezometric surface, changes more or less obviously whenever one or more hydrogeologic elements such as hydraulic conductivity and the position of boundaries are changed. Essentially,

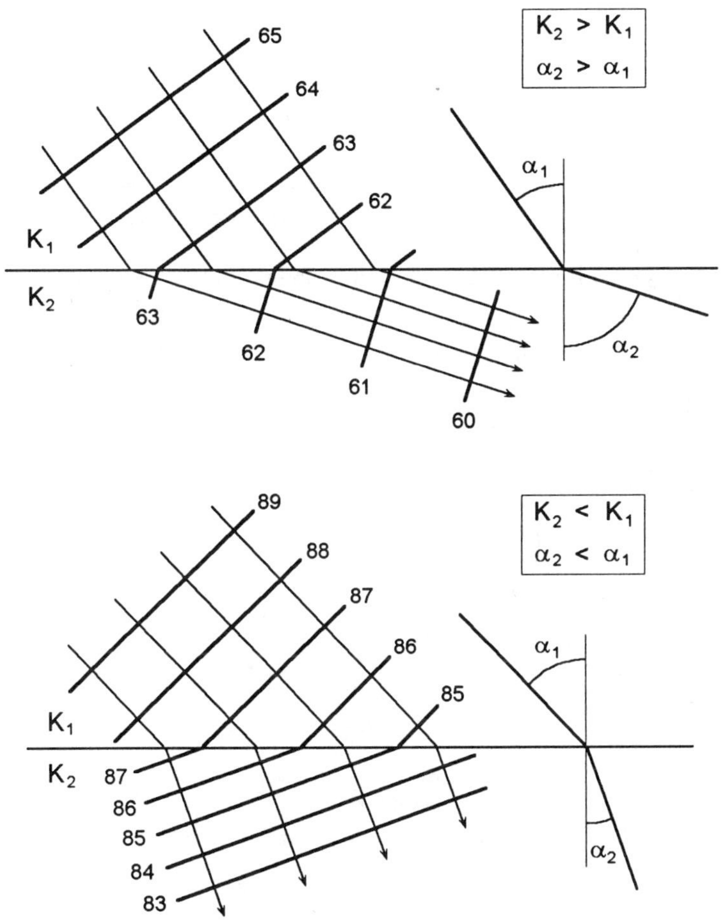

Figure 17.4 Refraction of streamlines at a boundary of higher hydraulic conductivity (top) and a boundary of lower hydraulic conductivity (bottom).

the model calibration is knowing what to expect from the changes and how to "manipulate" them to obtain a satisfactory contour map.

17.3 ESTIMATED PATH OF CONTAMINATION'

In general, the path of possible contamination is represented by two streamlines that bound the outer limits of the spill. Since the spill is localized in our case (less than 20m in diameter), it is not unreasonable to analyze just one streamline drawn through the spill's center (see Figure 17.5). The actual spread of pollution will depend largely on the amount of lateral dispersion and will have more or less diverging outer limits.

The streamline in the first hydrogeologic unit (silty sand) in the spill area is determined as shown in Figure 17.2. After reaching the first boundary, the streamline refracts with the angle of refraction calculated using equation (17.5):

$$\tan \alpha_2 = \frac{K_2 \cdot \tan \alpha_1}{K_1} \tag{17.6}$$

$$\alpha_2 = \arctan\left(\frac{K_2 \cdot \tan \alpha_1}{K_1}\right) \tag{17.7}$$

where

α_1 is the angle of incidence measured on the map (see Figure 17.5),

K_1 and K_2 are the hydraulic conductivities of the two porous media.

The angle of refraction is found after substituting known values into equation (17.7):

$$\alpha_2 = \arctan\left(\frac{1.45 \times 10^{-5}\, m/s \cdot \tan 16^o}{4.30 \times 10^{-6}\, m/s}\right)$$

$$\alpha_2 = \arctan 0.96569 \;=\; 44^o$$

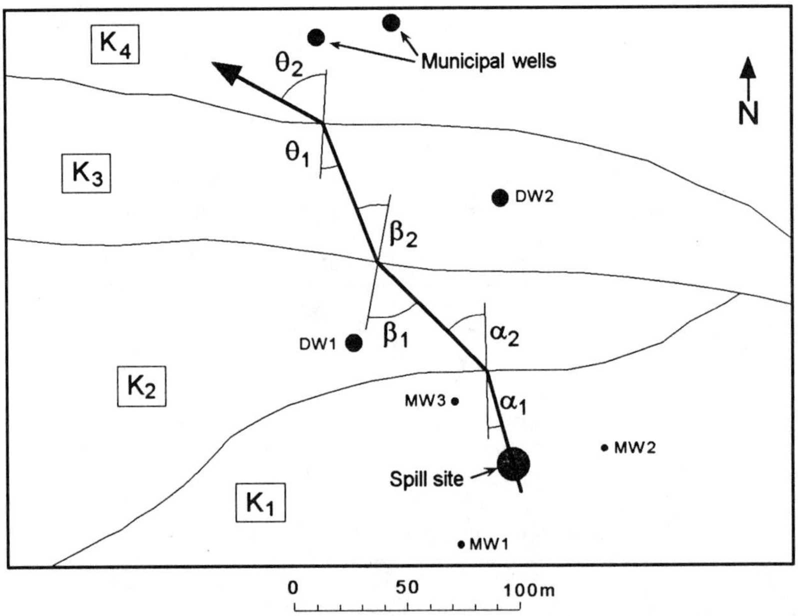

Figure 17.5 Streamline representing the path of possible contamination from the spill site. The angles of incidence at the boundaries (α_1, β_1, and θ_1) are measured on the map, and the angles of refraction (α_2, β_2, and θ_2) are calculated using equations (17.7) through (17.9). Note that an assumption was made that the groundwater flow is not disturbed by pumping wells or other sinks or sources.

The streamline is extended at a 44° angle until the next boundary. Next, the refraction angle β_2 is found by applying equation (17.5) again (note that the angle of incidence, β_1, is measured on the map):

$$\beta_2 = \arctan\left(\frac{K_3 \cdot \tan \beta_1}{K_2}\right) \tag{17.8}$$

$$\beta_2 = \arctan\left(\frac{6.11 \times 10^{-5}\, m/s \cdot \tan 56^o}{1.45 \times 10^{-5}\, m/s}\right)$$

$$\beta_2 = 32^o$$

Finally, after further extending the streamline, the angle of refraction at the last boundary is found as (θ_1 is measured on the map):

$$\theta_2 = \arctan\left(\frac{K_4 \cdot \tan\theta_1}{K_3}\right) \tag{17.9}$$

$$\theta_2 = \arctan\left(\frac{2.81 \times 10^{-5}\, m/s \cdot \tan 25^o}{6.11 \times 10^{-6}\, m/s}\right)$$

$$\theta_2 = 65^o$$

Note that the streamline shown in Figure 17.6 is drawn assuming no disturbance of groundwater flow by pumping wells or other sinks or sources. If there was a significant groundwater withdrawal in the area such as a continuous pumping in two municipal wells, this streamline and the corresponding equipotential lines would look different

17.4 FLOW VELOCITY

The velocity of groundwater flow (v), as defined by Darcy's law, is the product of the hydraulic conductivity (K) and the hydraulic gradient (i):

$$v = K \cdot i \tag{17.10}$$

However, this velocity, called Darcy's velocity, is not the real velocity at which water particles move through the porous medium. Darcy's law, derived experimentally (see Problem 3), assumes that the groundwater flow occurs through the entire cross-sectional area of a sample (porous medium) including both voids and grains. (Adequately, Darcy's velocity is often called "spread velocity" in Russian literature.) As illustrated with Figure 17.6, the volumetric flow rate in Darcy's law (Q) is the product of the Darcy's velocity and the total cross-sectional area of flow (A):

$$Q = v \cdot A \tag{17.11}$$

Since the flow takes place only through voids, the actual cross-sectional area is the sum of the individual cross-sectional areas of voids:

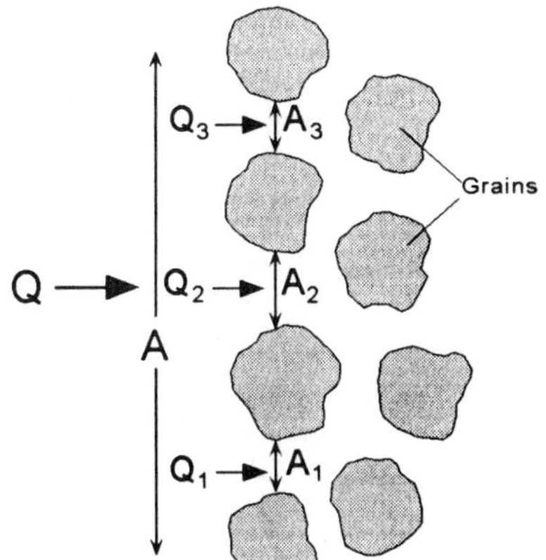

Figure 17.6 Schematic presentation of volumetric flow rates and the cross-sectional areas of groundwater flow through intergranular porous media.

$$A_v = A_1 + A_2 + A_3 \tag{17.12}$$

The total volumetric flow is the sum of the individual flows:

$$Q = Q_1 + Q_2 + Q_3 \tag{17.13}$$

In order to preserve the principle of the continuity of flow, a new equivalent velocity that describes flow through the cross-sectional area of voids has to be introduced:

$$v \cdot A = v_L \cdot A_v \tag{17.14}$$

where v_L is this new flow velocity often called the *linear velocity*.

Since the actual cross-sectional area of flow is smaller than the total area ($A_v<A$), it follows that the linear velocity must be greater than Darcy's velocity: $v_L>v$. Equation (17.14) can be written as:

$$v = v_L \cdot \frac{A_v}{A} \tag{17.15}$$

The fraction at the right-hand side can be multiplied by the unit aquifer (sample) width which gives the following relationship:

$$v = v_L \cdot \frac{V_v}{V} \tag{17.16}$$

where

V_v is the volume of the interconnected voids that allow groundwater flow,
V is the total aquifer (sample) volume.

These two volumes define the effective porosity of a porous medium (see Problem 1):

$$n_{ef} = \frac{V_v}{V} \tag{17.17}$$

Combining equations (17.16) and (17.17) gives the following expression for the linear flow velocity:

$$v_L = \frac{v}{n_{ef}} \tag{17.18}$$

or, when inserting equation (17.10) into equation (17.18):

$$v_L = \frac{K \cdot i}{n_{ef}} \tag{17.18a}$$

The linear groundwater velocity is accurate when used to estimate the average travel time of groundwater, and Darcy's velocity is accurate for calculating flow rates. Neither, however, is the real groundwater velocity which is the time of travel of a water particle along its actual path through the voids. It is obvious that, for practical purposes, the real velocity can not be measured or calculated.

As seen from equation (17.8), there are three elements needed to calculate the linear groundwater velocity in each of the four hydrogeologic sections (see Figure 17.1): the hydraulic gradient (i), the hydraulic conductivity (K), and the effective porosity (n_{ef}). In our case the hydraulic conductivities are known, the hydraulic gradients can be determined from the drawn streamline and the water table contours, while the effective porosities have to be estimated.

The *hydraulic gradient* in Section 1 is (see Figure 17.7):

$$i_1 = \frac{h_A - h_B}{L_{AB}}$$

$$i_1 = \frac{467.78m - 467.47m}{40m} = 0.00775$$

where

h_A and h_B are the hydraulic heads at points A and B respectively, whose values are determined from the water table contour map (Figure 17.7),

L_{AB} is the distance between points A and B measured on the map.

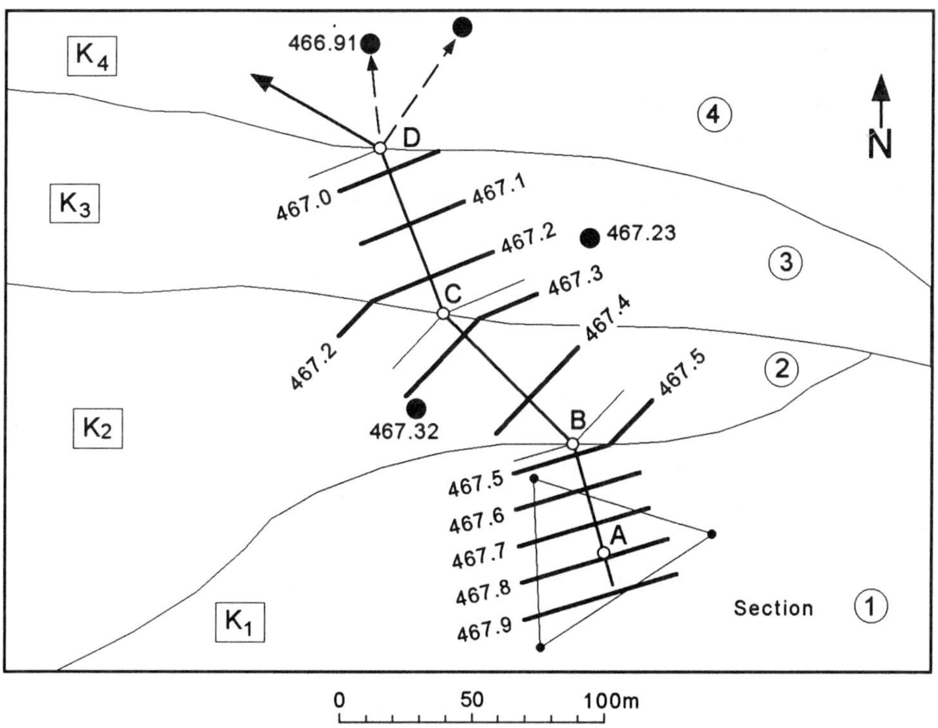

Figure 17.7 The streamline representing the path of potential contamination and the water table contour lines drawn perpendicular to it. The contours are constructed by the exact linear interpolation using field data and following the refraction rules. Contour interval is 0.1 meters.

As discussed in Problem 1, the *effective porosity* is always smaller than the overall porosity of a porous medium. This difference is up to several percent for clean, uniform sediments that are medium-to-coarse grained. When a sedimentary deposit is a mixture of various grain sizes this difference becomes larger, especially if there is a significant amount of fine sediments such as silt and clay. Some uniform clays have the overall porosity above 80% while their effective porosity is only 1–2%.

The overall porosity of various rocks is given in Appendix A and similar tables can be found throughout literature. Because of the wide possible ranges and much less experimental data available, tables of the effective porosity are not that common and are less accurate. Simple rules can be followed when estimating the effective porosity from the tables of overall porosity:

- for clean uniform sands and gravels the difference between the two porosities is up to 5%, usually less than that;
- for non-uniform sands and gravels, and their mixtures, the effective porosity is up to 10% smaller than the overall porosity;
- presence of fines, *i.e.*, silts and clays, rapidly reduces the effective porosity: a fifty-fifty mixture of uniform sand and clay can have the overall porosity of 50%, while the effective porosity can be less than 5%;
- the effective porosity of clean clays is always smaller than 5%, often only 1% or less;
- if the effective porosity is used to find travel time of contaminated groundwater (e.g., in a risk analysis), it is better to estimate a higher value.

Using these rules and data from Appendix A, the following values of the effective porosity are estimated for the hydrogeologic units in our case:

silty sand	20%
sand	35%
silty coarse sand	15%
sandy gravel	25%

Finally, the linear groundwater velocity in Section 1 is determined by inserting all known, calculated, and estimated values into equation (17.18):

$$v_{L1} = \frac{K_1 \cdot i_1}{n_{ef1}}$$

$$v_{L1} = \frac{4.30 \times 10^{-6} \, m/s \cdot 0.00775}{0.20} = 1.67 \times 10^{-7} \, m/s$$

$$v_{L1} = 0.0144 \, m/day$$

The linear velocity in Section 2 is:

$$V_{L2} = \frac{K_2 \cdot \dfrac{h_B - h_C}{L_{BC}}}{n_{ef2}}$$

$$V_{L2} = \frac{1.45 \times 10^{-5} \, m/s \cdot \dfrac{467.47m - 467.27m}{67m}}{0.35} = 1.24 \times 10^{-7} \, m/s$$

$$v_{L2} = 0.0107\ m / day$$

The linear velocity in Section 3 is:

$$V_{L3} = \frac{K_3 \cdot \dfrac{h_C - h_D}{L_{CD}}}{n_{ef3}}$$

$$V_{L3} = \frac{6.11 \times 10^{-6}\ m / s \cdot \dfrac{467.27m - 466.96m}{63m}}{0.15} = 2.00 \times 10^{-7}\ m / s$$

$$v_{L3} = 0.0173\ m / day$$

17.5 TRAVEL TIME

The time needed for the potential groundwater contamination caused by the spill to reach the sandy gravel aquifer (Section 4) is determined using the basic velocity equation:

$$v_L = \frac{L}{t} \tag{17.19}$$

where v_L is the linear velocity, L is the distance traveled, and t is the travel time. The total time is the sum of travel times in each of the three sections:

$$t = t_1 + t_2 + t_3 \tag{17.20}$$

or

$$t = \frac{L_{AB}}{v_{L1}} + \frac{L_{BC}}{v_{L2}} + \frac{L_{CD}}{v_{L3}} \tag{17.21}$$

$$t = \frac{40m}{0.0144\ m / day} + \frac{67m}{0.0107\ m / day} + \frac{63m}{0.0173\ m / day}$$

$$t = 7.6\ years\ +\ 17.2\ years\ +\ 10.0\ years\ =\ 34.8\ years$$

Note that the time of approximately 35 years is calculated on the assumption that the flow pattern shown in Figure 17.7 is not influenced by the groundwater withdrawal at the two municipal wells. A significant pumping in the wells would alter the shape of the guiding streamline and the contour lines. It would also induce steeper hydraulic gradients and reduce the travel time.

Chapter 7

WATER WELLS, STEADY STATE FLOW

PROBLEM 18
CONFINED AQUIFER, THREE-STEP PUMPING TEST, WELL PERFORMANCE

An exploitation well and three monitoring wells were used to conduct a pumping test in a confined aquifer. The location of the wells and a simplified hydrogeologic cross-section are shown in Figures 18.1 and 18.2, respectively. Monitoring wells MW1, MW2, and MW3 are 8, 35, and 120 meters from the pumping well PW, respectively. The radius of the pumping well in its screen section is 0.3 meters. The test was a three-step pumping test with each step lasting 6 hours. The pumping rates were 10, 15, and 20 liters per second.

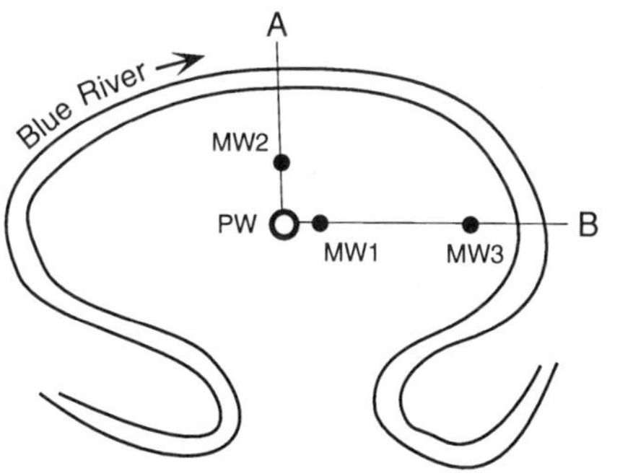

Figure 18.1 Map view of the locations of the pumping well PW and three monitoring wells used for aquifer testing. Hydrogeologic cross-section in Figure 18.2 is along AB.

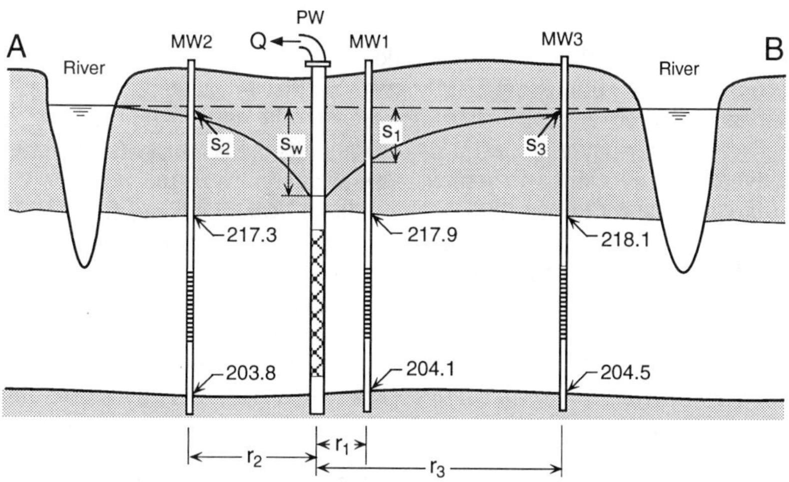

Figure 18.2 Simplified hydrogeologic cross-section AB.

Figure 3 shows registered drawdown in all wells. From the available data determine the transmissivity and the hydraulic conductivity of the aquifer. Also determine the following characteristics of the pumping well: well loss (well efficiency), specific capacity and optimum pumping rate for a long-term exploitation.

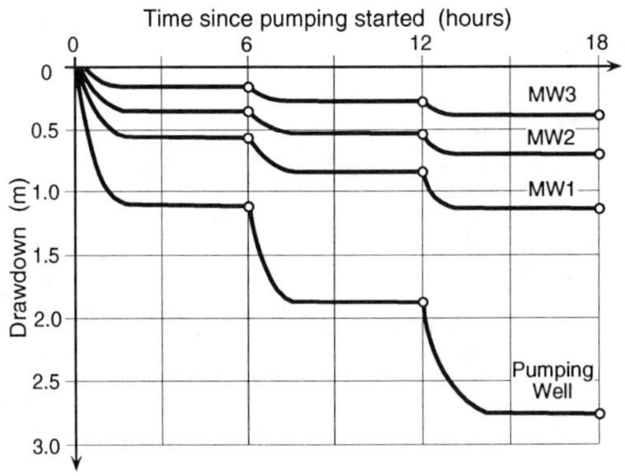

Figure 18.3 Drawdown in all wells registered during the three-step pumping test.

18.1 FLOW TOWARD A FULLY PENETRATING WELL

The equation describing time-independent (steady state) groundwater flow toward a well in a confined aquifer, also known as Theim equation (or Theim-Forchheimer equation), is based on the following assumptions:

- The well fully penetrates the confined aquifer and receives water from the entire thickness of the aquifer.
- The well is pumping water from the aquifer at a constant rate.
- The flow toward the well is radial, horizontal and laminar, *i.e.,* the flow lines are parallel along each radial cross-section.
- The aquifer is homogeneous and isotropic, and has uniform thickness and a horizontal base.
- After a certain period of pumping, the drawdown does not increase anymore (it is stabilized), and the steady state flow conditions are established.

Although these conditions are hardly ever completely satisfied in reality, there are several situations when a steady state approach to the well pumping test analysis is justified. These include pumping near a strong equipotential boundary (such as at an island in a lake/river), or at a locality partly surrounded by a large river (the meander in our case). The boundary and the confined aquifer are in a direct hydraulic contact, the radius of well influence reaches the boundary soon after the beginning of pumping (e.g., after several hours) and the drawdown remains constant afterwards.

Figure 18.3 shows that the steady state conditions were indeed established soon after the test began and then again after each consecutive increase in the pumping rate. These increases, called steps, are a common practice when the pumping test is designed to provide data on well efficiency and future exploitation. However, if determining aquifer parameters such as transmissivity and hydraulic conductivity is the only interest, a one step pumping test is sufficient.

Figures 18.4 and 18.5 show the elements for deriving the differential equation of groundwater flow toward a fully penetrating well in a confined aquifer. As is the case with

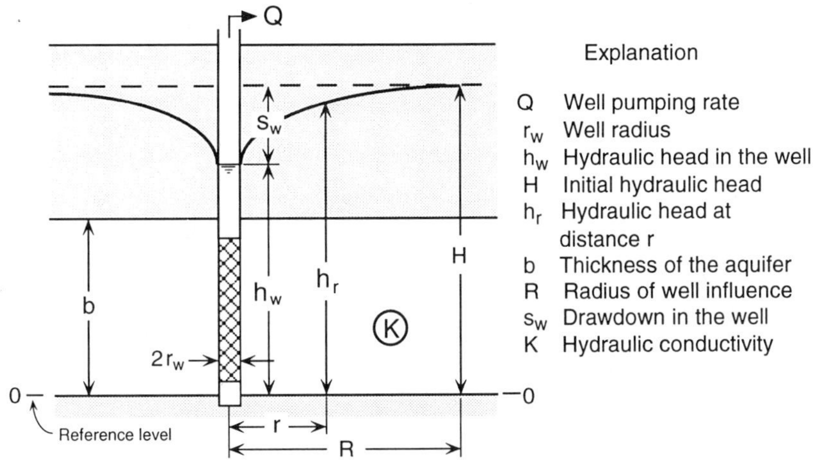

Figure 18.4 Elements of groundwater flow toward a fully penetrating well in a confined aquifer.

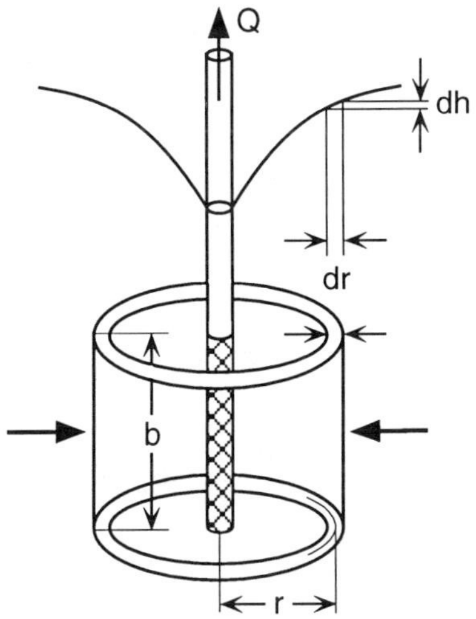

Figure 18.5 Scheme for deriving the differential equation of groundwater flow toward a fully penetrating well in a confined aquifer.

most other equations in hydrogeology, the flow rate (Q) is given as the product of the cross-sectional area of flow (A) and the flow velocity (v):

$$Q = A \cdot v \tag{18.1}$$

The cross-sectional area of flow at distance r from the pumping well is the side of the cylinder with the radius r and the thickness b (which is the thickness of the confined aquifer):

$$A = 2\pi rb \tag{18.2}$$

The velocity of flow at distance r is given as the product of the aquifer's hydraulic conductivity (K) and the hydraulic gradient which is an infinitesimally small drop in hydraulic head (dh) over an infinitesimally small distance (dr):

$$v = K\frac{dh}{dr} \tag{18.3}$$

Note that, unlike in the case of planar flow, there is no minus sign in equation (18.3). This is because the flow is toward the well, i.e., in the negative direction of the radial coordinate system which cancels out the negative sign in front of the hydraulic gradient.

$$Q = 2\pi rbK\frac{dh}{dr} \tag{18.4}$$

To solve this differential equation the variables h (hydraulic head) and r (radial distance from the well) first need to be separated:

$$dh = \frac{Q}{2\pi T}\frac{dr}{r} \tag{18.5}$$

Equation (18.5) is then integrated with the following boundary conditions:

- at distance r_w (which is the well radius) the hydraulic head is h_w (head in the well),
- at distance r from the well the hydraulic head is h_r:

$$\int_{h_w}^{h_r} dh = \frac{Q}{2\pi T}\int_{r_w}^{r}\frac{dr}{r} \tag{18.6}$$

The integrals on both sides of the equation are readily solved since they are *elementary* integrals (see Appendix G):

$$h_r - h_w = \frac{Q}{2\pi T}\ln\frac{r}{r_w} \tag{18.7}$$

Finally, the hydraulic head at any distance from the pumping well is given as:

$$h_r = h_w + \frac{Q}{2\pi T}\ln\frac{r}{r_w} \tag{18.8}$$

Equation (18.5) can also be integrated with the following boundary conditions:

- at distance r_w (which is the well radius) the hydraulic head is h_w (head in the well),
- at distance R from the well (which is the radius of well influence) the hydraulic head is H (which is the undisturbed head, equal to the initial head before the pumping started).

These boundary conditions enable introduction of the *drawdown in the well* (s_w) into equation (18.8):

$$H - h_w = s_w = \frac{Q}{2\pi T}\ln\frac{R}{r_w} \tag{18.9}$$

The pumping rate of the well is:

$$Q = \frac{2\pi T s_w}{\ln \frac{R}{r_w}} \tag{18.10}$$

18.2 TRANSMISSIVITY AND HYDRAULIC CONDUCTIVITY

Equation (18.9) can be rewritten to give the drawdown at any radial distance from the pumping well:

$$s_r = \frac{Q}{2\pi T} \ln \frac{R}{r} \tag{18.11}$$

Noting all constant terms in equation (18.11), the drawdown can be expressed as the function of distance only:

$$s_r = \frac{Q}{2\pi T} \ln R - \frac{Q}{2\pi T} \ln r \tag{18.12}$$

or, when the constant terms are replaced with general numbers:

$$s_r = a - b \ln r \tag{18.13}$$

Equation (18.13) means that the recorded data of drawdown versus the radial distance from the well would form a straight line when plotted on a semilogarithmic graph paper.

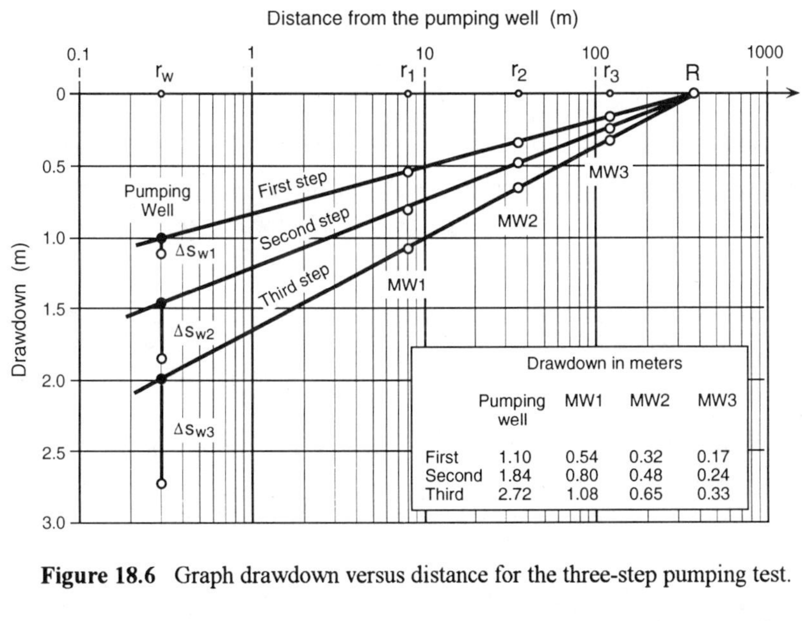

Figure 18.6 Graph drawdown versus distance for the three-step pumping test.

Figure 18.6 shows that data for the monitoring wells MW1, MW2, and MW3 form straight lines for all three steps. Note that the drawdown recorded in the pumping well does not fall on the straight line in any of the steps; it is below the straight lines indicating that there is an additional drawdown in the well because of the *well loss*. This well loss increases for each step due to the increasing pumping rates. The well loss is explained in detail at the end of this problem.

The transmissivity of the aquifer can be determined from the graph drawdown versus distance for any of the three steps. This is done by relating the coordinates of any two points on the same straight line (line of the same step):

$$s_1 = const - \frac{Q}{2\pi T}\ln r_1 \tag{18.14}$$

$$s_2 = const - \frac{Q}{2\pi T}\ln r_2 \tag{18.15}$$

$$s_1 - s_2 = \Delta s = \frac{Q}{2\pi T}\ln\frac{r_2}{r_1} \tag{18.16}$$

The transmissivity is then found from this difference in drawdown for the chosen points:

$$T = \frac{Q}{2\pi\Delta s}\ln\frac{r_2}{r_1} \tag{18.17}$$

For practical purposes it is better to express transmissivity in terms of the logarithm to the base 10 (log) since the corresponding graph is easier to use. Knowing that

$$\log x = 0.4343\ln x \tag{18.18}$$

and replacing "pi" with number 3.14, equation (18.17) becomes

$$T = \frac{0.366Q}{\Delta s}\log\frac{r_2}{r_1} \tag{18.19}$$

If two points on the straight line are chosen to be one log cycle apart (*i.e.*, the ratio of the distance coordinates r_2 and r_1 is 10), equation (18.19) reduces to the following simple form:

$$T = \frac{0.366Q}{\Delta s} \tag{18.20}$$

First step

For the monitoring wells MW1 and MW2, registered drawdown at the end of the first step are 0.54 m and 0.32 m, respectively. The transmissivity is calculated using equation (18.19) and knowing that the pumping rate for the first step is 10 l/s (0.01 m^3/s):

$$T = \frac{0.366 \cdot 0.01\ m^3/s}{(0.54m - 0.32m)}\log\frac{35m}{8m}$$

$$T = 0.0107\ m^2/s = 1.07\times 10^{-2}\ m^2/s$$

The hydraulic conductivity is calculated from the equation

$$K = \frac{T}{b} \tag{18.21}$$

and knowing that the average thickness of the aquifer determined from the logs for MW1, MW2, and MW3 (see Figure 18.2) is 13.6 meters:

$$K = \frac{1.07 \times 10^{-2}\ m^2/s}{13.6\ m} = 7.87 \times 10^{-4}\ m/s$$

Choosing any two points on the straight line, other than the actual measurements, and applying equation (18.19) or (18.20), would yield the same result for T in this case since the actual measurements fall exactly on the straight line. However, if the actual data do not all *exactly* fall along the straight line (which is usually the case, especially if there are more monitoring wells), choosing two points on the best-fit line is more appropriate.

Second step

For the monitoring wells MW2 and MW3, registered drawdown at the end of the second step are 0.48 m and 0.24 m respectively (Figure 18.6). The transmissivity is calculated using equation (18.19) and knowing that the pumping rate for the second step is 15 l/s (0.015 m^3/s):

$$T = \frac{0.366 \cdot 0.015\ m^3/s}{(0.48m - 0.24m)} \log \frac{120m}{35m}$$

$$T = 0.0122\ m^2/s = 1.22 \times 10^{-2}\ m^2/s$$

$$K = 9.04 \times 10^{-4}\ m/s$$

Third step

For the monitoring wells MW1 and MW3, registered drawdown at the end of the third step are 1.08 m and 0.33 m, respectively (Figure 18.6). The transmissivity is calculated using equation (18.19) and knowing that the pumping rate for the third step is 20 l/s (0.02 m^3/s):

$$T = \frac{0.366 \cdot 0.02\ m^3/s}{(1.08m - 0.33m)} \log \frac{120m}{8m}$$

$$T = 0.0115\ m^2/s = 1.15 \times 10^{-2}\ m^2/s$$

$$K = 8.46 \times 10^{-4}\ m/s$$

All three steps together (drawdown/pumping rate method)

Another graphical method for determining aquifer transmissivity and hydraulic conductivity combines all three steps/drawdown into a single straight line on a semilogarithmic paper. This method is preferred over individual steps since it averages the response of the aquifers to several different stresses. The equation that gives the drawdown at any radial distance from the pumping well can be divided by the pumping rate:

$$s_r = \frac{Q}{2\pi T}\ln\frac{R}{r} \tag{18.22}$$

$$\frac{s_r}{Q} = \frac{1}{2\pi T}\ln\frac{R}{r} \tag{18.23}$$

Equation (18.23) rewritten to include logarithm to the base 10 (note that *lnx=2.3logx*, and *"pi"=3.14*) is:

$$\frac{s_r}{Q} = \frac{0.366}{T}\log\frac{R}{r} \tag{18.24}$$

The ratio between the drawdown and the pumping rate can be expressed as the function of distance only since R and T are constant terms (assuming steady-state conditions and homogeneity of the aquifer):

$$\frac{s_r}{Q} = \frac{0.366}{T}\log R - \frac{0.366}{T}\ln r \tag{18.25}$$

or

$$\frac{s_r}{Q} = a - b\log r \tag{18.26}$$

The graph that shows points with the coordinates r (distance from the pumping well) and s/Q (registered drawdown over pumping rate) for all the wells and all three drawdowns is shown in Figure 18.7. As can be seen, the points for the individual monitoring wells have practically the same s-Q ratio for all three steps. However, this is not the case with the pumping well because of the increasing well loss. Note that, because of the well loss, the actual points for the pumping well are below the straight line formed by the monitoring wells.

The transmissivity is obtained by relating coordinates of any two points on the straight line formed by the monitoring wells:

$$\left(\frac{s}{Q}\right)_1 = \frac{0.366}{T}\log\frac{R}{r_1} \tag{18.27}$$

$$\left(\frac{s}{Q}\right)_2 = \frac{0.366}{T}\log\frac{R}{r_2} \tag{18.28}$$

$$T = \frac{0.366}{\left(\frac{s}{Q}\right)_2 - \left(\frac{s}{Q}\right)_1}\log\frac{r_2}{r_1} \tag{18.29}$$

Again, if two points are chosen so that the ratio between their distance coordinates is 10 (note that *log10* equals 1), equation (18.29) becomes

$$T = \frac{0.366}{\Delta \left(\frac{s}{Q}\right)} \tag{18.30}$$

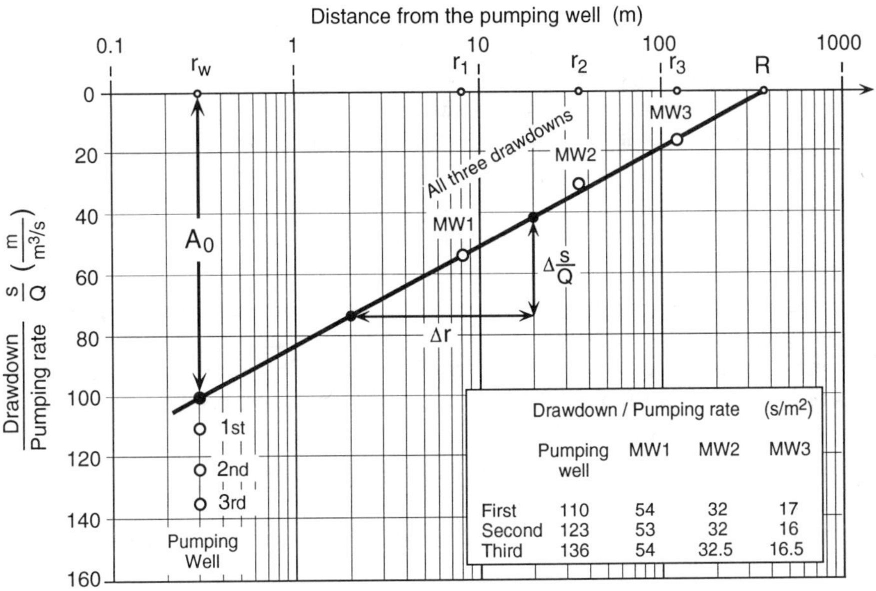

Figure 18.7 Graph distance versus drawdown/pumping rate ratio for the three-step pumping test.

In our example the two points chosen have distance coordinates 2 and 20 meters, respectively (see Figure 18.7) which gives the ratio of 10. Reading from the graph that $(s/Q)_2-(s/Q)_1=31.5\ s/m^2$, the transmissivity is calculated using equation (18.3):

$$T = \frac{0.366}{31.5\ s/m^2} = 1.16 \times 10^{-2}\ m^2/s$$

The hydraulic conductivity is $K = T/b$, or

$$K = 8.53 \times 10^{-4}\ m/s$$

18.3 RADIUS OF WELL INFLUENCE

The radius of well influence is determined from the graphs in Figures 18.6 and/or 18.7. It is the intercept of the straight line formed by the monitoring wells and the zero drawdown. As seen from the graph in Figure 18.6, this value remains constant for all three steps indicating the position of the meandering river (strong equipotential boundary). Although the pumping well's location may not be in the center of the meander, and the meander itself may be of a more or less irregular shape, the radius of well influence determined this way corresponds to a circle hydraulically equivalent to the actual meander.

The radius of well influence determined from both graphs is R=375 m.

18.4 WELL LOSS

Well loss is the difference between the actual measured drawdown in the pumping well and the theoretical drawdown expressed by equation (18.9) (also called the *formation loss*). It is the result of various factors, such as an inevitable disturbance of the porous medium near the well during drilling, an improper well development (e.g., drilling fluid is left in the formation, mud cake along the borehole is not removed), a poorly designed gravel pack and/or well screen, turbulent flow through the pack and well screen, etc. Well loss is always present in pumping wells and its evaluation is a very important part in deciding if the well performance is satisfactory or not.

The total measured drawdown at a well is a combination of the linear losses and turbulent losses:

$$s_w = AQ + BQ^2 \tag{18.31}$$

where A is the coefficient of the linear losses, B is the coefficient of turbulent losses, and Q is the pumping rate. The turbulent losses are usually assumed to be quadratic, but other powers may be examined as well. The linear losses include both formation loss (equation 18.9) and linear loss in the near-screen zone. Their respective coefficients are A, A_0 and A_1:

$$A = A_0 + A_1 \tag{18.32}$$

The formation loss, or the theoretical drawdown in the well (s_0) can be determined using the following equation:

$$s_0 = \frac{Q}{2\pi T}\ln\frac{R}{r_w} \tag{18.33}$$

The coefficient of linear formation loss (A_0) can be calculated as

$$A_0 = \frac{1}{2\pi T}\ln\frac{R}{r_w} \tag{18.34}$$

or determined graphically as shown in Figure 18.7: it is the point at distance r_w on the straight line formed by the monitoring wells (distance r_w is the radius of the pumping well).

The coefficients of the total linear loss (A) and the quadratic loss (B) can be determined from a graph pumping rate (Q) versus drawdown-pumping rate ratio (s/Q) as shown in Figure 18.8. The graph is a straight line of the following form:

$$\frac{s_w}{Q} = A + BQ \tag{18.35}$$

where A is the intercept and B is the slope of the best-fit straight line drawn through the experimental data (in our case all the points fall exactly along the straight line). After substituting values of A and B determined from the graph into equation (18.31), it is possible to calculate the total (*i.e.*, expected to be actually recorded) drawdown in the pumping well for any pumping rate:

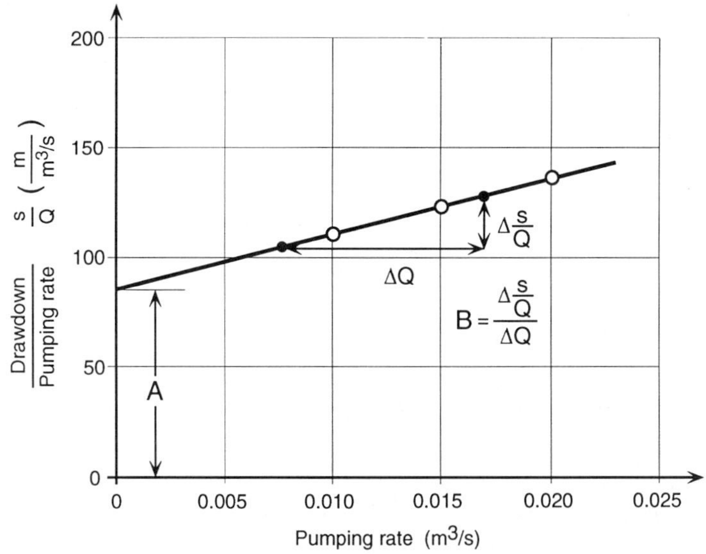

Figure 18.8 Graph pumping rate versus drawdown-pumping rate ratio used to determine the components of the drawdown in the pumping well PW (A: coefficient of the linear loss, B: coefficient of the turbulent loss).

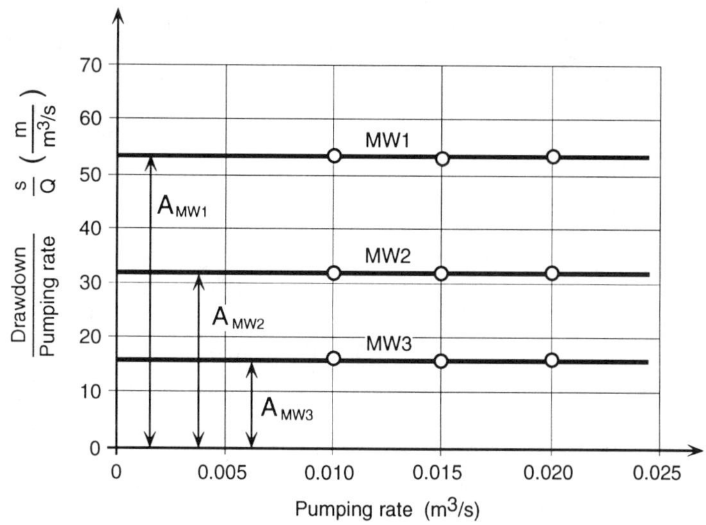

Figure 18.9 Graph pumping rate (Q) versus drawdown-pumping rate ratio (s/Q) used to determine the coefficient of linear loss in the monitoring wells.

$$s_w = 87\frac{s}{m^2}\cdot Q + 2600\frac{s^2}{m^5}\cdot Q^2 \tag{18.36}$$

The coefficient of turbulent (quadratic) well loss smaller than 2500 to 3000 s^2/m^5 is usually considered acceptable. Larger coefficients indicate improper well design and/or development, clogging of the well screen and other deterioration of the well.

The coefficients of the linear losses for the monitoring wells can also be found by constructing a graph similar to the graph in Figure 18.8. Since there is no pumping from the monitoring wells there are no turbulent losses. The total drawdown is equal to the linear loss (AQ) which, in turn, is equal to the theoretical loss of the hydraulic head due to groundwater flow through the porous medium (*i.e.*, the *formation loss*):

$$s_{MW} = AQ = A_0 Q \tag{18.37}$$

The drawdown in the monitoring wells, for any pumping rate, can be calculated using equation (18.22), or finding the coefficient A graphically as shown in Figure 18.9 and then applying equation (18.37). The latter is more convenient since it is not necessary to know the transmissivity and the radius of well influence.

18.5 WELL EFFICIENCY

Well efficiency is the ratio between the theoretical drawdown and the actual drawdown measured in the well. It is expressed in percent:

$$\text{Well Efficiency} = \frac{\text{Theoretical Drawdown } (s_0)}{\text{Measured Drawdown } (s_w)} \times 100\% \tag{18.38}$$

The theoretical drawdown is easily determined from the graph in Figure 18.6. It is the drawdown at distance r_w (which is the pumping well radius) read from the straight line formed by the monitoring wells. The values for the first, second, and third steps are 0.98, 1.45, and 1.96 m, respectively. As can be seen from the graph, the difference between the theoretical drawdown and the measured drawdown increases for each step, *i.e.*, it increases with the increasing pumping rate. *Consequently, the well efficiency decreases with the increasing pumping rate.* Its values for the first, second, and third steps are 89, 79, and 72%, respectively. The example below illustrates determination of the well efficiency for the first step:

$$\text{Well Efficiency} = \frac{0.98\ m}{1.10\ m} \times 100\% = 89.1\%$$

Note: A well efficiency of 70% or more is usually considered acceptable. If a newly developed well has less than 65% efficiency, it should not be approved.

18.6 WELL SPECIFIC CAPACITY AND OPTIMUM PUMPING RATE

The specific capacity of the well is given as:

$$\text{Well Specific Capacity} = \frac{\text{Pumping Rate } (Q)}{\text{Measured Drawdown } (s_w)} \tag{18.39}$$

and is expressed as the pumping rate per unit drawdown (e.g., liters per second per one meter of drawdown). Similarly to the well loss, it also decreases with the increasing pumping rate: 9.1 l/s/m, 8.1 l/s/m, and 7.3 l/s/m for the first, second, and third steps respectively.

Choosing an optimum rate at which a well will be pumped is a decision based on numerous factors. For example, if the well (or wells) will be used for a short-term construction dewatering, maintaining a desired drawdown may be the only relevant criterion. In some cases, where there are no alternatives, a certain pumping rate is all that matters. A graph showing drawdown versus pumping rate for several pumping steps, such as the one shown in Figure 18.10 for PW, is a simple tool for examining these options.

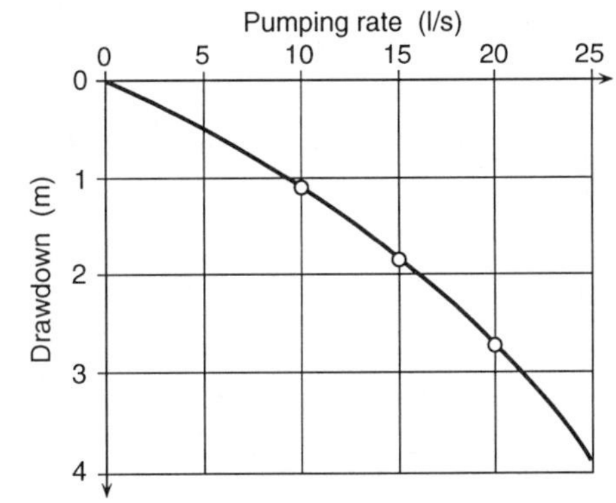

Figure 18.10 Graph pumping rate versus drawdown for PW. Coordinates of the data points are the pumping rates for three steps and the drawdown registered at the end of each step (see Figure 18.3).

On the other hand, if the well is designed for a long-term exploitation, in addition to the energy cost of pumping, the hydraulic criteria are the most important in deciding which pumping rate is the optimum one. This includes a comparative study of well losses and well efficiency. Graphs in Figures 18.11 and 18.12 are the final results of this analysis for PW. The procedure consists of the following steps:

1) For up to ten different pumping rates find drawdown which is expected to be actually recorded in the well (*i.e.*, the drawdown that includes well losses) using equation (18.31):

$$s_w = AQ + BQ^2$$

As determined from the graph in Figure 18.8, this equation for PW is:

$$s_w = 87\frac{s}{m^2}\cdot Q + 2600\frac{s^2}{m^5}\cdot Q^2$$

For example, the total expected drawdown in PW for the pumping rate of 25 l/s is:

$$s_w = 87\frac{s}{m^2}\cdot 0.025\frac{m^3}{s} + 2600\frac{s^2}{m^5}\cdot(0.025\frac{m^3}{s})^2$$

$$s_w = 3.80\ m$$

2) For the same pumping rates find theoretical drawdown which is the result of groundwater flow through the undisturbed portion of the aquifer only (the *formation loss*) using equation (18.33):

$$s_0 = \frac{Q}{2\pi T}\ln\frac{R}{r_w}$$

Substituting previously determined values of the transmissivity ($T=1.16\text{x}10^{-2}$ m^2/s), and the radius of influence (R=375 meters), this theoretical drawdown in PW (whose radius is 0.3 meters) for the pumping rate of 25 l/s is

$$s_0 = \frac{0.025\ m^3/s}{2 \cdot 3.14 \cdot 1.16 \times 10^{-2}\ m^2/s} \ln \frac{375m}{0.3m}$$

$$s_0 = 2.45\ m$$

3) Find the well efficiency for all chosen pumping rates using equation 18.38. In our example for PW and the pumping rate of 25 l/s, the well efficiency is

$$Well\ Efficiency = \frac{2.45\ m}{3.80\ m} \times 100\% = 64.5\%$$

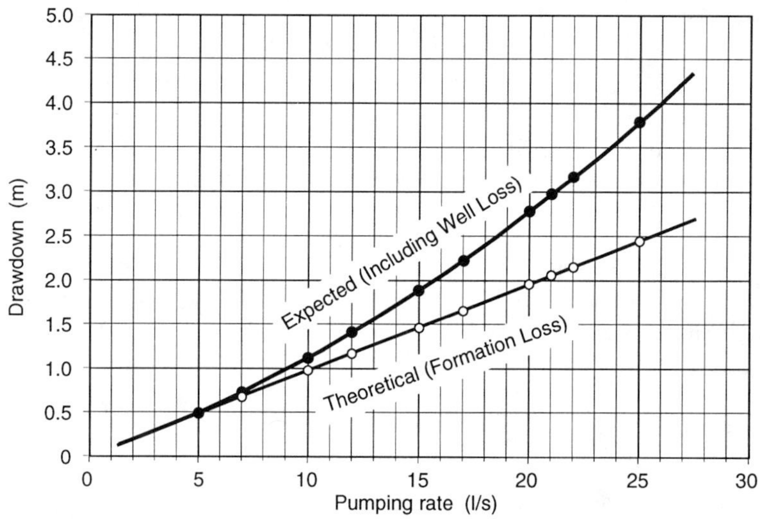

Figure 18.11 Pumping rate at PW versus 1) the formation loss calculated using equation (18.33), and 2) the expected total drawdown calculated using equation (18.36).

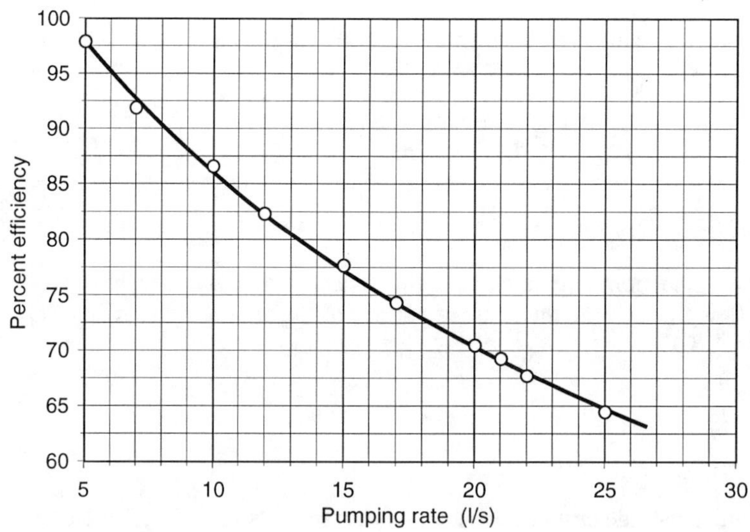

Figure 18.12 Pumping rate at PW versus the well efficiency.

The graph in Figure 18.12 is drawn after repeating the described procedure for ten different pumping rates. Accepting that the well efficiency of 70% is the minimum for an optimal long-term exploitation, the pumping rate of 20 l/s is recommended.

Note: Due to normal aging of pumping wells, the well efficiency is expected to decrease in time. Thus the step-pumping tests should be performed every few years in order to adjust the exploitation (pumping) regime and apply well rehabilitation measures if necessary.

PROBLEM 19
UNCONFINED AQUIFER, HYDRAULIC CONDUCTIVITY RADIUS OF WELL INFLUENCE, POSITION OF WATER TABLE

A fully penetrating well is pumping from an unconfined aquifer as shown in Figure 19.1. Drawdown registered in the pumping well (PW) and two monitoring wells (MW1 and MW2) reflects steady state conditions and is 7.73, 2.74 and 1.63 meters, respectively. Find the hydraulic conductivity of the aquifer and the radius of well influence if the pumping rate is 17.5 l/s (0.0175 m^3/s). Diameter of the pumping well is 300 mm and the monitoring wells MW1 and MW2 are at 8 and 30 m from PW1 respectively. Also find the approximate position of the water table in the vicinity of the well, as well as the distance from the well at which Dupuit's assumption is valid.

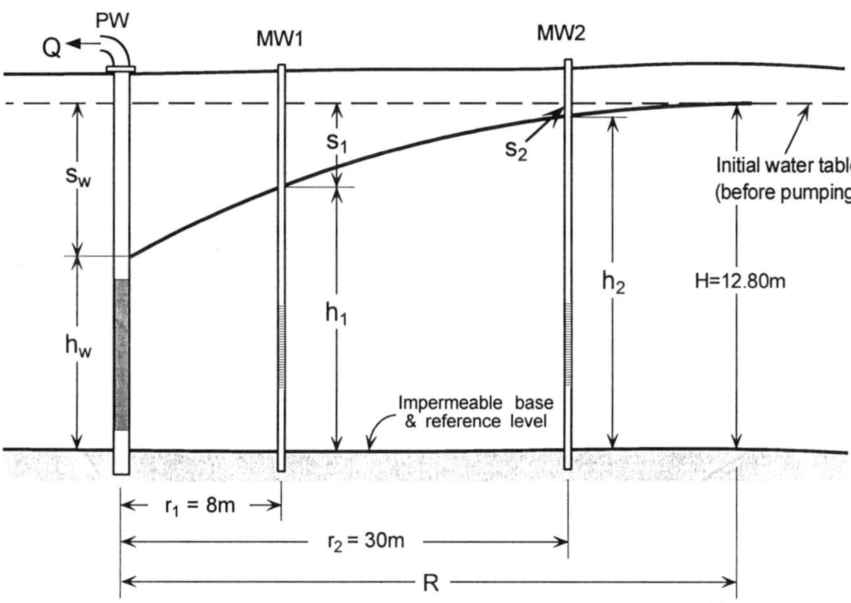

Figure 19.1 Hydrogeologic cross-section of the unconfined aquifer with the pumping well PW and two monitoring wells MW1 and MW2.

19.1 FLOW TOWARD A FULLY PENETRATING WELL IN AN UNCONFINED AQUIFER

Analysis of wells pumping from unconfined aquifers is more complex than that of confined aquifers since the top of the aquifer corresponds to the water table and its position changes due to pumping. In other words, the thickness of the aquifer (saturated zone) changes within the radius of the well influence being the smallest at the well perimeter (see Figure 19.1). If the aquifer impermeable base is horizontal, and the reference level is set at the base, the hydraulic head equals the water table which simplifies the derivation of the flow equation.

Applying Dupuit's hypothesis (see also Problem 12) allows the exact calculation of the flow rate, whereas finding the accurate position of the water table is more complicated and is based on various experimental/approximate equations. Figure 19.2 illustrates the concept of Dupuit's hypothesis together with the actual distribution of velocities around a pumping well. Dupuit's hypothesis states that:

- equipotential lines are vertical;
- velocity is constant along any given vertical (*i.e.*, equipotential line);
- the velocity vector has only the horizontal component (*i.e.*, the stream lines are horizontal and parallel).

However, the velocity vector has a vertical component which increases closer to the well and the water table. The actual position of the water table is above the calculated one (Dupuit's parabola) and this difference at the well perimeter corresponds to the seepage face. At a certain distance from the well (denoted r^* in Figure 19.2) the difference between the actual and the calculated water table becomes very small and can be ignored for practical purposes.

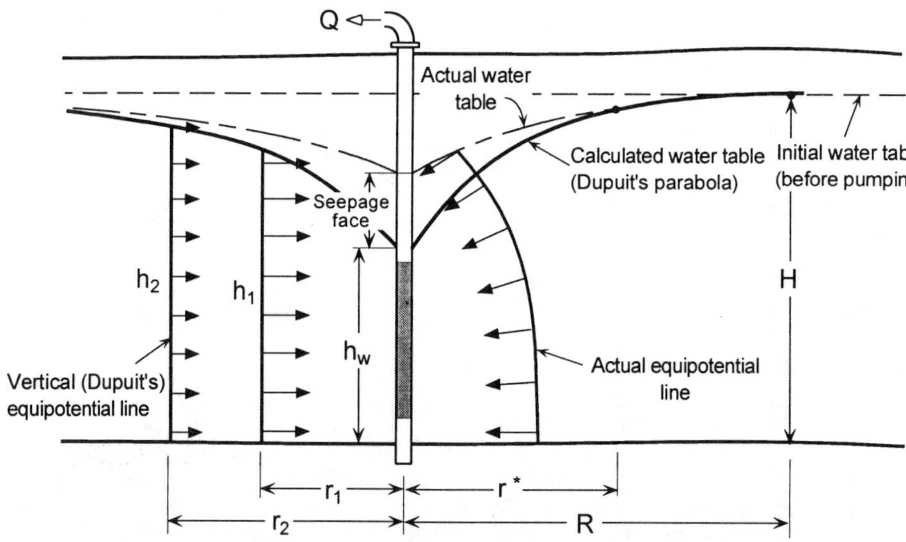

Figure 19.2 Illustration of Dupuit's hypothesis (left) and the actual flow (right) toward a fully penetrating well in an unconfined aquifer. Note that the actual equipotential line is curvilinear, and that the velocity vector has both horizontal and vertical components.

The flow rate (Q) is given as the product of the cross-sectional area of flow (A) and the flow velocity (v):

$$Q = A \cdot v \tag{19.1}$$

The cross-sectional area of flow at distance r from the pumping well is the side of the cylinder with radius r and thickness h (which is the thickness of the saturated zone; see Figure 19.3):

$$A = 2r\pi h \tag{19.2}$$

The velocity of flow at distance r is given as the product of the aquifer's hydraulic conductivity (K) and the hydraulic gradient (i) which is an infinitesimally small drop in hydraulic head (dh) over an infinitesimally small distance (dr):

$$v = K \cdot \frac{dh}{dr} \tag{19.3}$$

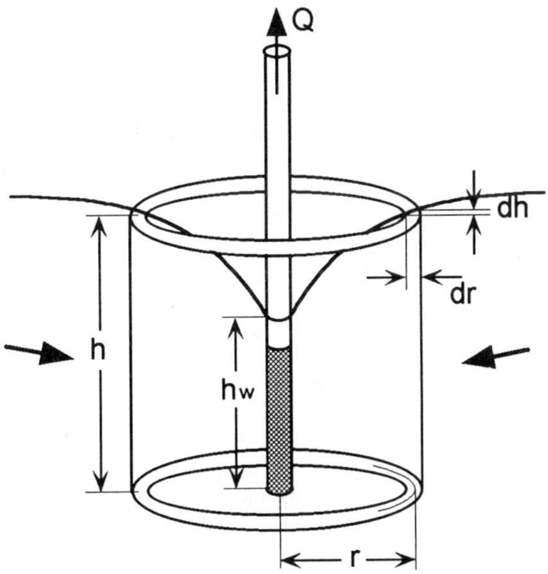

Figure 19.3 Scheme for deriving the differential equation of groundwater flow toward a fully penetrating well in an unconfined aquifer.

Note that there is no minus sign in equation (19.3) as is the case with Darcy's law in the Cartesian (rectangular) coordinate system (see Problem 7). This is because the decrease in hydraulic head is in the negative direction of the r axis (toward the well) and the two minus signs cancel each other out:

$$v = -K \cdot i = -K \cdot \left(-\frac{dh}{dr}\right) = K \cdot \frac{dh}{dr} \tag{19.3a}$$

The flow rate is then:

$$Q = 2r\pi hK\frac{dh}{dr} \tag{19.4}$$

To solve this differential equation the variables h (hydraulic head) and r (radial distance from the well) have to be separated:

$$hdh = \frac{Q}{2\pi K} \cdot \frac{dr}{r} \tag{19.5}$$

Equation (19.5) is then integrated for known boundary conditions (see Figure 19.3):

- at distance r_w (which is the well radius) the hydraulic head is h_w (head in the well);
- at distance r from the well the hydraulic head is h:

$$\int_{h_w}^{h} hdh = \frac{Q}{2\pi K} \int_{r_w}^{r} \frac{dr}{r} \tag{19.6}$$

The integrals on both sides are elementary and are easily solved (see Appendix G):

$$\frac{1}{2}(h^2 - h_w^2) = \frac{Q}{2\pi K} \ln \frac{r}{r_w} \tag{19.7}$$

which gives the following common expression for the pumping rate of a fully penetrating well in an unconfined aquifer:

$$Q = \pi K \frac{(h^2 - h_w^2)}{\ln \frac{r}{r_w}} \tag{19.8}$$

Equation (19.5) can be integrated for any other pair of known hydraulic heads and the corresponding distances from the well. For example:

- at distance r_w (which is the well radius) the hydraulic head is h_w (head in the well);
- at distance R from the well (which is the radius of well influence) the hydraulic head is H (which is the initial, undisturbed head before the pumping; see Figures 19.1 and 19.2):

$$Q = \pi K \frac{(H^2 - h_w^2)}{\ln \frac{R}{r_w}} \tag{19.9}$$

Equation (19.9) can be rewritten as:

$$Q = \pi K \frac{(H - h_w)(H + h_w)}{\ln \frac{R}{r_w}} \tag{19.10}$$

Knowing that the drawdown in the well (s_w) is the difference between the initial hydraulic head (H) and the head registered in the well during pumping (h_w), *i.e.*,:

$$s_w = H - h_w$$

and introducing the *average aquifer thickness* (thickness of the saturated zone) as:

$$b^* = \frac{H + h_w}{2}$$

equation (19.10) can be written in the following form:

$$Q = \frac{2\pi K b^* s_w}{\ln \frac{R}{r_w}} \tag{19.11}$$

which is similar to the equation describing flow toward a fully penetrating well in a *confined* aquifer (see Problem 18):

$$Q = \frac{2\pi b K s_w}{\ln \frac{R}{r_w}}$$

where b is the thickness of the confined aquifer.

For the analysis of steady-state groundwater flow toward wells in unconfined aquifers one should always apply equation (19.10) rather than equation (19.11) since the aquifer thickness is represented more accurately.

19.2 HYDRAULIC CONDUCTIVITY

As mentioned earlier, equation (19.5) can be integrated for any pair of known hydraulic heads and their corresponding distances from the pumping well. If the heads (or drawdown) in the two monitoring wells are recorded, the boundary conditions are as follows:

- at distance r_1 (the distance between the pumping well and the monitoring well MW1 - see Figure 19.1) the hydraulic head is h_1 (head in MW1);
- at distance r_2 (the distance between the pumping well and the monitoring well MW2) the hydraulic head is h_2 (head in MW2);

For these boundary conditions equation (19.5) becomes:

$$h_2^2 - h_1^2 = \frac{Q}{\pi K} \ln \frac{r_2}{r_1} \tag{19.12}$$

which, solved for the hydraulic conductivity, gives the following expression:

$$K = \frac{Q \ln \frac{r_2}{r_1}}{\pi (h_2^2 - h_1^2)} \tag{19.13}$$

The hydraulic heads in MW1 and MW2 to be substituted into equation (19.13) can be found from the data on recorded drawdown and the initial hydraulic head (see Figure 19.1):

$$h_2 = H - s_2 = 12.80m - 1.63m = 11.17m$$

$$h_1 = H - s_1 = 12.80m - 2.74m = 10.06m$$

and the hydraulic conductivity is then:

$$K = \frac{0.0175m^3/s \cdot \ln\frac{30m}{8m}}{3.14 \cdot [(11.17m)^2 - (10.06m)^2]} = 1.34 \times 10^{-3}\, m/s$$

If there are more than two monitoring wells, or there are several pumping tests, and it is required to find an average hydraulic conductivity representative of the tested aquifer, the graphoanalytical method presented in the next section is recommended.

19.3 RADIUS OF WELL INFLUENCE

The general equation of groundwater flow toward a fully penetrating well in an unconfined aquifer can be written in the following form:

$$H^2 - h^2 = \frac{Q}{\pi K}\ln\frac{R}{r} \tag{19.14}$$

where h is the hydraulic head at distance r from the pumping well, H is the initial hydraulic head and R is the radius of well influence. Knowing the basic rules of logarithmic calculus (see Appendix G), equation (19.14) can also be written as:

$$H^2 - h^2 = \frac{Q}{\pi K}\ln R - \frac{Q}{\pi K}\ln r \tag{19.15}$$

Since all the factors of the first term on the right-hand side are constant values (note that the radius of well influence does not change in time for steady-state conditions), equation (19.15) can be expressed as:

$$H^2 - h^2 = const - \frac{Q}{\pi K}\ln r \tag{19.16}$$

or, when the natural logarithm is replaced with the logarithm to the base 10 (note that lnx=2.3026*logx):

$$H^2 - h^2 = const - \frac{Q}{\pi K} \cdot 2.303 \cdot \log r \tag{19.17}$$

$$H^2 - h^2 = const - \frac{0.733 \cdot Q}{K}\log r \tag{19.17a}$$

From equation (19.17) it can be seen that the recorded cone of depression would appear as a straight line if plotted on a semi-logarithmic graph, *i.e.*, $H^2 - h^2 = f(\log r)$. Figure 19.4 shows data for PW, MW1, and MW2 plotted on such a graph. Coordinates of the points are:

PW: $H^2 - h_w^2 = (12.80m)^2 - (5.28m)^2 = 135.96m^2$ $\quad r_w = 0.150m$

MW1: $H^2 - h_w^2 = (12.80m)^2 - (10.06m)^2 = 62.6m^2$ $\quad r_1 = 8m$

MW2: $H^2 - h_w^2 = (12.80m)^2 - (11.17m)^2 = 39.07m^2$ $\quad r_2 = 30m$

The radius of well influence is determined by the interception of the straight line drawn through the monitoring well points with the line of zero drawdown, *i.e.*, for $H^2-h^2=0$. In our case the radius of well influence is R = 270m. Note that the straight line should be drawn only through the monitoring well points, since the actual drawdown recorded in the pumping well includes an additional undetermined component and the pumping well point in most cases is below the theoretical straight line. This additional drawdown is a consequence of well losses, as well as the formation of a seepage face along the well perimeter (see Section 19.1 and Figure 19.2).

Graphs, such as one shown in Figure 19.4, can also be used to determine the aquifer's hydraulic conductivity. Figure 19.5 shows the results of a hypothetical pumping test in an unconfined aquifer monitored by 4 piezometers. The coordinates of any two points on the straight line can be used to calculate the hydraulic conductivity using equation (19.17a):

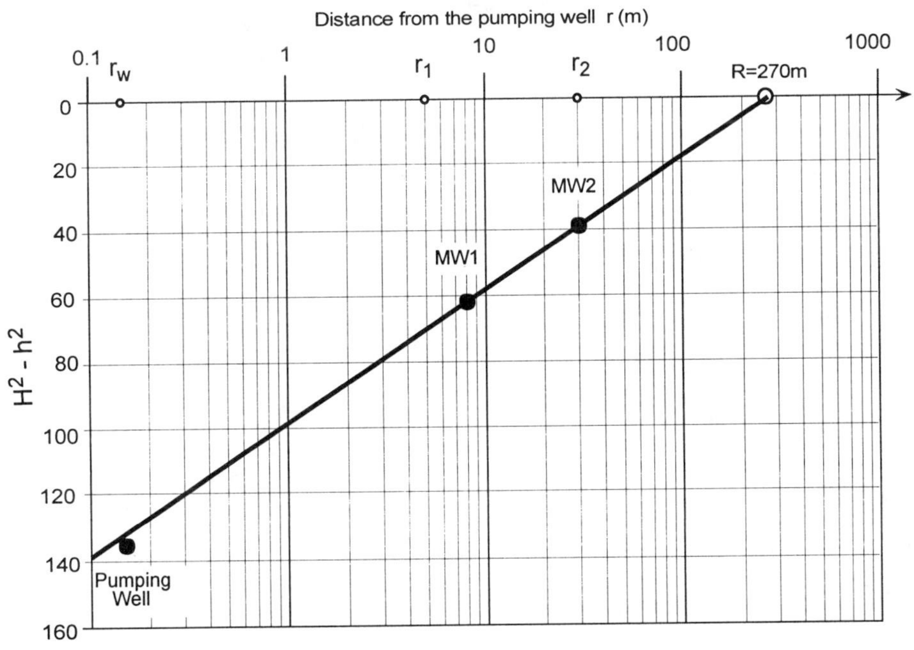

Figure 19.4 Semi-logarithmic graph H^2-h^2 versus radial distance from a pumping well (r) for determining radius of well influence.

$$(H^2 - h^2)_1 = const - \frac{0.733 \cdot Q}{K} \log r_1 \tag{19.18}$$

$$(H^2 - h^2)_2 = const - \frac{0.733 \cdot Q}{K} \log r_2 \tag{19.18a}$$

$$(H^2 - h^2)_1 - (H^2 - h^2)_2 = \Delta(H^2 - h^2) = \frac{0.733 \cdot Q}{K} \log \frac{r_2}{r_1} \tag{19.19}$$

$$K = \frac{0.733 \cdot Q}{\Delta(H^2 - h^2)} \log \frac{r_2}{r_1} \tag{19.20}$$

If two points on the straight line are chosen to be one log cycle apart (*i.e.*, the ratio of the distance coordinates r_2 and r_1 is 10), equation (19.20) reduces to:

$$K = \frac{0.733 \cdot Q}{\Delta(H^2 - h^2)} \tag{19.21}$$

For the pumping test shown in Figure 19.5, and the pumping rate of 0.015 m³/s (15 l/s or 238 gpm) the hydraulic conductivity is:

$$K = \frac{0.733 \cdot 0.015 m^3 / s}{38.5 m^2} = 2.86 \times 10^{-4} m/s$$

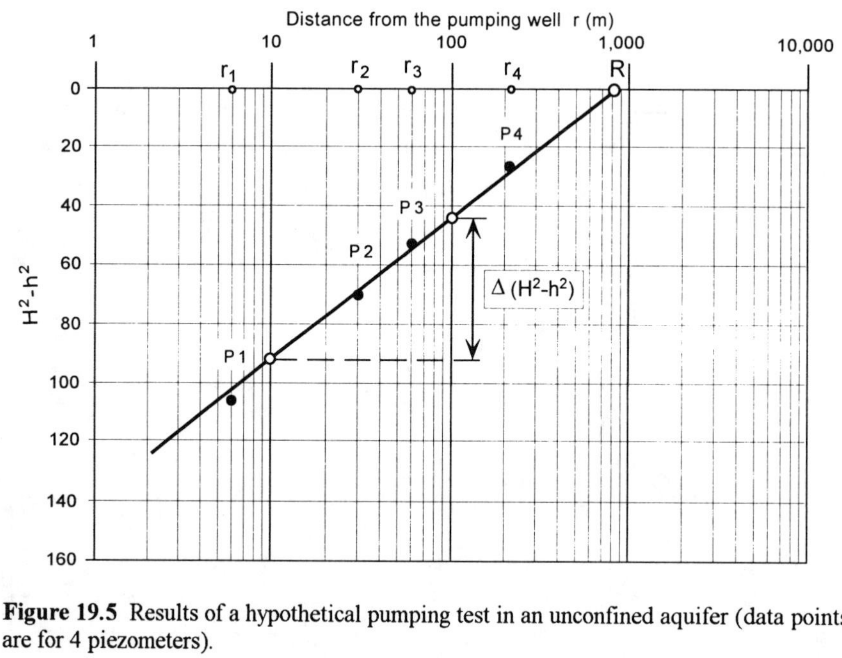

Figure 19.5 Results of a hypothetical pumping test in an unconfined aquifer (data points are for 4 piezometers).

19.4 POSITION OF THE WATER TABLE

An equation, based on Dupuit's assumption, that describes the position of the water table (h) at a distance r from the pumping well is derived from equation (19.8):

$$h = \sqrt{h_w^2 + \frac{Q}{\pi K} \ln \frac{r}{r_w}} \tag{19.22}$$

where h_w is the hydraulic head recorded in the well and r_w is the well radius. This equation is also known as *Dupuit's parabola*. The well pumping rate is given as (equation 19.9):

$$Q = \pi K \frac{(H^2 - h_w^2)}{\ln \dfrac{R}{r_w}} \tag{19.23}$$

By inserting equation (19.23) into equation (19.22), the hydraulic head is expressed only in terms of other known heads and distances at which they are recorded. This is a more convenient method of calculating Dupuit's parabola since one does not need to know the hydraulic conductivity and the well pumping rate:

$$h = \sqrt{h_w^2 + (H^2 - h_w^2)\frac{\ln \dfrac{r}{r_w}}{\ln \dfrac{R}{r_w}}} \tag{19.24}$$

where H is the initial hydraulic head (before pumping) and R is the radius of well influence.

As mentioned earlier, the position of the actual water table is above the Dupuit's parabola and several experimental (approximation) equations have been proposed for its calculation. One of the more common is the Babbitt-Caldwell equation (1948, adapted from Kashef, 1987) which estimates the actual hydraulic head at the well perimeter as:

$$h_0 = H - \frac{0.6}{H} \cdot \frac{H^2 - h_w^2}{\ln \dfrac{R}{r_w}} \cdot \ln \frac{R}{0.1 \cdot H} \tag{19.25}$$

In our case this gives the following value:

$$h_0 = 12.8m - \frac{0.6}{12.8m} \cdot \frac{(12.8m)^2 - (5.07m)^2}{\ln \dfrac{270m}{0.15m}} \cdot \ln \frac{270m}{0.1 \cdot 12.8m} = 8.18m$$

which means that the seepage face at the well perimeter is:

$$\Delta h = h_0 - h_w = 8.18m - 5.07m = 3.11m$$

Another equation, proposed by Schneebeli, is also used for determining the height of the seepage face (Δh) at the well perimeter (adapted from Vukovic and Soro, 1984):

$$\Delta h = \sqrt{h_w^2 + \frac{Q}{\pi K}\left[0.4343 \ln \frac{Q}{\pi K r_w^2} - 0.4\right]} - h_w \tag{19.26}$$

or

$$\Delta h = \sqrt{h_w^2 + \frac{H^2 - h_w^2}{\ln \dfrac{R}{r_w}}\left[0.4343 \ln \frac{H^2 - h_w^2}{r_w^2 \cdot \ln \dfrac{R}{r_w}} - 0.4\right]} - h_w \tag{19.27}$$

which in our case gives 3.41 m after substituting known values for the hydraulic heads and the radial distances. The position of the actual water table (h_0) at the well perimeter is then:

$$h_0 = h_w + \Delta h \;=\; 5.07m + 3.41m = 8.48m \tag{19.28}$$

where h_w is the recorded water level in the pumping well. Equations (19.26) and (19.27) should be used when the following condition is satisfied:

$$\frac{Q}{Kr_w^2} > 8$$

When this fraction is less than 8, the seepage face is very small and can be disregarded for practical purposes.

The position of the water table in the vicinity of the pumping well is estimated based on the calculated seepage face and a distance from the well at which Dupuit's assumption is valid, *i.e.*, where Dupuit's parabola and the actual water table can be considered close enough for practical purposes. This distance (r^*) is found from equation (19.22) for the condition that $dh/dr = 0.2$, which, for most practical purposes, provides that the actual water table and the Dupuit's parabola are close "enough" to each other (Vukovic and Soro, 1984):

$$\frac{2.5 \cdot (H^2 - h_w^2)}{\ln \dfrac{R}{r_w}} = r^* \cdot \sqrt{h_w^2 + (H^2 - h_w^2)\frac{\ln \dfrac{r^*}{r_w}}{\ln \dfrac{R}{r_w}}} \tag{19.29}$$

or

$$Const = r^* \cdot \sqrt{h_w^2 + (H^2 - h_w^2)\frac{\ln \dfrac{r^*}{r_w}}{\ln \dfrac{R}{r_w}}} \tag{19.29a}$$

As can be seen from equation (19.29), distance r^* is expressed implicitly and its value of 4.9 m (in our case) was found by trial (see Table 19.1). As practice shows, distance r^* is usually within the radius of 10 m from the pumping well. In cases of extremely high well losses r^* can be placed at greater distances.

Table 19.1 Calculation of distance r^* using equation (19.29)

r^* (m)	30	20	10	5	4.5	***4.9***	4.8
Calculated constant	333	215	101.5	47.52	42.31	***46.47***	45.43
Actual constant	46.07	46.07	46.07	46.07	46.07	***46.07***	46.07

Table 19.2 shows calculations of the water table around pumping well PW using equation (19.24) (Dupuit's parabola). The results are plotted in Figure 19.6 together with the estimated position of the actual water table. The latter one is constructed by connecting the upper limit of the seepage face (found as the average value of equations 19.25 and 19.28: 8.33 m) with the point on the Dupuit's parabola at distance $r^* = 4.9\ m$. This line should follow the trend of Dupuit's parabola beyond the radius of 4.9 m. Also plotted in Figure 19.6 is the recorded water table in MW1 which is 8 m from the pumping well. As can be seen, the actual water table is very close to the line calculated using equation (19.24). Again, this confirms common experience that a significant difference between the

actual water table and Dupuit's parabola occurs only in the immediate vicinity of a pumping well. Finding (calculating) this difference is important in cases of construction dewatering, drainage works and other "microspace"-related well operations.

Table 19.2 Values of the calculated water table using equation (19.24) (Dupuit's parabola) in the vicinity of PW

Distance from PW	Dupuit's parabola (m)	Measured in the field
0.15	5.07	5.07
1	7.79	
2	8.57	
3	8.99	
4	9.28	
5	9.50	
6	9.68	
7	9.82	
8	9.95	
9	10.06	
10	10.15	10.06
11	10.24	
12	10.32	
13	10.39	
14	10.45	
15	10.51	

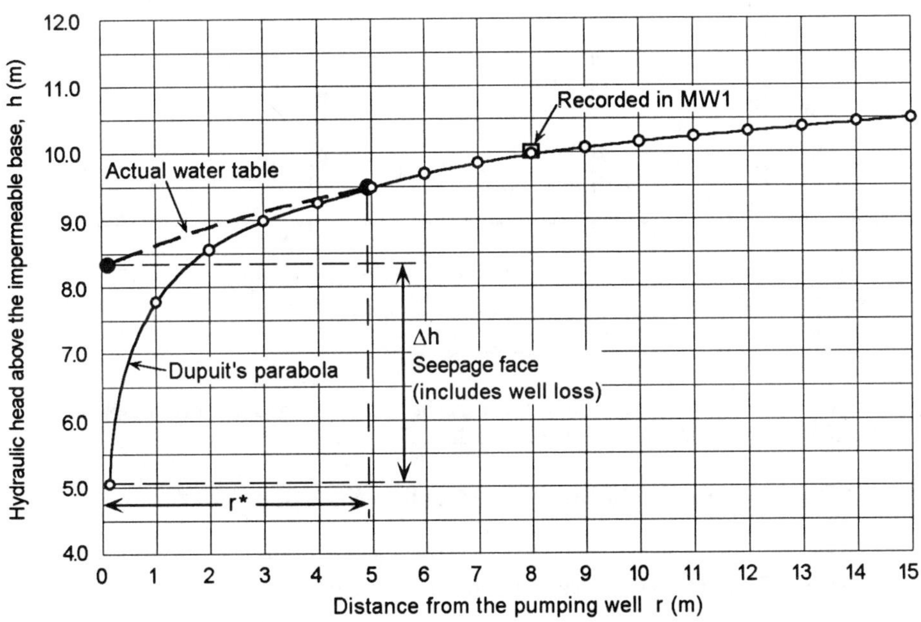

Figure 19.6 Calculated (Dupuit's parabola) and estimated actual water table in the vicinity of pumping well PW.

PROBLEM 20
UNCONFINED AQUIFER WITH INFILTRATION

A water supply well at a paper factory is tapping an unconfined aquifer which is exposed to significant infiltration from precipitation. An incomplete record shows a fairly constant drawdown in the well and the two nearby abandoned wells (see Figure 20.1) of 5.80, 1.65, and 0.95m, respectively. At the time of well completion, the static water table was at 100.80 m above mean sea level. The diameter of the well is 400 mm and the hydraulic conductivity of the aquifer was determined to be 9.3×10^{-5} m/s. According to the production record, the installed water supply system has been consistently using 137,500 gallons of water per day for the last three years. Although there is no exact data on the well pumping rate, it is suspected that there is a loss of water somewhere in the system. The average annual precipitation in the area was 29 inches for the last three years.

From the available data find the average pumping rate of the well and the average annual infiltration rate.

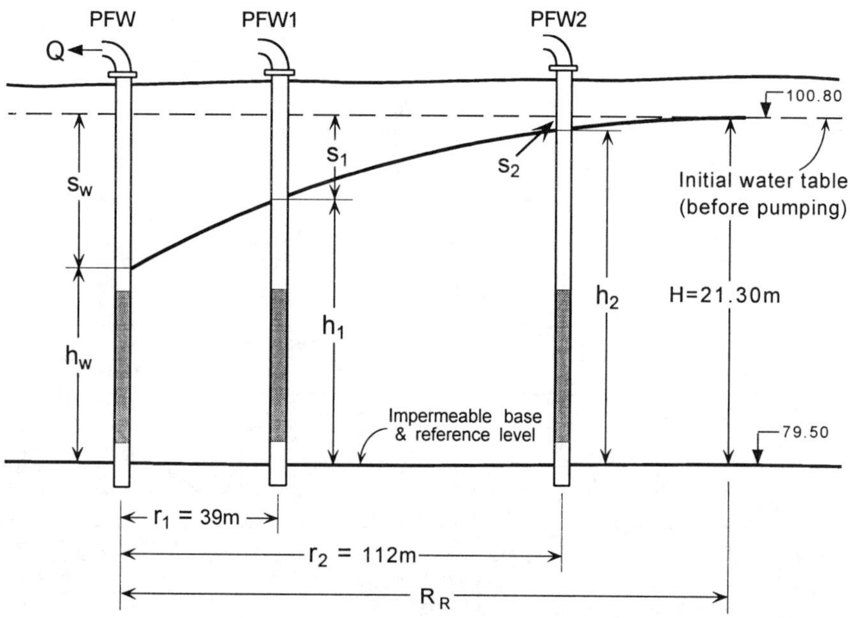

Figure 20.1 Hydrogeologic cross-section of the unconfined aquifer with the pumping well PFW and two abandoned wells PFW1 and PFW2.

20.1 RADIUS OF WELL INFLUENCE ACCORDING TO DUPUIT

The general equation of groundwater flow toward a fully penetrating well in an unconfined aquifer is (see also Problem 19 and equation 19.14):

$$H^2 - h^2 = \frac{Q}{\pi K} \ln \frac{R_D}{r} \tag{20.1}$$

where Q is the well pumping rate, h is the hydraulic head at distance r from the pumping well, H is the initial hydraulic head and R_D is the radius of well influence. In our case this formula cannot be applied directly since the pumping rate is not known. Also, although the infiltration plays an important role in the system, equation (20.1) does not include it. However, one can assume that the pumping rate is generated entirely by the infiltration rate since the system is in equilibrium, *i.e.*, in steady state. This means that equation (20.1) can be used to determine the radius of well influence as explained with equations (20.2) through (20.5), *i.e.*, using a graph $H^2 - h^2$ vs. distance r (see Figure 20.2). *Note, however, that this radius of well influence is not the real (actual) one. Rather, it is calculated, i.e., a graphically determined radius of influence according to Dupuit's assumptions* (see Problem 19, Section 19.1). The real radius of influence is somewhat larger and it depends on the infiltration.

Knowing the basic rules of logarithmic calculus (see Appendix G), equation (20.1) can be written as:

$$H^2 - h^2 = \frac{Q}{\pi K}\ln R_D - \frac{Q}{\pi K}\ln r \tag{20.2}$$

Since all the factors of the first term on the right-hand side are constant values (note that the radius of well influence does not change in time for steady-state conditions), equation (20.2) can be expressed as:

$$H^2 - h^2 = const - \frac{Q}{\pi K}\ln r \tag{20.3}$$

or, when the natural logarithm is replaced with the logarithm to the base 10 (note that $\ln x = 2.3026 \times \log x$):

$$H^2 - h^2 = const - \frac{Q}{\pi K}\cdot 2.303 \cdot \log r \tag{20.4}$$

$$H^2 - h^2 = const - \frac{0.733 \cdot Q}{K}\log r \tag{20.5}$$

From equation (20.5) it can be seen that the recorded cone of depression would appear as a straight line if plotted on a semi-logarithmic graph, *i.e.*, $H^2 - h^2 = f(\log r)$. Figure 20.2 shows data for PFW, PFW1, and PFW2 plotted on such a graph. Coordinates of the points are:

PFW: $H^2 - h_w^2 = (21.30m)^2 - (15.50m)^2 = 213.4m^2$ $r_w = 0.200m$

PFW1: $H^2 - h_w^2 = (21.30m)^2 - (19.65m)^2 = 67.6m^2$ $r_1 = 39m$

PFW2: $H^2 - h_w^2 = (21.30m)^2 - (20.35m)^2 = 39.6m^2$ $r_2 = 112m$

The radius of well influence according to Dupuit (R_D) is determined by the interception of the straight line drawn through the data points with the line of zero drawdown, *i.e.*, for $H^2 - h^2 = 0$. In our case Dupuit's radius of well influence is $R_D = 460\ m$. Note that the straight line should be drawn only through the monitoring well points, since the actual drawdown recorded in the pumping well includes an additional undetermined component and the pumping well point in most cases is below the theoretical straight line. This additional drawdown is a consequence of well losses, as well as the formation of a seepage face along the well perimeter (see Problem 19, Section 19.1 and Figure 19.2). It appears, however, that in our case the pumping well loss is not significant.

Once the radius of influence according to Dupuit is determined, it is possible to find the actual radius of influence and apply the differential equation of groundwater flow toward a fully penetrating well in an unconfined aquifer under the influence of infiltration.

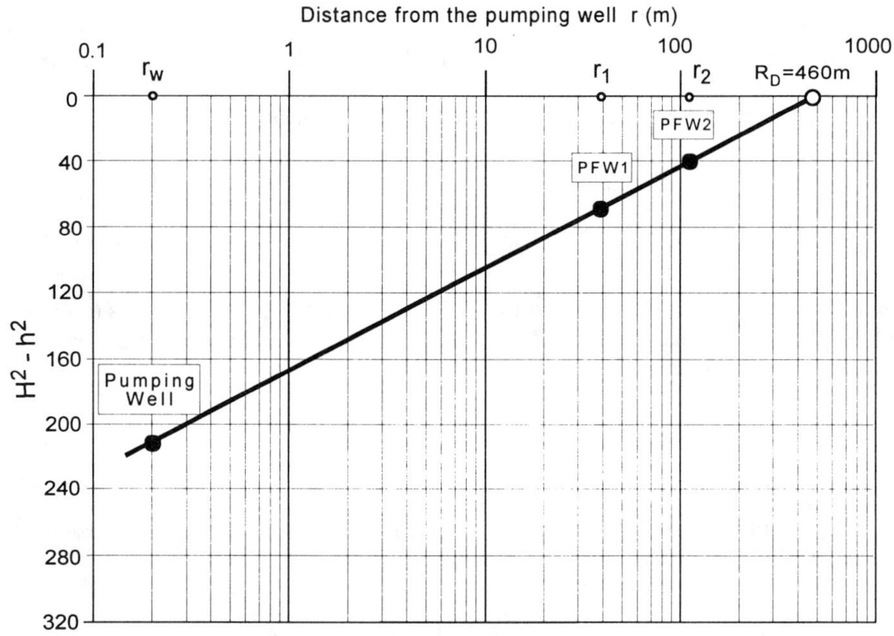

Figure 20.2 Semi-logarithmic graph H^2–h^2 versus radial distance from a pumping well (r) for determining radius of well influence according to Dupuit.

20.2 FLOW TOWARD A FULLY PENETRATING WELL UNDER THE INFLUENCE OF INFILTRATION

Figure 20.3 shows elements of groundwater flow toward a fully penetrating well in an unconfined aquifer that is exposed to infiltration from the surface. The flow, or pumping rate at the well (Q_w), is the sum of two flows:

- the flow from the aquifer (Q_r) that enters the side of the cylinder with radius r, and
- the flow generated by the infiltration (Q_{inf}) over the area with radius r.

(Note that the flow from the aquifer is also generated by the infiltration but, for the purpose of deriving the differential equation, it is considered simply as a flow through porous medium due to the existing hydraulic gradient.)

According to Dupuit's assumptions, the flow through the porous medium at distance r from the well is described with the following equation (see also Problem 19 and equation 19.4):

$$Q_r = 2r\pi hK\frac{dh}{dr} \qquad (20.6)$$

where h is the hydraulic head at distance r. This head is also the height of the cylinder, *i.e.*, the thickness of the saturated zone if the reference level is placed at the base of the aquifer.

The flow generated by the infiltration over the area with radius r is:

$$Q_{inf} = r^2 \cdot \pi \cdot w \tag{20.7}$$

where w is the infiltration rate.

As already mentioned, the well pumping rate is the sum of two flows:

$$Q_w = Q_r + Q_{inf} \tag{20.8}$$

or

$$Q_w = 2\pi r h K \frac{dh}{dr} + r^2 \pi w \tag{20.9}$$

In order to solve this differential equation, the unknowns have to be separated:

$$hdh = \frac{Q_w}{\pi K}\frac{dr}{r} - \frac{w}{2K} rdr \tag{20.10}$$

and then integrated for the given boundaries. In our case (Figure 20.3) the boundary conditions are:

- at distance r_w (which is the well radius) the hydraulic head is h_w,
- at distance r the hydraulic head is h, *i.e.,*:

$$\int_{h_w}^{h} hdh = \frac{Q_w}{\pi K}\int_{r_w}^{r}\frac{dr}{r} - \frac{w}{2K}\int_{r_w}^{r} rdr \tag{20.11}$$

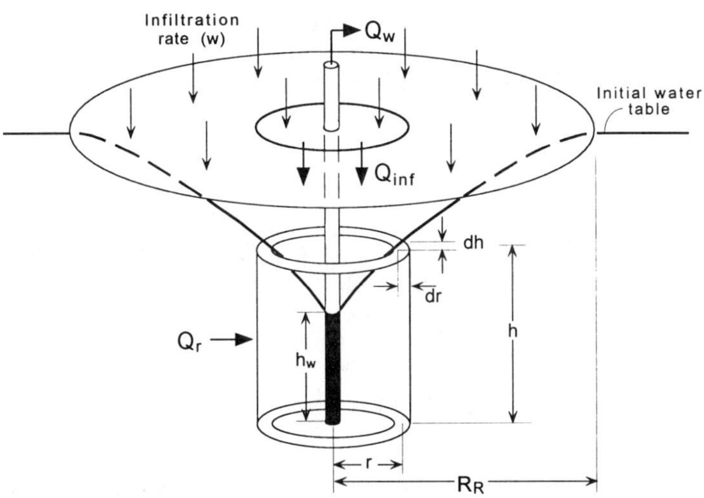

Figure 20.3 Scheme for deriving the differential equation of groundwater flow toward a fully penetrating well in an unconfined aquifer under the influence of infiltration.

All the integrals in equation (20.11) are elementary and are easily solved (see Appendix G):

$$\frac{1}{2}h^2\Big]_{h_w}^{h} = \frac{Q_w}{2\pi K}\cdot \ln r\Big]_{r_w}^{r} - \frac{w}{2K}\cdot\frac{1}{2}r^2\Big]_{r_w}^{r} \tag{20.12}$$

or:

$$h^2 - h_w^2 = \frac{Q_w}{\pi K} \cdot \ln \frac{r}{r_w} - \frac{w}{2K} \cdot (r^2 - r_w^2) \tag{20.12a}$$

This is the final form of the equation describing flow toward a fully penetrating well in an unconfined aquifer under the influence of infiltration.

For the following boundaries:

- at distance R_R (which is the real radius of well influence) the hydraulic head is H (which is the initial head before pumping)
- at distance r_w the hydraulic head is h_w,

equation (20.10) is solved as:

$$H^2 - h_w^2 = \frac{Q_w}{\pi K} \cdot \ln \frac{R_R}{r_w} - \frac{w}{2K} \cdot (R_R^2 - r_w^2) \tag{20.13}$$

Since the squared well radius (r_w^2) is much smaller than the squared radius of well influence (R_R^2), it can be ignored and equation (20.13) becomes:

$$H^2 - h_w^2 = \frac{Q_w}{\pi K} \cdot \ln \frac{R_R}{r_w} - \frac{w}{2K} \cdot R_R^2 \tag{20.13a}$$

It can be seen that if there is no infiltration, equation (20.13a) is the same as equation (20.1).

20.3 REAL RADIUS OF WELL INFLUENCE

For the equilibrium conditions assumed in our case, the flow at the well is generated entirely by the infiltration:

$$Q_w = R_R^2 \cdot \pi \cdot w \tag{20.14}$$

As shown in Figure 20.3, this flow is the product of the area exposed to infiltration and the infiltration rate which has dimensions of velocity.

The real radius of well influence is then:

$$R_R = \sqrt{\frac{Q_w}{\pi w}} \tag{20.15}$$

If the radius of well influence in equation (20.13a) is replaced by equation (20.15), it gives:

$$H^2 - h_w^2 = \frac{Q_w}{\pi K} \cdot \ln \frac{\sqrt{\frac{Q_w}{\pi w}}}{r_w} - \frac{w}{2K} \cdot \frac{Q_w}{\pi w} \tag{20.16}$$

or, after simplification:

$$H^2 - h_w^2 = \frac{Q_w}{\pi K} \cdot \ln \frac{\sqrt{\frac{Q_w}{\pi w}}}{r_w} - \frac{Q_w}{2\pi K} \tag{20.17}$$

Equation (20.17) can also be written as:

$$H^2 - h_w^2 = \frac{Q_w}{\pi K} \cdot \left[\ln \frac{\sqrt{\frac{Q_w}{\pi w}}}{r_w} - \frac{1}{2}\right] \tag{20.18}$$

Knowing that (see Appendix G):

$\ln e = 1$ and $\frac{1}{2} = \ln e^{1/2} = \ln \sqrt{e}$

equation (20.18) becomes:

$$H^2 - h_w^2 = \frac{Q_w}{\pi K} \cdot \left(\ln \frac{\sqrt{\frac{Q_w}{\pi we}}}{r_w}\right) \tag{20.19}$$

When equation (20.19) is compared with equation (20.1), it is apparent that, for the equivalent pumping rate and drawdown, the radius of well influence according to Dupuit ("Dupuit's radius of well influence") is expressed as:

$$R_D = \sqrt{\frac{Q_w}{\pi we}} \tag{20.20}$$

As shown earlier, the real radius of well influence is given as:

$$R_R = \sqrt{\frac{Q_w}{\pi w}}$$

which means that the two radii are connected through the following equation:

$$R_R = R_D \cdot \sqrt{e} \tag{20.21}$$

Dupuit's radius of well influence determined graphoanalytically (see Figure 20.2) is 460 m and the real radius of well influence is:

$$R_R = 460m \cdot 1.6487 = 758m$$

20.4 RATE OF INFILTRATION

The expression for the infiltration rate (w) is obtained by combining equations (20.13) and (20.14):

$$w = \frac{K}{R_R^2} \cdot \frac{(H^2 - h_w^2)}{\ln \dfrac{R_R}{r_w \cdot \sqrt{e}}} \tag{20.22}$$

which gives the following value when the values for all knowns are inserted:

$$w = \frac{9.3 \times 10^{-5}\, m/s}{(758m)^2} \cdot \frac{(21.30m)^2 - (15.50m)^2}{\ln \dfrac{758m}{0.2m \cdot 1.6487}}$$

$$w = 4.4635 \times 10^{-9}\, m/s \;=\; 0.141\, m/year$$

The average annual infiltration is calculated from the infiltration rate and the average annual precipitation in the area (note that 29 inches of rain equal 0.737m):

$$I = \frac{0.141m}{0.737m} \times 100\% \;=\; 19.1\%$$

20.5 WELL PUMPING RATE

The well pumping rate can be calculated from equation (20.13a) by solving for Q_w, or using equation (20.14) directly:

$$Q_w = R_R^2 \cdot \pi \cdot w$$

$$Q_w = (758m)^2 \cdot 3.14 \cdot 4.4635 \times 10^{-9}\, m/s$$

$$Q_w = 8.05 \times 10^{-3}\, m^3/s$$

which gives the total of 695.5 cubic meters or 183,730 gallons of pumped water per day. This calculated value of the withdrawn groundwater is considerably larger than the amount of water used in the production (136,500 gallons per day).

The efficiency of the water supply system is:

$$Efficiency \;=\; \frac{Used\ volume}{Withdrawn\ volume} \times 100\% \;=\; \frac{136500gal}{183730gal} \times 100\% \;=\; 74.3\%$$

which indicates that there indeed may be a significant loss of water in the system requiring a thorough analysis. This analysis should, in addition to possible mechanical/distribution failures, include testing of the well efficiency.

PROBLEM 21
CONFINED-UNCONFINED AQUIFER TRANSITION

An extraction well on a large river island (Figure 21.1) has been considered for a temporary lowering of the piezometric surface for construction purposes. The island's sandy aquifer, which is naturally confined, will have to undergo a change to unconfined conditions at the construction site because of the requirement that the piezometric surface must be kept at 1.5 meters below the aquifer top during the construction (see Figure 21.2).

Find the pumping rate at the well that will satisfy this requirement for safe construction. The well diameter is 500 mm. The hydraulic conductivity of the aquifer is 3.73×10^{-5} m/s. Other elements needed for the calculation are shown in Figures 21.2 through 21.4.

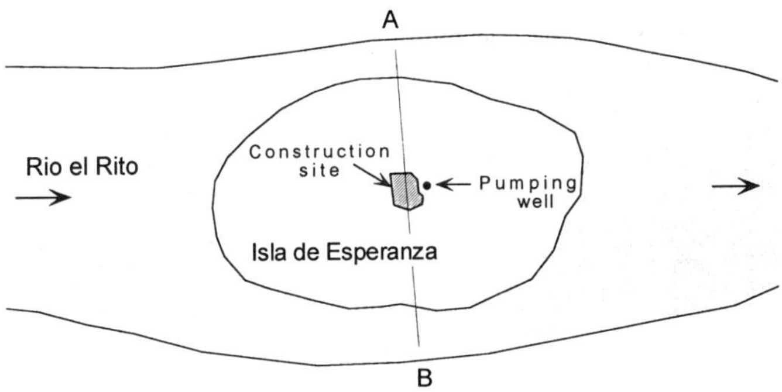

Figure 21.1 Map view of the location of the construction site and the dewatering well on an island (Isla de Esperanza) in a large river (Rio el Rito).

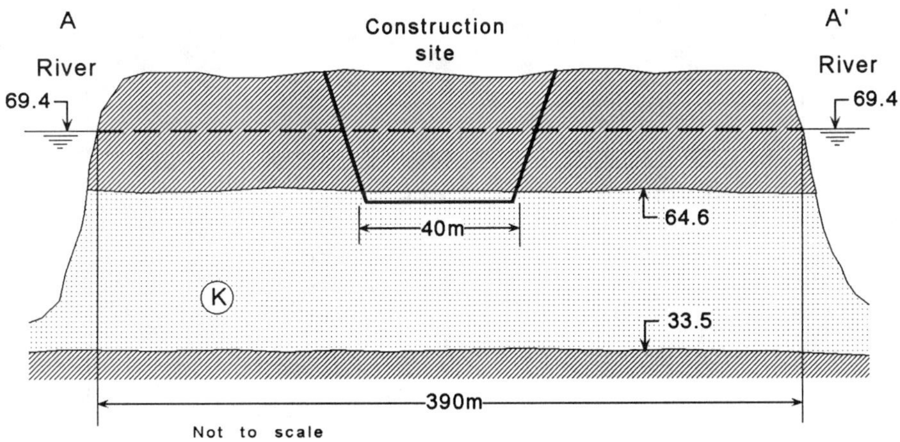

Figure 21.2 Hydrogeologic cross-section of the confined aquifer with the outline of the excavation pit.

21.1 EQUATION OF GROUNDWATER FLOW TOWARD THE WELL

The differential equation of groundwater flow toward a fully penetrating well in a confined aquifer is (see also Problem 18 and equation18.4):

$$Q = 2\pi rbK\frac{dh}{dr} \tag{21.1}$$

where

Q	is the well pumping rate,
r	is the radial distance from the well,
K	is the hydraulic conductivity,
h	is the hydraulic head at distance r.

In our case the confined flow is in part of the aquifer with the following boundary conditions (Figure 21.3):

- at distance r_u (which is the radial distance from the well at which the flow changes from confined to unconfined), the hydraulic head is h_u and is equal to the aquifer thickness b,
- at distance R (which is the radius of well influence, *i.e.*, the distance between the well and the river) the hydraulic head is H.

When integrated with these boundary conditions equation (21.1) is solved as:

$$Q = \frac{2\pi Kb\cdot(H-b)}{\ln\frac{R}{r_u}} \tag{21.2}$$

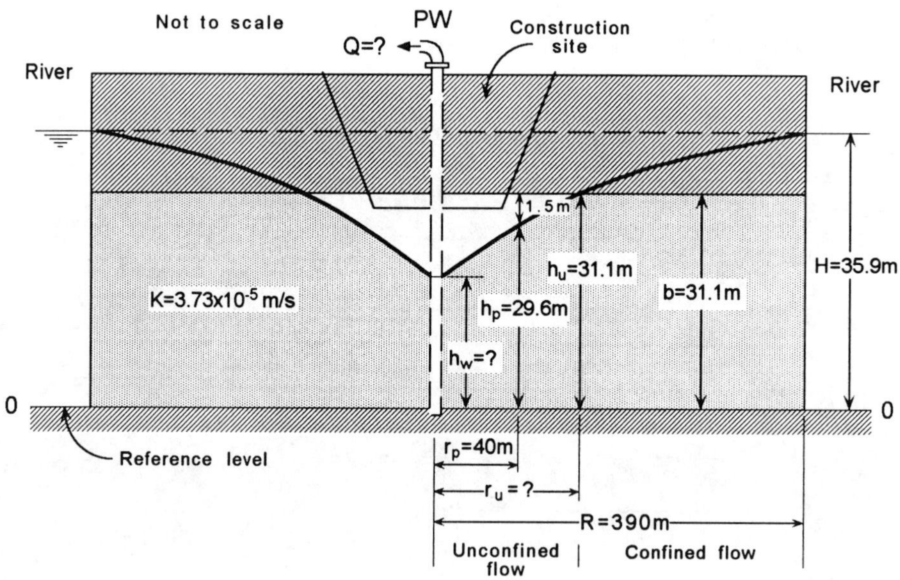

Figure 21.3 Calculation scheme for the problem.

The differential equation of groundwater flow toward a fully penetrating well in an unconfined aquifer is (see also Problem 19 and equation 19.5):

$$hdh = \frac{Q}{2\pi K} \cdot \frac{dr}{r} \tag{21.3}$$

In our case the unconfined flow is described with the following boundary conditions:

- at distance r_p (which is the protected area - see Figure 21.4) the hydraulic head is h_p,
- at distance r_u (which is the radial distance from the well at which the flow changes from confined to unconfined), the hydraulic head is h_u and is equal to the aquifer thickness b.

The solution of equation (21.3) for these boundary conditions is:

$$Q = \frac{\pi K \cdot (b^2 - h_p^2)}{\ln \frac{r_u}{r_p}} \tag{21.4}$$

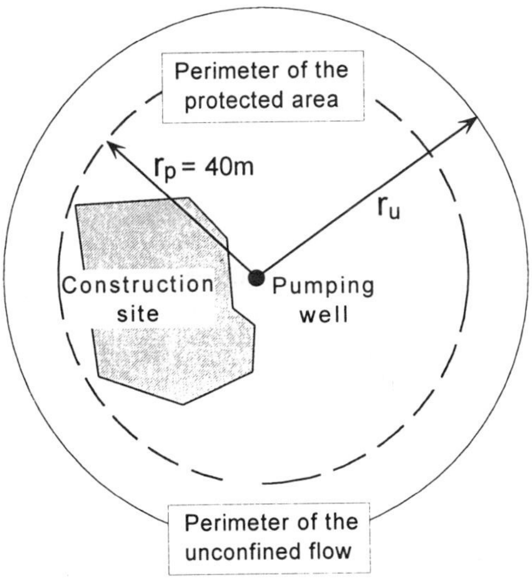

Figure 21.4 Map view of the construction site with the location of the extraction well. The dashed circle is the area of the drainage requirement (protected area), and the solid circle is the area of the aquifer with unconfined flow conditions during the well operation.

Equation (21.2) can be written as:

$$\ln R - \ln r_u = \frac{2\pi K b \cdot (H - b)}{Q} \tag{21.5}$$

Similarly, equation (21.4) is:

$$\ln r_u - \ln r_p = \frac{\pi K \cdot (b^2 - h_p^2)}{Q} \tag{21.6}$$

or

$$\ln r_u = \frac{\pi K \cdot (b^2 - h_p^2)}{Q} - \ln r_p \tag{21.6a}$$

Because of the principle of continuity of flow, equations (21.2) and (21.4) are equal, *i.e.*, they both give the same pumping rate at the well. This means that equation (21.6a) can be inserted into equation (21.5) which leads to the following:

$$\ln R - \ln r_p = \frac{2\pi Kb(H-b)}{Q} + \frac{\pi K(b^2 - h_p^2)}{Q} \tag{21.7}$$

or, after solving for Q:

$$Q = \frac{\pi K \cdot (2Hb - b^2 - h_p^2)}{\ln \frac{R}{r_p}} \tag{21.8}$$

As seen in Figure 21.3, the well pumping rate can be calculated directly since all the elements in equation (21.8) are known. The hydraulic head at the perimeter of the protected area, h_p, is found from the requirement that the piezometric surface must be at least 1.5m below the aquifer top within the excavation pit:

$$h_p = b - 1.5m = 31.1m - 1.5m = 29.6m$$

The well pumping rate that will meet the requirement for safe construction is:

$$Q = \frac{3.14 \cdot 3.73 \times 10^{-5}\, m/s \cdot [2 \cdot 35.9m \cdot 31.1m - (31.1m)^2 - (29.6m)^2]}{\ln \frac{390m}{40m}}$$

$$Q = 0.02004\ m^3/s \ = \ 20.04\ l/s$$

21.2 RADIUS OF UNCONFINED FLOW (r_u)

Any of equations (21.5) through (21.6a) can be used to find the radius of the area with unconfined flow conditions during the well operation. For example, from equation (21.6) it follows that

$$\ln \frac{r_u}{r_p} = \frac{\pi K(b^2 - h_p^2)}{Q} \tag{21.9}$$

and

$$\frac{r_u}{r_p} = e^{\frac{\pi \cdot K \cdot (b^2 - h_p^2)}{Q}} \tag{21.10}$$

which gives the following expression for the radius of unconfined flow:

$$r_u = r_p \cdot e^{\frac{\pi \cdot K \cdot (b^2 - h_p^2)}{Q}} \tag{21.11}$$

and, after inserting all known values (note that e=2.7183):

$$r_u = 40m \cdot e^{\frac{3.14 \cdot 3.73 \times 10^{-5} m/s \cdot [(31.1m)^2 - (29.6m)^2]}{0.02m^3/s}}$$

$$r_u = 68.2m$$

21.3 DRAWDOWN IN THE WELL

The hydraulic head at the well, which is needed to calculate the drawdown, can be found by solving equation (21.3) for the following boundary conditions:

- at distance r_w (which is the well radius) the hydraulic head is h_w,
- at distance r_p (which is the radius of the protected area), the hydraulic head is h_p,

which gives:

$$Q = \frac{\pi K \cdot (h_p^2 - h_w^2)}{\ln \frac{r_p}{r_w}} \tag{21.12}$$

Solved for the hydraulic head, equation (21.12) becomes:

$$h_w = \sqrt{\frac{Q}{\pi K} \ln \frac{r_p}{r_w} - h_p^2} \tag{21.13}$$

$$h_w = \sqrt{\frac{0.02m^3/s}{3.14 \cdot 3.73 \times 10^{-5} m/s} \cdot \ln \frac{40m}{0.25m} - (29.6m)^2}$$

$$h_w = 2.80m$$

The calculated drawdown in the pumping well is the difference between the initial hydraulic head (before pumping) and the hydraulic head during pumping:

$$s = H - h_w = 35.9m - 2.80m = 33.1m$$

Such a large drawdown (92.2% of the initial head) is not feasible because it does not leave enough saturated thickness for the screen and pump placement. *(Note that the actual drawdown in the well would be higher than the calculated one because of the inevitable well loss.)* This means that drainage of the construction site would have to be performed with two or more wells placed around the excavation perimeter. This would result in a smaller drawdown in individual wells and safer operation and maintenance of the dewatering system.

PROBLEM 22
GROUP OF WELLS

Two pumping wells (PW1 and PW2) are planned for containment and cleanup of a groundwater pollutant plume shown in Figure 22.1. The plume is in an unconfined aquifer with the horizontal impermeable base at 185 meters above mean sea level. The average water table elevation is at 205 m amsl. An earlier aquifer test at PW1, with the constant pumping rate of 2.5 l/s, indicated that quasi steady-state conditions were quickly established at the 3.2-meter drawdown. The radius of well influence, determined from observations in nearby domestic wells, was approximately 200 meters. From the available data find the following:

- the drawdown at point P within the plume if only PW1 and PW2 are operating with the pumping rates of 6.0 and 4.0 l/s respectively;
- the drawdown at point P if two recharge wells RW1 and RW2, with injection rates of 5.0 l/s each, were added to the system.

All wells are designed to have the same radius of 100 mm.

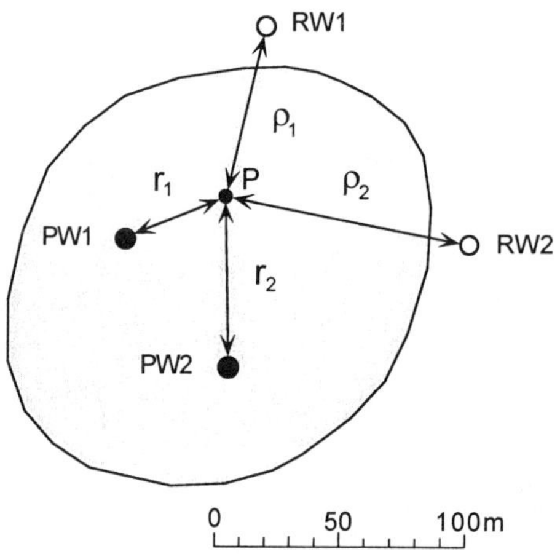

Figure 22.1 Map view of the proposed pump-and-treat well system. PW - pumping wells, RW - recharge wells.

22.1 SUPERPOSITION OF FLOWS

The drawdown at any point in the aquifer affected by a simultaneous pumping of two or more wells is the algebraic sum of drawdowns caused by each well. This principle of the flow superposition allows a simple application of the well-known equations describing groundwater flow toward a single well in a confined or an unconfined aquifer.

Figure 22.2 illustrates the principle for two extraction wells in an unconfined aquifer where the resulting drawdown is the sum of two individual drawdowns. Note the formation of a local groundwater divide between the two wells.

For calculation purposes, a recharge well is considered to be a well with a negative pumping rate since water is added to the aquifer. It means that the drawdown calculated using common equations is also negative, *i.e.*, the water table "builds up" around a recharging well. Figure 22.3 shows the resulting drawdown when one recharging and one extraction well operate simultaneously.

In hydrogeological practice it is far more common to analyze situations with more than one well operating. It is also advisable to check for the possible existence and influence of wells that are not in the immediate vicinity of the well(s) being currently studied.

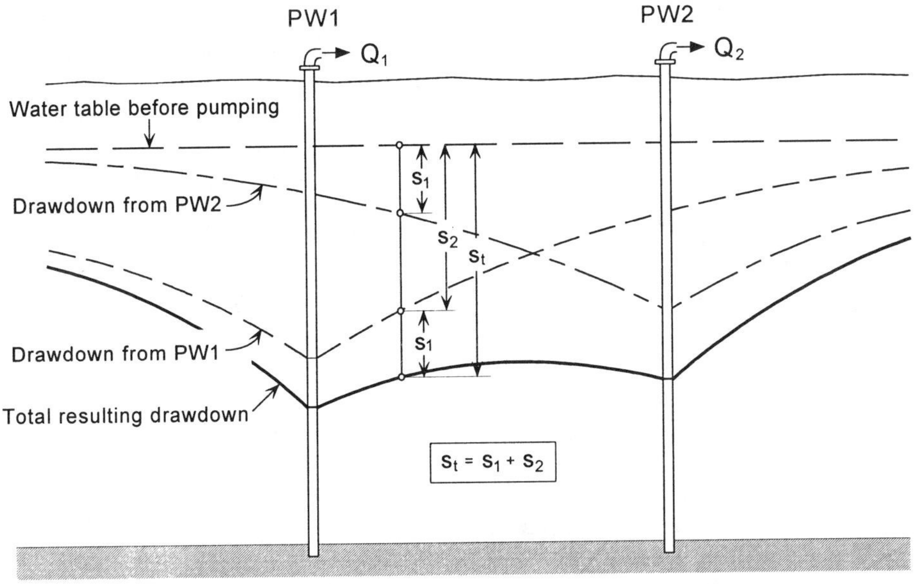

Figure 22.2 Superposition of flows toward two pumping wells. The resulting drawdown is the algebraic sum of the drawdowns caused by individual wells.

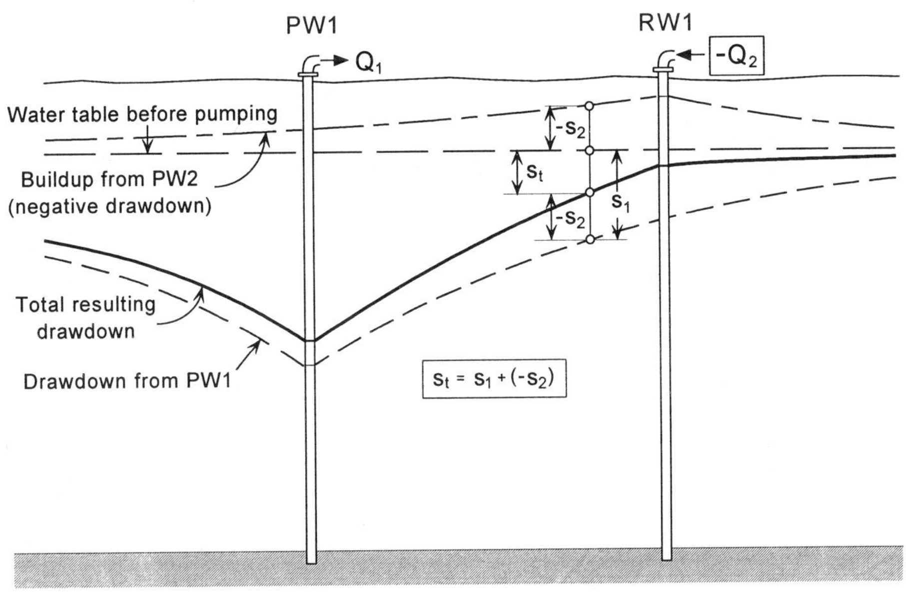

Figure 22.3 Superposition of flows for one pumping well (left) and one recharging well (right) operating simultaneously.

22.2 DRAWDOWN CALCULATION

Two Extraction Wells

The drawdown in a pumping well is the difference between the initial hydraulic head (before pumping) and the hydraulic head during pumping. When the drawdown has to be calculated using common equations derived from Dupuit's assumptions (as in our case), the result will need further adjustment to account for the well loss. The well loss, however, is a unique characteristic for each individual well and can be determined only after an appropriate test is performed in the real well (see Problem 18). Because of the (inevitable) well loss, the actual hydraulic head in the immediate vicinity of a pumping well is above the calculated Dupuit's parabola (see Problem 19). This difference decreases rapidly with the radial distance from the well and, for most practical purposes, it is safe to assume that the calculated hydraulic head equals the actual head beyond the 10-meter radius from the well.

As mentioned earlier, the drawdown at point P is the sum of two drawdowns caused by the pumping wells PW1 and PW2:

$$s_P = s_1 + s_2 \tag{22.1}$$

Equation (22.1) can also be written as:

$$s_P = (H - h_1) + (H - h_2) \tag{22.2}$$

where

- H is the initial hydraulic head,
- h_1 is the hydraulic head at point P due to the operation of PW1,
- h_2 is the hydraulic head at point P due to the operation of PW2.

The general equation of groundwater flow toward a fully penetrating well in an unconfined aquifer is (see also Problem 19 and equation 19.14):

$$H^2 - h^2 = \frac{Q}{\pi K} \ln \frac{R}{r} \tag{22.3}$$

where h is the hydraulic head at distance r from the well, and R is the radius of well influence. This equation can be applied to find the hydraulic heads h_1 and h_2 at point P:

$$h_1 = \sqrt{H^2 - \frac{Q_1}{\pi K} \ln \frac{R}{r_1}} \tag{22.4}$$

$$h_2 = \sqrt{H^2 - \frac{Q_2}{\pi K} \ln \frac{R}{r_2}} \tag{22.5}$$

Except for hydraulic conductivity, all factors in equations (22.4) and (22.5) are known (r_1 and r_2 are the distances between point P and PW1 and PW2 respectively, and can be measured on the map in Figure 22.1).

The *hydraulic conductivity* is determined from the PW1 aquifer test data using equation (22.2):

$$K = \frac{Q \ln \frac{R}{r_w}}{\pi (H^2 - h_w^2)} \tag{22.6}$$

Note, however, that the hydraulic conductivity calculated using the pumping well data is only approximate since the measured drawdown includes an undetermined well loss.

The initial hydraulic head is found by subtracting the elevation of the aquifer base from the elevation of the water table:

$$H = 205m - 185m = 20m$$

The steady-state hydraulic head at PW during the test is found from the initial hydraulic head and the recorded drawdown of 3.2 m:

$$h_w = 20m - 3.2m = 16.8m$$

Finally, the hydraulic conductivity is found by substituting this and other known values into equation (22.6):

$$K = \frac{0.0025m^3 / s \cdot \ln\frac{200m}{0.1m}}{3.14 \cdot [(20m)^2 - (16.8m)^2]}$$

$$K = 5.14 \times 10^{-5} m/s$$

The hydraulic head at point P due to the operation of PW1 is found by inserting the hydraulic conductivity and other known values into equation (22.4):

$$h_1 = \sqrt{(20m)^2 - \frac{0.006m^3 / s}{3.14 \cdot 5.14 \times 10^{-5} m/s} \ln\frac{200m}{43m}} = 18.52m$$

(Note that the pumping rate at PW1 is designed to be 6 l/s and the distance between point P and PW1 is 43 m as measured on the map.)

The hydraulic head at point P due to the operation of PW2 is:

$$h_2 = \sqrt{(20m)^2 - \frac{0.004m^3 / s}{3.14 \cdot 5.14 \times 10^{-5} m/s} \ln\frac{200m}{67m}} = 19.31m$$

Finally, the total drawdown at point P due to the simultaneous operation of both pumping wells is calculated using equation (22.2):

$$s_P = (20m - 18.52m) + (20m - 19.31m) = 2.17m$$

Two Extraction and Two Recharge Wells

The addition of two recharge wells to the system will increase the hydraulic head at point P. The resulting smaller total drawdown is calculated by adding two new, negative drawdowns to the already calculated value of 2.17 m.

The hydraulic head at point P due to the operation of RW1 is (note the negative flow rate):

$$h_{R1} = \sqrt{(20m)^2 - \frac{-(0.005m^3 / s)}{3.14 \cdot 5.14 \times 10^{-5} m/s} \ln\frac{200m}{69m}} = 20.8m$$

which gives the following negative drawdown:

$$s_{R1} = 20.0m - 20.8 = -0.8m$$

Due to the operation of RW2, the hydraulic head at point P is:

$$h_{R1} = \sqrt{(20m)^2 - \frac{-(0.005m^3/s)}{3.14 \cdot 5.14 \times 10^{-5} m/s} \ln \frac{200m}{102m}} = 20.5m$$

and the drawdown is −0.5m. The resulting total drawdown at point P within the plume when all four wells are operating is:

$$s_P = 2.17m - 0.8m - 0.5m = 0.87m$$

PROBLEM 23
WELLS NEAR BOUNDARIES

A proposed system for the containment and cleanup of contaminated groundwater consists of a fully penetrating grout curtain and an extraction well as shown in Figure 23.1. The plume is in the confined aquifer which is in direct hydraulic connection with the Black River, a large perennial stream. The aquifer transmissivity is 5.04×10^{-4} m/s, and the average water table elevation is at 87 meters above sea level. From the aquifer test at RW (real pumping well within the plume), it was determined that the radius of well influence is 75 m (this is an equivalent steady-state hydraulic radius which is very close to the radius of the river bend around RW).

From the available data find the pumping rate at RW that would create a 1.3-meter drawdown at the piezometer P1.

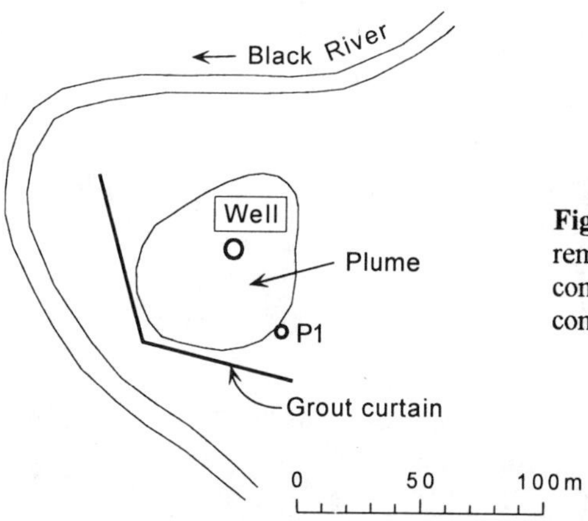

Figure 23.1 Map view of the proposed remediation of contaminated groundwater in the confined aquifer which is in direct hydraulic connection with the Black River.

23.1 EQUIPOTENTIAL BOUNDARY

Although, for various reasons (the most important being simplicity), it is not uncommon to analyze aquifers as if they were of infinite extent, in reality all aquifers have more or less clearly defined boundaries. When a well is completely surrounded by an equipotential boundary (e.g., when it is located on an island or in a meander - see Problem

18), application of the steady-state formulas described so far is justified. In all other cases one either has to assume "quasi steady-state" conditions without boundaries (if their influence is not apparent), or account for the boundaries by introducing appropriate hydraulic conditions. (*Ultimately, however, the most correct approach is to use transient equations with corresponding boundary conditions and apply the image theory.*)

The method of images is a convenient and effective way to account for the physical aquifer boundaries by replacing them with hydraulic entities that allow application of the equations describing flow toward wells. These new hydraulic entities are also wells, either discharging or recharging depending on the nature of the boundaries.

Figure 23.2 illustrates the method of images in case of a discharging (extraction) well near a large perennial river. If the river stage is not affected by the well, the river is considered a strong equipotential boundary equivalent to a line source having a constant head. This is the case with both fully and partially (as in our case) penetrating streams.

The equipotential boundary is simulated with an imaginary recharging well which is the exact image of the real extraction well, only with the opposite (negative) sign for the flow rate. This *image well*, which operates simultaneously with the real well, is placed on the other side of the boundary at the same distance from the boundary as the real well. The line connecting the two wells is perpendicular to the boundary (see Figure 23.3).

The recharge from the image well simulates the flow from the stream toward the real well since it builds up the water table. This additional head along the boundary line cancels the drawdown caused by the extraction well so that there is no change in the hydraulic head along the boundary. Elsewhere in the aquifer the resulting drawdown is the algebraic sum of the drawdown from the real well and the buildup (negative drawdown) from the image well.

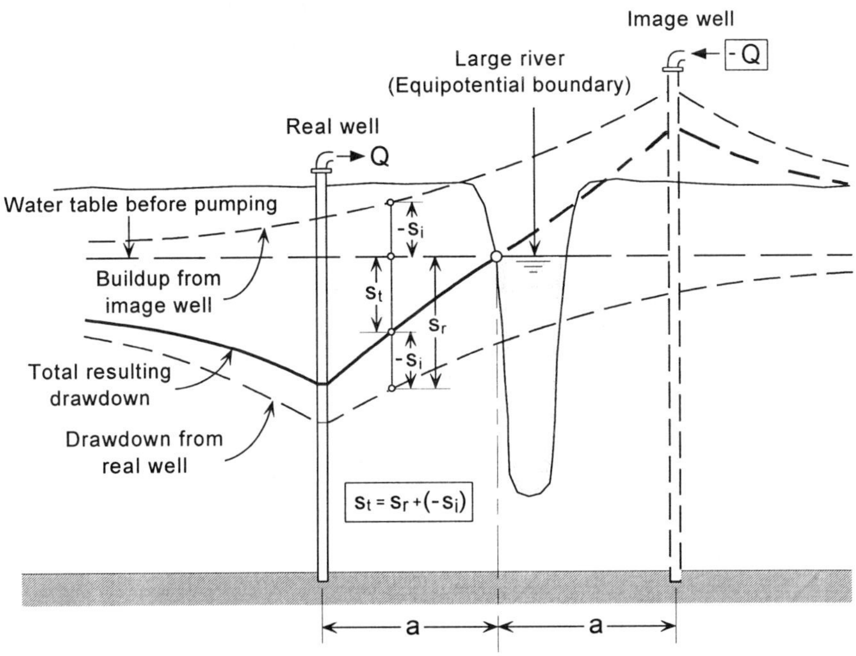

Figure 23.2 Schematic cross-sectional presentation of the resulting drawdown when an equipotential boundary near the extraction well is simulated with the imaginary recharge well of the same flow rate. Note that the drawdown at the boundary is (correctly) zero as the result of this substitution of the physical entity (river) with the new hydraulic entity (recharge well).

It is important to remember an important rule of the image theory: the image well on the other side of the equipotential boundary changes the character - it has the same flow rate as the real well but with the opposite sign. If the real well is extracting water, its image is adding water to the aquifer. If the real well is recharging the aquifer, the image well is extracting water.

Drawdown at the Real Well

The drawdown at the real well is the algebraic sum of the drawdown caused by groundwater withdrawal, and the buildup (negative drawdown) caused by the image well. This can be expressed with the following equation when the resulting drawdown is calculated for point A at the well perimeter (see Figure 23.3):

$$s_A = \frac{+Q}{2\pi T}\ln\frac{R}{r_w} + \frac{-Q}{2\pi T}\ln\frac{R}{2a - r_w} \tag{23.1}$$

where

Q	is the flow rate at the wells,
R	is the radius of well influence,
T	is the aquifer transmissivity,
r_w	is the well radius,
$2a - r_w$	is the distance between point A and the image well.

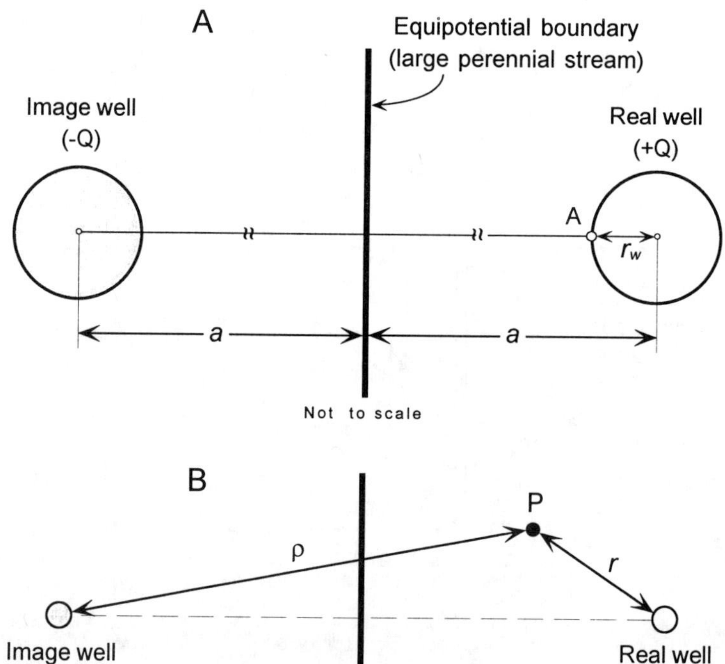

Figure 23.3 Scheme for calculating the drawdown at the extraction well (A) and at any point in the aquifer (B) influenced by the equipotential boundary.

(Note that the actual recorded drawdown in the real well will be larger because of the well loss which must be determined by testing -- see Problem 18).

Equation (23.1) can be written as:

$$s_A = \frac{Q}{2\pi T} \ln \frac{2a - r_w}{r_w} \tag{23.2}$$

Since the well radius (r_w) is much smaller than twice the distance between the wells ($2a$), it can be ignored and equation (23.2) becomes

$$s_A = \frac{Q}{2\pi T} \ln \frac{2a}{r_w} \tag{23.3}$$

When equation (23.3) is compared with the general equation of groundwater flow toward a fully penetrating well in a confined aquifer of infinite extent:

$$s = \frac{Q}{2\pi T} \ln \frac{R_D}{r_w}$$

where

R_D is "Dupuit's radius of well influence" (*i.e.*, calculated but not the real radius of influence),

it becomes apparent that the hydraulically equivalent radius of influence of the well near the equipotential boundary is:

$$R_D = 2a \tag{23.4}$$

where a is the distance between the real well and the equipotential boundary.

Drawdown in the Aquifer

As already mentioned, the drawdown at any point in the aquifer is the algebraic sum of the drawdown from the pumping well and the buildup (negative drawdown) from the injection well (see Figure 23.3):

$$s_P = \frac{+Q}{2\pi T} \ln \frac{R}{r} + \frac{-Q}{2\pi T} \ln \frac{R}{\rho} \tag{23.5}$$

where

r is the distance between the point and the real well,
ρ is the distance between the point and the image well.

Equation (23.5) can also be written as:

$$S_P = \frac{Q}{2\pi T} \ln \frac{\rho}{r} \tag{23.6}$$

As can be seen, the drawdown at any point in the aquifer is dependent on the distances between that point and the two wells (one real and one image). It also means that, in order to find the drawdown (*i.e.*, the hydraulic head), one does not need to know the real radius of well influence.

23.2 IMPERMEABLE BOUNDARY

An impermeable boundary near a discharging well is simulated by another discharging well placed on the other side of the boundary. This new well is an exact image of the real well and has the same pumping rate. It is placed at the same distance from the boundary as the real well (see Figure 23.4). The two wells are on the same line perpendicular to the boundary and operate simultaneously. The resulting drawdown in the aquifer is the algebraic sum of their individual drawdowns. The formation of a groundwater divide along the boundary, due to the operation of two extracting wells, satisfies the hydraulic condition that there is no flow across the impermeable boundary. Note that the actual drawdown along the boundary line is two times larger than it would be if the real well was operating in an infinite aquifer without the boundary.

Drawdown at the Real Well

The drawdown at the real well is calculated for a point placed at the well perimeter as in the case of the equipotential boundary. It is the algebraic sum of the drawdown caused by groundwater withdrawal at two extraction wells - the real and the image wells. This is expressed with the following equation:

$$s_A = \frac{+Q}{2\pi T}\ln\frac{R}{r_w} + \frac{+Q}{2\pi T}\ln\frac{R}{2a - r_w} \tag{23.7}$$

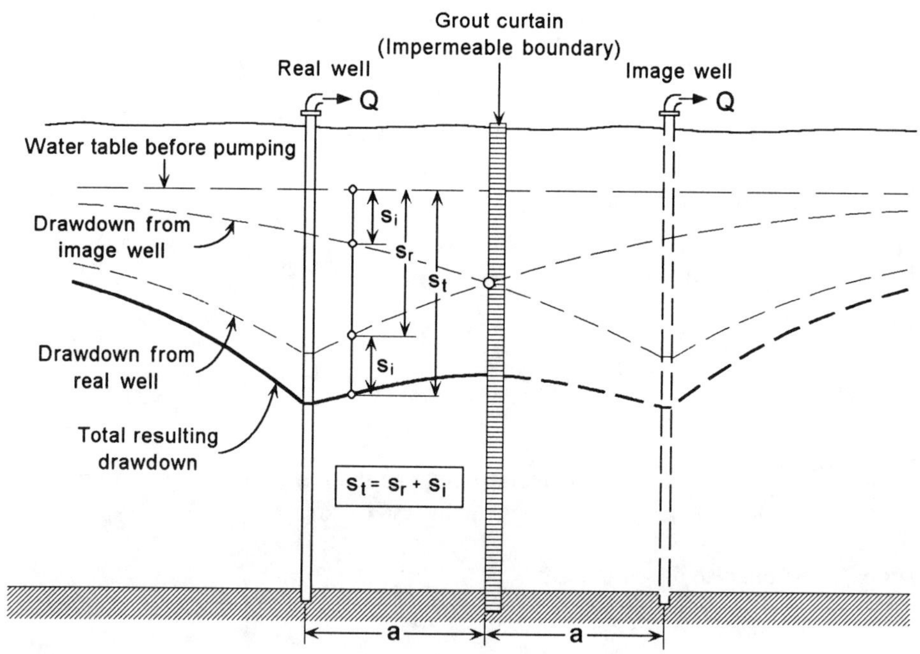

Figure 23.4 Schematic cross-sectional presentation of the resulting drawdown when an impermeable boundary near the extraction well is simulated with the imaginary extraction well of the same flow rate. Note that the water table between the real well and the boundary is much flatter.

where

Q	is the flow rate at the wells,
R	is the radius of well influence,
T	is the aquifer transmissivity,
r_w	is the well radius,
$2a - r_w$	is the distance between point A and the image well.

(Again, note that the actual recorded drawdown in the real well will be larger because of the well loss which must be determined by testing.)

Equation (23.7) can be written as:

$$s_A = \frac{Q}{2\pi T} \ln \frac{R^2}{2ar_w - r_w^2} \tag{23.8}$$

Since the squared well radius (r_w^2) is much smaller than the product $2ar_w$, it can be ignored and equation (23.8) becomes

$$s_A = \frac{Q}{2\pi T} \ln \frac{R^2}{2ar_w} \tag{23.9}$$

or

$$s_A = \frac{Q}{2\pi T} \ln \frac{\frac{R^2}{2a}}{r_w} \tag{23.10}$$

When equation (23.10) is compared with the general equation of groundwater flow toward a fully penetrating well in a confined aquifer of infinite extent:

$$s = \frac{Q}{2\pi T} \ln \frac{R_D}{r_w}$$

where

R_D	is "Dupuit's radius of well influence" (*i.e.*, the calculated but not the real radius of influence),

it becomes apparent that hydraulically equivalent radius of influence of the well near the impermeable boundary is:

$$R_D = \frac{R^2}{2a} \tag{23.11}$$

where a is the distance between the real well and the impermeable boundary and R is the real radius of influence of a single well in an infinite aquifer.

As can be seen, in order to find the drawdown at the well one must know the real radius of well influence (unlike in case of the equipotential boundary where R_D=$2a$). Since the real radius of well influence in an infinite aquifer depends on time, it follows that flow toward a well near an impermeable boundary has to be analyzed as transient (time-dependent).

Drawdown in the Aquifer

As previously stated, the drawdown at any point in the aquifer is the algebraic sum of the individual drawdowns from two pumping wells - the real and the image wells:

$$s_P = \frac{+Q}{2\pi T}\ln\frac{R}{r} + \frac{+Q}{2\pi T}\ln\frac{R}{\rho} \tag{23.12}$$

where

r	is the distance between the point and the real well,
ρ	is the distance between the point and the image well.

Equation (23.13) can also be written as:

$$S_P = \frac{Q}{2\pi T}\ln\frac{R^2}{r \cdot \rho} \tag{23.13}$$

As can be seen, the drawdown at any point in the aquifer depends upon the distances between that point and the two wells (one real and one image), as well as on the real, time-dependent radius of well influence.

23.3 MULTIPLE BOUNDARIES

The principle of hydraulic simulation of two or more boundaries that intersect, or are parallel, remains the same as in the case of a single boundary: the boundaries are replaced with image wells whose character depends upon the boundary type. An image well over an impermeable boundary has the flow rate with the same sign as the real well (e.g., the image of an extraction well is also an extraction well, and the image of a recharge well is also a recharge well). When the imaging is done for an equipotential boundary, the well flow rate changes the sign. For both types of boundaries the image well and the real well are on the same line perpendicular to the boundary and at the same distance from the boundary.

Figure 23.5 illustrates, with map views, some of the possible relationships between aquifer boundaries. It is important to remember that, when boundaries intersect, they have to be extended beyond their actual physical limits. The image well(s) from the real well are then imaged over these extended boundaries as well as over the real boundaries creating "images of the images". This process should continue until the loop is closed, *i.e.*, until there is no room for another image (see Figure 23.5E and F).

In the case of parallel boundaries, every new image well is imaged back over the next "available" boundary (*i.e.*, the boundary that was not used for its imaging -- see Figure 23.5B and D) thus creating an infinite number of image wells. In practice, it will be enough to create several images since the solution quickly converges. If a more accurate analysis is needed, it is recommended to draw a graph drawdown vs. number of wells included in the calculation. This will provide a convenient way to choose the right number of wells for a given level of accuracy. The graph is particularly useful in the case of one equipotential and one impermeable parallel boundary since the solution oscillates along a straight line.

For multiple boundaries, the drawdown at any point in the aquifer is calculated by adding drawdown components from each real and image well which are considered simply as a group of wells (see Problem 22). Again, it is very important to remember that the individual drawdowns can be positive (from extraction wells) or negative (from recharge wells).

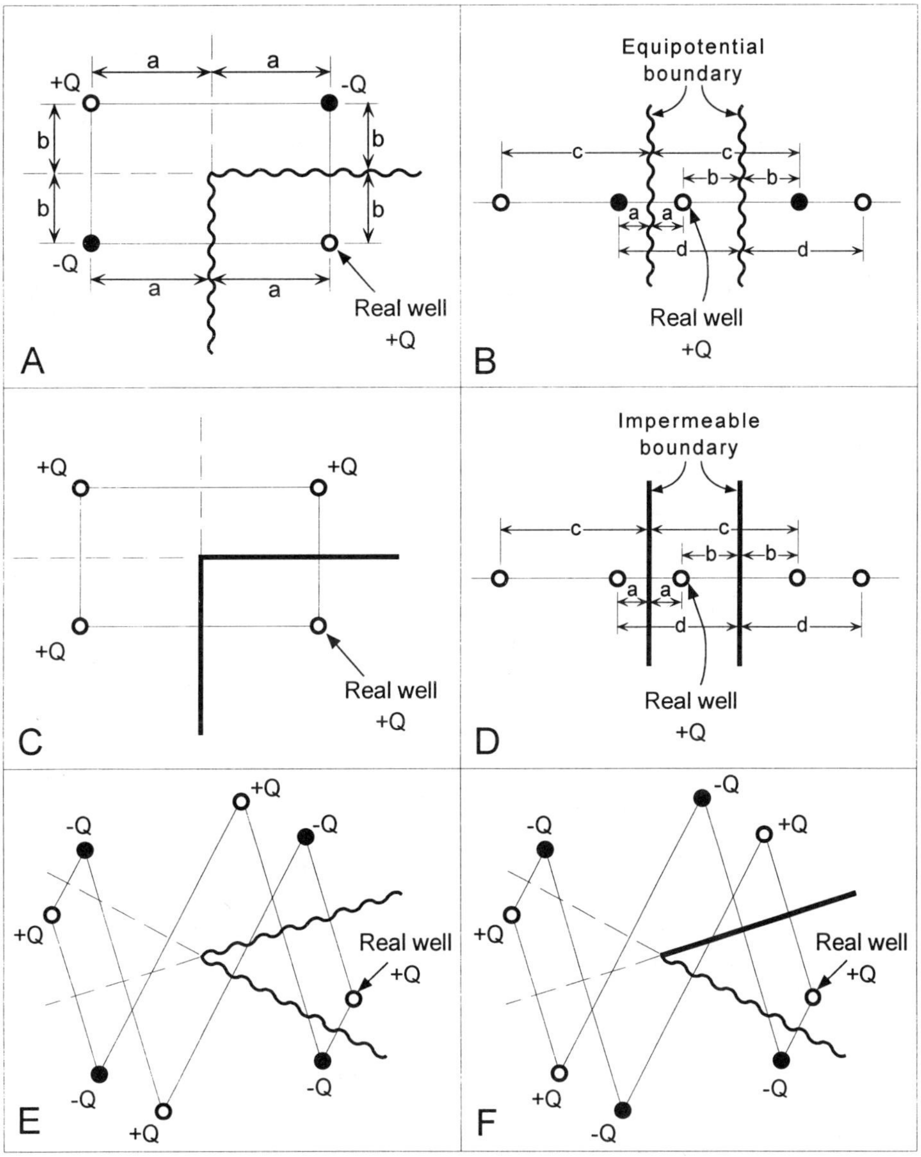

Figure 23.5 Map views for various combinations of aquifer boundaries. The extraction wells are shown with empty circles and (+Q), and the recharge wells are shown with filled circles and (–Q). A: Two equipotential boundaries intersecting at a right angle. B: Two parallel equipotential boundaries (note that the number of image wells is theoretically infinite). C: Two impermeable boundaries intersecting at a right angle. D: Two parallel impermeable boundaries. E: Two equipotential boundaries intersecting at a sharp angle. F: One equipotential and one impermeable boundary intersecting at a sharp angle.

23.4 SOLUTION FOR THIS PROBLEM

Figure 23.6 shows the real extraction well (RW) and its images which simulate the aquifer boundaries. The image wells IW1 and IW2 are created over the impermeable boundary and are also extracting water from the aquifer. The image well IW3 is a recharge well, *i.e.*, with the opposite flow rate sign than the real well because it is created over the equipotential boundary. Well IW4 is the image of the image well IW3 and is also a recharge well -- it is created over the extension of the impermeable boundary thus keeping its sign.

Drawdown at piezometer P1 is the algebraic sum of the individual drawdowns caused by the real well and four image wells shown in Figure 23.6:

$$s_P = s_r + s_{i1} + s_{i2} + s_{i3} + s_{i4} \tag{23.14}$$

or

$$s_P = \frac{Q}{2\pi T}\ln\frac{R}{r_r} + \frac{(+Q)}{2\pi T}\ln\frac{R}{r_{i1}} + \frac{(+Q)}{2\pi T}\ln\frac{R}{r_{i2}} + \frac{(-Q)}{2\pi T}\ln\frac{R}{r_{i3}} + \frac{(-Q)}{2\pi T}\ln\frac{R}{r_{i4}} \tag{23.15}$$

where the distances between P1 and the wells (r) are shown in Figure 23.7.

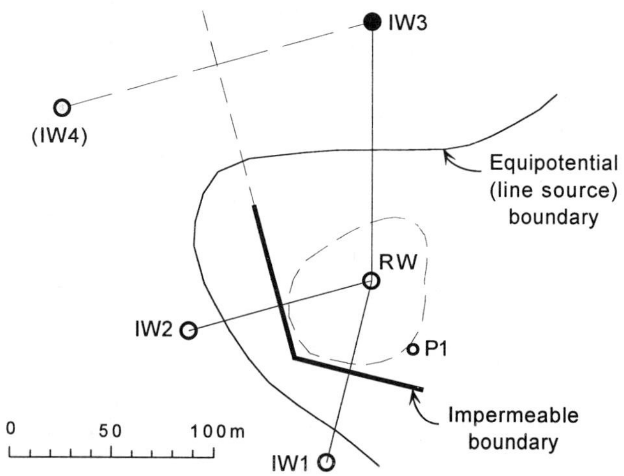

Figure 23.6 The real extraction well (RW) and its images (IW). Note that the recharge image well IW3 has its own image well IW4 created over the extended portion of the impermeable boundary (grout curtain).

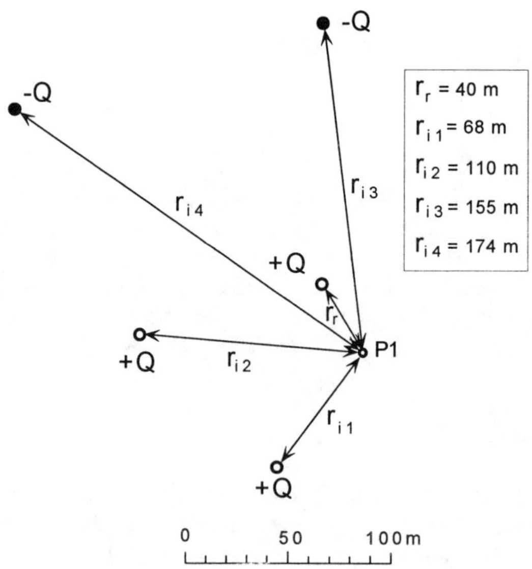

Figure 23.7 Calculation scheme for the problem.

Equation (23.15) can be written as:

$$s_P = \frac{Q}{2\pi T}\ln\frac{R^3}{r_r \cdot r_{i1} \cdot r_{i2}} - \frac{Q}{2\pi T}\ln\frac{R^2}{r_{i3} \cdot r_{i4}} \tag{23.16}$$

or, after further simplifying it:

$$s_P = \frac{Q}{2\pi T}\ln\frac{R \cdot r_{i3} \cdot r_{i4}}{r_r \cdot r_{i1} \cdot r_{i2}} \tag{23.17}$$

Solved for Q (the pumping rate needed to create the drawdown of 5 meters at P1) it is:

$$Q = \frac{s_P 2\pi T}{\ln\frac{R \cdot r_{i3} \cdot r_{i4}}{r_r \cdot r_{i1} \cdot r_{i2}}} \tag{23.18}$$

or, after inserting all known values:

$$Q = \frac{5m \cdot 2 \cdot 3.14 \cdot 5.04 \times 10^{-4}\, m^2 / s}{\ln\frac{75m \cdot 155m \cdot 174m}{40m \cdot 68m \cdot 110m}}$$

$$Q = 8.28 \times 10^{-3}\, m^3 / s = 8.3\ l / s$$

Chapter 8

WATER WELLS, TRANSIENT FLOW

PROBLEM 24
CONFINED, NON-LEAKY AQUIFER, PUMPING TESTS

A 24-hour well pumping test in a confined aquifer was conducted in order to determine its hydrogeologic parameters. Use information from the schematic hydrogeologic cross-section in Figure 24.1 and pumping data from Table 24.1 to calculate aquifer transmissivity, hydraulic conductivity and storage coefficient. Apply several methods of pumping test analysis. Also determine well loss and its acceptability (*i.e.*, efficiency of the well).

The pumping rate was kept constant at 8 l/s (0.008 m^3/s) during all 24 hours of the test. The drilling diameter of the well screen portion is 350 mm.

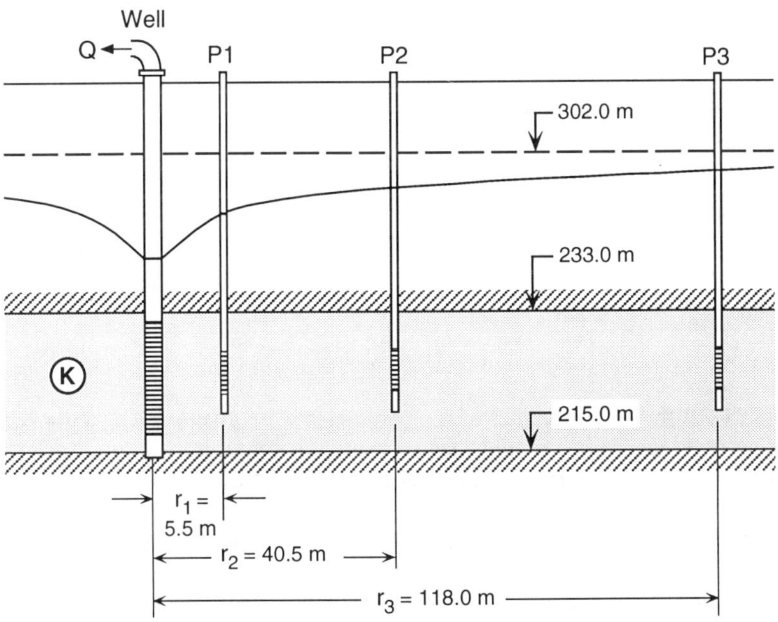

Figure 24.1 Schematic hydrogeologic cross-section of the well pumping test conditions. Drawdown data measured in the pumping well and piezometers are given in Table 24.1.

24.1 THE THEIS METHOD (CURVE MATCHING METHOD)

The Theis equation, which describes transient (non-equilibrium) groundwater flow toward a fully penetrating well in the confined aquifer, is the basis for practically all methods of pumping test analysis. Using the equation, transmissivity and storage can be determined from the drawdown measurements without having to wait for the stabilization of pumping water levels as in the case of steady state methods. In addition, only one observation well (piezometer), or sometimes only the pumping well itself, is enough to determine aquifer hydrogeologic parameters as opposed to steady state calculations where at least two observation wells are needed.

The Theis equation gives drawdown (s) at any time after the beginning of pumping:

$$s = \frac{Q}{4\pi T} W(u) \tag{24.1}$$

where Q is pumping rate kept constant during the test, T is transmissivity, and $W(u)$ is called *well function of u*, also known as *The Theis function*, or simply *well function*. Dimensionless parameter u is given as:

$$u = \frac{r^2 S}{4Tt} \tag{24.2}$$

where r is the distance from the pumping well where the drawdown is recorded, S is the storage coefficient, and t is the time since the beginning of pumping.

The well function of u is the exponential integral

$$W(u) = -E(-i) = -\int_u^\infty \frac{e^{-u}}{u} du \tag{24.3}$$

which can also be expressed with the following series:

$$W(u) = -0.5772 - \ln u + u - \frac{u^2}{2 \cdot 2!} + \frac{u^3}{3 \cdot 3!} - \frac{u^4}{4 \cdot 4!} + \ldots \tag{24.4}$$

or as

$$W(u) = \ln \frac{0.5615}{u} + \sum_{n=1}^{n=\infty} (-1)^{n+1} \frac{u^n}{n \cdot n!} \tag{24.5}$$

Values of $W(u)$ for various values of the parameter u are given in Appendix E and can be readily found elsewhere in groundwater literature. The graph in Figure 24.2, which is prepared as a part of the Theis method, also shows this relationship in form of $W(u)$ versus $1/u$.

Theis derived his equation based on quite a few assumptions and it is very important to understand its limitations. If the aquifer tested and the test conditions significantly deviate from these assumptions (which, in fact, is very often the case in reality), other methods of analysis using appropriate modifications of the Theis equation should be used. The equation (24.1) is valid for the following conditions (adapted from Driscoll, 1986, and Domenico and Schwartz, 1990):

- The aquifer is homogeneous and isotropic.
- The aquifer is uniform in thickness and the pumping never affects its exterior boundary (the aquifer can be considered infinite in areal extent).
- The aquifer is confined and it does not receive any recharge.
- Well discharge is derived entirely from aquifer storage.
- The water removed from storage is discharged instantaneously when the head is lowered.
- The pumping rate is constant.

Table 24.1 Results of the pumping test

Time Since Test Started		Drawdown in Meters			
Minutes	Seconds	P1	P2	P3	Well
1	60	1.875	0.412	0.041	6.407
2	120	2.288	0.684	0.105	6.831
3	180	2.495	0.862	0.175	7.038
4	240	2.651	0.992	0.241	7.197
5	300	2.757	1.092	0.302	7.302
6	360	2.839	1.172	0.357	7.384
7	420	2.906	1.240	0.407	7.451
8	480	2.965	1.298	0.451	7.509
9	540	3.016	1.348	0.492	7.560
10	600	3.060	1.394	0.530	7.605
12	720	3.146	1.471	0.594	7.691
14	840	3.212	1.535	0.650	7.757
16	960	3.267	1.592	0.700	7.812
18	1080	3.316	1.641	0.745	7.861
20	1200	3.360	1.686	0.786	7.904
25	1500	3.436	1.771	0.870	7.979
30	1800	3.534	1.854	0.940	8.078
35	2100	3.579	1.913	1.002	8.121
40	2400	3.651	1.974	1.005	8.194
45	2700	3.702	2.024	1.104	8.247
50	3000	3.730	2.064	1.147	8.273
55	3300	3.782	2.108	1.186	8.325
60	3600	3.821	2.145	1.222	8.364
70	4200	3.868	2.202	1.283	8.410
80	4800	3.945	2.268	1.338	8.489
90	5400	3.975	2.309	1.387	8.517
100	6000	4.036	2.361	1.431	8.579
110	6600	4.076	2.401	1.471	8.621
120	7200	4.115	2.437	1.507	8.658
150	9000	4.175	2.513	1.593	8.717
180	10800	4.254	2.590	1.667	8.795
210	12600	4.319	2.655	1.731	8.861
240	14400	4.408	2.730	1.792	8.950
270	16200	4.429	2.765	1.839	8.971
300	18000	4.498	2.824	1.887	9.040
330	19800	4.537	2.862	1.926	9.080
360	21600	4.574	2.989	1.963	9.117
390	24400	4.607	2.931	1.996	9.151
420	25200	4.637	2.961	2.027	9.181
450	27000	4.666	2.989	2.056	9.210
480	28800	4.692	3.016	2.083	9.245
540	32400	4.718	3.054	2.128	9.259
600	36000	4.791	3.177	2.178	9.332
660	39600	4.822	3.151	2.217	9.364
720	43200	4.861	3.188	2.253	9.403
780	46800	4.894	3.220	2.285	9.437
840	50400	4.924	3.250	2.315	9.467
900	54000	4.952	3.277	2.343	9.495
960	57600	4.977	3.303	2.370	9.521
1020	61200	5.002	3.328	2.394	9.545
1080	64800	5.024	3.351	2.418	9.568
1140	68400	5.046	3.372	2.440	9.590
1200	72000	5.066	3.393	2.461	9.610
1260	75600	5.085	3.412	2.481	9.630
1320	79200	5.105	3.434	2.501	9.647
1380	82800	5.124	3.453	2.520	9.666

- The radius of the well is infinitesimally small, *i.e.*, the storage in the well can be ignored.
- The pumping well is fully penetrating: it receives water from the entire thickness of the aquifer and is 100 % efficient (there are no well losses).
- The initial potentiometric surface (before pumping) is horizontal.

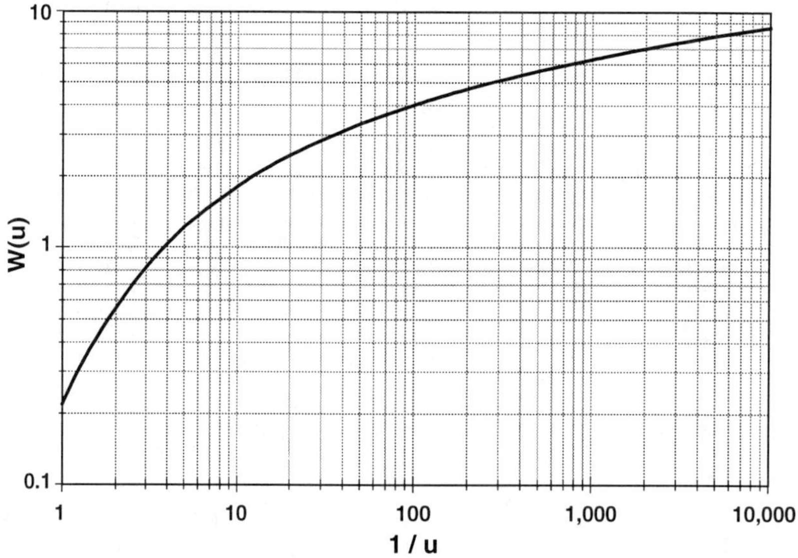

Figure 24.2 Theoretical curve ("type curve") of $W(u)$ versus $1/u$ used for the curve matching method of well pumping test analysis (the Theis method). (Vertical scale slightly exaggerated.)

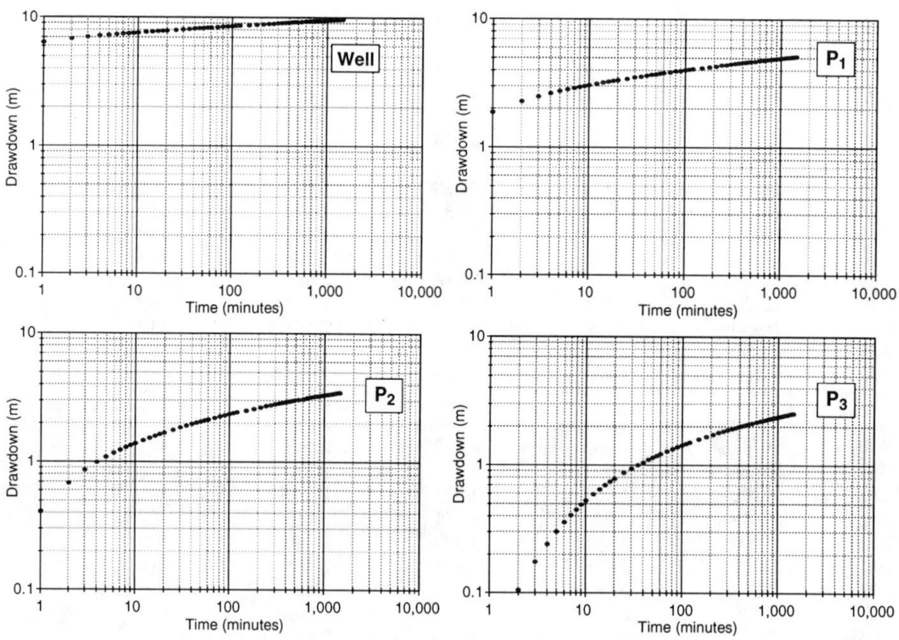

Figure 24.3 Field data of drawdown versus time for the pumping well and piezometers P1, P2 and P3 plotted on a log-log graph paper.

Theis equation can be used to determine aquifer transmissivity and storage if frequent measurements of drawdown versus time are performed in one or more observation wells. Equation (24.1) has no explicit solution and Theis introduced a graphical method which gives T and S if other terms are known. Theoretical curve *W(u)* versus *1/u* is plotted on a graph paper with logarithmic scales as shown in Figure 24.2. Field data of *drawdown (s)* versus *time (t)* are plotted separately for each observation well (Figure 24.3). Field curve for one observation well at a time is superimposed on the theoretical curve, also called *type curve. It is essential that both graphs/curves have identical logarithmic scales and cycles as shown in Figures 24.4 through 24.6.* Keeping coordinate axes of the curves *parallel*, field data are matched to the type curve.

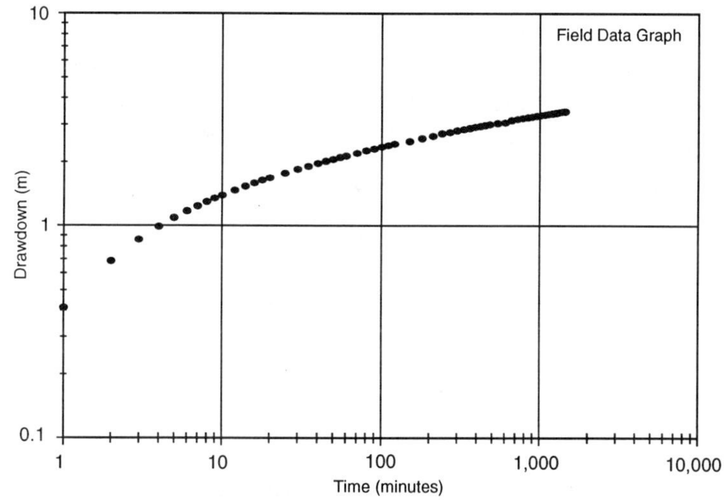

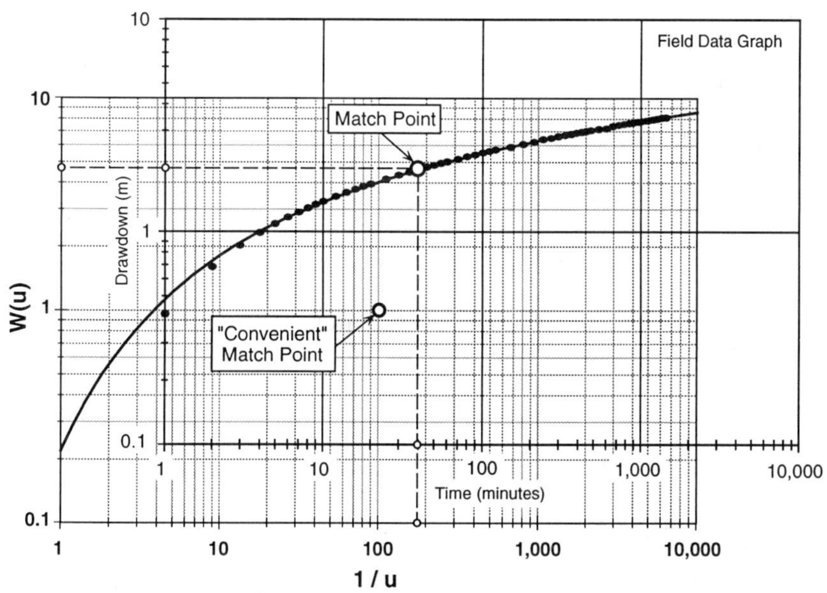

Figure 24.4 Top: Field data of drawdown versus time for the piezometer P2. **Bottom**: Curve matching for the piezometer P2. The coordinates of the match point are: *W(u)*=4.75, *1/u*=180, *s*=2.0 m, *t*=39 minutes=2440 seconds.

Once a satisfactory match is found, a *match point* on the overlapping graphs is selected. The match point is defined by four coordinates, the values of which are read on two graphs: *W(u)* and *1/u* on the type curve graph, *s* and *t* on the field graph. The match point can be any point on the overlapping graphs, *i.e.*, it does not have to be on the matching curves. Figure 24.4 shows two possibilities of choosing the match point: one is on the theoretical curve/measured data (which looks somewhat more logical), and one is chosen outside to obtain convenient values of *W(u)* and *1/u*: 1 and 100 respectively. Both matching points should lead to the same values of *T* and *S* if the coordinates are read carefully.

The transmissivity is calculated using equation (24.1) and the match point coordinates *s* and *W(u)*:

$$T = \frac{Q}{4\pi s} W(u) \tag{24.6}$$

The storage coefficient is calculated using equation (24.2), the match point coordinates *1/u* and *t*, and the transmissivity value already determined:

$$S = \frac{4Ttu}{r^2} \tag{24.7}$$

Calculation for piezometer P2

The match point coordinates are (see Figure 24.4):

$W(u) = 4.75$ $\quad 1/u = 180$ $\quad u = 0.0055$ $\quad s = 2.0\ m$ $\quad t = 39\ minutes = 2440\ seconds$

The transmissivity is calculated using equation (24.6):

$$T = \frac{0.008\ m^3/s}{4 \cdot \pi \cdot 2.0\ m} \cdot 4.75 = 1.51 \times 10^{-3}\ m^2/s$$

The storage coefficient is calculated using equation (24.7):

$$S = \frac{4 \cdot 1.51 \times 10^{-3}\ m^2/s \cdot 2340\ sec \cdot 0.0055}{(40.5\ m)^2} = 4.7 \times 10^{-5}$$

The hydraulic conductivity (*K*) is calculated from the transmissivity (*T*) and the aquifer thickness (*b*):

$$K = \frac{T}{b} = \frac{1.51 \times 10^{-3}\ m^2/s}{18\ m} = 8.39 \times 10^{-5}\ m/s$$

Calculation for piezometer P3

The match point coordinates are (see Figure 24.5):

$W(u) = 2.35$ $\quad 1/u = 18.5$ $\quad u = 0.054$ $\quad s = 1.0\ m$ $\quad t = 37.5\ minutes = 2250$ sec

The transmissivity is calculated using equation (24. 6):

$$T = \frac{0.008\ m^3/s}{4 \cdot \pi \cdot 1.0\ m} \cdot 2.35 = 1.50 \times 10^{-3}\ m^2/s$$

The storage coefficient is calculated using equation (24.7):

$$S = \frac{4 \cdot 1.50 \times 10^{-3}\ m^2/s \cdot 2250\ sec \cdot 0.054}{(118\ m)^2} = 5.2 \times 10^{-5}$$

The hydraulic conductivity is calculated from the transmissivity and the aquifer thickness:

$$K = \frac{T}{b} = \frac{1.50 \times 10^{-3}\ m^2/s}{18\ m} = 8.33 \times 10^{-5}\ m/s$$

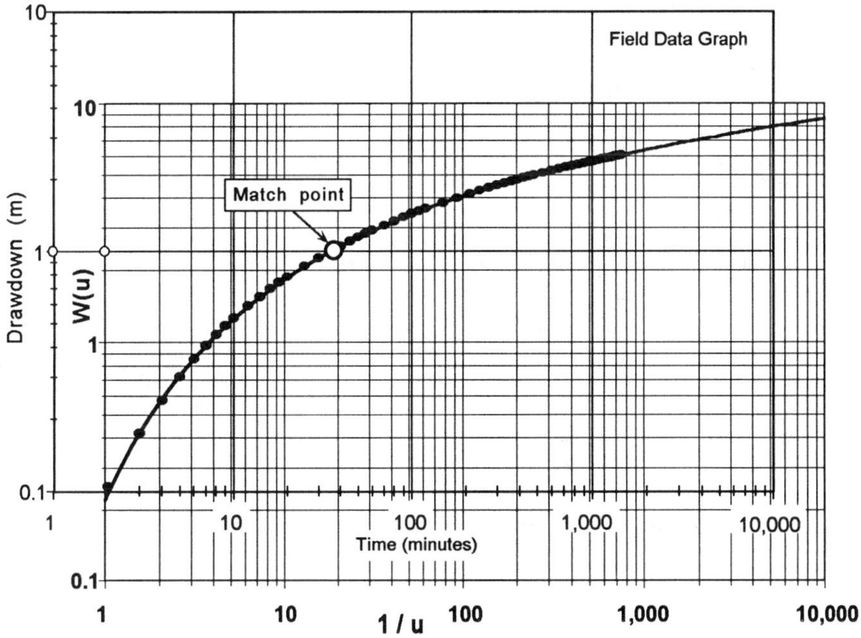

Figure 24.5 Curve matching for the piezometer P3. Coordinates of the match point are: *W(u)*=2.35, *1/u*=18.5, *s*=1.0 m, *t*=37.5 minutes = 2250 seconds.

Calculation for piezometer P1

The match point coordinates (Figure 24.6):

$W(u) = 8.0 \quad 1/u = 6000 \quad u = 1.66x10^{-4} \quad s = 3.7\ m \quad t = 48\ minutes = 2880\ seconds$

The transmissivity is calculated using equation (24.6):

$$T = \frac{0.008\ m^3/s}{4 \cdot \pi \cdot 3.7\ m} \cdot 8.0 = 1.38 \times 10^{-3}\ m^2/s$$

The storage coefficient is calculated using equation (24.7):

$$S = \frac{4 \cdot 1.38 \times 10^{-3}\ m^2/s \cdot 2880\ sec \cdot 1.66 \times 10^{-4}}{(5.5\ m)^2} = 8.7 \times 10^{-5}$$

The hydraulic conductivity is calculated from the transmissivity and the aquifer thickness:

$$K = \frac{T}{b} = \frac{1.38 \times 10^{-3}\ m^2/s}{18\ m} = 7.67 \times 10^{-5}\ m/s$$

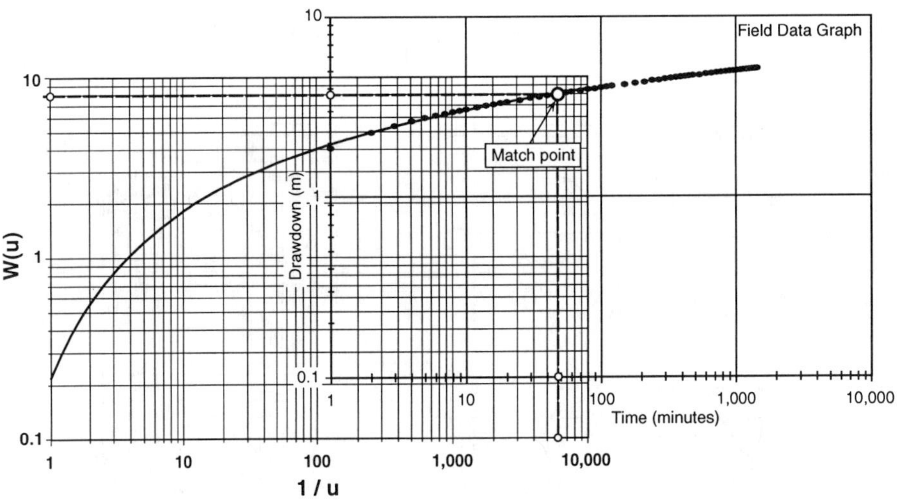

Figure 24.6 Curve matching for the piezometer P1. The coordinates of the match point are: $W(u)$=8.0, $1/u$=6000, s=3.7 m, t=48 minutes=2880 seconds.

The transmissivity value calculated from the P1 data is somewhat different because the recorded data are much less curved than for P2 and P3 and matching the theoretical Theis curve was therefore more arbitrary. The Theis curve in Figure 24.3 could have been extended to cover higher values of W. This addition would be a near-straight line and the resulting matching might have been more accurate. The pumping well data were not used for matching because they fall entirely along a near-straight line (Figure 24.3) that could not be matched to the theoretical curve shown in Figure 24.2.

If field data of a well pumping test cannot be matched to the theoretical Theis curve because of their "odd" shapes, and the test has been properly conducted, the reason may be one or more of the following:

- the aquifer is not confined,
- the aquifer is heterogeneous and/or anisotropic,
- there is a source of recharge/discharge,
- there is one or more hydraulic boundaries.

In such cases a hydrogeological assessment of the possible causes should be made and the pumping test data should be analyzed with an appropriate method (see the following Problems).

24.2 THE COOPER-JACOB METHOD (DRAWDOWN VERSUS TIME)

Theis equation can be modified, as pointed out by Cooper and Jacob (1946), to simplify calculations and determination of the aquifer parameters using well pumping data. For small values of parameter u, *i.e.*, sufficiently large values of pumping time t, and/or small values of r (*i.e.*, when a monitoring well is close to the pumping well), the well function $W(u)$ can be approximated by the first two members of the series in equation (24.4):

$$W(u) = -0.5772 - \ln u = \ln 0.5615 - \ln u \tag{24.8}$$

$$W(u) = \ln \frac{0.5615}{u} \tag{24.9}$$

For values of u equal to 0.05, equation (24.9) approximates the well function with an error of 2% (errors for u=0.03 and u=0.07 are 1% and 3.2% respectively). When u is equal or less than 0.05, the well function expressed in the following form is considered acceptable for practical purposes:

$$W(u) = \ln \frac{2.25Tt}{r^2 S} \tag{24.10}$$

and the Theis equation (24.1) becomes:

$$s = \frac{Q}{4\pi T} \ln \frac{2.25Tt}{r^2 S} \tag{24.11}$$

or

$$s = \frac{0.183Q}{T} \log \frac{2.25Tt}{r^2 S} \tag{24.11a}$$

when the logarithm to the base e or "*natural logarithm*" (ln) is replaced by the logarithm to the base 10 (log). The logarithmic form of Theis equation is much easier to work with and enables the usage of semi-logarithmic graphs for well pumping test analyses.

If well discharge (Q) is kept constant, and it is assumed that transmissivity (T) and storage coefficient (S) do not change in time, equation (24.11) can be rewritten to separate these constant terms from the logarithm of time (t) which is the only variable on the right-hand side:

$$s = \frac{Q}{4\pi T} \ln \frac{2.25T}{r^2 S} + \frac{Q}{4\pi T} \ln t \tag{24.12}$$

Simplification of equation (24.12), *i.e.*, replacement of all constants with new terms a and b, gives the following function:

$$s = a + b \ln t \tag{24.13}$$

Equation (24.13) shows that, if the condition that $u < 0.05$ is satisfied, recorded data of drawdown versus time would form a straight line when plotted on semi-logarithmic graph paper. As shown in Figure 24.7, this is the case with our data for both the pumping well and the piezometers (Table 24.1). There is a deviation from the straight line for P3 for the initial period of pumping which indicates that the parameter u is still greater than 0.05. This is because piezometer P3 is relatively far from the pumping well (r=118 m) compared to P1 and P2 and more time is needed for the condition regarding parameter u to be satisfied. This time can be determined using the following equation:

$$t = \frac{r^2 S}{4Tu} \tag{24.14}$$

$$t = \frac{(118m)^2 \cdot 5.2 \times 10^{-5}}{4 \cdot 1.5 \times 10^{-3} m^2 / s \cdot 0.05} = 2413 \ sec \approx 40 \ min$$

In other words, data recorded in P3 form a straight line on the graph drawdown (s) versus logarithm of time (t) after approximately 40 minutes of pumping (see Figure 24.7).

Drawdown and time coordinates of any two points on a straight line drawn through the recorded (field) data are used to determine aquifer *transmissivity* as shown with the following equations:

$$s_1 = \frac{Q}{4\pi T} \ln \frac{2.25T}{r^2 S} + \frac{Q}{4\pi T} \ln t_1 \tag{24.15}$$

$$s_2 = \frac{Q}{4\pi T} \ln \frac{2.25T}{r^2 S} + \frac{Q}{4\pi T} \ln t_2 \tag{24.16}$$

$$s_2 - s_1 = \Delta s = \frac{Q}{4\pi T} (\ln t_2 - \ln t_1) \tag{24.17}$$

Finally, the transmissivity is given as:

$$T = \frac{Q}{4\pi \Delta s} \ln \frac{t_2}{t_1} \tag{24.18}$$

For practical purposes it is better to express transmissivity in terms of logarithm to the base 10 since the corresponding graph papers are easier to use:

$$T = \frac{0.183 \cdot Q}{\Delta s} \log \frac{t_2}{t_1} \tag{24.18a}$$

If two points on the straight line are chosen to be one log cycle apart (*i.e.*, ratio of time coordinates t_2 and t_1 is 10), equation (24.18a) reduces to:

$$T = \frac{0.183 \cdot Q}{\Delta s} \tag{24.19}$$

The straight lines representing recorded data in our example are all parallel, *i.e.*, with the same slope (Δs) which confirms the assumption that the aquifer is homogeneous and isotropic. Therefore any of the piezometers or the pumping well itself can be used to determine aquifer transmissivity since the result would be the same. If, however, the lines were with significantly different or varying slopes it would indicate that the aquifer may be heterogeneous or with some other disturbance (e.g., source of recharge/discharge, a boundary, etc.).

Calculation for piezometer P1

Knowing the pumping rate value (0.008 m^3/s) and finding from the graph in Figure 24.7 that Δs=0.92m for the log cycle 10–100, the *transmissivity* is calculated using equation (24.19):

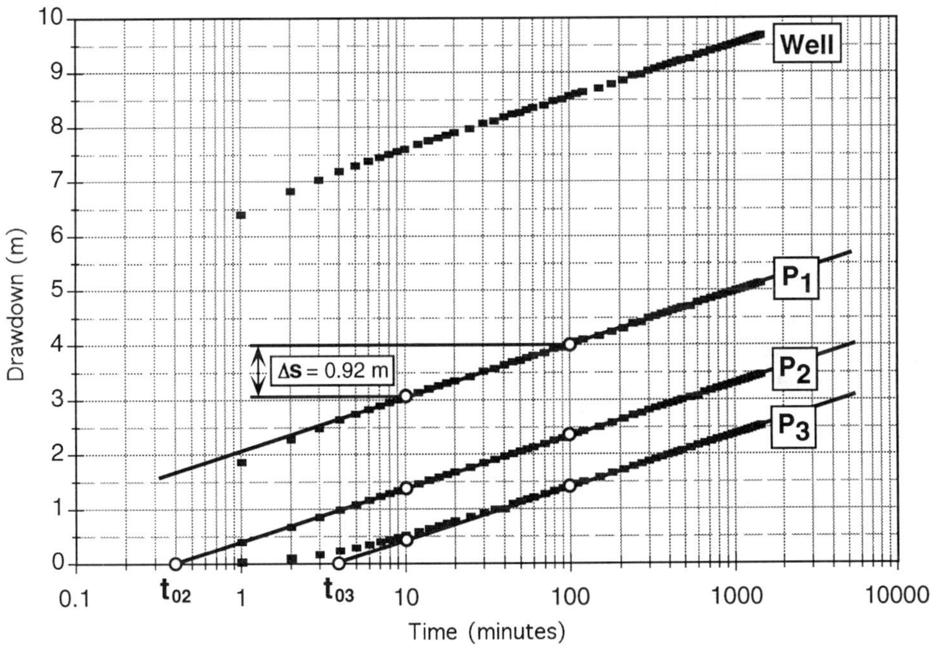

Figure 24.7 Semilog graph drawdown versus time for the pumping well and three piezometers. The graph is used for the calculation of aquifer transmissivity and storage coefficient by the Cooper-Jacob method (straight line method).

$$T = \frac{0.183 \cdot 8.0 \times 10^{-3}\ m^3/s}{0.92\ m} = 1.59 \times 10^{-3}\ m^2/s$$

In addition to transmissivity, the Cooper-Jacob method is used to determine the *storage coefficient* of the aquifer. This is done graphically using the same graph shown in Figure 24.7. The straight line for any of the piezometers is extended to intercept the zero-drawdown. The time of the intercept (t_0) is related to the zero drawdown through the following equation:

$$s = 0 = \frac{0.183Q}{T}\log\frac{2.25Tt_0}{r^2S} \tag{24.20}$$

For equation (24.20) to be valid the term under the logarithm must be equal to 1:

$$\frac{2.25Tt_0}{r^2S} = 1 \tag{24.21}$$

Knowing that $log1 = 0$, *the storage coefficient* is then expressed as:

$$S = \frac{2.25Tt_0}{r^2} \tag{24.22}$$

Calculation for piezometer P2

The storage coefficient is calculated by substituting the already determined transmissivity value ($T = 1.59 \text{x} 10^{-3}$ m²/s) and the time of the zero-drawdown intercept (t_{02} = 0.4 minutes = 24 seconds, see Figure 24.7) into equation (24.22):

$$S = \frac{2.25 \cdot 1.59 \times 10^{-3}\ m^2 / s \cdot 24 \text{ sec}}{(40.5\ m)^2}$$

$$S = 0.0000523 = 5.2 \times 10^{-5}$$

Calculation for piezometer P3

The zero-drawdown intercept is t_{03}=3.9 minutes or 244 seconds and the distance between the pumping well and P3 is r=118 meters:

$$S = \frac{2.25 \cdot 1.59 \times 10^{-3}\ m^2 / s \cdot 234 \text{ sec}}{(118\ m)^2}$$

$$S = 6 \times 10^{-5}$$

Only the data for the piezometers can be used to determine the storage coefficient because the actual drawdown measured in the pumping well includes a well loss. This additional drawdown is not known beforehand and neither is the theoretical position of the pumping well data (without the well loss). Therefore the zero-drawdown intercept of the straight line for the pumping well data is not the theoretical value of t_0 from equations (24.20) to (24.22) and would lead to an erroneous result.

24.3 DRAWDOWN VERSUS RATIO OF TIME AND SQUARE DISTANCE

This method is an obligatory part of every well pumping test analysis because, in addition to aquifer transmissivity and storage coefficient, it allows the determination of well loss and efficiency. It is also based on the application of the modified Theis equation (24.11) introduced by Cooper and Jacob (1946):

$$s = \frac{Q}{4\pi T} \ln \frac{2.25Tt}{r^2S}$$

This equation can be rewritten in the following form:

$$s = \frac{Q}{4\pi T}\ln\frac{2.25T}{S} + \frac{Q}{4\pi T}\ln\frac{t}{r^2} \tag{24.23}$$

which, similarly to equation (24.12), separates the constant terms (*i.e.*, transmissivity and storage coefficient) from the t/r^2 ratio which now becomes a variable related to drawdown (s). By replacing all the constant terms with new parameters a and b, equation (24.24) is further simplified:

$$s = a + b\ln\frac{t}{r^2} \tag{24.24}$$

Equation (24.24) shows that, if the condition that $u < 0.05$ is satisfied, the recorded data of drawdown (s) versus t/r^2 ratio would form a straight line if plotted on a semi-logarithmic graph paper. Figure 24.8 shows separate plots of this function for all three piezometers and the pumping well. Again, except for the initial stage of pumping for P3, the recorded data form straight lines confirming the assumption about the parameter u.

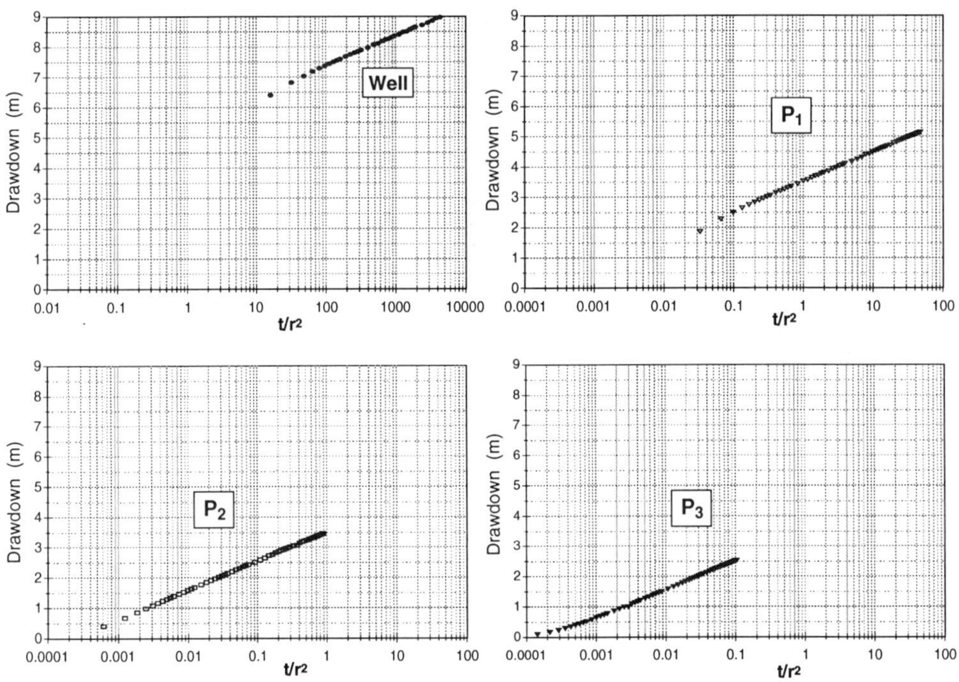

Figure 24.8 Semilog graphs drawdown versus t/r^2 for the pumping well and the three piezometers.

When plotted together, as shown in Figure 24.9, data for all piezometers form a single straight line indicating that, for practical purposes, the aquifer can be considered as homogeneous and isotropic, without recharge, and without boundaries influenced by the pumping.

Coordinates of any two points on the straight line can be used to determine aquifer transmissivity:

$$s_2 - s_1 = \Delta s = \frac{Q}{4\pi T}\left[\ln\left(\frac{t}{r^2}\right)_2 - \ln\left(\frac{t}{r^2}\right)_1\right] \tag{24.25}$$

Solving equation (24.25) for *transmissivity* and introducing logarithm to the base 10 gives:

$$T = \frac{0.183Q}{\Delta s}\left[\log\left(\frac{t}{r^2}\right)_2 - \log\left(\frac{t}{r^2}\right)_1\right] \tag{24.26}$$

If two points on the straight line are chosen to be one log cycle apart (*i.e.*, ratio of coordinates $(t/r^2)_2$ and $(t/r^2)_1$ is 10), equation (24.26) reduces to a simple-to-use form:

$$T = \frac{0.183Q}{\Delta s} \tag{24.27}$$

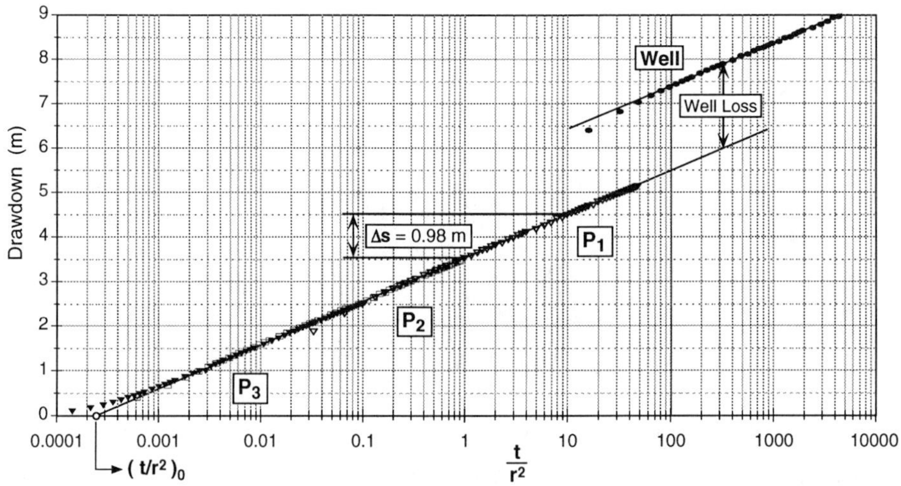

Figure 24.9 Semilog graph drawdown versus t/r^2 used for the determination of aquifer transmissivity, storage coefficient, and pumping well loss.

In our example (Figure 24.9), the difference in drawdown for the log cycle 1–10 is 0.98 m. The *transmissivity* is then easily calculated knowing that the pumping rate is 8 l/s (8×10^{-3} m^3/s):

$$T = \frac{0.183 \cdot 8 \times 10^{-3}\ m^3/s}{0.98\ m}$$

$$T = 0.001494\ m^2/s = 1.49 \times 10^{-3}\ m^2/s$$

The *storage coefficient* is, similarly to the Cooper-Jacob method, determined by substituting the previously calculated transmissivity and the zero-drawdown intercept of the straight line into equation (24.28).

$$S = 2.25T\left(\frac{t}{r^2}\right)_0 \qquad (24.28)$$

Time in the ratio (t/r^2) in Figure 24.9 is given in minutes (see Table 24.1) and the zero-drawdown intercept is 0.00025 min/m^2. When time is expressed in seconds (which is needed for the consistency of units since T is given in m^2/s), this value is 0.00025 min/m^2 x 60 = 0.015 s/m^2.

$$S = 2.25 \cdot 1.49 \times 10^{-3}\ m^2 / s \cdot 0.015\ s / m^2$$

$$S = 0.0000503 = 5 \times 10^{-5}$$

Well loss is the difference between the actual measured drawdown in the pumping well and the theoretical drawdown which is expressed by the Theis equation and is the result of groundwater flow through the aquifer undisturbed zone only. The additional drawdown, or well loss, which is always present in pumping wells, is created by a combination of various factors such as improper well development (drilling fluid is left in the formation, mud cake along the borehole walls is not removed, fines from the formation are not removed), and a poorly designed gravel pack and/or well screen.

Well loss is determined directly from the graph in Figure 24.9. It is the vertical distance, in meters of drawdown, between the line of piezometers and the pumping well line (if there were not any well losses, the well data would fall along the line of piezometers). In our case the well loss is 1.9 meters.

Well efficiency, expressed in percent, is the ratio between the theoretical drawdown and the actual drawdown measured in the well:

$$Well\ Efficiency = \frac{Theoretical\ Drawdown}{Measured\ Drawdown} \times 100\% \qquad (24.29)$$

Values of theoretical and measured drawdown, as depicted from the graph in Figure 24.9, are 6 m and 7.9 m respectively, and the well efficiency is:

$$Well\ Efficiency = \frac{6\ m}{7.9\ m} \times 100\% = 76\%$$

Note: An efficiency of 70 to 80 % is considered good. If a newly developed well has less than 65% efficiency it should not be accepted.

24.4 DRAWDOWN VERSUS DISTANCE

The drawdown-distance analysis can be performed when there are two or more piezometers (monitoring wells) available. The method allows calculation of both the aquifer transmissivity and storativity.

Assuming that the aquifer is homogeneous and that parameter u is less than 0.05 (*i.e.*, the Cooper-Jacob equation is valid), the drawdown at distance r_1 and at time t after the start of pumping is given as (see also equations 24.11 and 24.11a):

$$s_1 = \frac{Q}{4\pi T} \ln \frac{2.25Tt}{r_1^2 S} \tag{24.30}$$

or

$$s_1 = \frac{0.183Q}{T} \cdot \log \frac{2.25Tt}{r_1^2 S} \tag{24.31}$$

At the same time t, the drawdown at distance r_2 from the pumping well is:

$$s_2 = \frac{0.183Q}{T} \cdot \log \frac{2.25Tt}{r_2^2 S} \tag{24.32}$$

The difference between two drawdowns is:

$$\Delta s = s_1 - s_2 = \frac{0.183Q}{T} \log \frac{r_2^2}{r_1^2} \tag{24.33}$$

knowing that

$$\log \frac{r_2^2}{r_1^2} = \log \frac{1}{r_1^2} - \log \frac{1}{r_2^2}$$

and that

$$\log \frac{1}{r^2} = 2 \log \frac{1}{r}$$

Equation (24.33) can also be written as:

$$\Delta s = \frac{0.366Q}{T} \log \frac{r_2}{r_1} \tag{24.34}$$

Equation (24.34) shows that our drawdown data recorded at 3 piezometers, which are at various distances from the pumping well, can be plotted on a semilog graph drawdown vs. distance forming a straight line. Figure 24.10 shows such a graph with the drawdown data recorded at 6 different times. As can be seen, all the lines drawn through the piezometer data are indeed straight and parallel to each other thus confirming the assumption that the aquifer is homogeneous and isotropic.

The drawdown registered in the pumping well is above the straight line formed by the piezometers for each time period. This 1.9 meter difference between the theoretical and the actual drawdown in the well (as seen in Figure 24.10) is the result of the well loss.

From equation (24.34) the aquifer *transmissivity* is found as:

$$T = \frac{0.366Q}{\Delta s} \cdot \log \frac{r_2}{r_1} \tag{24.35}$$

The drawdown difference (Δs) can be found for any two points on the straight line drawn through the piezometers data for one time period. Considering this difference for one log cycle simplifies equation (24.35):

$$T = \frac{0.366Q}{\Delta s} \tag{24.36}$$

In our case the transmissivity is:

$$T = \frac{0.366 \cdot 0.008 m^3/s}{1.87 m} = 1.57 \times 10^{-3} m^2/s$$

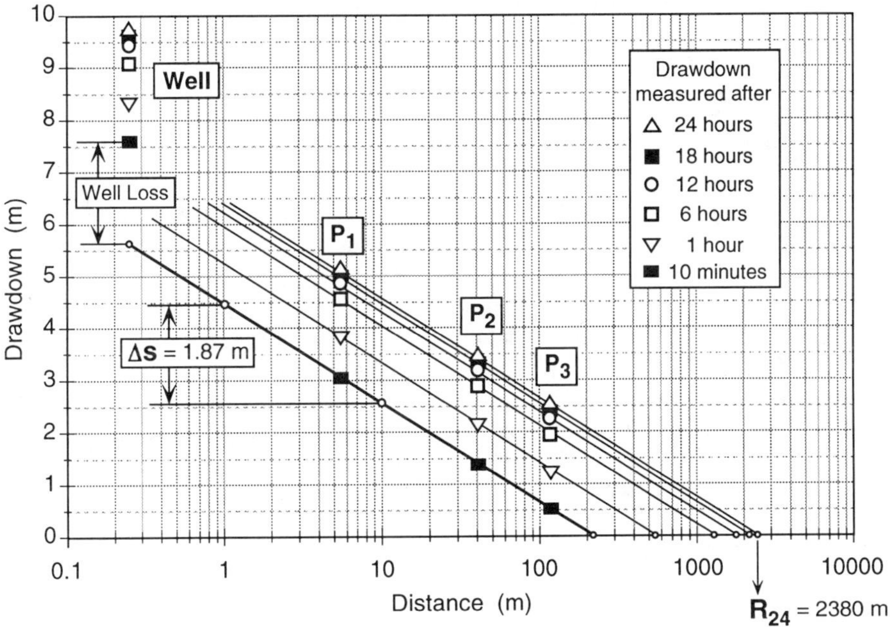

Figure 24.10 Semilog graph drawdown versus distance for determining aquifer transmissivity and the radius of well influence in time.

The drawdown-distance method is also used to calculate the *storage coefficient* of the aquifer: any of the straight lines is extended to intersect the zero-drawdown axis and find the corresponding value of r_0. By definition, this distance for which the drawdown equals zero is the *radius of well influence*. As can be seen in Figure 24.10, the radius of well influence increases in time (after 10 minutes of pumping it is approximately 215m, and after one day of pumping it is 2380m).

From the condition that the drawdown equals zero, *i.e.*:

$$s = 0 = \frac{0.183Q}{T} \log \frac{2.25Tt}{r_0^2 S} \tag{24.37}$$

it follows that the term under the logarithm must be equal to 1:

$$\frac{2.25Tt}{r_0^2 S} = 1 \tag{24.38}$$

Knowing that *log1 = 0, the storage coefficient* is then expressed as:

$$S = \frac{2.25Tt}{r_0^2} \qquad (24.39)$$

In our case the intercept of the straight line for a period of 24 hours is 2380 m. Inserting this value, and the previously calculated value of transmissivity into equation (24.39), gives:

$$S = \frac{2.25 \cdot 1.57 \times 10^{-3} m^2 / s \cdot 86400s}{(2380m)^2} = 5.4 \times 10^{-5}$$

PROBLEM 25
METHOD OF RECOVERY, VARYING PUMPING RATE

A 110-hour pumping test was conducted at a fully penetrating well in a confined aquifer considered for water supply purposes. The pumping (drawdown) lasted 94 hours and the recovery was recorded for an additional 16 hours. The pumping rate at the well was kept constant at 17 l/s. The water level during the test was recorded in a piezometer at 38 meters from the pumping well. Figure 25.1 shows the graph of the drawdown at the piezometer. The residual drawdown recorded in the piezometer during the recovery is given in Table 25.1. Find the aquifer hydrogeologic parameters (transmissivity and storage coefficient) using the recovery data. Compare these results with the results obtained from the calculated recovery (Table 25.2) and the drawdown data (Table 25.3).

Describe necessary adjustments in the recovery analysis in case of a varying pumping rate.

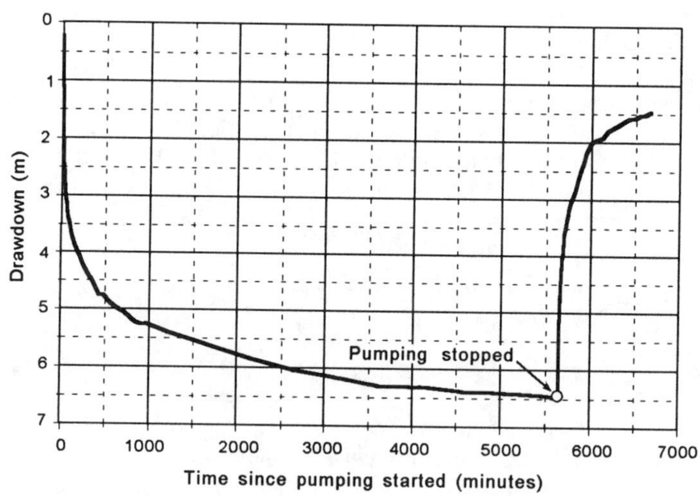

Figure 25.1 Drawdown recorded in the piezometer during the 110-hour pumping test.

25.1 HYDRAULIC CHARACTER OF THE RECOVERY

Recording and analysis of the drawdown recovery is an obligatory part of every aquifer pumping test. Pumping at a test well lowers the piezometric surface around the well and creates the *drawdown*. After pumping stops there is an inflow of water from the aquifer into the previously created *cone of depression* in the piezometric surface. This inflow causes a rise in the depleted hydraulic head, *i.e.*, the drawdown *recovery* (see Figure 25.2).

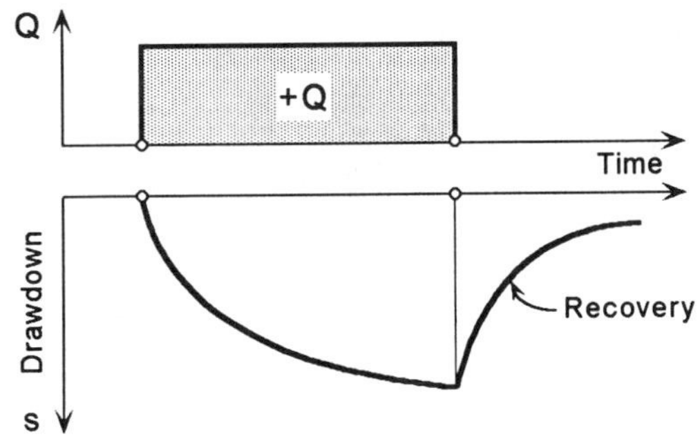

Figure 25.2 Drawdown versus time, and the corresponding pumping rate hydrograph for a common well pumping test.

The inflow of water from the aquifer during the recovery can be simulated with an imaginary well injecting water into the aquifer. This well has the same pumping rate as the real test well, only with the negative sign. Accordingly, the entire aquifer test can be simulated as a superposition of flows at two wells. This principle is illustrated in Figure 25.3: at the time when actual pumping stops, the imaginary well starts adding water to the aquifer and the real well continues extracting it at the same rate. Regarding the aquifer's water balance, the effect of the two wells on the aquifer is the same as in case of the one real well.

The resulting drawdown at any time after the actual pumping stops is the algebraic sum of the drawdown from the extraction well (that continues to operate) and the buildup (negative drawdown) from the introduced imaginary injection well (see Figure 25.3):

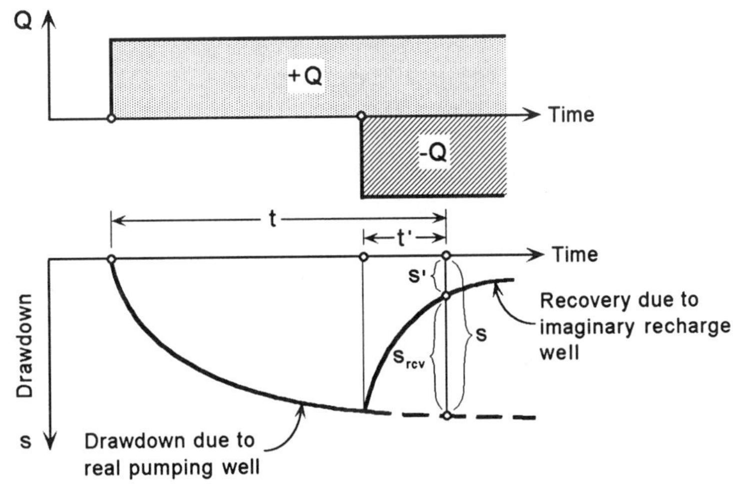

Figure 25.3 The principle of superposition of flows used to simulate the drawdown recovery during a well pumping test.

$$s' = s + s_{rcv} \tag{25.1}$$

where s' is the resulting drawdown measured at the well (*residual drawdown*), s is the drawdown from the extraction well, and s_{rcv} is the *recovery* (negative drawdown) from the recharge well.

When the above drawdown components are expressed with the Theis equation (see Problem 24, equation 24.1), it follows that:

$$s' = \frac{+Q}{4\pi T} \cdot W(u) + \frac{-Q}{4\pi T} \cdot W(u_{rcv}) \tag{25.2}$$

The parameter u for the extraction well is:

$$u = \frac{r^2 S}{4Tt} \tag{25.3}$$

where

- r is the radial distance from the pumping well where the drawdown is recorded,
- S is the storage coefficient,
- T is the transmissivity,
- t is the time since the actual pumping started.

The parameter u_{rcv} for the recharge (imaginary) well is:

$$u_{rcv} = \frac{r^2 S}{4Tt'} \tag{25.4}$$

where t' is the time since the actual pumping stopped.

When parameter u is less than 0.05, the Theis equation can be simplified as suggested by Jacob and Cooper (see Problem 24, equation 24.11), and equation 25.2 becomes:

$$s' = \frac{Q}{4\pi T}\left(\ln\frac{2.25Tt}{r^2 S} - \ln\frac{2.25Tt'}{r^2 S}\right) \tag{25.5}$$

or

$$s' = \frac{Q}{4\pi T} \cdot \ln\frac{t}{t'} \tag{25.6}$$

Knowing that $lnx=2.3logx$, equation (25.6) can be written as:

$$s' = \frac{0.183Q}{T} \cdot \log\frac{t}{t'} \tag{25.7}$$

25.2 RESIDUAL DRAWDOWN VERSUS RATIO *t/t'*

As seen from equation (25.7), if plotted versus the t/t' ratio on semilog graph paper, the residual drawdown data recorded in the piezometer during the recovery would form a straight line. This is the case for our data for the most part as shown in Figure 25.4. The residual drawdown measurements (s') and the calculated ratio t/t' used to draw the graph are given in Table 25.1.

Table 25.1 The residual drawdown (s') at the piezometer, and calculation of the t/t' ratio.

Time since start t (min)	Time since stop t' (min)	Ratio t/t'	Residual draw-down s' (m)
6540	0	-	6.47
5641	1	5641.0	6.29
5642	2	2821.0	5.97
5643	3	1881.0	5.75
5644	4	1411.0	5.54
5645	5	1129.0	5.39
5646	6	941.0	5.29
5647	7	806.7	5.19
5648	8	706.0	5.09
5649	9	627.0	5.02
5650	10	565.0	4.94
5652	12	471.0	4.80
5654	14	403.8	4.72
5656	16	353.5	4.64
5658	18	314.3	4.49
5660	20	283.0	4.42
5665	25	226.6	4.28
5670	30	189.0	4.17
5675	35	162.1	4.07
5680	40	142.0	3.96
5685	45	126.3	3.86
5690	50	113.8	3.76
5695	55	103.5	3.64
5700	60	95.0	3.59
5715	75	76.2	3.49
5730	90	63.7	3.33
5745	105	54.7	3.20
5760	120	48.0	3.07
5790	150	38.6	2.95
5820	180	32.3	2.84
5850	210	27.8	2.65
5880	240	24.5	2.48
5910	270	21.9	2.35
5940	300	19.8	2.19
5970	330	18.1	2.07
6000	360	16.7	2.00
6060	420	14.4	1.96
6120	480	12.7	1.93
6180	540	11.4	1.82
6240	600	10.4	1.76
6300	660	9.5	1.71
6360	720	8.8	1.65
6420	780	8.2	1.60
6480	840	7.7	1.60
6540	900	7.3	1.54
6600	960	6.9	1.53
6660	1020	6.5	1.49

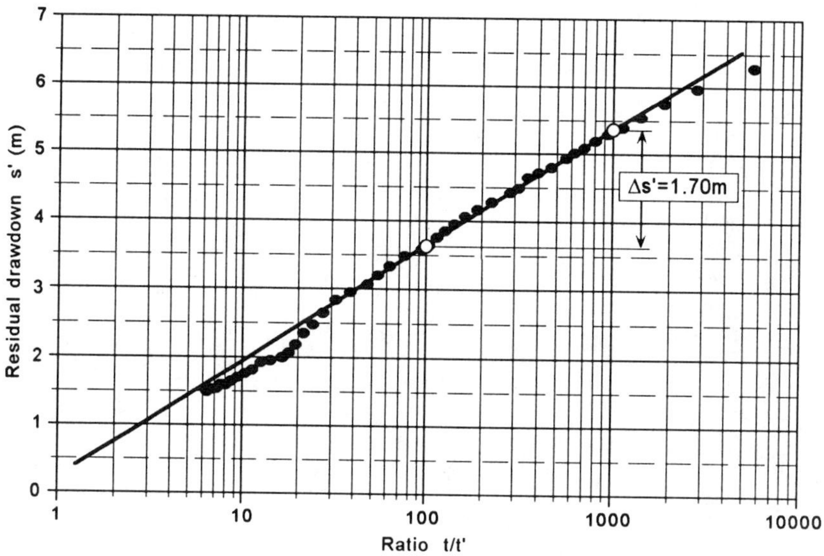

Figure 25.4 Semilog graph residual drawdown versus t/t' ratio for determining the aquifer transmissivity.

The residual drawdown coordinates of any two points on the straight line are:

$$s_1' = \frac{0.183Q}{T} \cdot \log\left(\frac{t}{t'}\right)_2$$

$$s_2' = \frac{0.183Q}{T} \cdot \log\left(\frac{t}{t'}\right)_2$$

and their difference is

$$\Delta s' = \frac{0.183Q}{T} \cdot \log \frac{(t/t')_2}{(t/t')_1} \tag{25.8}$$

If the two points are chosen over one log cycle, and equation (25.8) is solved for the aquifer transmissivity, it follows that

$$T = \frac{0.183Q}{\Delta s'} \tag{25.9}$$

or in our case:

$$T = \frac{0.183 \cdot 0.017 m^3/s}{1.68m} = 1.85 \times 10^{-3} m^2/s$$

Note that the residual drawdown method of the recovery analysis does not allow calculation of the storage coefficient, neither in the pumping well nor in piezometers. This is obvious from the absence of the storage coefficient in the basic equation describing the method (equation 25.6).

For an infinite, homogeneous and isotropic aquifer, and the groundwater flow that satisfies Theis assumptions (see Problem 24), the straight line on the semilog graph s' versus t/t' intersects the zero-drawdown axis at $t/t'=1$. This theoretical value is obtained from the condition that the residual drawdown equals 0 in equation (25.7):

$$s' = 0 = \frac{0.183Q}{T} \cdot \log\frac{t}{t'}$$

For this equation to be valid, the logarithm of t/t' must equal 0 and the ratio t/t' is then 1.

Deviation of the experimental straight line from the theoretical position, *i.e.*, when the t/t' intercept is different than 1, indicates that the aquifer is not "ideal" and may suggest a possible cause. For example, the t/t' intercept of 2 or more suggests that there is a source of recharge resulting in a full recovery in a shorter time than the corresponding time of drawdown (the imaginary well adds more water to the aquifer than what is being extracted by the real well). The t/t' intercept noticeably smaller than 1 means that the recovery is slow and incomplete thus indicating an aquifer of limited extent.

25.3 CALCULATED RECOVERY VERSUS t'

The residual drawdown recorded in the well after the end of pumping is the result of the combined influence of one extraction and one recharge (imaginary) well. Although the individual influence of each well cannot be found explicitly, it is possible to estimate it using a graphoanalytical approach as shown in Figure 25.5. Once the influence of the imaginary well is separated, it can be considered a simple case of the time-drawdown method of pumping test analysis (drawdown here corresponds to the negative drawdown,

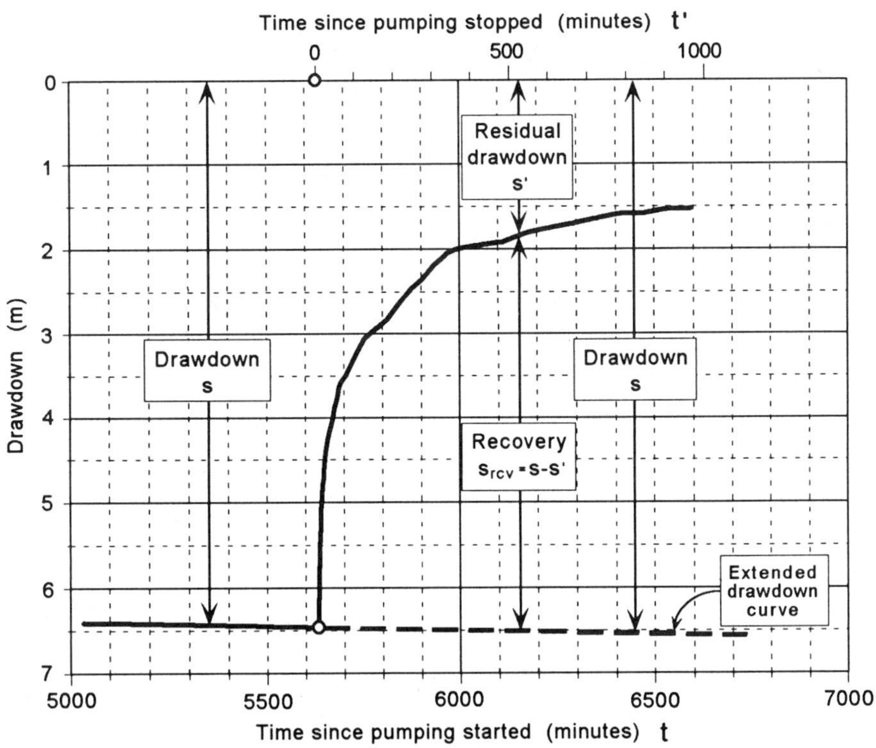

Figure 25.5 Graphoanalytical determination of the recovery from the extended drawdown curve.

Table 25.2 Calculated recovery in the piezometer.

Time since stop t' (min)	Extended drawdown (m)	Residual drawdown (m)	Calculated recovery (m)
0	6.47	6.47	0
1	6.47	6.29	0.18
2	6.47	5.97	0.50
3	6.47	5.75	0.72
4	6.47	5.54	0.93
5	6.47	5.39	1.08
6	6.47	5.29	1.18
7	6.47	5.19	1.28
8	6.47	5.09	1.38
9	6.47	5.02	1.45
10	6.47	4.94	1.53
12	6.47	4.80	1.67
14	6.47	4.72	1.75
16	6.47	4.64	1.83
18	6.47	4.49	1.98
20	6.47	4.42	2.05
25	6.47	4.28	2.19
30	6.48	4.17	2.31
35	6.48	4.07	2.41
40	6.48	3.96	2.52
45	6.48	3.86	2.62
50	6.48	3.76	2.72
55	6.48	3.64	2.84
60	6.48	3.59	2.89
75	6.48	3.49	2.99
90	6.48	3.33	3.15
105	6.48	3.20	3.28
120	6.48	3.07	3.41
150	6.48	2.95	3.53
180	6.48	2.84	3.64
210	6.48	2.65	3.83
240	6.48	2.48	4.01
270	6.49	2.35	4.14
300	6.49	2.19	4.30
330	6.49	2.07	4.42
360	6.50	2.00	4.50
420	6.51	1.96	4.55
480	6.51	1.93	4.58
540	6.52	1.82	4.71
600	6.52	1.76	4.76
660	6.53	1.71	4.83
720	6.53	1.65	4.88
780	6.54	1.60	4.95
840	6.54	1.60	4.95
900	6.55	1.54	5.01
960	6.56	1.53	5.03

or the recovery from the imaginary well). The method requires that the drawdown curve from the actual well be extended through the monitored recovery period. The recovery is then found as the difference between the extended (estimated) drawdown and the residual drawdown recorded at the well:

$$s_{rcv} = s - s' \tag{25.10}$$

Although relatively simple, this procedure is often lengthy (the analyst must manually define "enough" points) and there is subjectivity in extending the drawdown line.

The calculated recovery for our case is shown in Table 25.2. Figure 25.6 is the semilog plot of the recovery versus time since the pumping stopped - t' (which is the time relevant for the imaginary well).

The aquifer transmissivity and the storage coefficient are found from the semilog graph in Figure 25.6 as in case of the common time-drawdown method using the modified Theis equation (the Cooper-Jacob method; see Problem 24, Section 24.2).

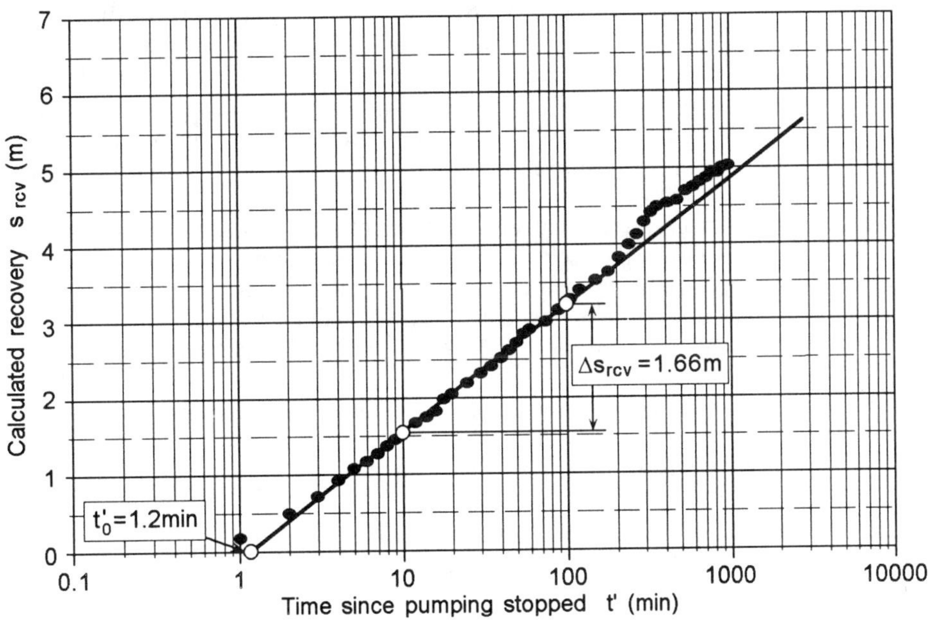

Figure 25.6 Semilog graph of the calculated recovery (s_{rcv}) versus time since pumping stopped (t') for finding the aquifer transmissivity and the storage coefficient.

When two points on the straight line are chosen over one log cycle, the transmissivity is calculated as:

$$T = \frac{0.183Q}{\Delta s_{rcv}} \tag{25.11}$$

$$T = \frac{0.183 \cdot 0.017 m^3 / s}{1.66 m} = 1.87 \times 10^{-3} m^2 / s$$

The storage coefficient is:

$$S = \frac{2.25 T t_0'}{r^2} \tag{25.12}$$

where t_0' is the intersection of the straight line and the zero-drawdown axis (1.2 minutes in our case -- see Figure 25.6). Knowing that the piezometer is 38 meters from the pumping well (*r=38m*), and inserting the calculated transmissivity into equation (25.12) gives

$$S = \frac{2.25 \cdot 1.87 \times 10^{-3} m^2 / s \cdot 1.2 \times 60s}{(38m)^2} = 2.1 \times 10^{-4}$$

One very important advantage of both recovery methods described is that, during the "operation" of the imaginary well, the well loss is practically minimal since the pump is turned off. Although there will still be some linear resistance to the flow in the near-well zone due to aquifer disturbance during the well completion, the turbulent well loss is absent. In addition, the water level recorded in the pumping well is much more accurate when the pump is not working. This all means that the recovery data from the pumping well itself can be used to determine the aquifer parameters with reasonable accuracy. (*Note, however, that for more reliable results the recovery analyses should always be performed with the piezometer data when available.*)

When the recovery is correctly calculated, values of the transmissivity and the storage coefficient should be close to those obtained using the actual drawdown data from the first part of the test (Cooper-Jacob straight line method). Table 25.3 gives values of the actual drawdown and times of measurements in our case. Plotted on a semilog graph, the data look quite similar to the plot recovery versus t' (Figures 25.7 and 25.6 respectively).

Table 25.3 Time-drawdown data for the piezometer.

Time (min)	Drawdown (m)	Time (min)	Drawdown (m)	Time (min)	Drawdown (m)
1	0.24	40	2.96	420	4.75
2	0.57	45	3.03	480	4.77
3	0.77	50	3.11	540	4.87
4	1.01	55	3.18	600	4.95
5	1.20	60	3.26	660	5.02
6	1.36	75	3.43	720	5.07
7	1.53	90	3.53	780	5.16
8	1.62	105	3.67	840	5.24
9	1.75	120	3.75	900	5.27
10	1.86	150	3.89	960	5.26
12	1.99	180	4.03	1200	5.40
14	2.08	210	4.11	2160	5.85
16	2.20	240	4.24	2640	6.04
18	2.30	270	4.32	3120	6.16
20	2.38	300	4.40	3600	6.30
25	2.57	330	4.47	4080	6.33
30	2.75	360	4.54	4560	6.39
35	2.88	390	4.66	5040	6.41

The aquifer transmissivity from the graph time-drawdown (Figure 25.7) is:

$$T = \frac{0.183Q}{\Delta s}$$

$$T = \frac{0.183 \cdot 0.017 m^3 / s}{1.62 m} = 1.92 \times 10^{-3} m^2 / s$$

The coefficient of storage is:

$$S = \frac{2.25 T t_0}{r^2}$$

$$S = \frac{2.25 \cdot 1.92 \times 10^{-3} m^2 / s \cdot 0.87 \times 60 s}{(38 m)^2} = 1.6 \times 10^{-4}$$

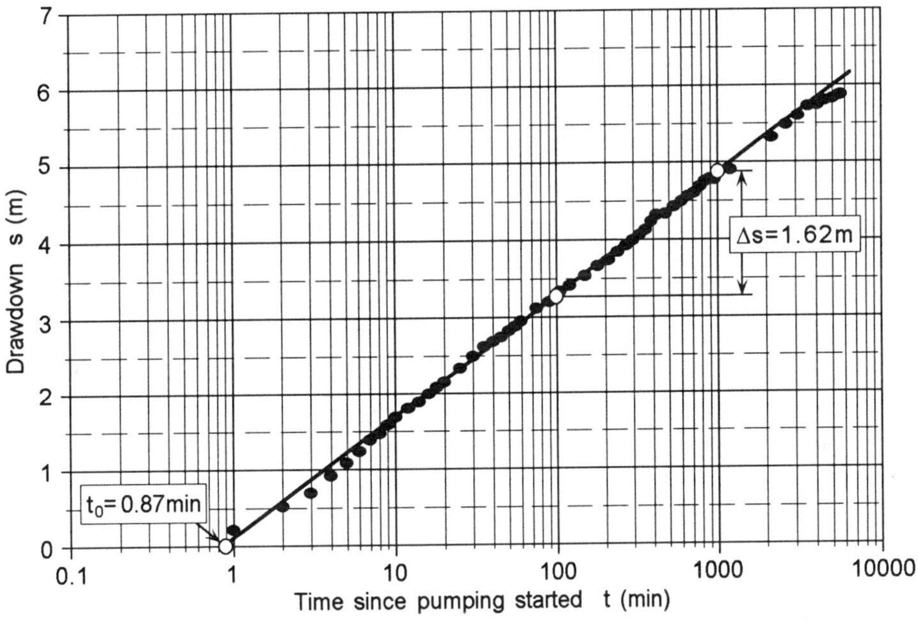

Figure 25.7 Time-drawdown plot for the data recorded in the piezometer during the pumping period (Cooper-Jacob straight line method).

As can be seen, both values are very close to the results obtained by the recovery method. Often, however, the storage coefficient will change during the test and will be smaller for the recovery period. This is because the solid matter in the aquifer is not perfectly elastic and it will tend to rebound at a slower rate during pressure recovery as compared to the rate of compression during the preceding pumping (Driscoll, 1986).

25.4 VARYING PUMPING RATE

One of the advantages of the recovery analysis is that it can be used to calculate aquifer parameters with reasonable accuracy when the pumping rate was not kept constant during the test. The pumping rate can vary for various reasons, intentional (such as during a step test) or unintentional, which is more common. A frequent problem during pumping tests in newly constructed wells is an improper (insufficient) well development which results in erratic surges and drops of the pumping rate. Another example is a constant decrease of the pumping rate due to lowering of the hydraulic head in the well and the consequent decrease of the pump capacity. In both cases an attempt to sporadically adjust the pumping rate may sometimes further aggravate the problem.

Figure 25.8 shows the hydrograph of a pumping test initially designed to have the pumping rate of 8 liters per second and duration of 8 hours. Due to various problems, the pumping rate (recorded every 30 minutes) varied between 7.8 and 6.6 l/s. Practice shows that when the pumping rate varies less than 10 to 15 % during the test, the recovery data can be used for the transmissivity determination after necessary adjustments. In our case this variation is 15%, just "enough" to consider the result acceptable.

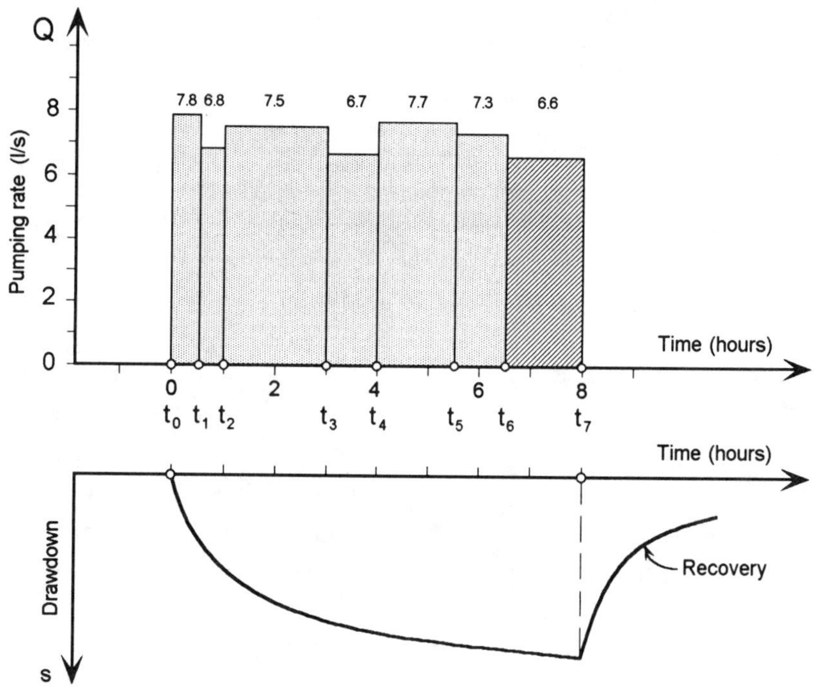

Figure 25.8 Hydrograph of the varying pumping rate and the corresponding drawdown-recovery curve at the aquifer test well.

The actual varying pumping rate is replaced with an equivalent constant pumping rate which will be used in calculations. The new pumping rate is not the average of all the recorded rates. Rather, it is the last pumping rate of a significant duration - 6.6 l/s in our case (see Figure 25.8). This last recorded pumping rate has the most influence on the future development of the cone of depression (drawdown) around the pumping well.

Introduction of the new pumping rate requires an adjustment to the hydrograph time base so that the aquifer water balance does not change. In other words, the volume of water actually withdrawn must equal the volume of water that would have been withdrawn with the new adjusted pumping rate. The duration of pumping with the adjusted pumping rate, *i.e.*, the new time base of the hydrograph (t^*), is found using the following general formula:

$$t^* = \frac{Q_1 \cdot t_1 + Q_2 \cdot (t_2 - t_1) + ... + Q_n \cdot (t_n - t_{n-1})}{Q_n} \tag{25.13}$$

where

t_1, t_2, t_{n-1}, t_n	are the times of the pumping rate change,
Q_1, Q_2, Q_{n-1}, Q_n	are the recorded pumping rates,
Q_n	is the last recorded pumping rate.

In our case this time is (see Figures 25.8 and 25.9):

$$t^* = \frac{7.8\,l/s \cdot 0.5h + 6.8\,l/s \cdot 0.5h + 7.5\,l/s \cdot 2h + 6.7\,l/s \cdot 1h + 7.7\,l/s \cdot 1.5h + 7.3\,l/s \cdot 1h + 6.6\,l/s \cdot 1.5h}{6.6\,l/s}$$

$$t^* = 8.75\ hours$$

As can be seen, the new time base is longer than the time of actual pumping. The adjusted time at the beginning of pumping is set 8.75 hours backwards from the time at which the actual pumping stopped. Note that the point in time at which the actual pumping stopped remains the same, *i.e.*, it is "fixed". The new time-zero (t_0), which is to the left of the initial beginning of the actual pumping, is now the reference point for the recovery analysis (see Figure 25.9).

When the pumping rate varies during the test, the drawdown data cannot be used for the transmissivity calculation using the graphoanalytical approaches already described. However, the residual drawdown recorded at the well during recovery can be used since it is assumed that the aquifer supplies water to the well at a constant rate (*i.e.*, the imaginary recharge well injects water to the aquifer at a constant rate). Again, this constant rate equals the last recorded actual pumping rate of a significant duration (6.6 l/s in our case).

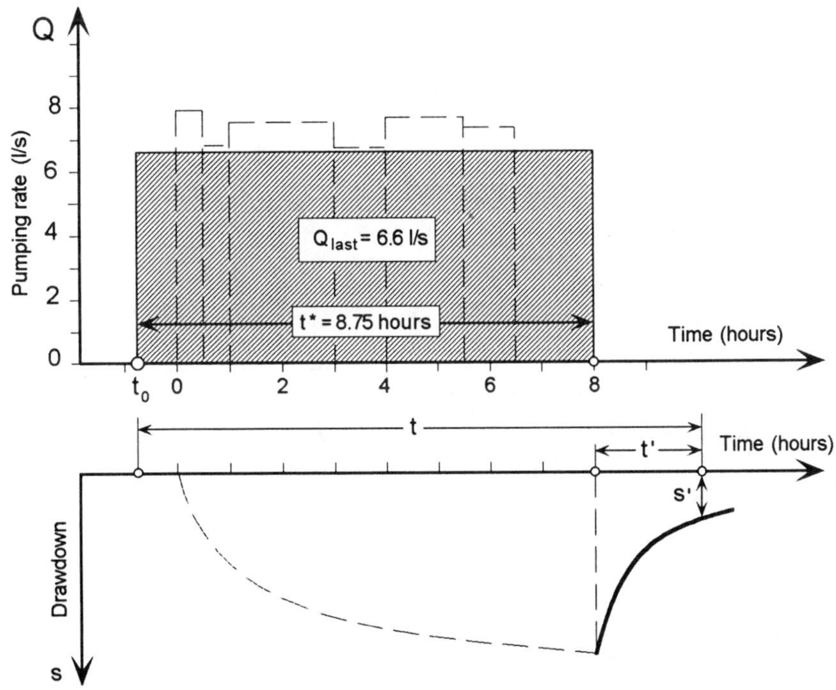

Figure 25.9 New adjusted hydrograph with the constant pumping rate of 6.6 l/s and the new time base of 8.75 hours. Note that the new beginning time of the test is shifted to the left.

The transmissivity is found using a semilog graph residual drawdown (s') versus t/t' ratio and choosing two points on the straight line over one log cycle as explained in Section 25.2 and in Figure 25.4. Note that time t is now changed -- it is the time since the new adjusted beginning of the test as shown in Figure 25.9. This means that the actual time data have to be corrected accordingly. Time since the pumping stopped (t') remains the same and does not require correction. The transmissivity is then calculated using equation (25.9) where Q_{last} is the last recorded pumping rate of significant duration during the test:

$$T = \frac{0.183 Q_{last}}{\Delta s'}$$

PROBLEM 26
THREE-STEP PUMPING TEST, WELL LOSS

A three-step pumping test at a fully penetrating well in a confined aquifer was conducted in order to determine the well characteristics. The graph in Figure 26.1 shows the drawdown at the well versus the time for all three steps, as well as the pumping rate hydrograph. The drawdown recorded during the first step at a domestic well 92 meters from the test well is given in Table 26.1. The residual drawdown after the end of pumping at the test well is given in Table 26.2. From the available data find the aquifer hydrogeologic parameters and the well loss.

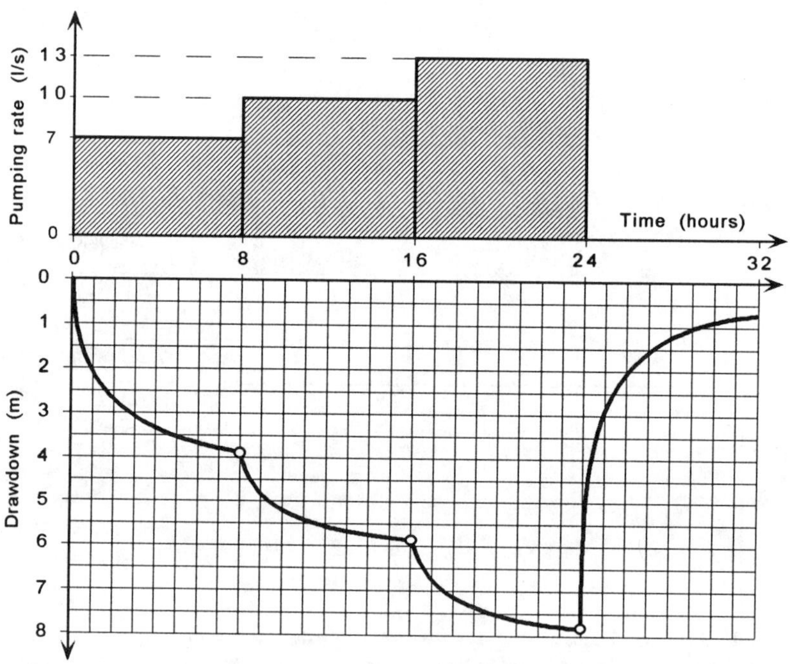

Figure 26.1 Drawdown at the pumping well versus time for the three-step pumping test.

Table 26.1 Drawdown (in meters - m) versus time (in minutes - min) at 92m from the pumping well. The pumping rate during the first step was 7 liters per second.

min	m	min	m	min	m	min	m
1	-	8	0.02	40	0.26	180	0.58
2	-	9	0.03	50	0.31	240	0.64
3	0.002	10	0.04	60	0.35	300	0.69
4	0.004	15	0.08	75	0.39	360	0.72
5	0.006	20	0.12	90	0.42	420	0.75
6	0.008	25	0.16	120	0.49	480	0.78
7	0.01	30	0.20	150	0.54		

26.1 HYDROGEOLOGIC PARAMETERS

The Cooper-Jacob Straight Line Method

Aquifer hydrogeologic parameters, transmissivity and storage coefficient, can be determined using the data recorded during the first step at a non-pumping well near the test well (e.g., a monitoring well, a piezometer, or a domestic well as in our case), and using the recovery data at both the pumping (test) well and nearby non-pumping wells. It is very important to remember that the data from the second and the third steps cannot be used because of the influence of the preceding steps.

Figure 26.2 shows a semilog graph drawdown versus time recorded at a domestic well 92 meters from the test well. The data (Table 26.1) are for the first step. As can be seen from the graph, the data fall along a straight line roughly 20 minutes after pumping. This also means that after 20 minutes the parameter u in the well function (the Theis equation) is smaller than 0.05 and the Cooper-Jacob straight line method can be used to find the transmissivity and the storage coefficient (see Problem 24, Section 24.2).

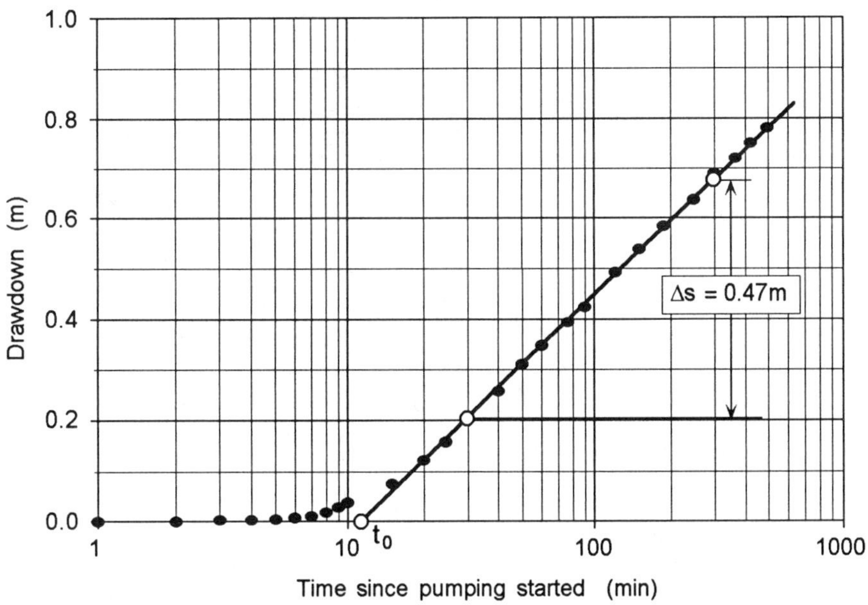

Figure 26.2 Time-drawdown graph for the first step at 92 meters from the test well (recordings in a domestic well).

The transmissivity is (see also equation 24.19):

$$T = \frac{0.183Q}{\Delta s} \tag{26.1}$$

$$T = \frac{0.183 \cdot 0.007 m^3 / s}{0.47 m} = 2.725 \times 10^{-3} m^2 / s$$

The storage coefficient is (see also equation 24.22):

$$S = \frac{2.25 T t_0}{r^2} \tag{26.2}$$

$$S = \frac{2.25 \cdot 2.725 \times 10^{-3} m^2 / s \cdot 11.5 \times 60 s}{(92 m)^2} = 0.0005$$

Residual Drawdown Method

When the pumping rate varies during the test, which in the case of a step test is done deliberately, the residual drawdown data can be used for the transmissivity analysis only after the pumping rate has been analytically adjusted to a new time base. This new pumping rate and the new time base preserve aquifer water balance: the volume of water actually withdrawn equals the volume of water that would have been withdrawn with the new constant pumping rate (see Problem 25, Section 25.4). As mentioned earlier, this adjusted pumping rate is not the average of the three rates. Rather, it is the pumping rate of

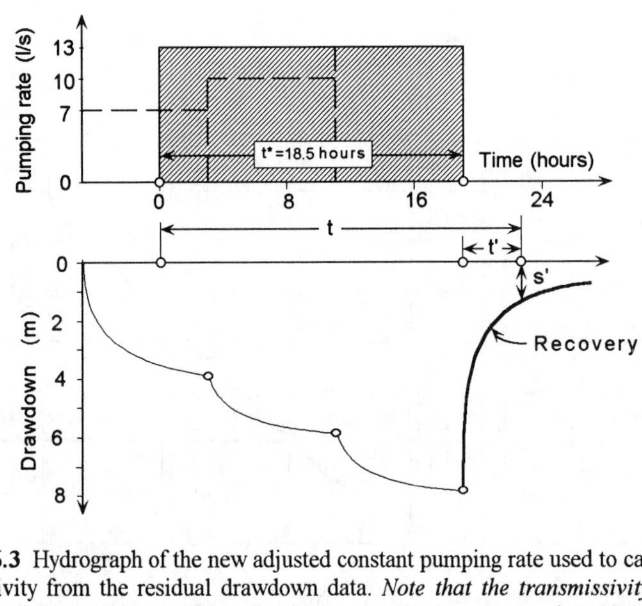

Figure 26.3 Hydrograph of the new adjusted constant pumping rate used to calculate the transmissivity from the residual drawdown data. *Note that the transmissivity found in this way is only approximate if the actual pumping rate varies more than 15% during the test (as in our case).*

the last step (13 l/s in our case) which has the most influence on the future development of the cone of depression (drawdown) around the pumping well. The adjusted constant pumping rate is also the rate at which an imaginary recharge well simulating the recovery starts adding water to the aquifer after the actual pumping stopped (Section 25.1).

During operation of the imaginary well, well loss is practically minimal since the pump is turned off. Although there will still be some linear resistance to the flow in the near-well zone due to aquifer disturbance during the well completion, turbulent well loss is absent. In addition, the water level recorded in the pumping well is much more accurate when the pump is not working. This means that both the test well and the monitoring well(s) data can be used for transmissivity analysis. Practice shows that, if the pumping rate varies less than 10 to 15 % during the test, the residual drawdown method provides reasonable accuracy. However, in case of a three-step pumping test, this variation is commonly greater than 15% (often more than 100%) which means that the recovery data at the test well should be used only if there is no data for the first step at monitoring (non-pumping) wells.

The residual drawdown analysis for the varying pumping rate is presented for our data as another illustration of the method. The duration of pumping with the adjusted pumping rate is found using the following formula (see Figure 26.3):

$$t^* = \frac{Q_1 \cdot t_1 + Q_2 \cdot (t_2 - t_1) + Q_3 \cdot (t_3 - t_2)}{Q_3} \qquad (26.3)$$

$$t^* = \frac{0.007 m^3/s \cdot 8h + 0.010 m^3/s \cdot 8h + 0.013 m^3/s \cdot 8h}{0.013 m^3/s} = 18.46\ hours \approx 1108 \text{ min}$$

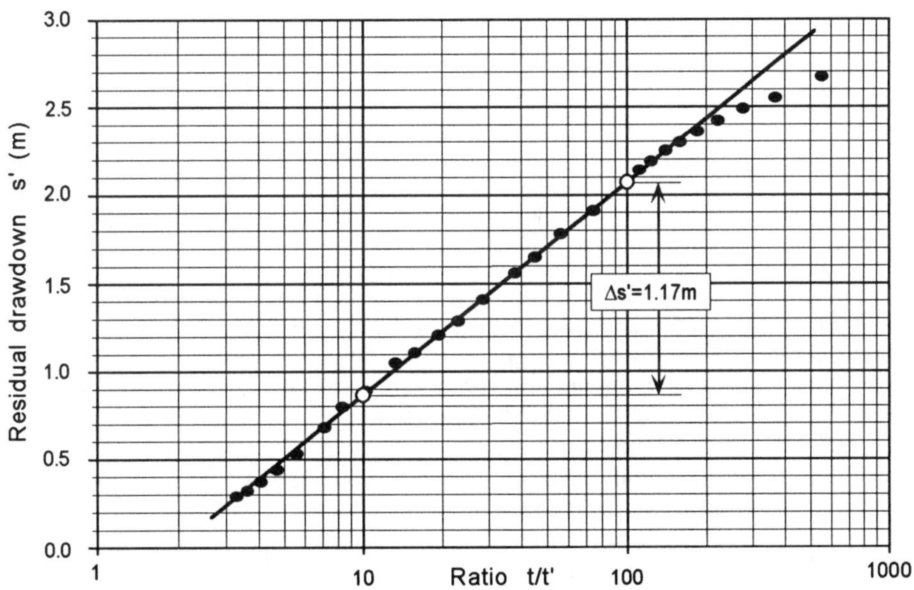

Figure 26.4 Residual drawdown at the test well versus ratio t/t'.

Calculation of the t/t' ratio needed to plot the semilog graph shown in Figure 26.4 is given in Table 26.2. The transmissivity is found as:

$$T = \frac{0.183Q}{\Delta s'}$$

$$T = \frac{0.183 \cdot 0.013m^3 / s}{1.17m} = 2.03 \times 10^{-3} m^2 / s$$

Table 26.2 Residual drawdown (s') at the pumping well versus t/t' ratio. *Note that the time since pumping started is adjusted for the last pumping rate (see equation 26.3, and Figure 26.3), i.e., it is not the actual time since the beginning of the test.*

Time since stopped t' (min)	Time since started t (min)	Ratio t/t'	Residual drawdown s' (m)
1	1109	1109	2.81
2	1110	555	2.67
3	1111	370	2.55
4	1112	278	2.49
5	1113	227	2.42
6	1114	186	2.36
7	1115	159	2.30
8	1116	139	2.26
9	1117	124	2.19
10	1118	112	2.14
15	1123	75	1.91
20	1128	56	1.78
25	1133	45	1.65
30	1138	38	1.56
40	1148	29	1.41
50	1158	23	1.29
60	1168	19	1.21
75	1183	16	1.11
90	1198	13	1.05
120	1228	10	0.89
150	1258	8.4	0.80
180	1288	7.2	0.68
240	1348	5.6	0.53
300	1408	4.7	0.44
360	1468	4.1	0.37
420	1528	3.6	0.32
480	1588	3.3	0.29

As can be seen, this value is 25% smaller than the transmissivity determined from the first step data recorded at 92 m from the test well.

Figure 26.5 shows hydrographs and the corresponding drawdown for an "ideal" pumping test. The first part of the test, which has three steps, is designed to determine the well characteristics such as well loss and an optimum pumping rate. The duration of each step should be the same, usually not more than 6 to 8 hours. Drawdown recorded during the first step can also be used to find the transmissivity and the storage coefficient. The second part of the test should be performed after complete recovery and with a maximum feasible pumping rate. Duration of this part of the test, which is designed to determine the overall aquifer transmissivity for an extensive radius of influence, depends on specific project requirements and may vary from 24 hours to several weeks in case of a major water supply development. Long pumping with a maximum rate uncovers various aquifer irregularities that may be less obvious after a short test: distant boundaries, leakage, changes in storage. Both drawdown and recovery data should be used to find the aquifer parameters.

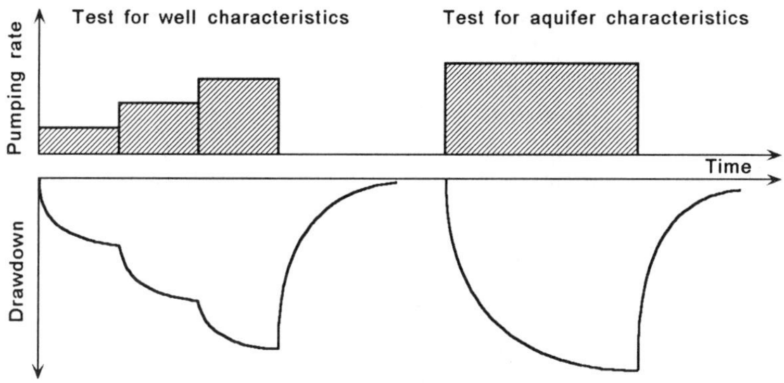

Figure 26.5 Pumping rate hydrographs and corresponding drawdown for an ideal pumping test designed to determine characteristics of both the test well and the aquifer.

26.2 WELL LOSS

The well loss is the difference between the theoretical drawdown at the well, which is the consequence of groundwater flow through the porous media, and the actual (recorded) drawdown at the well. The actual drawdown (s_w) is thus the sum of the theoretical drawdown (also called the formation loss -- s) and the well loss (Δs):

$$s_w = s + \Delta s \tag{26.4}$$

As explained in detail in Problem 18, the well loss consists of linear and turbulent parts due to various factors such as well imperfections, drilling-caused disturbances in the near-well zone, poor gravel pack and/or screen design, operation of the pump. According to Dupuit's hypothesis, it is assumed that formation loss is linear since the groundwater flow velocities are small and the flow is laminar.

The recorded drawdown can also be expressed as:

$$s_w = A_0 Q + A_1 Q + BQ^2 \tag{26.5}$$

where

- A_0 is the coefficient of linear formation loss,
- A_1 is the coefficient of linear well loss,
- B is the coefficient of turbulent well loss.

The coefficient of formation loss is easily found by applying the Theis equation:

$$A_0 = \frac{1}{2\pi T} W(u) \tag{26.6}$$

Parameter u for the test well is given as:

$$u = \frac{r_w^2 \cdot S}{4Tt} \tag{26.7}$$

where

r_0	is well radius,
S	is storage coefficient,
T	is aquifer transmissivity,
t	is time since the pumping started.

As shown by Cooper and Jacob (see Problem 24, Section 24.2), for small values of parameter u (u<0.05), *i.e.*, sufficiently long pumping time, the well function $W(u)$ is:

$$W(u) = \frac{2.25Tt}{r_w^2 S} \tag{26.8}$$

and the formation loss (*i.e.*, theoretical drawdown - s) can be written as:

$$s = \frac{Q}{2\pi T} \cdot \frac{1}{2} \ln \frac{2.25Tt}{r_w^2 S} \tag{26.9}$$

$$s = \frac{Q}{2\pi T} \cdot \ln \sqrt{\frac{2.25Tt}{r_w^2 S}} \tag{26.10}$$

$$s = \frac{Q}{2\pi T} \cdot \ln \frac{\sqrt{\dfrac{2.25Tt}{S}}}{r_w} \tag{26.11}$$

$$s = \frac{Q}{2\pi T} \cdot \ln \frac{1.5 \cdot \sqrt{Tt/S}}{r_w} \tag{26.12}$$

Notice that equation (26.12) looks similar to the steady state equation describing groundwater flow toward a fully-penetrating well in a confined homogeneous aquifer:

$$s = \frac{Q}{2\pi T} \ln \frac{R_D}{r_w} \tag{26.13}$$

where R_D is the radius of well influence which does not change in time (steady-state flow) and is also called Dupuit's radius of well influence (see Problems 18 and 19). From the analogy between equations (26.12) and (26.13) it is apparent that, in transient conditions, Dupuit's radius of well influence is time-dependent and is expressed as:

$$R_D = 1.5 \cdot \sqrt{Tt/S} \tag{26.14}$$

Theoretically, for an infinite confined aquifer, the groundwater flow forms in infinity and reaches the well pumping rate at the well perimeter (r_0). The corresponding radius of well influence also approaches infinity for a long pumping period ($t \to \infty$) which means that Dupuit's radius of well influence does not have a real physical meaning. For most practical purposes, however, Dupuit's radius of well influence given with equation (26.14) will yield satisfactory results in various analytical calculations involving the Theis equation.

Again, it should be noted that a definite real radius of well influence cannot be formed in a homogeneous confined aquifer unless there is a source of recharge, such as from a boundary or from leakage.

Using the expression for Dupuit's radius of well influence, the coefficient of the linear formation loss is:

$$A_0 = \frac{1}{2\pi T}\ln\frac{1.5\cdot\sqrt{Tt/S}}{r_w} \tag{26.15}$$

Similarly to steady state conditions (see Problem 18), the coefficients of linear and turbulent well losses are found graphoanalytically from a graph s/Q versus Q. However, since the radius of well influence in transient conditions is not constant (it increases with time), the drawdown data recorded during the three-step test must be modified in order to plot a correct graph. Figure 26.6 shows the components of drawdown recorded at the end of each step and the error made if the three drawdowns (s_1, s_2, and s_3) are used to draw a graph s/Q vs. Q without necessary corrections.

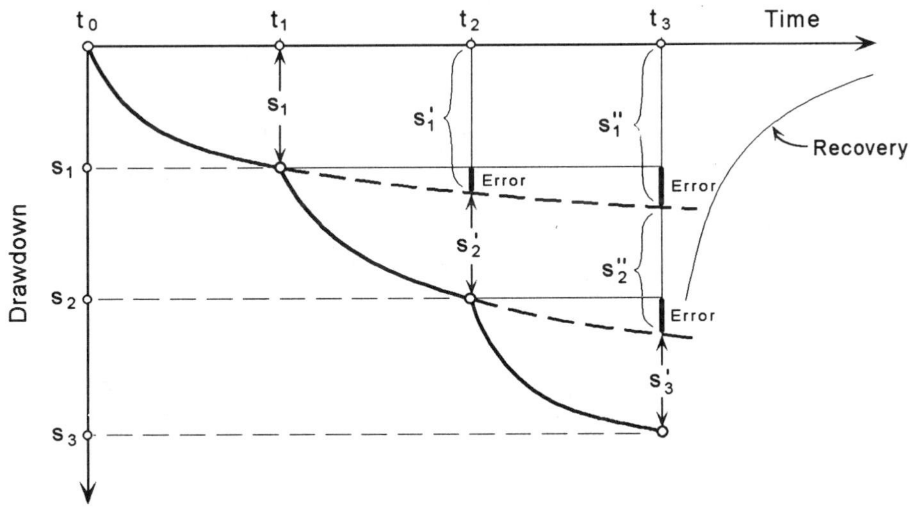

Figure 26.6 Components of drawdown recorded at the end of each pumping step. Note the error made if drawdowns s_1, s_2, and s_3 are used directly to draw graph s/Q vs. Q

The drawdown recorded at the end of the first step is:

$$s_1 = \frac{Q_1}{2\pi T}\ln\frac{1.5\cdot\sqrt{T\cdot t_1/S}}{r_w} \quad (+well\ loss) \tag{26.16}$$

or, after inserting all known values:

$$3.87m = \frac{0.007m^3/s}{2\cdot 3.14\cdot 2.725\times 10^{-3}m^2/s}\ln\frac{1.5\cdot\sqrt{(2.725\times 10^{-3}m^2/s\cdot 8hours)/0.0005}}{0.3m} \quad (+well\ loss)$$

$$3.87m = 3.10m\ +\ well\ loss$$

This drawdown of 3.87m does not need adjustments and can be used directly to draw the first point on the graph with the coordinates s_1/Q_1 and Q_1 (553s/m^2 and 0.007m^3/s).

The drawdown at the end of the second step (s_2) has two components: the drawdown caused by the increase in pumping rate (s_2') and the drawdown caused by the extended influence of the first pumping rate/step (s_1'). From the principle of superposition of flows it follows that:

$$s_2 = s_1' + s_2' \quad (+well\ loss) \tag{26.17}$$

or

$$s_2 = \frac{Q_1}{2\pi T}\ln\frac{1.5\cdot\sqrt{T\cdot t_2 / S}}{r_w} + \frac{(Q_2 - Q_1)}{2\pi T}\ln\frac{1.5\cdot\sqrt{T\cdot(t_2 - t_1)/S}}{r_w} \quad (+well\ loss) \tag{26.18}$$

The adjusted drawdown at the end of the second step, which excludes the error caused by the extended influence of the first step, is:

$$s_2^* = s_2 - error \tag{26.19}$$

where *error* is the difference between the theoretical (calculated) drawdown at the end of the second step (t_2) caused by the first pumping rate (Q_1), and the theoretical drawdown at the end of the first step (t_1) caused by the same pumping rate (Q_1):

$$error = \frac{Q_1}{2\pi T}\ln\frac{1.5\cdot\sqrt{(T\cdot t_2)/S}}{r_w} - \frac{Q_1}{2\pi T}\ln\frac{1.5\cdot\sqrt{(T\cdot t_1)/S}}{r_w} \tag{26.20}$$

or, after simplifying:

$$error = \frac{Q_1}{2\pi T}\ln\sqrt{\frac{t_2}{t_1}} \tag{26.21}$$

If the time intervals of each step are the same, as in our case, equation (26.21) becomes:

$$error = \frac{Q_1}{2\pi T}\ln\sqrt{2} \tag{26.21a}$$

and the adjusted drawdown is:

$$s_2^* = s_2 - \frac{Q_1}{2\pi T}\ln\sqrt{2} \tag{26.22}$$

After substituting all known values, the adjusted drawdown for the second step is:

$$s_2^* = 5.84m - \frac{0.007m^3/s}{2\cdot 3.14\cdot 2.725\times 10^{-3}m^2/s}\ln 1.4142$$

$$s_2^* = 5.84m - 0.14m = 5.70m$$

This adjusted drawdown, which includes the well loss, is used to draw the second point on the graph s/Q vs. Q with the following coordinates: s_2^* / Q_2 and Q_2 (570s/m^2 and 0.010m^3/s).

The recorded drawdown at the end of the third step (s_3) has three components: $s_3^{'}$ which is the drawdown caused by the increase in pumping rate from Q_2 to Q_3, the influence of the first step ($s_1^{''}$), and the influence of the second step ($s_2^{''}$):

$$s_3 = s_1^{''} + s_2^{''} + s_3^{'} \tag{26.23}$$

or

$$s_3 = \frac{Q_1}{2\pi T}\ln\frac{1.5\cdot\sqrt{T\cdot t_3/S}}{r_w} + \frac{(Q_2-Q_1)}{2\pi T}\ln\frac{1.5\cdot\sqrt{T\cdot(t_3-t_1)/S}}{r_w} + \frac{(Q_3-Q_2)}{2\pi T}\ln\frac{1.5\cdot\sqrt{T\cdot(t_3-t_2)/S}}{r_w}$$

(+ *well loss*) (26.24)

The adjusted drawdown at the end of the third step (s_3^*) excludes errors introduced by the influences of the first and second steps (see Figure 26.6):

$$s_3^* = s_3 - error_I - error_{II} \tag{26.25}$$

The error from the influence of the first step is:

$$error_I = \frac{Q_1}{2\pi T}\ln\frac{1.5\cdot\sqrt{(T\cdot t_3)/S}}{r_w} - \frac{Q_1}{2\pi T}\ln\frac{1.5\cdot\sqrt{(T\cdot t_1)/S}}{r_w} \tag{26.26}$$

or, after simplifying (note that $t_3=3t_1$ in our case):

$$error_I = \frac{Q_1}{2\pi T}\ln\sqrt{3} \tag{26.26a}$$

The error from the influence of the second step is:

$$error_{II} = \frac{Q_2-Q_1}{2\pi T}\ln\frac{1.5\cdot\sqrt{T\cdot(t_3-t_1)/S}}{r_w} - \frac{Q_2-Q_1}{2\pi T}\ln\frac{1.5\cdot\sqrt{T\cdot(t_2-t_1)/S}}{r_w} \tag{26.27}$$

or, after simplifying (note that $t_3=3t_1$ and $t_2=2t_1$ in our case):

$$error_{II} = \frac{Q_2-Q_1}{2\pi T}\ln\sqrt{2} \tag{26.28}$$

The corrected drawdown at the end of the third step is then:

$$s_3^* = s_3 - \frac{Q_1}{2\pi T}\ln\sqrt{3} - \frac{(Q_2-Q_1)}{2\pi T}\ln\sqrt{2} \tag{26.29}$$

$$s_3^* = 7.84m - 0.22m - 0.06m = 7.56m$$

The coordinates of the third adjusted point on the graph are then s_3^*/Q_3 and Q_3 (581s/m^2 and 0.013m^3/s).

By correcting the recorded drawdowns, the time-dependent influence of the preceding steps is eliminated and transient flow is converted into quasi steady state. This allows the use of the graphoanalytical method for determining well losses for steady state flow as explained earlier (see Problem 18, Section 18.4).

Figure 26.7 shows graph *s/Q* vs. *Q* for the three-step pumping test for both corrected and raw data. The intercept of the straight line drawn through the corrected points is the coefficient of the total linear loss at the well (*A*). It includes the coefficients of the linear formation loss (A_0) and the linear well loss (A_1):

$$A = A_0 + A_1 \tag{26.30}$$

Unlike for steady state flow, where these two components of linear loss can be separated, in transient conditions this separation is not possible since the coefficient of the linear formation loss is time-dependent (see equation 26.15):

$$A = \frac{1}{2\pi T} \ln \frac{1.5 \cdot \sqrt{Tt/S}}{r_w} + A_1 \tag{26.31}$$

From the graph in Figure 26.7, the coefficient of the total linear loss determined with our three-step pumping test is 521 s/m^2. This value would change for other pumping rates and pumping times.

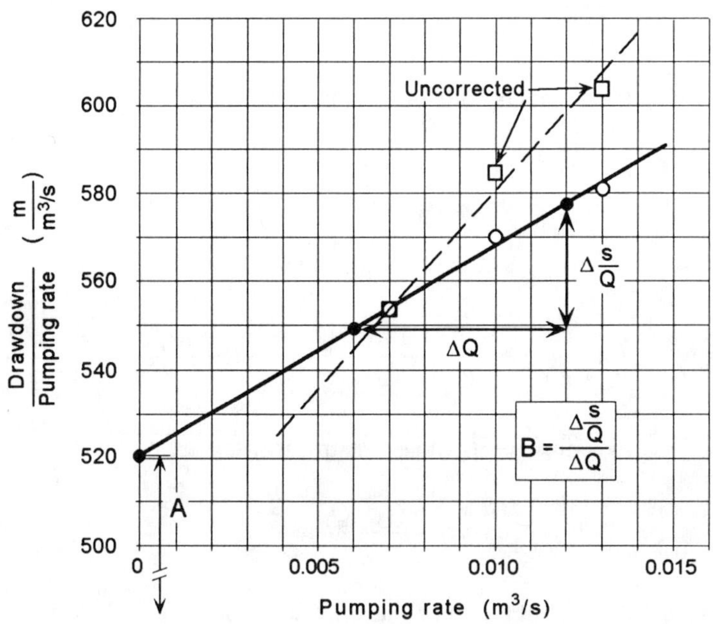

Figure 26.7 Graph pumping rate versus drawdown-pumping rate ratio used to determine the components of the drawdown in the pumping well (A: coefficient of the linear loss, B: coefficient of the turbulent loss).

The coefficient of the turbulent well loss (B) is the slope of the straight line drawn through the corrected points. Theoretically, it is not time dependent and should remain the same for different pumping rates. A common exception is pumping from karst and fissured aquifers where turbulent well loss may increase with an increasing pumping rate. In such a case, the points on the graph *s/Q* vs. *Q* would form a parabola rather than a straight line.

From the two chosen points (filled black circles) on the straight line in our case, the coefficient of turbulent well loss is found as:

$$B = \frac{\Delta \frac{s}{Q}}{\Delta Q} \tag{26.32}$$

$$B = \frac{29\ s\ /\ m^2}{0.006\ m^3\ /\ s} = 4833\ s^2\ /\ m^5$$

This value of B is high and indicates that the well should receive some corrective measures such as better development (if it is a new well), rehabilitation (if the well is old), or other appropriate treatment depending upon the cause of well loss. In general, if the coefficient of turbulent loss is less than 2500 to 3000 s^2/m^5, the well is considered to be in good condition.

PROBLEM 27
APPLICATION OF THE THEIS EQUATION

A homogeneous confined aquifer is tapped by a fully-penetrating well whose radius is 225 millimeters. The aquifer transmissivity is $8.35x10^{-4}$ m^2/s and the storage coefficient is 0.00065. For the given parameters determine the following:

1. Drawdown at 50 meters from the pumping well 10 minutes and 24 hours after the beginning of pumping with the rate of 12 liters per second (0.012 m^3/s).
2. Time after which the drawdown at 30 meters from the pumping well would be 2.5 meters if the pumping rate was 9 liters per second.
3. Pumping rate needed to produce 1.5-meter drawdown at 28 meters from the pumping well after 10 days of pumping.
4. Distance at which the drawdown is 0.85 meters after one week of pumping with the rate of 5 liters per second.
5. The radius of well influence after 5 days and 6 months of pumping with the rate of 15 liters per second.

27.1 DRAWDOWN

Drawdown at any time and at any distance from the pumping well is calculated using the well known Theis equation:

$$s = \frac{Q}{4\pi T} W(u) \tag{27.1}$$

where

Q is well pumping rate,
T is aquifer transmissivity,
u is dimensionless parameter of the Theis (well) function given as

$$u = \frac{r^2 S}{4Tt} \tag{27.2}$$

where

r is distance from the pumping well,
S is storage coefficient,
t is time since pumping started.

The value of $W(u)$ can be found either in tables for the known parameter u (see Appendix E) or calculated using the Cooper-Jacob approximation and assuming that $u<0.05$ (which is commonly the case after 10–20 minutes of pumping for distances that are not very far from the pumping well). In our case parameter u after 10 minutes and at 50 meters from the pumping well is:

$$u = \frac{(50m)^2 \cdot 0.00065}{4 \cdot 8.35 \times 10^{-4} m^2 / s \cdot 10 \cdot 60s} = 0.811$$

and the value of corresponding well function is (Appendix E):

$$W(u) = 0.305$$

The drawdown is

$$s = \frac{0.012m^3 / s}{4 \cdot 3.14 \cdot 8.35 \times 10^{-4} m^2 / s} \cdot 0.305 = 0.35m$$

Well function calculated using the Cooper-Jacob approximation is

$$W(u) = \ln \frac{2.25Tt}{r^2 S} \tag{27.3}$$

$$W(u) = \ln \frac{2.25 \cdot 8.35 \times 10^{-4} m^2 / s \cdot 600s}{(50m)^2 \cdot 0.00065} = -0.366$$

Since the value is negative (which would lead to a negative drawdown), it is obvious that equation (27.3) cannot be used to find the well function after 10 minutes of pumping. The time after which this approximate function can be used (accepting an error of 2% or less), *i.e.*, time at which parameter u is equal to 0.05, is found using the following equation (see also equation 24.14):

$$t = \frac{r^2 S}{4Tu} \tag{27.4}$$

In our case, after substituting all known values into equation (27.4), this time equals 9730 seconds (162 minutes or 2.7 hours).

Drawdown after 24 hours of pumping at 50 meters from the well using the Theis equation and table *W(u) – u* (Appendix E) is:

$$s = \frac{Q}{4\pi T} \cdot W\left[\frac{(50m)^2 \cdot 0.00065}{4 \cdot 8.35 \times 10^{-4} m^2 / s \cdot 86400s}\right]$$

$$s = \frac{0.012m^3 / s}{4 \cdot 3.14 \cdot 8.35 \times 10^{-4} m^2 / s} \cdot W(0.00563)$$

$$s = 1.14m \cdot 4.61 = 5.255m$$

When using the approximate logarithmic formula, the drawdown is:

$$s = \frac{Q}{4\pi T} \ln \frac{2.25T \cdot 24h}{r^2 S} = 5.26m$$

which is practically identical to the result obtained by the Theis equation.

27.2 TIME

From the Theis equation, well function for a given drawdown is:

$$W(u) = \frac{4s\pi T}{Q} \tag{27.5}$$

or, in our case:

$$W(u) = \frac{4 \cdot 2.5m \cdot 3.14 \cdot 8.35 \times 10^{-4} m^2 / s}{0.009m^3 / s} = 2.913$$

For this value of well function, parameter *u* is 0.0315 as determined from the table in Appendix E. The corresponding time is found using equation (27.4):

$$t = \frac{r^2 S}{4Tu}$$

$$t = \frac{(30m)^2 \cdot 0.00065}{4 \cdot 8.35 \times 10^{-4} m^2 / s \cdot 0.0315} = 5560s \text{ or approximately 1.5 hours}$$

27.3 PUMPING RATE

From the Theis equation, the pumping rate needed to produce the given drawdown is

$$Q = \frac{4\pi sT}{W(u)} \tag{27.6}$$

Well function for the given aquifer parameters, time and distance is:

$$W(u) = W(\frac{r^2S}{4Tt}) = W(\frac{(28m)^2 \cdot 0.00065}{4 \cdot 8.35 \times 10^{-4} m \ / s \cdot 864000s}) = W(0.0001766) = 8.0684$$

After inserting all known values into equation (27.6), the pumping rate is:

$$Q = \frac{4 \cdot 3.14 \cdot 1.5m \cdot 8.35 \times 10^{-4} m^2 \ / s}{8.0684} = 0.0013m^3 \ / s \quad \text{or 1.3 liters per second}$$

27.4 DISTANCE

The distance at which a given drawdown is measured is part of the expression for parameter u in the Theis equation, *i.e.*, it is part of the well function:

$$W(u) = W(\frac{r^2S}{4Tt}) \tag{27.7}$$

The well function is (from the Theis equation):

$$W(u) = \frac{4\pi sT}{Q} \tag{27.8}$$

For the given values this function is:

$$W(u) = \frac{4 \cdot 3.14 \cdot 0.85m \cdot 8.35 \times 10^{-4} m^2 \ / s}{0.005m^3 \ / s} = 1.783$$

The corresponding parameter u is found from the table in Appendix E (note that it will almost always be necessary to choose an interpolated value for u; in our case this value is between 0.1 and 0.11 for the corresponding table values of $W(u)$ – 1.8229 and 1.7371 respectively):

$u=0.104$

From the expression for parameter u (equation 27.2), the distance at which the drawdown is measured at a given time is:

$$r = \sqrt{\frac{4uTt}{S}} \tag{27.9}$$

or in our case:

$$r = \sqrt{\frac{4 \cdot 1.04 \cdot 8.35 \times 10^{-4} \cdot 7 \cdot 86400s}{0.00065}} = 1798m$$

27.5 RADIUS OF WELL INFLUENCE

As mentioned earlier, the radius of well influence in a homogeneous, confined aquifer of infinite extent and without recharge theoretically forms in infinity. However, it is often necessary, for practical purposes, to find some approximate ("realistic") value for the radius of well influence. This radius is defined by a given small drawdown that can still be registered in the field (say, 5 cm or 0.05 m).

The Theis equation for such a small drawdown (s_R) is:

$$s_R = \frac{Q}{4\pi T} W(u_R) \tag{27.10}$$

where parameter u_R is defined by the distance from the pumping well where this small drawdown is registered, *i.e.*, by the "realistic" radius of well influence (R_R):

$$u_R = \frac{R_R^2 S}{4Tt} \tag{27.11}$$

After finding the value for $W(u_R)$ from the following expression:

$$W(u_R) = \frac{4\pi T s_R}{Q}$$

and the corresponding value of u_R from the table in Appendix E, the radius of well influence is calculated by inserting all known values into:

$$R_R = \sqrt{\frac{4Ttu_R}{S}}$$

The radius of well influence for sufficiently long pumping periods (*i.e.*, when $u<0.05$) can also be found from the Cooper-Jacob approximation of the Theis equation (see Problem 24, Section 24.2):

$$s_R = \frac{Q}{4\pi T} \ln \frac{2.25Tt}{R_R^2 S} \tag{27.12}$$

which gives:

$$\frac{2.25Tt}{R_R^2 S} = e^{\frac{4\pi T s_R}{Q}} \tag{27.13}$$

and

$$R_R = \sqrt{\frac{2.25Tt}{S \cdot e^{\frac{4\pi T s_R}{Q}}}} \tag{27.14}$$

Note that (in general notation) if $\ln x = a$, *then* $e^a = x$ *(see also Appendix G).*

Assuming a small drawdown of 0.05m at R_R, and inserting all known values into equation (27.14), gives the following value for the realistic radius of well influence after 5 days of pumping:

$$R_R = \sqrt{\frac{2.25 \cdot 8.35 \times 10^{-4} m^2 / s \cdot 5 \cdot 86400s}{0.00065 \cdot e^{\frac{4 \cdot 3.14 \cdot 8.35 \times 10^{-4} \cdot 0.05m}{0.015 m^3 / s}}}} = 1098m$$

After 6 months of pumping with the same pumping rate of 15 liters per second, the radius of well influence is 6643 meters.

PROBLEM 28
LEAKY CONFINED AQUIFER

A pumping test was conducted at a fully penetrating well completed in a homogeneous confined aquifer which, for practical purposes, can be considered to be of infinite extent. (It is assumed that the aquifer boundaries are distant and do not influence groundwater flow for the duration of the test.) The aquifer is overlain by an aquitard which, in turn, is overlain by an unconfined (water table) aquifer as shown in Figure 28.1. The hydraulic conductivity of the unconfined aquifer is 8.45×10^{-3} m^2/s. The drawdown of the piezometric surface of the confined aquifer was measured at a monitoring well 128 meters from the pumping well and the results are given in Table 28.1. From the available data find the hydrogeologic parameters of the confined aquifer and the aquitard. Calculate the drawdown at 10 meters from the pumping well after 10 hours of pumping.

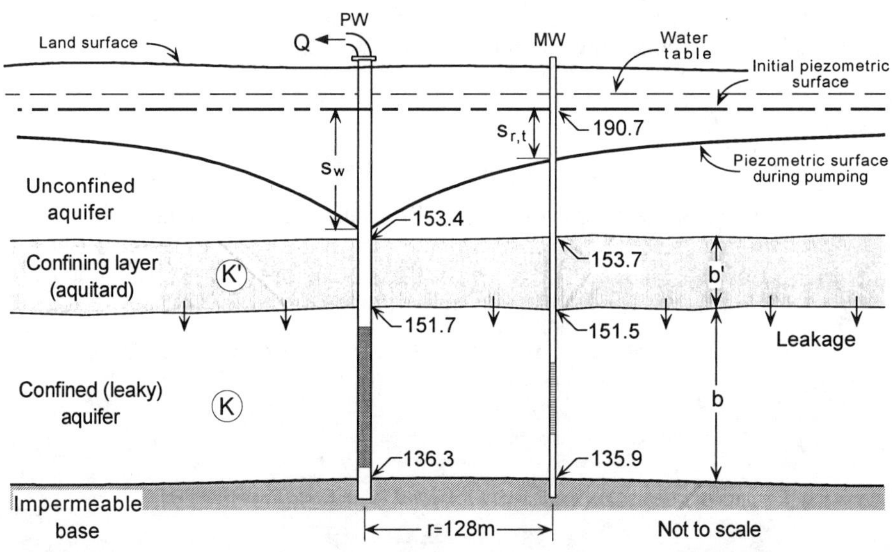

Figure 28.1 Simplified hydrogeologic cross-section of the aquifer test at pumping well PW.

28.1 FLOW TOWARD A WELL IN A LEAKY CONFINED AQUIFER

The Theis non-equilibrium equation describes flow toward a fully penetrating well in a confined aquifer and is based on numerous assumptions, one of which is that the aquifer receives no recharge for the duration of pumping (see Problem 24). However, this condition is seldom entirely satisfied since most confined aquifers receive recharge, either continuously or intermittently. One common source of recharge is leakage through an overlaying and/or underlying aquitard which separates the pumped aquifer from another aquifer. The rate of leakage can significantly increase during pumping due to an increased hydraulic gradient between the two aquifers. Eventually, if the rate of leakage balances the pumping rate, the drawdown will stabilize at a certain level and the radius of well influence will cease to expand. Figure 28.2, which is a log-log plot of the recorded

Table 28.1 Time versus drawdown recorded at 128 m from the pumping well in a leaky confined aquifer.

Time (min)	Drawdown (m)	Time (min)	Drawdown (m)
1	0.11	120	1.46
2	0.52	150	1.47
3	0.75	180	1.48
4	0.86	240	1.49
5	0.96	300	1.50
6	1.01	360	1.50
7	1.05	420	1.51
8	1.08	480	1.51
9	1.11	600	1.51
10	1.14	720	1.51
15	1.24	840	1.52
20	1.28	960	1.52
25	1.33	1080	1.53
30	1.36	1200	1.53
45	1.39	1320	1.53
60	1.43	1440	1.53
90	1.45		

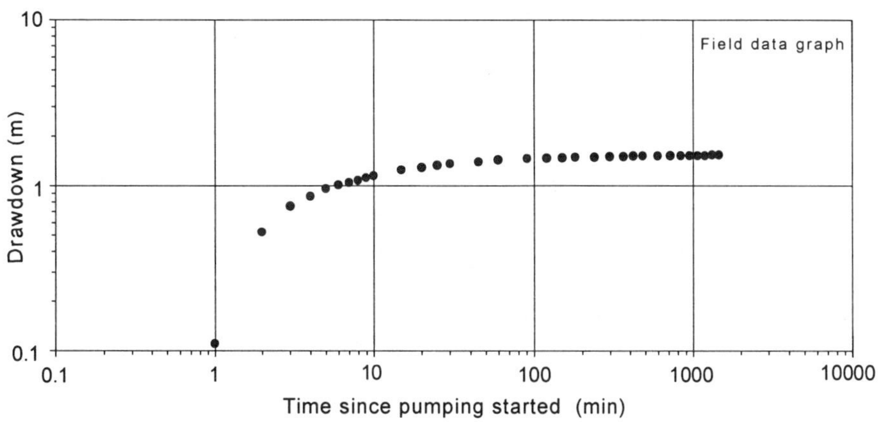

Figure 28.2 *Time* versus *drawdown* recorded at MW, 128 m from the pumping well (data in Table 28.1).

drawdown versus time in our case, shows this effect of leakage. Approximately 100 minutes after pumping started, the leakage balances the withdrawal of water from the confined aquifer and the drawdown remains practically unchanged for the duration of the test. (Note that the influence of an equipotential boundary would look very similar and interpreters should verify its absence before assuming leaky conditions.)

The equation describing flow toward a well in a leaky confined aquifer, which is separated from an unconfined aquifer by an aquitard, was derived in 1955 by Hantush and Jacob in the following form (Lohman, 1972):

$$\frac{s}{\frac{Q}{4\pi T}} = 2K_0(2v) - \int_{v^2/u}^{\infty} \frac{1}{y}\exp\left(\frac{-y-v^2}{y}\right)dy \qquad (dimensionless) \tag{28.1}$$

where

s	is drawdown in the confined aquifer,
Q	is pumping rate of the well,
T	is transmissivity of the confined aquifer,
K_0	is modified Bessel function of the second time of the zero order,
y	is variable of integration,
u	is the same parameter (dimensionless) as in the Theis equation:

$$u = \frac{r^2 S}{4Tt} \tag{28.2}$$

where

r	is distance from the pumping well to the monitoring well,
S	is storage coefficient of the confined aquifer,
t	is time since pumping started,

and

$$v = \frac{r}{2} \cdot \sqrt{\frac{K'}{b'T}} \tag{28.3}$$

where

K'	is vertical hydraulic conductivity of the confining bed,
b'	is thickness of the confining bed,
T	is transmissivity of the aquifer.

The authors gave two sets of equations for the solution of equation (28.1), both for large and small values of time. The method was later made more friendly by various authors who computed corresponding well function and/or derived graphoanalytical methods for solving equation (28.1) with the aid of type curves (see Hantush, 1956; Walton, 1960 and 1962; Cooper, 1963).

Presented here is the Walton graphoanalytical method, based on type curves, for solving the Hantush-Jacob equation in the following form:

$$s = \frac{Q}{4\pi T} \cdot W(u, r/B) \tag{28.4}$$

where

$W(u,r/B)$	is well function for the leaky confined aquifer,
B	is leakage factor given as:

$$B = \sqrt{\frac{Tb'}{K'}} \tag{28.5}$$

The values of the well function are given in Appendix E, and type curves for various values of *1/u* and *r/B* are shown in Figure 28.3.

28.2 HYDROGEOLOGIC PARAMETERS

The procedure for finding hydrogeologic parameters is as follows:

- Plot log-log graph *time* versus recorded *drawdown* at the same log-cycle scale as the graph with type curves (see Figure 28.2).
- Superimpose the field graph on the graph with type curves and, keeping the axes of both graphs parallel, match the field data with a type curve *r/B* (Figure 28.4).
- Choose a match point anywhere where the two graphs overlap and, noting the *r/B* value, read the four coordinates as follows: drawdown (*s*) and time (*t*) from the field graph, and *W(u,r/B)* and *1/u* from the type curve graph. It is recommended to choose

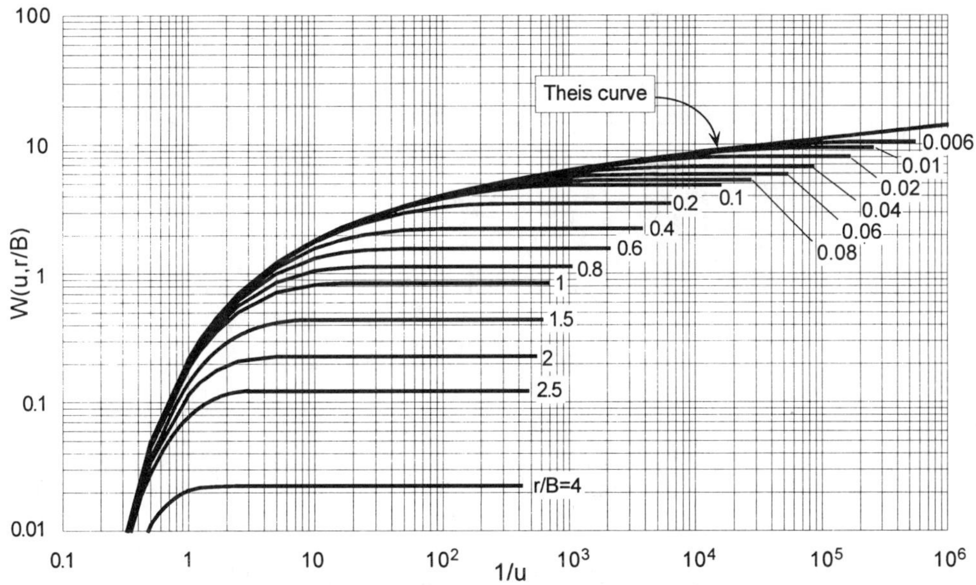

Figure 28.3 Type curves for leaky confined aquifer without the release of water from storage in confining layer. The curves are plotted using values computed by Hantush (1956; see Appendix E), and values given by Fetter (*Applied Hydrogeology*, Macmillan CPC, New York, 1994, p. 620)

rounded values for *1/u* and *W(u,r/B)* for easier calculations. In our case the coordinates of the match point are:

well function W(u,r/B)	*1*	
ratio 1/u	*10*	(and *u*=0.1)
drawdown	*0.32m*	
time	*2.5 min*	

- Calculate transmissivity by substituting known values into equation

$$T = \frac{Q}{4\pi s} \cdot W(u,\ r/B) \tag{28.6}$$

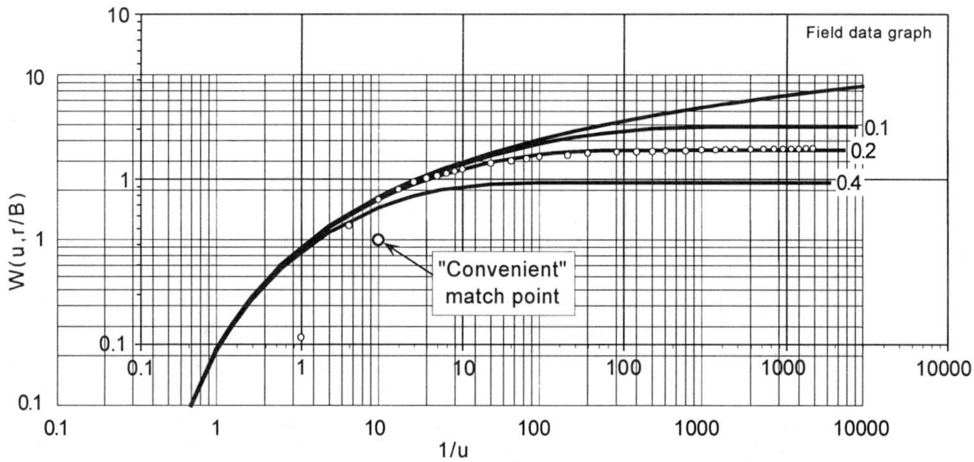

Figure 28.4 Matching field data to a type curve. The best fit is for the curve $r/B=0.2$. The match point is chosen to have convenient coordinates $W(u,r/B)=1$, and $1u=10$.

$$T = \frac{0.012m^3/s}{4 \cdot 3.14 \cdot 0.32m} \cdot 1 = 2.98 \times 10^{-3} m^2/s$$

- Find the coefficient of storage using equation

$$S = \frac{4Tut}{r^2} \tag{28.7}$$

$$S = \frac{4 \cdot 2.98 \times 10^{-3} m^2/s \cdot 0.1 \cdot 2.5 \cdot 60s}{(128m)^2} = 1.09 \times 10^{-5}$$

- Find the hydraulic conductivity of the confining layer (aquitard) from

$$K' = \frac{Tb'(r/B)^2}{r^2} \tag{28.8}$$

$$K' = \frac{2.98 \times 10^{-3} m^2/s \cdot 2m \cdot 0.2^2}{(128m)^2} = 1.45 \times 10^{-8} m/s$$

(Note that the average thickness of the aquitard is 2 m as determined from the cross-section in Figure 28.1.)

It is important to remember that equation (28.1) has limited application because of the following assumptions introduced as part of its solution (in addition to assumptions applied to the Theis equation – see problem 24):

- The water table in the unconfined aquifer does not change during pumping;
- There is no water released from the aquitard;
- The piezometric surface of the confined aquifer does not fall below top of the overlying aquitard.

The first assumption is valid if the following condition is satisfied (after Neuman and Witherspoon, 1969):

$$T_{UA} = 100 \times T_{CA} \tag{28.9}$$

where

T_{UA} is transmissivity of the unconfined aquifer,
T_{CA} is transmissivity of the confined (pumped) aquifer.

In our case the transmissivity of the unconfined aquifer is

$$T_{UA} = K_{UA} \times b_{UA} = 8.45 \times 10^{-3} m/s \cdot 37m = 3.21 \times 10^{-1} m^2/s$$

and the first assumption is valid since

$$3.21 \times 10^{-1} m^2/s > 100 \cdot 2.98 \times 10^{-3} m^2/s$$

The rate of flow attributed to the release of water from storage in the aquitard can be ignored if the following condition is satisfied (Hantush, 1960):

$$t > \frac{0.036 \cdot b' \cdot S'}{K'} \tag{28.10}$$

where S' is storage coefficient of the aquitard. Since in our case this storage coefficient is not known, it has to be estimated. If S' is at least 10 times smaller than S ($S'=1.1x10^{-6}$), equation (28.10) gives:

$$36000s > \frac{0.036 \cdot 2m \cdot 1.1 \times 10^{-6}}{1.45 \times 10^{-8} m/s} > 5.5s$$

which means that the second assumption is valid and equation (28.6) can be applied to our data.

28.3 DRAWDOWN AT GIVEN DISTANCE AND TIME

Drawdown at 50 m from the pumping well after 10 hours of pumping is found using equation (28.6), and table values or type curves of the well function – $W(u,r/B)$ after calculating u and r/B:

$$u = \frac{r^2 S}{4Tt}$$

$$u = \frac{(50m)^2 \cdot 0.000011}{4 \cdot 2.98 \times 10^{-3} m^2/s \cdot 36000s} = 6.41 \times 10^{-5} \qquad \text{and} \qquad 1/u = 15600$$

$$\frac{r}{B} = \frac{r}{\sqrt{\frac{Tb'}{K'}}}$$

$$\frac{r}{B} = \frac{50m}{\sqrt{\dfrac{2.98 \times 10^{-3} m^2 / s \cdot 2m}{1.45 \times 10^{-8} m / s}}} = 0.078$$

For these values of $1/u$ and r/B, the well function is approximately 5.3 as determined from Figure 28.3. Finally, the drawdown is calculated by inserting all known values into equation (28.6) solved for s:

$$s = \frac{Q}{4\pi T} \cdot W(u, r / B)$$

$$s = \frac{0.012 m^3 / s}{4 \cdot 3.14 \cdot 2.98 \times 10^{-3} m / s} \cdot 5.3 = 1.70m$$

PROBLEM 29
UNCONFINED AQUIFER WITH DELAYED GRAVITY RESPONSE

An unconfined aquifer, with the average initial saturated thickness of 24 meters, was tested with a fully penetrating well pumping 35 liters per second constantly for 24 hours. The drawdown recorded at a monitoring well 31 meters from the pumping well is given in Table 29.1. Find hydrogeologic parameters of the aquifer.

Table 29.1 Drawdown (s) recorded at a monitoring well 31 m from the pumping well.

time (min)	s (m)	time (min)	s (m)	time (min)	s (m)
0.25	0.00	15	0.34	300	0.53
0.5	0.04	20	0.34	360	0.57
1	0.14	25	0.35	420	0.60
1.5	0.22	30	0.36	480	0.63
2	0.25	45	0.37	540	0.65
3	0.30	60	0.38	600	0.68
4	0.31	75	0.39	720	0.72
5	0.31	90	0.40	840	0.77
6	0.32	105	0.42	960	0.80
7	0.32	120	0.43	1080	0.86
8	0.33	150	0.45	1200	0.91
9	0.33	180	0.47	1320	0.99
10	0.33	240	0.50	1440	1.08

29.1 CHARACTERISTICS OF UNCONFINED FLOW TOWARD A WELL

The Theis equation, with all its assumptions, is derived for confined aquifers (see Problem 24). In general, if the drawdown in monitoring wells does not exceed 25% of the saturated thickness, the Theis equation can be applied to unconfined aquifers with certain adjustments. For drawdowns that are less than 10% of the aquifer's pre-pumping thickness, it is not necessary to adjust the recorded data since the error introduced by using the Theis equation is small. When the drawdown is between 10 and 25 percent, it is recommended to correct the measured values using the following equation derived by Jacob (1963):

$$s' = s - \frac{s^2}{2h} \tag{29.1}$$

where

- s' is corrected drawdown,
- s is measured drawdown in a monitoring well,
- h is saturated thickness of the unconfined aquifer before pumping started.

This correction is needed since the transmissivity of the aquifer changes during the test as the saturated thickness decreases (remember that, for unconfined aquifers, $T=Kb$ where b is the saturated thickness).

If the drawdown in a monitoring well is more than 25%, the Theis equation should not be used in the aquifer analysis.

The response of unconfined aquifers to well pumping tests is usually considerably different than the response of confined aquifers. One of Theis' assumptions in deriving the well equation is that water is withdrawn from storage instantaneously with a decrease in hydraulic head. Although this assumption, for practical purposes, may be considered "true enough" for most confined aquifers, it is not acceptable for most unconfined aquifers because of a lag in releasing water from storage. This lag is caused by slow gravity drainage of the porous media within the cone of depression and above the hydraulic head surface, especially at the beginning of pumping. As a result, the storage coefficient determined from the early drawdown data and using the (unmodified) Theis equation, will be underestimated. A more realistic value of the storage coefficient is obtained from the late drawdown data when the cone of depression spreads at a lower rate and the gravity drainage comes to an equilibrium with other pumping influences. Figure 29.1, which is a plot of the data given in Table 29.1, shows the typical delayed response of unconfined aquifers to pumping tests (note that the original data were not corrected using equation (29.1) since the maximum drawdown is less than 10% of the initial saturated thickness). When the porous medium has a high effective porosity, the aquifer's response will be less delayed and may look similar to confined aquifers in which case the methods described in Problem 24 can be used directly.

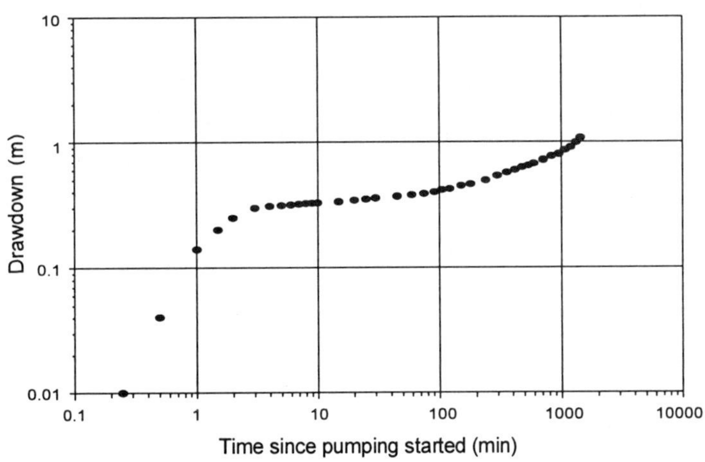

Figure 29.1 Log-log graph *time* versus *drawdown* recorded at 31 meters from a pumping well in an unconfined aquifer.

A log-log graph of the recorded drawdown versus time in Figure 29.1 indicates three distinct segments as the result of a delayed aquifer response:

1. The early segment, which in our case represents the first 4–5 minutes of pumping, shows a rapid drawdown similar to confined conditions. Almost all water supplied to the well comes from the aquifer storage in the saturated zone. Gravity water above the hydraulic head within the cone of depression still has not reached the saturated zone. The coefficient of storage during this stage is close to a confined aquifer of the same porous material as that of the unconfined aquifer.
2. The intermediate segment, between 5 and approximately 100 minutes, is a subhorizontal (flat) curve which indicates that the gravity water is reaching the saturated zone, but is still not in equilibrium with the saturated flow.
3. The third (late) segment represents an equilibrium between the gravity drainage and the saturated flow when the delayed response ceases. During this third period of the pumping test, the storage properties are those of a truly unconfined aquifer and are termed *specific yield.*

Transient flow toward a well in an unconfined aquifer with a delayed response to pumping was first analyzed by Boulton (1954a and b). Based on his theory, commonly called the Boulton method, an approach of analyzing water wells with delayed gravity response using type curves can be found in most textbooks on groundwater. Slight differences between circulating versions originate from modifications, simplifications, and graphical solutions that have been introduced to the method by various authors, including Boulton (e.g., Stallman, 1961a, 1961b; 1963, 1965; Boulton, 1963; 1970; Prickett, 1965; Boulton and Pontin, 1971; Neuman, 1972, 1975).

Presented here is the Neuman method which allows the determination of the hydrogeologic parameters of unisotropic unconfined aquifers with delayed response to pumping from field drawdown data, *i.e.*, the transmissivity, the horizontal and vertical hydraulic conductivities, the elastic storage coefficient, and the specific yield.

29.2 NEUMAN METHOD

During pumping from a fully penetrating well, the drawdown at a fully penetrating observation well at distance r and at time t is (Neuman, 1975):

$$s_{r,t} = \frac{Q}{4\pi T}\int_0^\infty 4yJ_0(y\beta^{1/2})[u_0(y)+\sum_{n=1}^{\infty}u_n(y)]dy \tag{29.2}$$

where

$$u_0(y) = \frac{\{1-\exp[-t_s\beta(y^2-\gamma_0^2)]\}\tanh(\gamma_0)}{\{y^2+(1+\sigma)\gamma_0^2-[(y^2-\gamma_0^2)^2/\sigma]\}\gamma_0} \tag{29.3}$$

$$u_n(y) = \frac{\{1-\exp[-t_s\beta(y^2+\gamma_n^2)]\}\tanh(\gamma_n)}{\{y^2-(1+\sigma)\gamma_n^2-[(y^2+\gamma_n^2)^2/\sigma]\}\gamma_n} \tag{29.4}$$

and the terms γ_0 and γ_n are the roots of the equations

$$\sigma\gamma_0\sinh(\gamma_0)-(y^2-\gamma_0^2)\cosh(\gamma_0)=0 \tag{29.5}$$

$$\text{for}\quad \gamma_0^2<y^2$$

$$\sigma\gamma_n \sin(\gamma_n) + (y^2 + \gamma_n^2)\cos(\gamma_n) = 0 \qquad (29.6)$$
$$\text{for} \quad (2n-1)(\pi/2) < \gamma_n < n\pi \quad \text{and} \quad n \geq 1$$

The notification in equations (29.2) through (29.6) is:

s	drawdown at observation well,
r	radial distance from pumping well,
t	time since pumping started,
Q	pumping rate,
T	transmissivity,
$J_0(x)$	zero-order Bessell function of the first kind,
y	a variable given as $y=x(K_D)^{1/2}$,
K_D	degree of anisotropy given as $K_D=K_z/K_r$,
K_z	vertical hydraulic conductivity,
K_r	horizontal hydraulic conductivity,
β	dimensionless parameter given as $\beta=K_D r^2/b^2$,
b	initial saturated thickness of the aquifer,
σ	dimensionless parameter given as $\sigma=S/S_y$,
S	storage coefficient given as $S=S_s b$,
S_s	specific (elastic) storage,
S_y	specific yield.

Equation (29.2) is expressed in terms of independent dimensionless parameters σ and β, and dimensionless time given either as time with respect to S_s (early data when the elastic storage is dominant):

$$t_s = \frac{Tt}{Sr^2} \qquad (29.7)$$

or as time with respect to S_y (specific yield is dominant):

$$t_y = \frac{Tt}{S_y r^2} \qquad (29.8)$$

For field drawdown data, equation (29.2) is solved with the aid of type curves shown in Figure 29.2. The corresponding numerical values computed by Neuman are given in Appendix E. The graph consists of two merged families of curves: type A curves lying to the left of the central (flat) graph area, and Type B curves lying to the right. The curves are plotted for various values of dimensionless parameter β, dimensionless times t_s and t_y, and dimensionless drawdown s_D which is given as:

$$s_D = \frac{4\pi T}{Q} s \qquad (29.9)$$

Note that type A curves are used with early drawdown data and correspond to the top scale expressed in terms of t_s (elastic storage is dominant). Type B curves are intended for use with late drawdown data when specific yield is dominant (true unconfined conditions without delayed response). They correspond to the bottom scale expressed in terms of s_y.

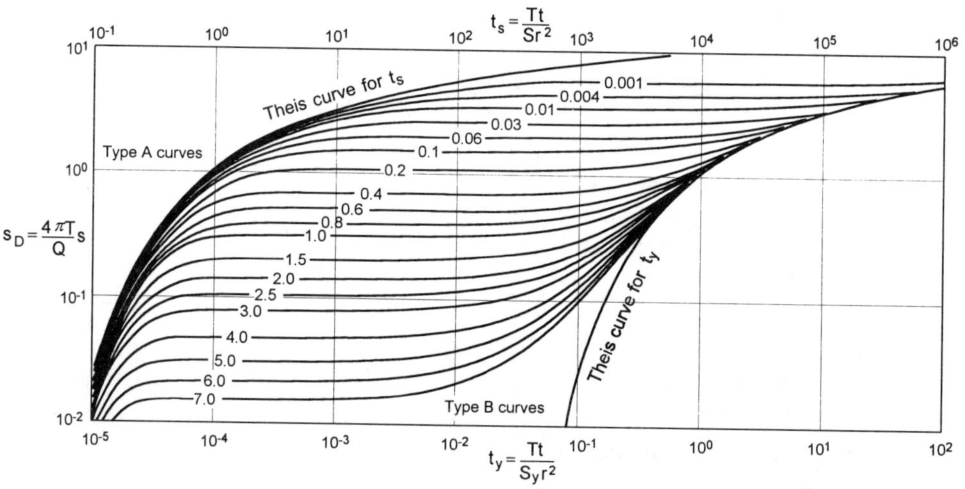

Figure 29.2 Type curves for fully penetrating wells in unconfined aquifers with delayed gravity response. (S. Neuman, 1975. Analysis of pumping test data from anisotropic unconfined aquifers considering delayed gravity response. *Water Resources Research*, 11(2), p. 329–342. Copyright by the American Geophysical Union.)

The procedure for calculating hydrogeologic parameters is as follows:

- Plot log-log graph *time* versus recorded *drawdown* at the same log-cycle scale as the graph with type curves (see Figure 29.1).
- Superimpose the field curve (data) on the type B curves and, keeping the axes of both graphs parallel, match as much of the late drawdown data as possible to a particular β curve (Figure 29.3).
- Choose a match point anywhere where the two graphs overlap and read the four coordinates as follows: s_D and t_y from the type graph, and s and t from the field graph. It is recommended to choose rounded values for s_D and t_y (bottom scale) for easier calculations. In our case these "B" coordinates are:

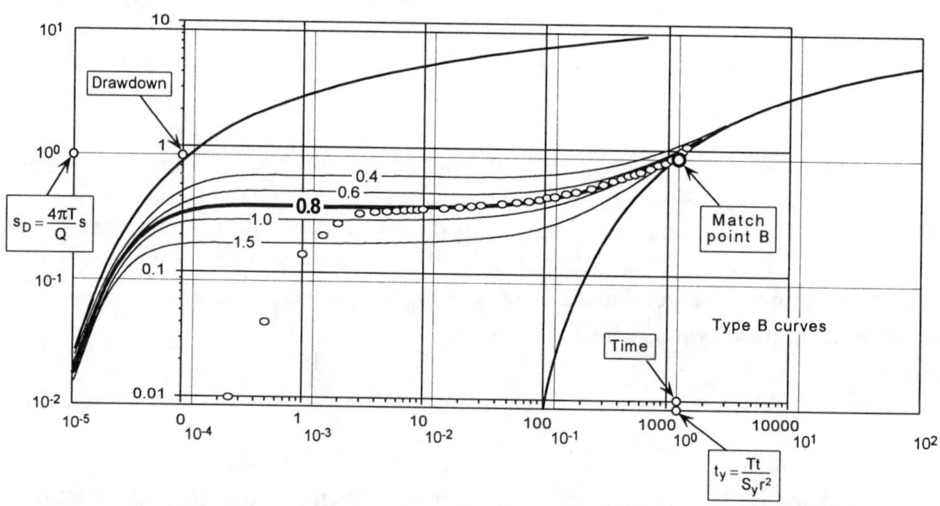

Figure 29.3 Matching late drawdown data to a type B curve. The best fit is for the curve β=0.8. The match point is chosen conveniently (s_D=*1* and t_y=*1*) where both graphs overlap. Note that the coordinate t_y is read on the lower axis of the type graph.

dimensionless drawdown s_D	10^0 *i.e., 1*
dimensionless time t_y	10^0 *i.e., 1*
drawdown s	*0.85m*
time t	*1200 min*

- Calculate the transmissivity using the following equation:

$$T = \frac{1}{4\pi} \cdot \frac{Q \cdot s_D}{s} \tag{29.10}$$

which in our case gives:

$$T = 0.0796 \cdot \frac{0.035m^3 / s \cdot 1}{0.85m} = 3.28 \times 10^{-3} m^2 / s$$

- Calculate the specific yield using equation:

$$S_y = \frac{Tt}{r^2 t_y} \tag{29.11}$$

or, in our case:

$$S_y = \frac{3.28 \times 10^{-3} m^2 / s \cdot 1200 \cdot 60s}{(31m)^2 \cdot 1} = 0.246$$

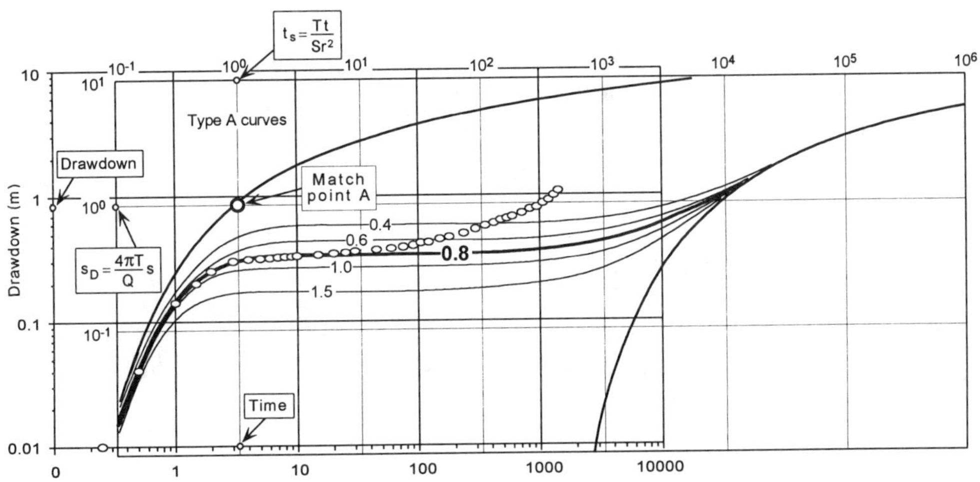

Figure 29.4 Matching early drawdown data to a type A curve. The curve must have the same value of β as the previously matched type B curve (0.8 in our case). The match point is again chosen conveniently ($s_D=1$ and $t_s=1$) where both graphs overlap. Note that the t_s coordinate is read on the upper axis of the type graph.

- Overlap the early field data (curve) and the A type curve which has the same value of β as the B curve (0.8 in our case). While doing this try to match as much data as possible to the A type curve keeping the axes on both graphs parallel (see Figure 29.4).

(Note that since data have to be matched to the same β curve, the graphs should move only in the horizontal direction).

- Choose a match point anywhere where the two graphs overlap and read the four coordinates as follows: s_D and t_s (lower axis) from the type graph, and s and t from the field graph. Again, it is recommended to choose rounded values for s_D and t_s for easier calculations. These "A" coordinates are:

dimensionless drawdown s_D	10^0 *i.e.,* 1
dimensionless time t_y	10^0 *i.e.,* 1
drawdown s	$0.85\ m$
time t	$3.3\ min$

- Calculate the transmissivity again using equation (29.10). The value should be the same as the one calculated for the late data:

$$T = 0.0796 \cdot \frac{0.035 m^3 / s \cdot 1}{0.85 m} = 3.28 \times 10^{-3} m^2 / s$$

- Calculate the elastic storage coefficient using equation:

$$S = \frac{Tt}{r^2 t_s} \tag{29.11a}$$

$$S = \frac{3.28 \times 10^{-3} m^2 / s \cdot 3.3 \cdot 60 s}{(31 m)^2 \cdot 1} = 0.00068$$

- Find the horizontal hydraulic conductivity knowing the average aquifer thickness (b=24m) and the previously calculated transmissivity:

$$K_r = \frac{T}{b} \tag{29.12}$$

$$K_r = \frac{3.28 \times 10^{-3} m^2 / s}{24 m} = 1.37 \times 10^{-4} m / s$$

- Find the degree of anisotropy (K_D), which is needed to calculate the vertical hydraulic conductivity, using equation:

$$K_D = \frac{\beta b^2}{r^2} \tag{29.13}$$

$$K_D = \frac{0.8 \cdot (24 m)^2}{(31 m)^2} = 0.48$$

- And finally, find the vertical hydraulic conductivity from:

$$K_z = K_D \cdot K_r \tag{29.14}$$

$$K_z = 0.48 \cdot 1.37 \times 10^{-4}\, m/s = 6.58 \times 10^{-4}\, m/s$$

If the saturated thickness of the aquifer decreases more than 10 % during the test, Neuman (1975) recommends that the recorded drawdown should be corrected using Jacobs formula (equation 29.1) for the late data only. Correcting early drawdown would lead to erroneous results since the response of the aquifer to pumping is primarily due to the elastic properties of the porous media and water (*i.e.*, the elastic storage is predominant).

PROBLEM 30
WELLS NEAR BOUNDARIES

A 24-hour aquifer pumping test with a constant rate of 15 liters per second was conducted in a fully penetrating well completed in a confined aquifer. The recorded drawdown in the three monitoring wells is given in Table 30.1. Monitoring wells MW1, MW2, and MW3 are 20, 60, and 80 meters from the pumping well PW respectively (Figure 30.1). A semi-log graph *time* vs. *drawdown* at the monitoring wells (Figure 30.2) suggests that there is a low-permeable boundary near the pumping well which causes an abrupt increase in drawdown at all three monitoring wells after approximately 30 to 90 minutes from the time pumping started. Find the probable location of the boundary.

30.1 SUPERPOSITION OF FLOWS

As explained earlier in Problem 23, the hydraulic effects of physical boundaries located near water wells can be simulated by introducing imaginary (image) wells and applying the theory of superposition of flows. If there are several monitoring wells at different distances and directions from the pumping well, it may be possible to locate a single boundary. Such a boundary would theoretically be a straight line representing an abrupt change in aquifer transmissivity, something not always true in reality. However, locating approximately a concealed (unnoticed) impermeable or low-permeable boundary using the image theory is useful: in many cases it can successfully redirect field investigations to confirm the real nature and position of the contact.

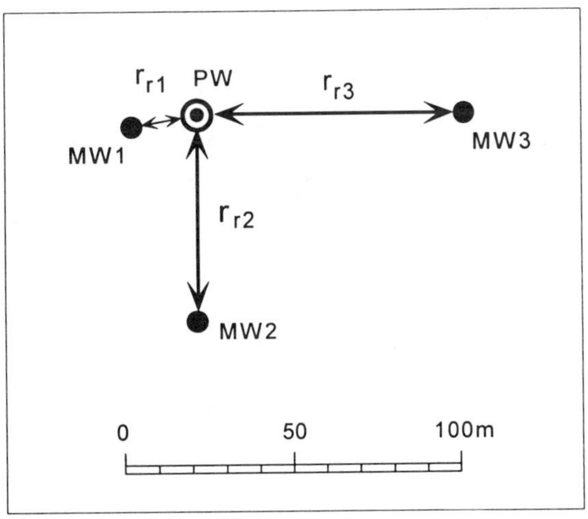

Figure 30.1 Map view of the locations of pumping well PW and monitoring wells MW1, MW2, and MW3.

Table 30.1 Drawdown, in meters, recorded in monitoring wells MW1, MW2, and MW3.

time (min)	MW1 (m)	MW2 (m)	MW3 (m)
1	0.027	0.001	0.001
2	0.098	0.014	0.001
3	0.142	0.034	0.005
4	0.181	0.058	0.011
5	0.209	0.085	0.019
10	0.298	0.167	0.058
15	0.367	0.210	0.097
20	0.399	0.257	0.119
25	0.431	0.296	0.156
30	0.453	0.320	0.175
45	0.602	0.370	0.218
60	0.746	0.430	0.264
75	0.828	0.517	0.288
90	0.917	0.610	0.310
105	0.986	0.700	0.370
120	1.020	0.750	0.422
150	1.163	0.867	0.519
180	1.201	0.942	0.593
210	1.288	1.032	0.675
240	1.331	1.052	0.707
300	1.442	1.150	0.850
360	1.500	1.214	0.904
420	1.587	1.283	0.976
480	1.635	1.347	1.040
540	1.678	1.382	1.083
600	1.747	1.428	1.128
720	1.836	1.515	1.203
840	1.935	1.605	1.269
960	1.992	1.693	1.340
1080	2.014	1.742	1.387
1200	2.072	1.802	1.425
1320	2.101	1.870	1.486
1440	2.160	1.921	1.541

In our case, the boundary is close to the pumping and monitoring wells since its effect is noticed in a relatively short time (see Figure 30.2). The drawdown recorded in MW1 changes the slope after approximately 30 to 45 minutes, followed by MW2 and MW3. This also means that MW1 is closest to the boundary and MW3 is the farthest.

An impermeable boundary is simulated with another discharging well located the same distance from the boundary as the real well, but on the opposite side. The real and image wells are on the same line perpendicular to the boundary. At the time when the cone of depression formed by the real well reaches the boundary, the image well starts pumping at the same rate. Superposition of flows (drawdowns) of the two wells creates the same effect as the boundary. The resulting drawdown at any distance from the real pumping well is the algebraic sum of two drawdowns -- that of the real well (s_r), and that of the image well (s_i):

$$s = s_r + s_i \tag{30.1}$$

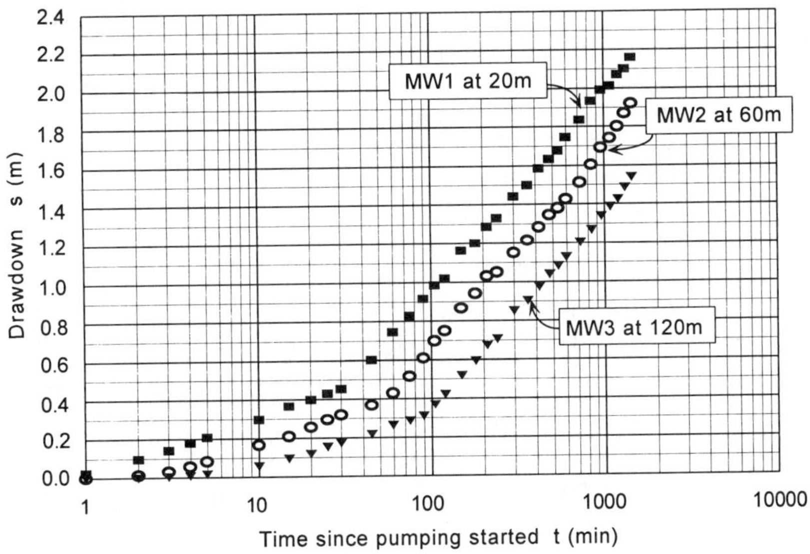

Figure 30.2 Semi-log graph *time* versus *drawdown* recorded in the monitoring wells.

Equation (30.1) can be expanded using either the Theis formula or the Cooper-Jacob formula. Consequently, there are two methods for analyzing the influence of the boundary and locating it.

30.2 USE OF COOPER-JACOB FORMULA

Use of the Cooper-Jacob approximation formula is recommended because change in the slope of a straight line drawn through field data is easier to notice than in a log-log curve (compare Figures 30.2 and 30.5). The formula is valid for values of $u<0.05$, *i.e.*, for sufficiently long pumping periods and/or small distances between the pumping well and the monitoring wells. The time-drawdown curve of the real well is represented by a straight line on a semi-log graph:

$$s_r = \frac{2.3Q}{4\pi T}\log\frac{2.25\cdot T\cdot t_r}{r_r^2 S} \tag{30.2}$$

where

t_r is the time since pumping started within the first time period when there is only the influence of the real well (the cone of depression did still not reach the boundary),

r_r is the distance between the real well and the monitoring well.

A straight line of the image well is:

$$s_i = \frac{2.3Q}{4\pi T}\log\frac{2.25\cdot T\cdot t_i}{r_i^2 S} \tag{30.3}$$

where

t_i is the time since pumping started within the time period after inclusion of the image well,

r_i is the distance between the image well and the monitoring well.

Figure 30.3 illustrates the process for finding the distance between monitoring well MW2 and the image well. The same is later repeated for the two other wells (or, in general, for all monitoring wells available):

- The straight line, drawn through data points representing the operation of only the real well, is extended through the duration of the test. The difference in drawdown anywhere between this line and the line formed after the inclusion of the image well (steep slope), represents the influence of the image well: s_i. The corresponding time t_i is noted.

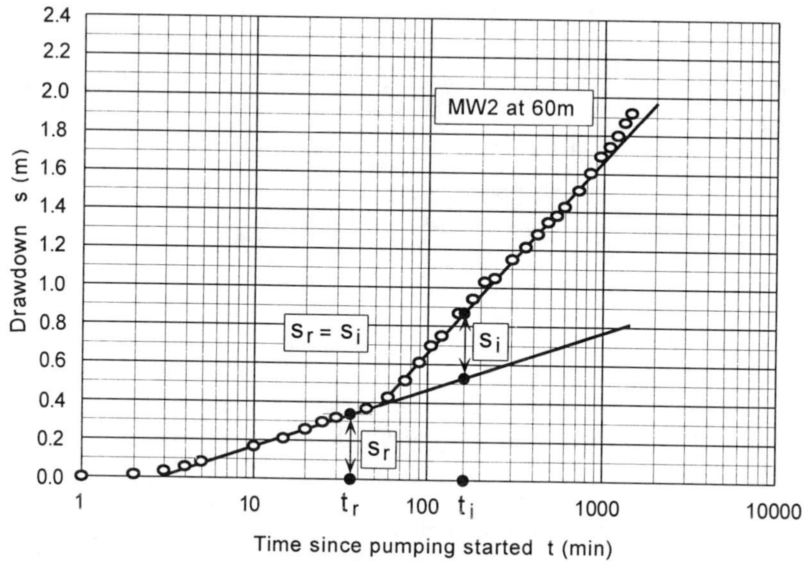

Figure 30.3 Elements for determining the distance between MW2 and the image well using the straight-line method (Cooper-Jacob formula).

- Time at which the real well produces the same drawdown as the image well (s_r=s_i) is noted for the first portion of the curve (first straight line): t_r.
- Since the two drawdowns are equal, from equations (30.2) and (30.3) it follows that

$$\frac{2.25T \cdot t_r}{r_r^2 S} = \frac{2.25T \cdot t_i}{r_i^2 S} \tag{30.4}$$

and

$$r_i = r_r \cdot \sqrt{\frac{t_i}{t_r}} \tag{30.5}$$

In our case, after inserting the values determined from the semi-log graph in Figure 30.3 into equation (30.5), the distance between MW2 and the image well is

$$r_i = 60m \cdot \sqrt{\frac{165min}{36min}} = 128m$$

After repeating this procedure for MW1 and MW3, the corresponding distances are 102 m and 163 m, respectively. Theoretically, if all assumptions were valid, the image well would be defined as the (only) point of intersection of all three circles having the corresponding calculated radii, and whose centers are at MW1, MW2, and MW3. Often, however, the circles will not all intersect in one point and an estimated weighted location would have to be chosen as in our case (see Figure 30.4).

Once the image well is located, the boundary is drawn at the midpoint and perpendicular to the line connecting the two pumping wells.

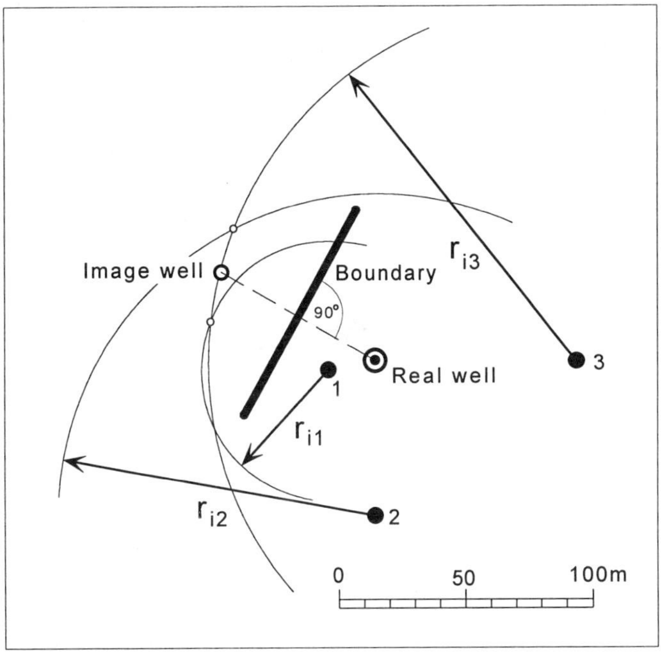

Figure 30.4 Location of the image well and the impermeable boundary using equation (30.5).

30.3 THEIS TYPE CURVE METHOD

Figure 30.5 is a log-log graph of time versus drawdown recorded in the three monitoring wells. Again, the shape of the three curves defined by data points changed between 30 and 90 minutes after pumping started, but less obviously than in the semi-log graph in Figure 30.2. Similar to the straight line method, the influence of the image well is found as the difference in drawdown between the extended theoretical Theis curve matched to the early data, and the curve formed by late data (see Figure 30.6). Time corresponding to this drawdown is t_i. Paying attention to the log scale, exactly the same drawdown (s_r) is scaled vertically on the early time (Theis) curve. The corresponding time is t_r.

Using the Theis equation, the drawdowns from the real and image wells are:

$$s_r = \frac{Q}{4\pi T} W(u_r) \qquad s_i = \frac{Q}{4\pi T} W(u_i) \tag{30.6}$$

where

$$u_r = \frac{r_r^2 S}{4Tt_r} \qquad u_i = \frac{r_i^2 S}{4Tt_i} \tag{30.7}$$

Since the two drawdowns are equal, from equation (30.7) it follows that

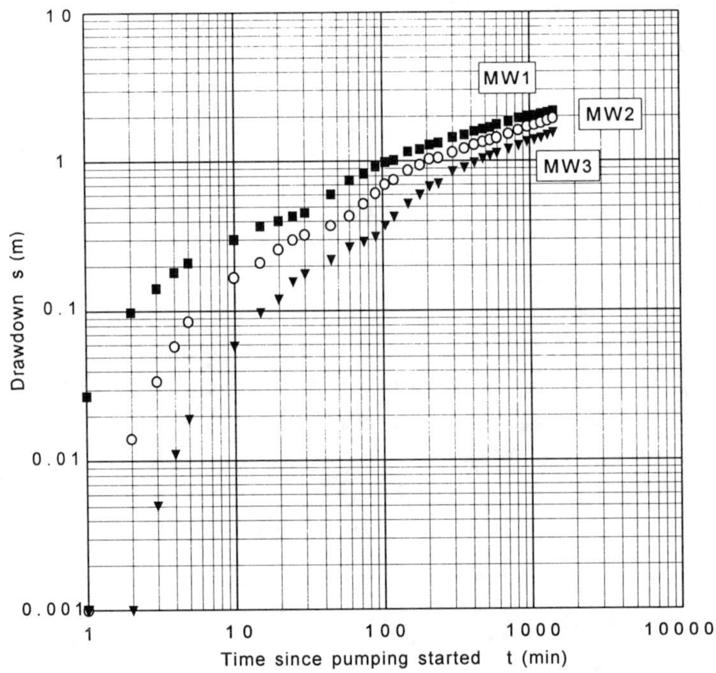

Figure 30.5 Log-log graph *time* versus *drawdown* recorded in the monitoring wells.

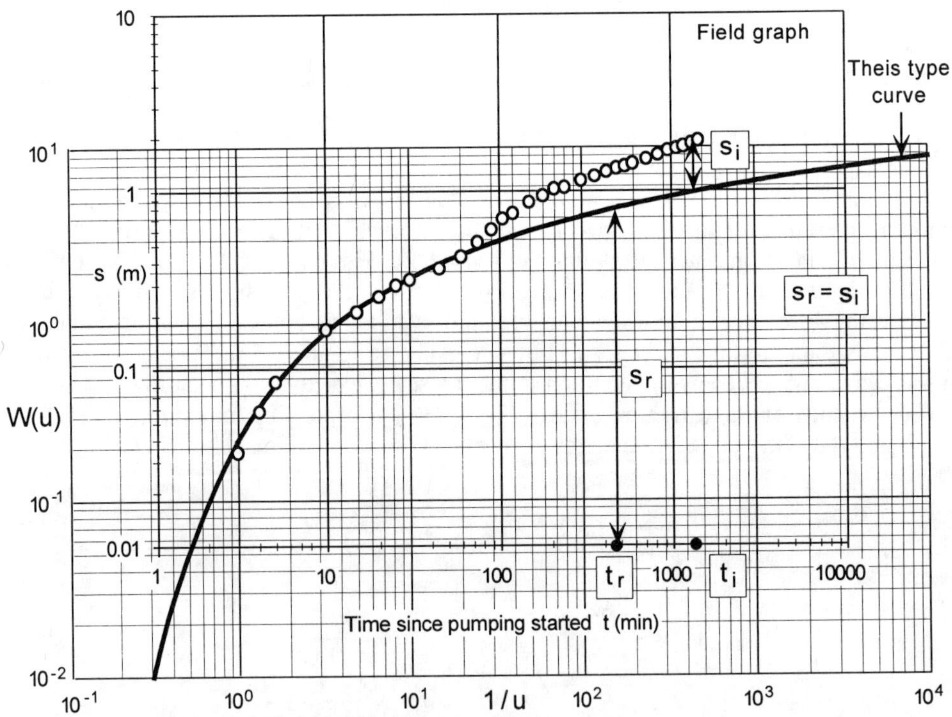

Figure 30.6 Elements for determining the distance between MW2 and the image well using the curve-matching method (Theis formula).

$$\frac{r_r^2}{t_r} = \frac{r_i^2}{t_i} \tag{30.8}$$

and

$$r_i = r_r \cdot \sqrt{\frac{t_i}{t_r}} \tag{30.9}$$

which is the same as equation (30.5) derived using the Cooper-Jacob approximation. In our case times t_i and t_r are 2350 min and 470 min, respectively as determined from the graph in Figure 30.6. The distance between MW2 and the image well is then:

$$r_i = 60m \cdot \sqrt{\frac{2350min}{470min}} = 134m$$

which is close to the value calculated using the Cooper-Jacob approximation. After finding the distance between the image well and the other two monitoring wells, the boundary is located as shown in Figure 30.4.

PROBLEM 31
SLUG TEST

The results of a slug test performed in a monitoring well completed in an unconfined aquifer are given in Tables 31.1 and 31.2. The test was for both rising and falling heads measured with a pressure transducer and recorded with a digital data logger. The monitoring well is partially penetrating with the total depth of 25 ft. The water table in the well before the test was at 24.3 ft above the aquifer base, and 9.2 ft below the land surface. A screen with diameter of 4 inches is placed in the bottom 9 ft of the well. The diameter of the borehole in the screen section is 9 in. Space between the side of the borehole and the screen is filled with a gravel pack. From the available data find the hydraulic conductivity of the aquifer.

31.1 CHARACTERISTICS OF A SLUG TEST

Slug tests are performed in exploratory boreholes and small-diameter monitoring wells to find an approximate hydraulic conductivity in their immediate vicinity. Despite this limitation, and the fact that most methods in use cannot determine the storage properties, slug test are widely applied in groundwater studies because of the following advantages:

- the cost is incomparably smaller than the cost of an average pumping test (not including the cost of a pumping well itself);
- the time needed for its preparation and the duration of the test are short;
- there is no need for observation wells;
- the test can be performed in exploratory boreholes (including inexpensive soil borings) that are only temporarily equipped with a screen;
- the results are not influenced by short-term or long-term changes such as variations of barometric pressure, or interference from nearby pumping wells.

Table 31.1 Data logger recordings of the rising head slug test: *start time* -- time since test started in minutes; *corr. time* -- time since the first displacement recorded; *dispmt.* – displacement in feet; *D.T.W.* – depth to water in feet (data courtesy Tom Fowler, Foster Wheeler Environmental Corporation).

start time in min	*corr. time* in min	*dispmt* in feet	*D.T.W.* in feet	*start time* in min	*corr. time* in min	*dispmt* in feet	*D.T.W.* in feet
0			11.49	0.75	0.25	0.2	11.69
0.0033			11.49	0.8333	0.3333	0.16	11.65
0.0066			11.49	0.9167	0.4167	0.13	11.62
0.0099			11.49	1	0.5	0.1	11.59
0.0133			11.49	1.0833	0.5833	0.09	11.58
0.0166			11.49	1.1667	0.6667	0.08	11.57
0.02			11.49	1.25	0.75	0.07	11.56
0.0233			11.49	1.3333	0.8333	0.06	11.55
0.0266			11.49	1.4166	0.9166	0.05	11.54
0.03			11.49	1.5	1	0.05	11.54
0.0333			11.49	1.5833	1.0833	0.04	11.53
0.05			11.49	1.6667	1.1667	0.04	11.53
0.0666			11.49	1.75	1.25	0.03	11.52
0.0833			11.49	1.8333	1.3333	0.03	11.52
0.1			11.49	1.9167	1.4167	0.03	11.52
0.1166			11.49	2	1.5	0.03	11.52
0.1333			11.49	2.5	2	0.02	11.51
0.15			11.49	3	2.5	0.02	11.51
0.1666			11.49	3.5	3	0.01	11.5
0.1833			11.49	4	3.5	0.01	11.5
0.2			11.49	4.5	4	0.01	11.5
0.2166			11.49	5	4.5	0.01	11.5
0.2333			11.49	5.5	5	0.01	11.5
0.25			11.49	6	5.5	0	11.49
0.2666			11.49	6.5	6	0	11.49
0.2833			11.49	7	6.5	0	11.49
0.3			11.49	7.5	7	0	11.49
0.3166			11.49	8	7.5	0	11.49
0.3333			11.49	8.5	8	0	11.49
0.4167			11.95	9	8.5	0	11.49
0.5	0	0.54	12.03	9.5	9	0	11.49
0.5833	0.0833	0.37	11.86	10	9.5	0	11.49
0.6667	0.1667	0.27	11.76				

Table 31.2 Data logger recordings of the falling head slug test: *start time* – time since test started in minutes; *corr. time* – time since the first displacement recorded; *dispmt.* – displacement in feet; *D.T.W.* – depth to water in feet (data courtesy Tom Fowler, Foster Wheeler Environmental Corporation).

start time in min	*corr. time* in min	*dispmt* in feet	*D.T.W.* in feet	*start time* in min	*corr. time* in min	*dispmt* in feet	*D.T.W.* in feet
0	0	0.4	11.09	0.75	0.75	0.08	11.41
0.0033	0.0033	0.4	11.09	0.8333	0.8333	0.07	11.42
0.0066	0.0066	0.39	11.1	0.9167	0.9167	0.06	11.43
0.0099	0.0099	0.39	11.1	1	1	0.05	11.44
0.0133	0.0133	0.38	11.11	1.0833	1.0833	0.05	11.44
0.0166	0.0166	0.38	11.11	1.1667	1.1667	0.04	11.45
0.02	0.02	0.37	11.12	1.25	1.25	0.04	11.45
0.0233	0.0233	0.37	11.12	1.3333	1.3333	0.04	11.45
0.0266	0.0266	0.36	11.13	1.4166	1.4166	0.04	11.45
0.03	0.03	0.36	11.13	1.5	1.5	0.03	11.46
0.0333	0.0333	0.35	11.14	1.5833	1.5833	0.03	11.46
0.05	0.05	0.33	11.16	1.6667	1.6667	0.03	11.46
0.0666	0.0666	0.32	11.17	1.75	1.75	0.03	11.46
0.0833	0.0833	0.37	11.12	1.8333	1.8333	0.03	11.46
0.1	0.1	0.31	11.18	1.9167	1.9167	0.03	11.46
0.1166	0.1166	0.27	11.22	2	2	0.03	11.46
0.1333	0.1333	0.26	11.23	2.5	2.5	0.02	11.47
0.15	0.15	0.25	11.24	3	3	0.02	11.47
0.1666	0.1666	0.23	11.26	3.5	3.5	0.01	11.48
0.1833	0.1833	0.22	11.27	4	4	0.01	11.48
0.2	0.2	0.21	11.28	4.5	4.5	0.01	11.48
0.2166	0.2166	0.2	11.29	5	5	0.01	11.48
0.2333	0.2333	0.2	11.29	5.5	5.5	0.01	11.48
0.25	0.25	0.19	11.3	6	6	0.01	11.48
0.2666	0.2666	0.18	11.31	6.5	6.5	0.01	11.48
0.2833	0.2833	0.18	11.31	7	7	0.01	11.48
0.3	0.3	0.17	11.32	7.5	7.5	0.01	11.48
0.3166	0.3166	0.16	11.33	8	8	0.01	11.48
0.3333	0.3333	0.16	11.33	8.5	8.5	0	11.49
0.4167	0.4167	0.13	11.36	9	9	0	11.49
0.5	0.5	0.11	11.38	9.5	9.5	0	11.49
0.5833	0.5833	0.1	11.39	10	10	0	11.49
0.6667	0.6667	0.08	11.41				

Since hydraulic conductivity calculated from a slug test is an estimate for a very small part of the aquifer around the borehole, it is recommended to test as many boreholes in the area of interest as possible to find an average hydraulic conductivity. (On the other hand, the transmissivity and the storage coefficient from a single well pumping test already represent weighted averages of a comparatively much larger area affected by pumping.)

In general, a slug test consists of either adding or extracting quickly a volume of water from the borehole (well) and then recording the change in the head at the borehole. When water is instantaneously added to the well, the head is displaced upwards and then starts to fall as water enters the aquifer (the hydraulic gradient is from the higher head in the well

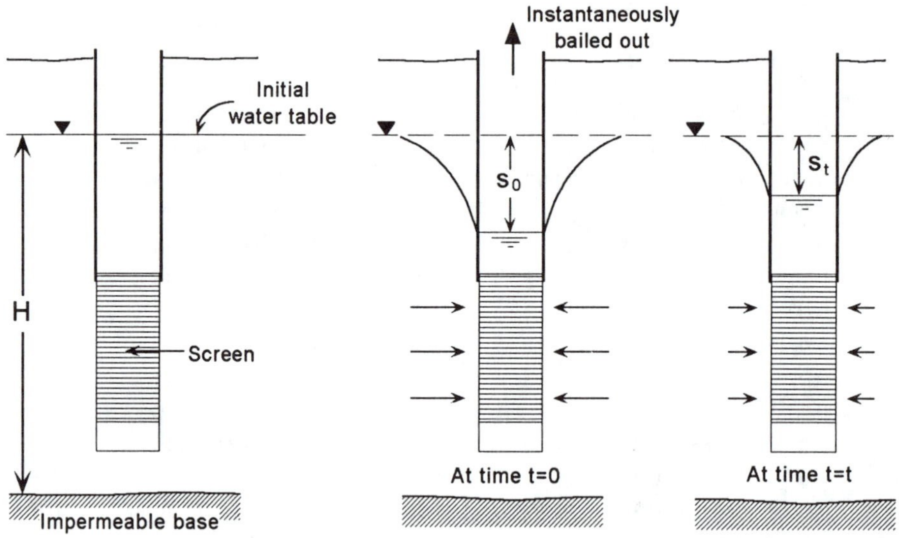

Figure 31.1 Elements of a rising head slug test. Water is instantaneously withdrawn from the well with a bailer which causes displacement in the hydraulic head at the well (s_0). Water from the aquifer starts to enter the well because of the created gradient, which causes the head in the well to rise.

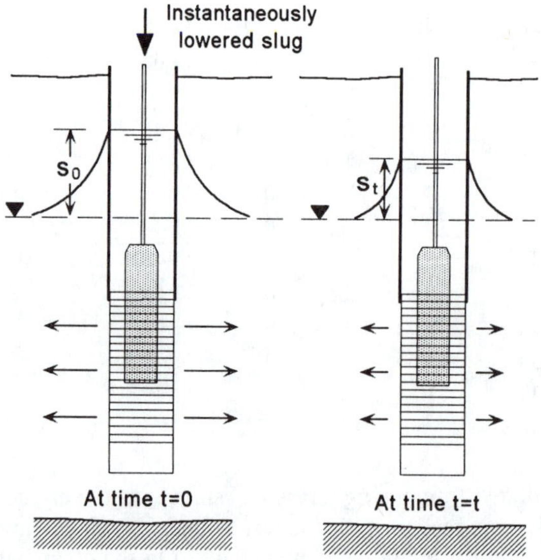

Figure 31.2 Elements of a falling head slug test. A water-tight metal cylinder (slug) on a rod is instantaneously lowered in the well which causes a displacement (rise) in the hydraulic head (s_0). The volume of displaced water is equal to the volume of the slug. Water from the well starts to enter the aquifer because of the created gradient, which causes the head in the well to fall.

toward a lower head in the surrounding porous medium). This type of slug test is called the *falling head test* (see Figure 31.1). Water does not have to be added to the well in order to displace the head. A more common practice is to instantaneously lower into the well a heavy water-tight cylinder (usually metal) on a rod which produces the same effect of head displacement.

When water is quickly extracted from the well, usually with a bailer, the head is displaced downwards and then starts to rise as water from the surrounding porous medium (aquifer) enters the well. This is called the *rising head test* (Figure 31.2).

Since the amount of water added to, or extracted from a borehole is relatively small, the head displacement is also small and the recovery is fast (usually not more than several minutes in moderately permeable aquifers). For this reason it is necessary to take as many measurements of the head in the well as possible in order to have enough data for a graphoanalytical analysis. This is most accurately done with the help of a pressure transducer and a digital data logger which can be programmed to take recordings every few seconds, or with even smaller intervals during the initial portion of the test (see Tables 31.1 and 31.2).

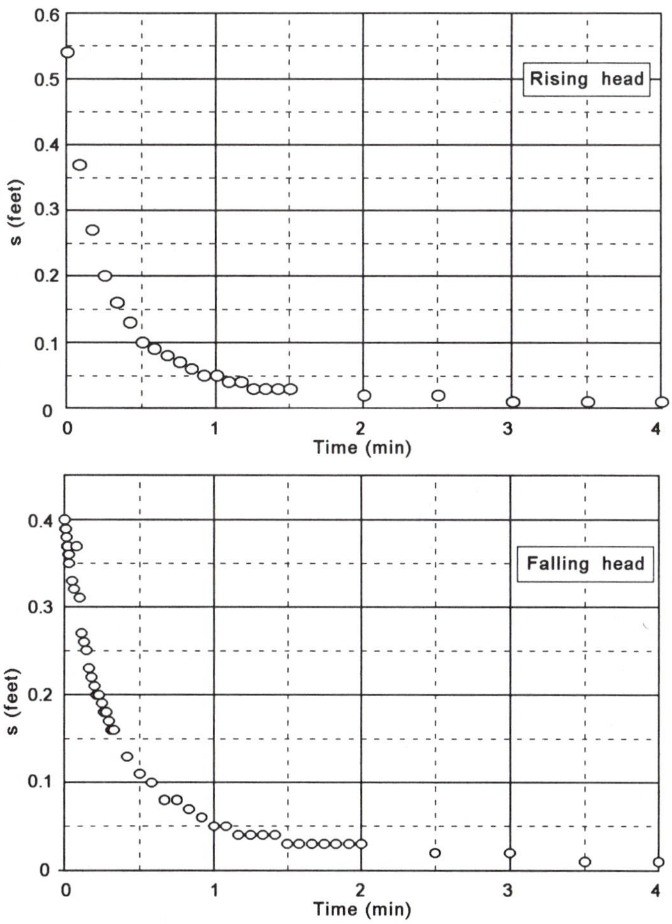

Figure 31.3 Arithmetic graphs of *time* versus *head displacement* (*s*) recorded during the two slug tests in the monitoring well.

Figure 31.3 shows the arithmetic plots *time* versus *displacement* (s) for both the falling head and the rising head slug tests in our case. As can be seen, the two graphs look very similar because in both cases the displacement is given as a difference between the water level (head) measured in the well at time t and the initial head measured before the test started (at time zero). These two elements -- *displacement* (in one form or another) and *time*, are used in practically all methods of slug test analysis.

31.2 BOUWER AND RICE METHOD

The Bouwer and Rice slug test can be used for fully or partially penetrating and partially screened, perforated, or otherwise open wells (Bouwer and Rice, 1976; Bouwer, 1989). Although initially developed for unconfined aquifers, the test has been successfully used for confined aquifers given that the top of the screened section is below the overlying confining layer (Bouwer, 1989).

The elements needed for determining hydraulic conductivity from the test data are shown in Figure 31.4. For reasons of consistency, the displacement of water level in the well is here denoted with s (equivalent to drawdown) rather than y as in the original work.

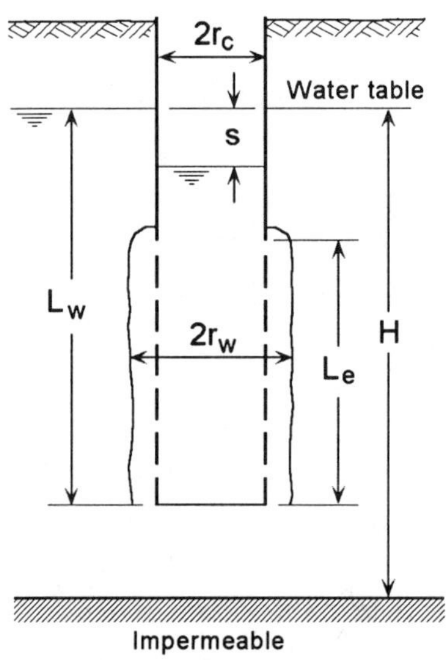

Figure 31.4 Geometry and symbols for the Bouwer and Rice slug test in an unconfined aquifer with gravel pack and/or developed zone around screen (Bouwer, H., 1989: The Bouwer and Rice slug test. *Ground Water*, 27(3), p. 304–309. Reprinted by permission of Ground Water Publishing Company. Copyright 1989. All rights reserved).

The rate of water flow into the well, when the head in the well is lower than the initial water table in the surrounding aquifer, is given by the Thiem equation:

$$Q = 2\pi K L_e \frac{s}{\ln \frac{R_e}{r_w}} \tag{31.1}$$

where

- Q is flow rate,
- K is hydraulic conductivity,
- L_e is length of screened (perforated, open) section of the well,
- s is vertical difference between water level inside the well and water level outside well (it is assumed that water table in the aquifer does not change for duration of the test),

R_e is effective radial distance over which s is dissipated,
r_w is radial distance of undisturbed portion of aquifer from centerline.

Values of R_e were determined experimentally by the authors and expressed in terms of the dimensionless ratio $\ln(R_e/r_w)$ which is calculated with the aid of the curves shown in Figure 31.5. The value of r_w is the radius of the screen plus the thickness of the gravel pack plus the thickness of the developed zone. In practice, the thickness of the developed zone is usually ignored since it is very difficult to define accurately.

The rate of water level rise in the well is:

$$\frac{ds}{dt} = -\frac{Q}{\pi r_c^2} \tag{31.2}$$

where r_c is the radius of the casing or other section of the well where the rise of water level is measured. If the water rises in the screened portion of the well and there is a gravel pack around the screen, r_c should be adjusted to account for the total free-water surface area in the well and the gravel pack:

$$r_c^* = [(1-n)\cdot r_c^2 + nr_w^2]^{1/2} \tag{31.3}$$

where n is effective porosity of the gravel pack.

Solving equation (31.2) for Q, equating the resulting expression to equation (31.1), then integrating, and solving for K gives the Bouwer-Rice equation (Bouwer, 1989):

$$K = \frac{r_c^2 \ln\frac{R_e}{r_w}}{2L_e} \cdot \frac{1}{t} \cdot \ln\frac{s_0}{s_t} \tag{31.4}$$

where

s_0 is displacement at time zero,
s_t is displacement at time t.

When the well is partially penetrating ($L_w<H$), the dimensionless ratio $\ln(R_e/r_w)$ is:

$$\ln\frac{R_e}{r_w} = \left[\frac{1.1}{\ln(L_w/r_w)} + \frac{A + B\cdot\ln[(H-L_w)/r_w]}{L_e/r_w}\right]^{-1} \tag{31.5}$$

For fully penetrating wells this ratio is:

$$\ln\frac{R_e}{r_w} = \left[\frac{1.1}{\ln(L_w/r_w)} + \frac{C}{L_e/r_w}\right]^{-1} \tag{31.6}$$

where A, B, and C are dimensionless numbers plotted in Figure 31.5 as a function of L_e/r_w.

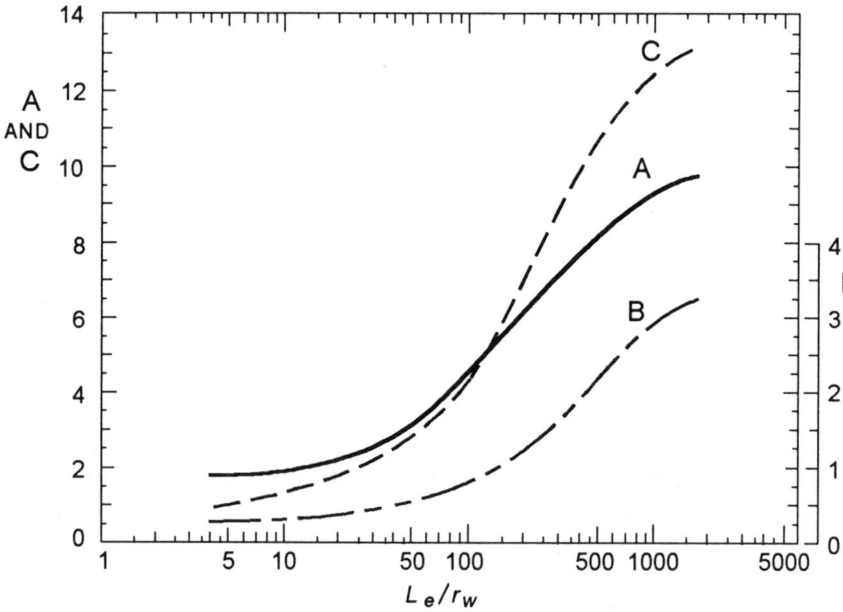

Figure 31.5 Dimensionless parameters A, B, and C as a function of L_e/r_w for calculation of $\ln(R_e/r_w)$ in the Bouwer and Rice slug test. (Bouwer, H., 1989: The Bouwer and Rice slug test. *Ground Water*, 27(3), p. 304–309. Reprinted by permission of Ground Water Publishing Company. Copyright 1989. All rights reserved.)

31.3 HYDRAULIC CONDUCTIVITY

Since time (t) and displacement (s) are the only variables in logarithmic equation (31.4), the plot of t versus s on a semi-log paper must show a straight line. However, because the drawdown of the water table in the aquifer becomes more significant during the later part of the test, the basic assumption of equation (31.4) does not hold any more and data points start to deviate from the straight line.

The slope of the best-fitting straight line through field data is found as:

$$\frac{\ln(s_1 / s_2)}{(t_2 - t_1)}$$

and then inserted into equation (31.4) with other known/calculated values to determine the hydraulic conductivity. s_1, t_1 and s_2, t_2 are the coordinates of any two points on the straight line. If the first point is chosen for time zero, the slope is then given as in equation (31.4).

$$\frac{\ln(s_0 / s_t)}{t - t_0} = \frac{1}{t} \ln \frac{s_0}{s_t}$$

As in our case (see Figure 31.6), field data may often show the so-called *double straight line effect*. The first straight line portion is explained by a highly permeable zone around the well (gravel pack or developed zone), which quickly sends water into the well immediately after the water level in the well has been lowered (Bouwer, 1989). The second straight line is more indicative of the flow from the undisturbed aquifer into the well and its slope should be used in equation (31.4).

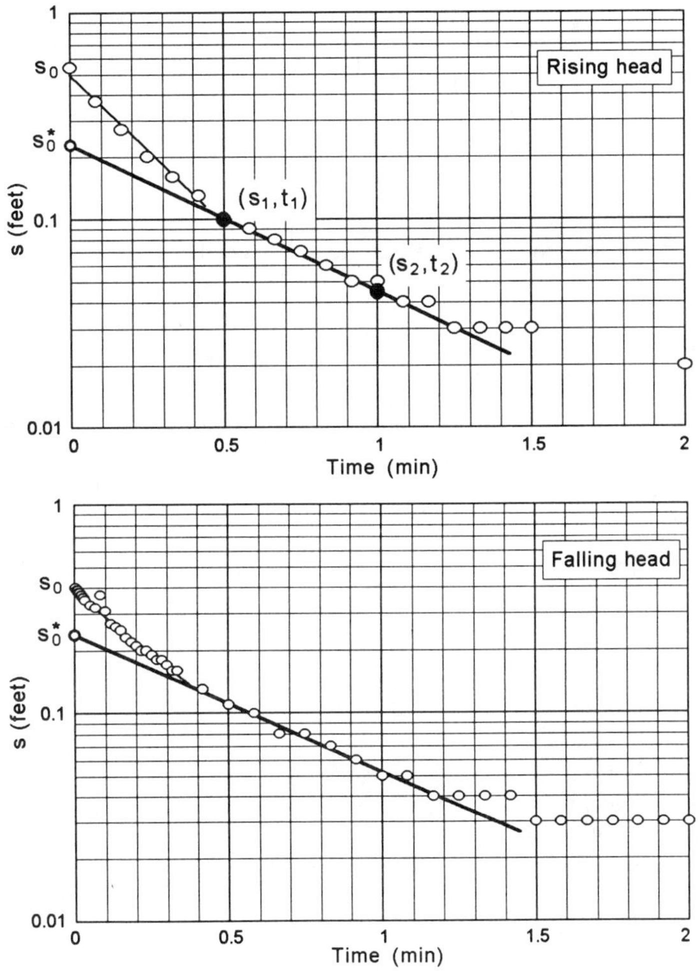

Figure 31.6 Semi-log graphs *time* versus *head displacement* (s) recorded during the two slug tests in the monitoring well.

Since the test well in our case is partially penetrating, the ratio $\ln(R_e/r_w)$ is found using equation (31.5) and curves A and B in Figure 31.5. For our well the ratio between the well depth (L_e) and the radius of developed zone (r_w; see Figure 31.4) is:

$$\frac{L_e}{r_{r_w}} = \frac{9\,ft}{4.5in} = \frac{2.74m}{0.114m} = 24 \qquad \text{(note that 1m=3.28ft=39.37in.)}$$

which gives the following values of the dimensionless parameters A and B (see Figure 31.6):

$$A = 2.45$$
$$B = 0.40$$

Inserting these values into equation (31.5) gives:

$$\ln\frac{R_e}{r_w} = \left[\frac{1.1}{\ln(L_w / r_w)} + \frac{A + B \cdot \ln[(H - L_w) / r_w]}{L_e / r_w}\right]^{-1}$$

$$\ln\frac{R_e}{r_w} = \left[\frac{1.1}{\ln(4.82m / 0.114m)} + \frac{2.45 + 0.40 \cdot \ln[(7.41m - 4.82m) / 0.114m]}{24}\right]^{-1} = 2.233$$

The slope of the second straight line for the rising head test (see Figure 31.6) is:

$$\frac{\ln(s_1 / s_2)}{(t_2 - t_1)} = \frac{\ln(0.1 ft / 0.045 ft)}{1 \min - 0.5 \min} = 1.597 \min^{-1}$$

After inserting this and other known values into equation (31.4), the hydraulic conductivity is found as:

$$K = \frac{r_c^2 \ln\frac{R_e}{r_w}}{2L_e} \cdot \frac{1}{t} \cdot \ln\frac{s_0}{s_t}$$

$$K = \frac{(0.102m)^2 \cdot 2.233}{2 \cdot 2.74m} \cdot 1.597 \times 60s^{-1} = 4.42 \times 10^{-5} m / s$$

As can be seen in Figure 31.6, the second straight line in the graph for the falling head test has almost the identical slope as the corresponding line for the rising head test. This means that the hydraulic conductivity calculated from the falling head test would be almost the same as K already calculated from the rising head test.

Chapter 9

SPRING FLOW

PROBLEM 32
RECESSION ANALYSIS

Analyze the recession period of the spring hydrograph shown in Figure 32.1. Find the volume of water stored in the aquifer at the beginning of recession. Calculate the spring discharge rate 3 months after the beginning of recession if there was no precipitation for the entire period.

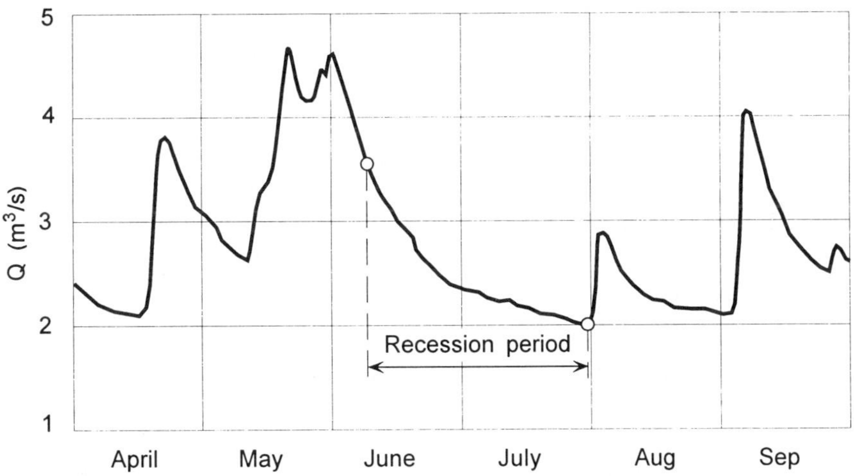

Figure 32.1 Part of the annual spring discharge hydrograph with the recession period.

32.1 DISCHARGE HYDROGRAPH

The hydrograph of a spring is the final result of various processes that govern the transformation of precipitation and other water inputs in the spring's drainage area into the flow at the point of discharge. In many cases, the discharge hydrograph of a spring closely resembles hydrographs of surface streams, particularly if the aquifer is unconfined and has a high transmissivity. In some low permeable media, such as silty and clayey sediments, or unaltered, unfractured metamorphic rocks, springs are weak and usually do not react visibly to daily, weekly, or even monthly (seasonal) water inputs. On the other hand, the reaction of large springs draining karst or intensely fissured aquifers to precipitation events is sometimes only a matter of hours. A thorough analysis of a spring discharge hydrograph provides useful information on aquifer characteristics such as the nature of its storage and transmissivity, and types and quantity of its groundwater reserves.

Although the processes that generate hydrographs of springs and surface streams are quite different, there is much that is analogous between them, and the hydrograph terminology is the same. Figure 32.2 shows the main elements of a discharge hydrograph. The beginning of discharge after a rainfall episode is marked with point A, and the time between the beginning of rainfall and the beginning of discharge, called *the starting time*, with t_s. The time in which the hydrograph raises to its maximum (point *C*), is called *the concentration time* – t_c. The time from maximum discharge until the end of the hydrograph, when the discharge theoretically equals zero (point *E*), is *the falling time* – t_f. Together, the concentration time and the falling time are called *the base time* – t_b. The time between the centroid of the precipitation episode (C_P) and the centroid of the hydrograph (C_H) is called *the retardation time* (t_r). The time interval for recording the amount of precipitation and the flow rate at the spring is Δt.

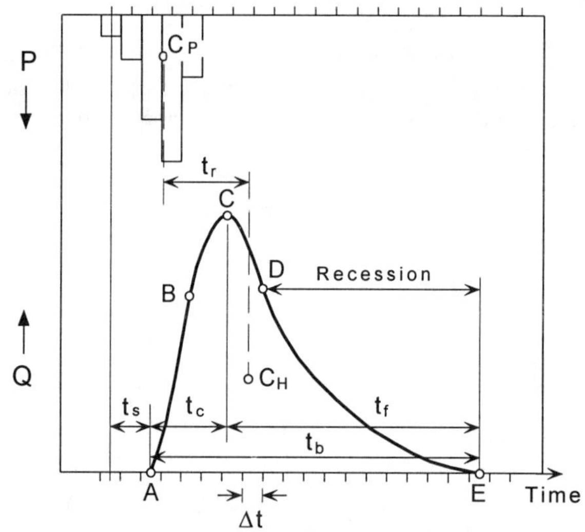

Figure 32.2 Components of a discharge hydrograph. Explanation in text.

The shape of the hydrograph is defined by its base (*AE*), the rising limb (*AB*), the crest (*BCD*) and the falling limb (*DE*). The falling limb also corresponds to the recession period. *B* and *C* are *inflexion points* where the hydrograph curve changes its shape from convex to concave and vice versa. For surface streams, point *D* is the end of direct runoff after the rain. In the case of a spring hydrograph, point *D* is the beginning of *recession* – discharge without direct influence of precipitation.

The shape of a discharge hydrograph depends upon the size and shape of the drainage area, as well as precipitation intensity. When a rainfall episode lasts longer, and the intensity is lower, the hydrograph has a bigger time base and vice versa: intensive short storm events cause sharp hydrographs with small time bases.

The area under the hydrograph is the volume of discharged water for the recording period. In reality, unless a spring or a surface stream is intermittent, the recorded hydrograph has a more complex shape which reflects the influence of antecedent precipitation and other possible water inputs. Such hydrographs are formed by the superposition of single hydrographs corresponding to separate precipitation events (see Figure 32.3).

32.2 HYDROGRAPH COMPONENT SEPARATION

The separation of a surface stream hydrograph is the determination of the individual elements that participate in flow formation. Theoretically, they are divided into: flow formed by direct precipitation over the surface stream, surface runoff collected by the stream, near-surface flow of the newly infiltrated water (also called underflow), and groundwater inflow. However, it is practically impossible to separate all these components of discharge from a real physical drainage area. In practice, the problem of component separation is therefore reduced to an estimation of the *base flow*, formed by groundwater, and *surface runoff*, which is the integration of all the other components. In the case of a spring hydrograph, the equivalents are the discharge formed by the draining of previously accumulated groundwater, and the discharge that is the result of newly infiltrated water inflow.

Base Flow of Surface Streams

The first method of hydrograph component separation shown in Figure 32.4 (line *ABC*) is commonly applied to surface streams with significant groundwater inflow. Assuming

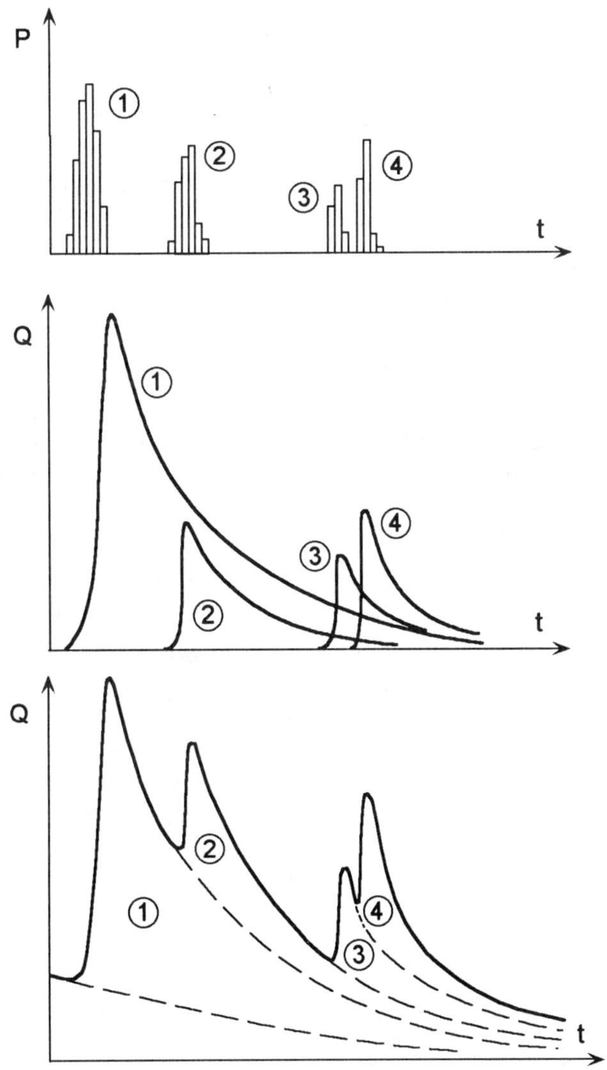

Figure 32.3 A complex discharge hydrograph (bottom) composed of single hydrographs (middle) as the result of isolated precipitation events (top) (after Jevdjevic, 1956, simplified).

that point C represents the end of all surface runoff, and the beginning of flow generated solely by groundwater drainage, the late near-straight line section of the hydrograph is extrapolated backwards until it intersects the ordinate of the maximum discharge (point B). Point A, representing the beginning of surface runoff after rainfall, and point B are then connected with the straight line. The area under the line ABC is the base flow, *i.e.*, the groundwater component of the surface stream flow. Point C is easily found from a log-log plot of time versus flow rate as a part of the hydrograph recession analysis (see Section 32.3).

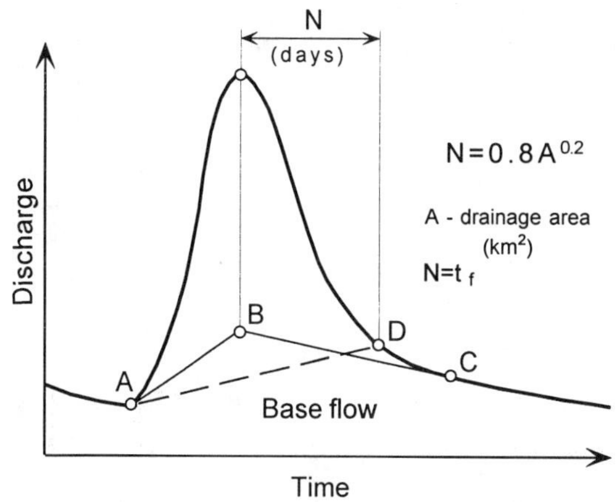

Figure 32.4 Two common methods of graphical separation of the components of surface stream hydrographs.

The second graphical method of base flow separation is used for surface streams in low permeable terrain without significant groundwater flow. It is highly conditional since point D (*i.e.*, the hydrograph falling time) is found by the following empirical formula (Linsley et al., 1975):

$$t_f = 0.8A^{0.2} \quad \text{(days)} \tag{32.1}$$

where A is the drainage area in square kilometers. In general, this method gives short falling times: for an area of 100 km^2 t_f is 2 days, and for 10,000 km^2 t_f is 5 days. Thus, the method should be applied cautiously after analyzing a sufficient number of single hydrographs and establishing of an adequate *area-time* relationship.

The portion of a spring hydrograph that corresponds to the base flow component of a surface stream hydrograph is attributed partially to a decreasing hydraulic gradient, and partially to a slow release of water from smaller voids (small fissures, matrix porosity).

The impact of new rainwater on spring discharge varies with respect to the predominant type of porosity and the stage of groundwater level. In any case, it should be noted that the first reaction of highly permeable karst or fissured aquifers to recharge, *i.e.*, the initial increase of discharge rate, is in most cases the consequence of pressure propagation through karst conduits/large fractures, and not the outflow of newly infiltrated water. The new water arrives at the spring with certain delay and its contribution to the overall flow rate is often less than 50%.

After a recharge episode and direct runoff are over, two cases may occur:

1) An increase in reserves of groundwater accumulated in storage, which is reflected in the shifting of the base flow to a higher hydrograph level (case a in Figure 32.5), and

2) New water is mostly transmitted through a well-developed network of fissures or karst channels/conduits. It is then discharged without significant accumulation of groundwater in the aquifer's matrix porosity (case *b* in Figure 32.5) and the base flow continues along the same extrapolated recession line present before the rainfall event.

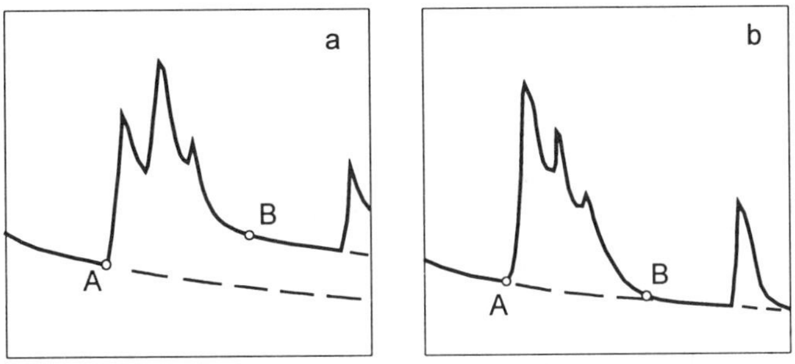

Figure 32.5 Possible reaction of aquifers to infiltration of recent precipitation as seen from spring discharge hydrographs. a) Lifting of the base flow to a higher hydrograph level as a consequence of storage increase; b) Discharge without storage increase.

The first case is characteristic for period(s) of the main aquifer recharge (March-June in moderate climates): both groundwater level and moisture content in the aquifer's unsaturated zone are high, while evapotranspiration loss is relatively small compared to other periods of the year. Newly infiltrated water adds to an already high piezometric pressure and groundwater is more easily injected into the aquifer's matrix porosity and narrow fissures. During the summer/autumn period the groundwater level and the hydraulic gradients are low, and new water is transmitted mainly through large fissures/conduits. This second case may also indicate lack of a significant matrix/fissure porosity in the aquifer.

32.3 RECESSION ANALYSIS

Analysis of the falling hydrograph limb (*DE* line in Figure 31.2), which corresponds to a period without significant precipitation, is called the *recession analysis*. Knowing that the spring discharge is without disturbances caused by a rapid inflow of new water into the aquifer, the recession analysis provides good insight into the aquifer structure. By establishing an appropriate mathematical relationship between spring discharge and time, it is possible to predict the discharge rate after a given period without precipitation, and to calculate the volume of discharged water. For these reasons, recession analysis has been a popular quantitative method in hydrogeological studies for a long time.

The shape and characteristics of a recession curve depend upon different factors such as type of aquifer porosity (the most important), position of piezometric surface, and recharge from other aquifers. The ideal recession conditions -- a long period of several months without precipitation, are rare in moderate/humid climates. Consequently, summer and early fall storms can cause various disturbances in the recession curve that cannot be removed unambiguously during analysis. It is therefore desirable to analyze as many recession curves from different years as possible. Larger sample allows for the derivation of the average recession curve as well as the envelope of minima (Figure 32.6). In addition, conclusions about the porosity structure, its accumulative ability, and expected long-term minimum discharge, are more accurate.

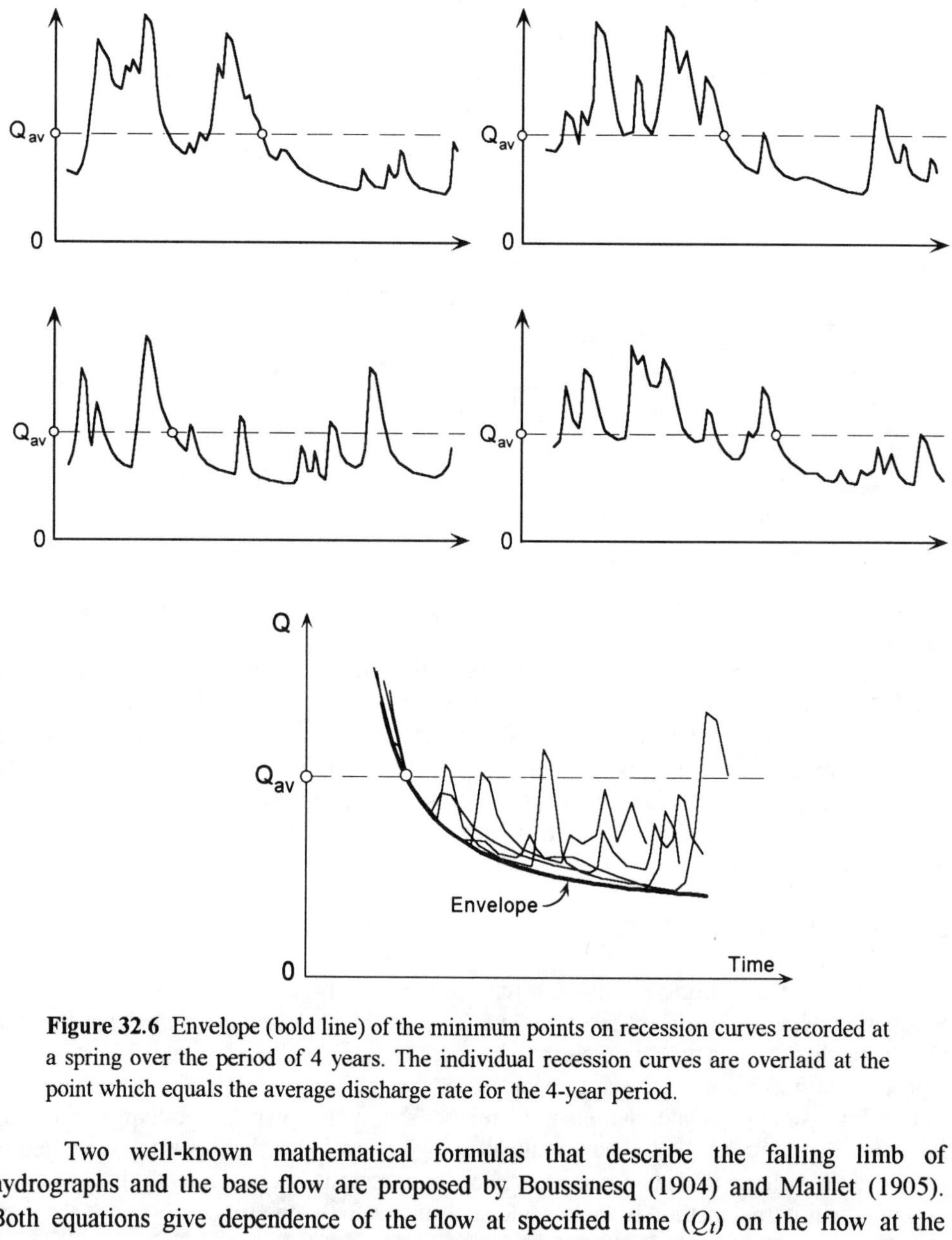

Figure 32.6 Envelope (bold line) of the minimum points on recession curves recorded at a spring over the period of 4 years. The individual recession curves are overlaid at the point which equals the average discharge rate for the 4-year period.

Two well-known mathematical formulas that describe the falling limb of hydrographs and the base flow are proposed by Boussinesq (1904) and Maillet (1905). Both equations give dependence of the flow at specified time (Q_t) on the flow at the beginning of recession (Q_0). The Boussinesq equation is of hyperbolic form:

$$Q_t = \frac{Q_0}{[1 + \alpha(t - t_0)]^2} \qquad (32.2)$$

where

t is time since the beginning of recession for which the flow rate is calculated,

t_0 is time at the beginning of recession usually (but not necessarily) set equal to zero.

The Maillet equation, which is more commonly used, is an exponential function:

$$Q_t = Q_0 \cdot e^{-\alpha(t-t_0)} \tag{32.3}$$

The dimensionless parameter α in both equations represents the *coefficient of discharge* (or *recession coefficient*) which depends upon the aquifer's transmissivity and specific yield. The Maillet equation, when plotted on a semi-log diagram, is a straight line with the coefficient of discharge (α) being its slope:

$$\log Q_t = \log Q_0 - 0.4343 \cdot \alpha \cdot \Delta t \qquad \Delta t = t - t_0 \tag{32.4}$$

$$\alpha = \frac{\log Q_0 - \log Q_t}{0.4343 \cdot (t - t_0)} \tag{32.5}$$

Introduction of the conversion factor (0.4343) is a convenience for expressing discharge in equation (32.5) in cubic meters per second, and time in days. Dimension of α is day^{-1}.

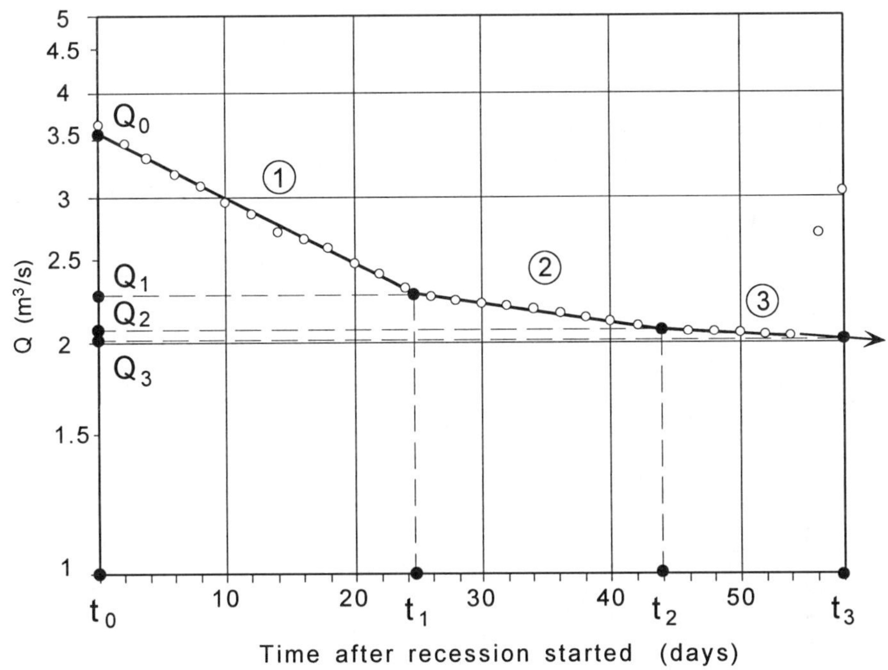

Figure 32.7 Semi-log graph of *time* versus *discharge* for the recession period shown in Figure 32.1. The duration of the recession period is 54 days.

Figure 32.7 is a semi-log plot of time versus discharge rate for the recession period shown in Figure 32.1. The recorded daily discharges form three straight lines which means that the recession curve can be approximated by three corresponding exponential functions with three different coefficients of discharge (α). The three lines correspond to three *micro-*

micro-regimes of discharge during the recession. The coefficient of discharge for the first micro-regime, using equation (32.5), is:

$$\alpha_1 = \frac{\log Q_0 - \log Q_1}{0.4343 \cdot (t_1 - t_0)}$$

$$\alpha_1 = \frac{\log(3.55 m^3/s) - \log(2.25 m^3/s)}{0.4343 \cdot 24.5\ days} = 0.019$$

The coefficient of discharge for the second micro-regime is (see Figure 32.6):

$$\alpha_2 = \frac{\log(2.25 m^3/s) - \log(2.06 m^3/s)}{0.4343 \cdot (44d - 24.5d)} = 0.0045$$

The third coefficient of discharge, or slope of the third straight line, is found by choosing a discharge rate anywhere on the line, including its extension if the actual line is short. In our case the discharge rate after 60 days is 2.01 m^3/s and α_3 is then:

$$\alpha_3 = \frac{\log(2.06 m^3/s) - \log(2.01 m^3/s)}{0.4343 \cdot (60d - 44d)} = 0.0015$$

After determining the coefficients of discharge, the flow rate at any given time after the beginning of recession can be calculated using the appropriate Maillet equation. For example, discharge of the spring 35 days after the recession started, when the second micro-regime is active, is calculated as:

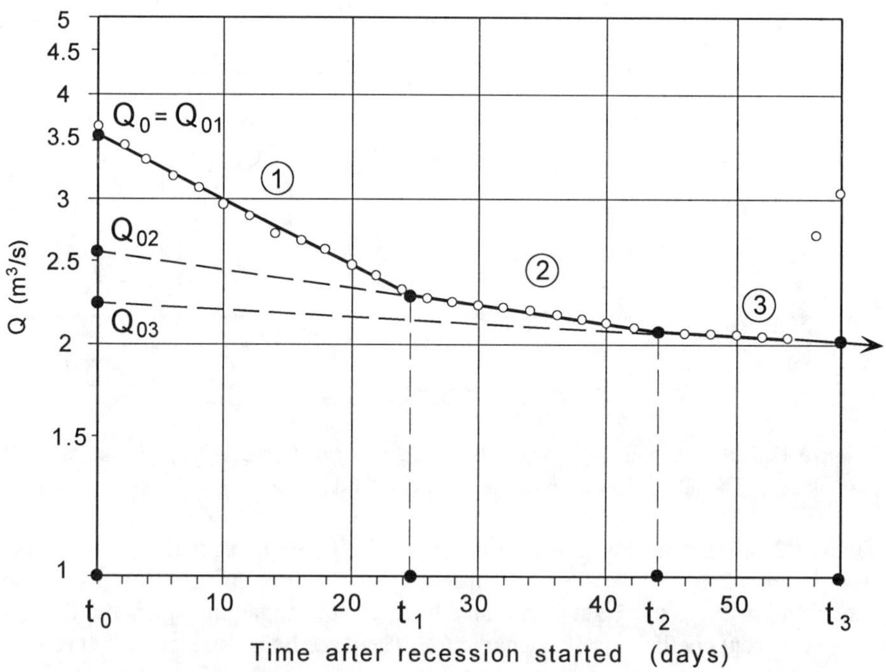

Figure 32.8 Alternative initial discharge rates for the three micro-regimes shown in Figure 32.6.

$$Q_{35} = Q_1 \cdot e^{-\alpha_2(35d - t_1)}$$

$$Q_{35} = 2.25m^3 / s \cdot e^{-0.0045(35d-24.5d)} = 2.146m^3 / s$$

Note that the initial discharge rate for the second micro-regime is 2.25 m^3/s and the corresponding time is 24.5 days.

Alternatively, discharge after 35 days may be determined by extending the second straight line backward to where it intersects the Q axis and then using that value in the calculation as the initial discharge rate (see Figure 32.8):

$$Q_{35} = Q_{02} \cdot e^{-\alpha_2(35d - t_0)}$$

$$Q_{35} = 2.55m^3 / s \cdot e^{-0.0045 \cdot 35d} = 2.178m^3 / s$$

Spring discharge 3 months after the beginning of recession, assuming no precipitation for the entire period, may be predicted by using the characteristic values for the third (or last-detected) micro-regime (see Figure 32.7):

$$Q_{90} = Q_2 \cdot e^{-\alpha_3(90d - t_2)}$$

$$Q_{90} = 2.06m^3 / s \cdot e^{-0.0015(90d-44d)} = 1.923m^3 / s$$

or, alternatively (see Figure 32.8):

$$Q_{90} = Q_{03} \cdot e^{-\alpha_3(90d - t_0)}$$

$$Q_{90} = 2.20m^3 / s \cdot e^{-0.0015(90d)} = 1.922m^3 / s$$

The variation of the coefficient of discharge has a physical explanation. It is accepted in practice that α of the order 10^{-2} indicates rapid drainage of well interconnected large fissures/fractures (or karst channels in case of karst aquifers), while milder slopes of the recession curve (α of the order 10^{-3}) represent slow drainage of small voids, *i.e.*, narrow fissures and aquifer matrix porosity. Accordingly, the main contribution to the spring discharge in our case is from storage in small voids.

The coefficient of discharge (α) and the volume of free gravitational groundwater stored in the aquifer above spring level (*i.e.*, groundwater that contributes to spring discharge), are inversely proportional:

$$\alpha = \frac{Q_t}{V_t} \tag{32.6}$$

where Q_t is the discharge rate at time t, and V_t is the volume of water stored in the aquifer above the level of discharge (spring level). This relationship is valid only for descending gravitational springs.

Equation (32.6) allows for the calculation of the volume of water accumulated in the aquifer at the beginning of recession, as well as the volume discharged during a given period of time. The calculated remaining volume of groundwater always refers to the reserves stored above the current level of discharge. The draining of an aquifer with three micro-regimes of discharge (as in our case), and the corresponding volumes of the discharged water are shown in Figure 32.9. The total initial volume of groundwater stored in the aquifer (above the level of discharge) at the beginning of the recession period is the sum of the three volumes that correspond to three different types of storage (effective porosity):

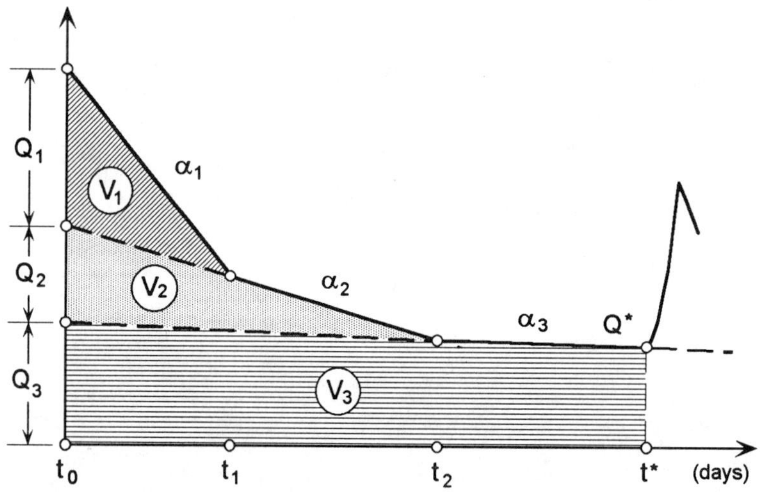

Figure 32.9 Schematic presentation of the recession with three micro-regimes of discharge, and the corresponding volumes of water discharged from the three different types of storage.

$$V_0 = V_1 + V_2 + V_3 = \left[\frac{Q_1}{\alpha_1} + \frac{Q_2}{\alpha_2} + \frac{Q_3}{\alpha_3}\right] \cdot 86400s \qquad [m^3] \qquad (32.7)$$

where discharge rates are given in m^3/s.

The volume of groundwater remaining in the aquifer at the end of the third micro-regime is the function of the discharge rate at time t^* and the coefficient of discharge α_3:

$$V^* = \frac{Q^*}{\alpha_3} \qquad (32.8)$$

The difference between volumes V_0 and V^* is the volume of all groundwater discharged during the period t^*-t_0.

In our case, the volume of groundwater stored in the aquifer at the beginning of recession is (see also Figure 32.8):

$$V_0 = \left[\frac{(Q_{01} - Q_{02})}{\alpha_1} + \frac{Q_{02} - Q_{03}}{\alpha_2} + \frac{Q_{03}}{\alpha_3}\right] \cdot 86400s \qquad [m^3]$$

$$V_0 = \left[\frac{(3.55m^3/s - 2.55m^3/s)}{0.019} + \frac{(2.55m^3/s - 2.20m^3/s}{0.0045} + \frac{2.20m^3/s}{0.0015}\right] \cdot 86400s$$

$$V_0 = 4.547 \times 10^6 m^3 + 6.720 \times 10^6 m^3 + 1.267 \times 10^8 m^3 \;=\; 1.380 \times 10^8 m^3$$

As can be seen, most water is stored in the fine storage which is the main contributor to discharge during the recession.

The volume of water remaining in the aquifer above the spring level, *i.e.*, the volume of free gravitational water in the fine storage at the end of recession is:

$$V^* = \frac{2.03m^3/s}{0.0015} \cdot 86400s = 1.169 \times 10^8 m^3$$

which gives the following volume of water discharged at the spring for the duration of recession (54 days):

$$V = V_0 - V^* = 1.380 \times 10^{-8} - 1.169 \times 10^8 m^3 = 21.1 \times 10^6 m^3$$

Recession periods of large perennial karstic springs, or springs draining highly permeable fissured aquifers, often have two or three micro-regimes of discharge as in our example. However, the first micro-regime rarely corresponds to the simple exponential expression of the Maillet's type, and is better explained by hyperbolic functions. Deviations from exponential dependence can be easily detected if recorded data plotted on a semi-log diagram do not form straight line(s). Usually the best approximation of the rapid (and often turbulent) drainage of large groundwater transmitters at the beginning of recession is the hyperbolic relation of the Boussinesq type. Its general form is:

$$Q_t = \frac{Q_0}{(1 + \alpha t)^n} \tag{32.9}$$

In many cases this function correctly describes the entire recession curve. On the basis of 100 analyzed recession curves of karstic springs in France, Drogue (1972) concludes that among 6 exponents studied, the best first approximations of exponent n are 1/2, 3/2, and 2.

The exact determination of exponent n and the discharge coefficient α for the function that best fits the measured data is performed graphically and by computation as follows:

- The minimum recorded discharge at the end of recession is noted (Q_2=0.057 m^3/s for the example in Figure 32.10);
- Any discharge Q_1 on the recession curve, which is not a consequence of (possible) recent rainfall, is chosen in the section between Q_2 and Q_0;
- The value of α that satisfies the following equation:

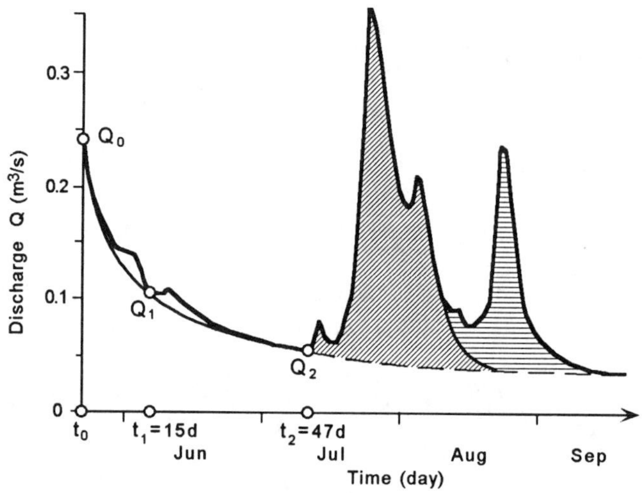

Figure 32.10 Recession curve of a spring with discharges used to determine the hyperbolic function. Hydrographs 1 and 2 are the result of summer storms.

$$\frac{\log(Q_0 / Q_1)}{\log(Q_0 / Q_2)} = \frac{\log(1 + \alpha t_1)}{\log(1 + \alpha t_2)} \tag{32.10}$$

is determined by trial adopting an initial value for α (usually 0.5). The result can be graphically checked as shown in Figure 32.11: the correct coefficient of discharge gives the straight line through the points determined by Q_0, Q_1, Q_2, and the corresponding times t_0, t_1, t_2. The exact value of α for our example shown in Figures 32.9 and 32.10 is 0.202.

$$\frac{\log(0.240m^3 / s \ / \ 0.105m^3 / s)}{\log(0.240m^3 / s \ / \ 0.060m^3 / s)} = \frac{\log(1 + 0.202 \cdot 15d)}{\log(1 + 0.202 \cdot 47d)}$$

$$0.5963 \approx 0.5929$$

- The exponent n is calculated by substituting the determined value for α into either of the following two equations:

$$n = \frac{\log(Q_0 / Q_1)}{\log(1 + \alpha t_1)} \tag{32.11}$$

$$n = -\frac{\log(Q_1 / Q_2)}{\log\left(\frac{1 + \alpha t_1}{1 + \alpha t_2}\right)} \tag{32.12}$$

For our example, equation (32.11) gives:

$$n = \frac{\log(0.240m^3 / s \ / \ 0.105m^3 / s)}{\log(1 + 0.202 \cdot 15d)} = 0.593$$

and the recession discharge equation is then:

$$Q_t = \frac{0.24 m^3 / s}{(1 + 0.202 \cdot t)^{0.593}} \qquad (32.13)$$

The coefficient of discharge and the exponent in equation (32.9) have the following general relationship:

$$\alpha = \frac{\sqrt[n]{Q_0} - \sqrt[n]{Q_t}}{t \cdot \sqrt[n]{Q_t}} \qquad (32.14)$$

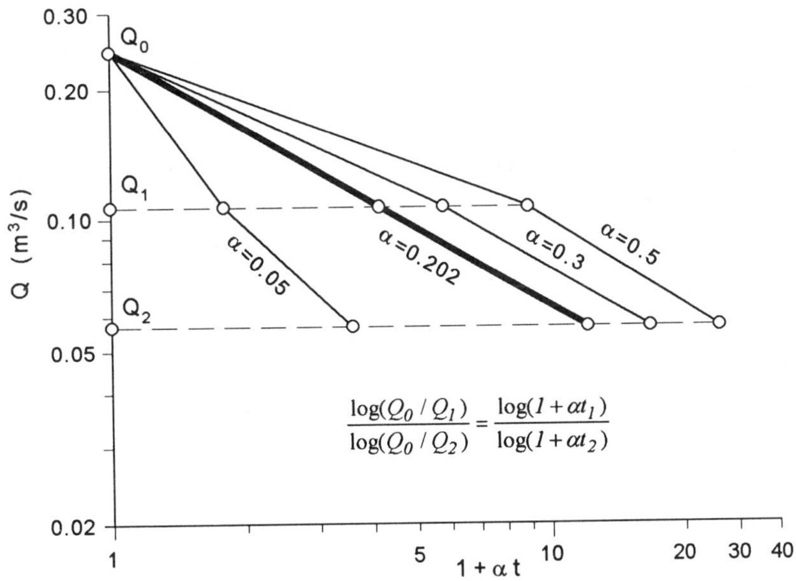

Figure 32.11 Graphical determination of discharge coefficient α for the example shown in Figure 32.10.

As in the case of the Maillet equation, the determined hyperbolic function can be used to calculate the volume of free gravitational water stored in the aquifer above the spring level. In general, this volume at any time t since the beginning of recession is:

$$V_t = \frac{Q_0}{\alpha \cdot (n-1)} \cdot \left[1 - \frac{1}{(1+\alpha t)^{n-1}}\right] \cdot 86400 s \qquad [m^3] \qquad (32.15)$$

PROBLEM 33
HYDROCHEMICAL SEPARATION OF THE SPRING HYDROGRAPH

Table 33.1 shows the results of daily monitoring of the spring discharge rate, the contents of the calcium ion in the spring water, and the precipitation in the drainage area of a permanent spring. Analyze the impact of precipitation on the spring discharge and find the percentage of newly infiltrated water in the overall spring discharge after major precipitation events.

Table 33.1 Elements monitored at the spring during an 89-day period with two major precipitation events. Q: spring discharge; Ca: concentration of the calcium ion; P: precipitation in the spring drainage area.

No.	Q (l/s)	Ca (mg/l)	P (mm)	No.	Q (l/s)	Ca (mg/l)	P (mm)	No.	Q (l/s)	Ca (mg/l)	P (mm)
1	35.2	95.8	0	31	74.4	97.0	0	60	135.0	98.6	0
2	35.9	96.2	0	32	70.6	97.8	0.2	61	139.7	99.8	0
3	36.7	97.0	0	33	66.9	97.8	0	62	131.5	97.4	0
4	35.9	97.0	0	34	63.3	96.2	2.9	63	128.1	97.4	0
5	37.4	96.2	0	35	65.1	95.0	0	64	121.3	97.4	0
6	35.9	96.2	0	36	63.3	94.6	0	65	105.9	96.6	0.9
7	36.7	97.0	0.3	37	57.9	95.8	0	66	122.4	96.6	0
8	34.4	96.6	2.55	38	57.1	96.2	1.2	67	119.0	96.2	0.4
9	34.4	96.6	0.3	39	55.3	97.0	1.1	68	110.2	96.2	2.75
10	33.7	96.2	0	40	51.0	97.0	0.9	69	114.6	95.8	5.55
11	34.4	97.0	0	41	55.3	96.6	6.8	70	100.6	96.2	1.55
12	34.4	97.0	0	42	161.4	96.2	16.1	71	112.4	96.2	4.8
13	33.7	96.2	0	43	301.0	95.4	10.6	72	114.6	95.8	0
14	31.5	96.2	0	44	280.0	95.4	0	73	112.4	95.4	0
15	30.8	97.0	0	45	283.0	72.7	25	74	105.9	95.0	0.5
16	31.5	97.4	0	46	346.2	69.3	11.6	75	102.7	96.2	2.2
17	68.8	97.8	17.1	47	330.4	66.5	0	76	103.8	94.2	0.2
18	55.3	97.0	6.1	48	268.2	73.7	0.9	77	97.4	93.0	0.6
19	93.3	96.3	11.8	49	260.9	76.5	2.6	78	83.2	93.8	0.9
20	97.4	95.4	4	50	243.7	85.0	0	79	101.6	93.8	19.1
21	437.4	91.4	18.8	51	202.5	85.8	0	80	137.3	95.0	7.8
22	330.4	79.4	0	52	73.5	90.6	0	81	110.2	96.2	0
23	232.5	83.0	0	53	70.6	92.8	0	82	91.2	95.4	0
24	87.2	85.0	0	54	62.4	95.8	0.9	83	91.2	95.8	0
25	87.2	85.8	0	55	71.6	96.6	0	85	93.3	95.8	0
26	51.0	87.4	0	56	151.6	96.6	0	86	95.3	95.8	1.2
27	64.2	88.6	0	57	149.2	96.6	0	87	104.8	96.2	10
28	66.0	91.4	1.5	58	156.5	97.0	8.3	88	99.5	96.2	11.1
29	114.6	95.4	14.5	59	140.9	97.8	8.7	89	99.5	97.0	2.1
30	79.2	96.2	0								

The simultaneous recording of spring discharge and chemical constitution of the spring water allows for a fairly accurate separation into "old" (pre-storm), and "new" (rain) water. This separation is based on the assumption that the constitution of the water entering the aquifer is considerably different than that already in it. When recharge by rain takes place, it is evident that the concentration of most cations characterizing groundwater, such as calcium and magnesium, is much less in rain water. Additional preconditions for the application of a hydrochemical hydrograph separation are (after Dreiss, 1989a) that:

- Concentrations of the chemical constituents in the rain water chosen for monitoring are uniform in both area and time;
- Corresponding concentrations in the pre-storm water are also uniform in area and time;
- The effects of other processes in the hydrologic cycle during the episode, including recharge by surface waters, are negligible;
- The concentration and transport of elements are not changed by chemical reactions in the aquifer.

This last condition is the most delicate, and assumes a minor dissolution of carbonate rocks during the flow of new water through porous medium. The first two conditions regarding calcium ion, monitored in our case, are acceptable.

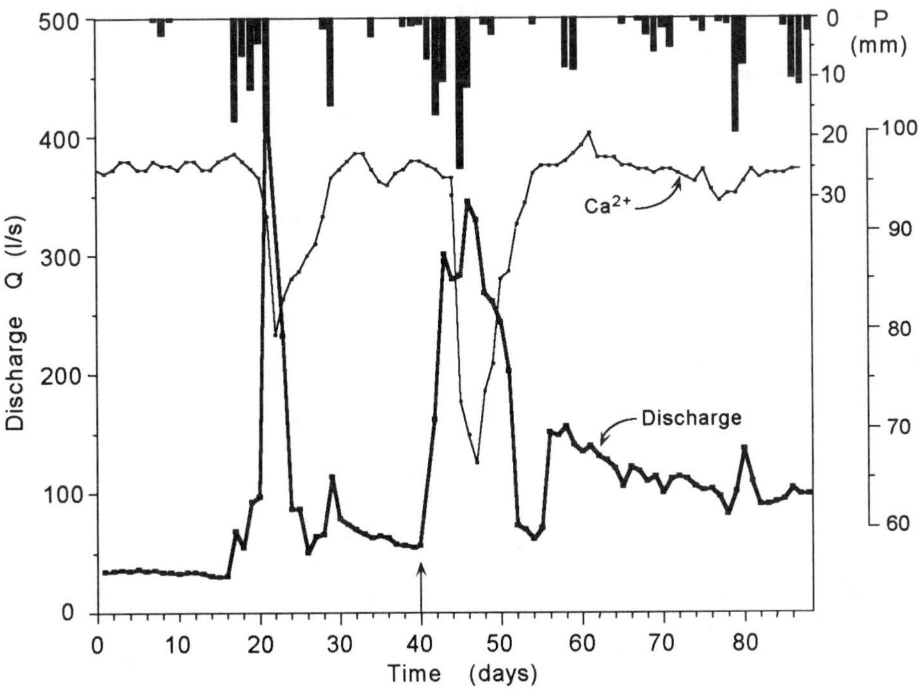

Figure 33.1 Elements monitored at the spring during an 89-day period. Plotted data are given in Table 33.1.

From the graph in Figure 33.1, it is clearly visible that the calcium ion concentration in the spring water drops rapidly after the heavy rains that cause an increase in discharge rate. Assuming a simple mixing of old aquifer water (Q_{old}) and newly infiltrated rain water (Q_{new}), the total recorded discharge of the spring is the sum of the two (after Dreiss, 1989a):

$$Q_{total} = Q_{old} + Q_{new} \tag{33.1}$$

If chemical reactions in the aquifer do not cause significant and rapid changes in the concentration of calcium ion in the infiltrating rain water (which is often true for unconfined karst and intensely fissured aquifers where the flow velocity is high), the calcium ion balance in the spring water is:

$$Q_{total} \cdot C_{total} = Q_{old} \cdot C_{old} + Q_{new} \cdot C_{new} \tag{33.2}$$

where

- Q_{total} is the recorded spring discharge,
- C_{rec} is the recorded concentration of calcium ion in the spring water,
- Q_{old} is the portion of the spring flow attributed to the "old" water (*i.e.*, water already present in the aquifer before the rain event),
- C_{old} is the recorded concentration of calcium ion in the spring water before the rain event,
- Q_{new} is the portion of the spring flow attributed to the "new" water,
- C_{new} is the concentration of calcium ion in the new water.

If C_{new} is much smaller than C_{old} (which is correct in our case since the concentration of calcium ion in rain water is usually less than 5 mg/l), the input mass of calcium ion is relatively small compared to its mass in the "old" aquifer water:

$$Q_{new} \cdot C_{new} << Q_{old} \cdot C_{old} \tag{33.3}$$

From equation (33.2) it follows, after excluding the (small) input mass, that:

$$Q_{old} = \cdot \frac{Q_{total} \cdot C_{total}}{C_{old}} \tag{33.4}$$

Combining equations (33.1) and (33.4) gives :

$$Q_{new} = Q_{total} - \frac{Q_{total} \cdot C_{total}}{C_{old}} \tag{33.5}$$

By applying equation (33.5) it is possible to estimate the discharge component formed by the inflow of new rain water if the spring discharge recordings and a continuous hydrochemical monitoring are performed before, during, and after the storm event.

The result of the hydrochemical hydrograph separation for the second major rainfall event, which started approximately 40 days after the beginning of monitoring (see Figure 33.1), is shown in Figure 33.2. For example, the flow rate attributed to "new water" 5 days after the first increase in spring discharge (day 45, see Table 33.1) is calculated by applying equation (33.5) as:

$$Q_{new} = 0.283m^3 / s - \frac{0.283m^3 / s \cdot 72.7mg / l}{96.4mg / l} = 0.070m^3 / s$$

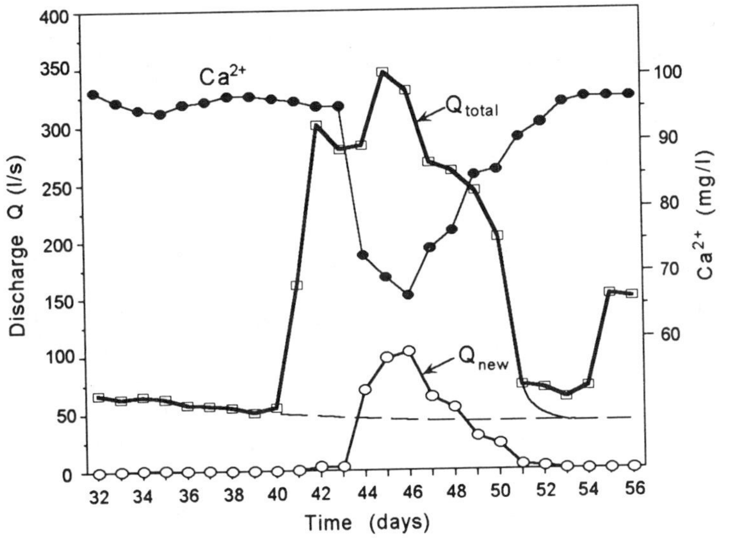

Figure 32.2 Hydrochemical separation of the newly infiltrated water discharging at the spring after a major rainfall event.

where 96.4 mg/l is the average concentration of calcium ion in the spring water before the rainfall event, *i.e.*, during the recession period when the flow is generated by "old" water only (day 32 – day 40, see Figures 33.1 and 33.2).

As can be seen from the graph in Figure 33.2, only a minor amount of "new" water started discharging at the spring one day after the major increase in the flow rate. This lag is 3 days for the first major increase in the discharge of "new" water. The lag between the maximum spring discharge and the maximum discharge of "new" water is again 1 day. New water stopped discharging at the spring 14 days after the rainfall started influencing the overall flow rate. Figure 33.3 shows these three components in the overall spring discharge:

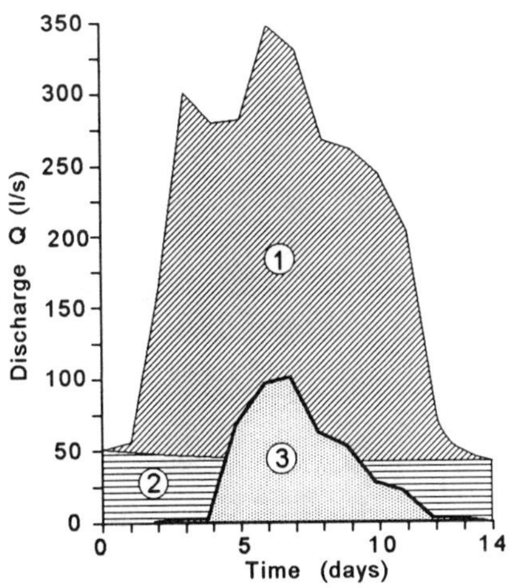

Figure 33.3 Components of flow at the spring during and after the second major rainfall event. 1: Overall increase in the discharge rate. 2: Discharge of pre-rainfall water that would have occurred if there was no precipitation. 3: Flow due to discharge of newly infiltrated water.

1. Flow which is the result of the rainfall event (the corresponding volume of the discharged water is $2.01x10^5$ m^3).
2. Flow that would have occurred if there was no precipitation. This component, equivalent to the base flow of surface streams, is separated by extending the pre-rainfall recession curve (the volume is $5.08x10^4$ m^3).
3. Flow due to discharge of newly infiltrated water ($3.67x10^4$ m^3).

Comparison of these three components and the corresponding volumes of discharged water show that "new" water contributes with only 18.3% to the overall flow increase at the spring (component 1). The percentage of new water in the total spring discharge for the 14-day period is 14.6% (note that the total spring discharge includes components 1 and 2.)

This analysis shows that the main effect of the rainfall event and the related infiltration is the displacement of the "old" water which is forced to discharge from the aquifer. This mechanism is attributed to the propagation of pressure through the well-interconnected conduits and fractures which often causes a rapid increase in spring discharge. Only a small amount of "new" water, with a certain delay, actually reaches the spring during and immediately after the rainfall. Similar mechanisms of spring discharge are common in karst and intensely fissured aquifers (e.g., see Dreiss 1989a; Kresic, 1991).

Chapter 10

TIME SERIES MODELING

PROBLEM 34
AUTO-REGRESSIVE CROSS-REGRESSIVE MODEL

Analyze stochastic properties and develop an auto-regressive cross-regressive model (ARCR) of a time series of spring discharge using daily measurements of the spring flow and the precipitation in its drainage area. Figure 34.1 shows characteristic spring hydrographs for both a dry and a wet year.

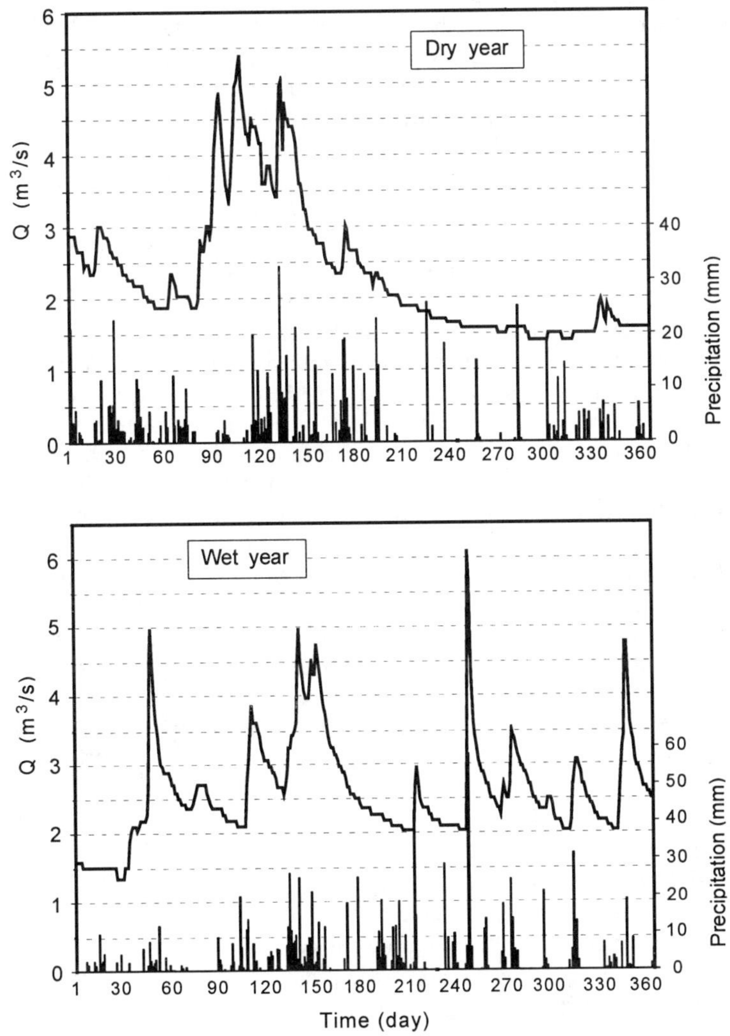

Figure 34.1 Characteristic hydrographs of the spring discharge for typical dry and wet years.

34.1 STOCHASTIC PROPERTIES OF TIME SERIES

Time series, a common term in hydrologic studies, is simply a series of a time-dependent hydrologic variable such as the flow rate in a surface stream or at a spring. When analyzing a time series in practice, we deal with a limited amount of recorded data – *a sample*. This sample, regardless of its size (e.g., one or one hundred years of data), consists of a limited number of realizations (one or hundred) of the same hydrologic process. All possible realizations of that process constitute a *population*. The goal of most hydrologic and hydrogeologic studies is to understand and quantitatively describe the population, as well as the process that generates it, based on a limited number of samples.

A time series can be *continuous* (such as the flow rate in a perennial stream) or *discrete* (such as daily precipitation). For practical and computational purposes, most continuous time series are converted into discrete time series by introducing the recording (or modeling) time interval – 1 day, 1 week, 6 hours or so.

When a time series is described with statistical and probabilistic parameters, it represents a probability of occurrence (realization) of one of its possible stages. A good example is a time series of monthly precipitation at a certain location in a moderate climate. Our long-term experience can tell us that, for example, April through June is the wet period, and July through September is the dry period of the year. Accordingly, we can, with a high probability, expect that in the near future (say, next year) these two periods will again last about the same time. However, no one can state with 100% accuracy that this will indeed happen (for example, June may be an unusually dry month next year). Also, it is impossible to accurately predict the annual or monthly amount of precipitation using some physical laws of nature. We can only apply tools of statistics and make predictions about the future using probabilistic models based on past data. A time series studied in this way is called a *stochastic time series* and its parameters described with statistic and probabilistic terms are called stochastic parameters (properties).

On the other hand, the stage of a deterministic process at time t is defined with certainty knowing its stage at some earlier time t_0. In other words, a deterministic process is described with physical laws rather than laws of probability. An example is the flow of groundwater from a point of recharge to a point of discharge when described with equations such as the Dupuit equation, the Laplace equation or the Theis equation, to name just a few. Quantitative hydrogeology is based on physical laws of groundwater flow as are traditional numeric models presented in Chapter 11.

Strictly speaking, most time series in hydrologic and hydrogeologic studies are stochastic since they depend on at least one random variable with precipitation being the most important. This also means that the result of a deterministic calculation, such as the drawdown in a well after 1 year of pumping, although given explicitly is actually just more or less *probable*.

In general, a time series has the following five components, all of which may or may not be present (adapted from McCuen and Snyder, 1986):

- *Trend*, which is a tendency to increase or decrease continuously for an extended period of time in a systematic manner. This component can often be described by fitting a functional form such as a line or polynomial. The coefficients of the equation are commonly evaluated using regression analysis. The trend is also referred to as a *deterministic component* even though its physical explanation may not always be clear.
- *Periodicity*, which is very common in hydrologic time series: annual and seasonal periodicity of precipitation, temperature, flows. The period(s) in time series can be identified using a moving-average analysis, an autocorrelation analysis, or a spectral analysis, after which it is described by one or more trigonometric functions.

- *Cycle*, which occurs with an irregular period and is hard to detect (for example, hydrometeorological time series are thought to be influenced by sunspot activity which has an irregular period).
- *Episodic variation*, which results from extremely rare or "one-shot" events such as hurricanes. Identification of this component requires supplementary information.
- *Random fluctuations*, which are often a dominant source of the variation in time series and are the main "target" of a probabilistic identification.

Stochastic models describe time series without (formally) considering their physical nature. Simply stated, they statistically/mathematically analyze the past of the time series, as system input, and then predict the present or the future as system output. They can also analyze the past of one time series and based on it predict the present and future of some other time series proven to be influenced by the first one. Or, they can combine several inputs and give one or several outputs. Examples would be a model that predicts water table elevation based on its position in the past, a model that predicts water table based on its own past and the antecedent precipitation, or a model that includes past stages of a nearby river as well.

General form of input-output stochastic models of a discrete time series (or a continuous time series transformed into a discrete one) with the same recording time interval is (Bozic, 1977):

$$y_t = f(x_t, x_{t-1}, x_{t-2}, \ldots; y_{t-1}, y_{t-2}, \ldots; \theta_1, \theta_2, \ldots) + \varepsilon_t \tag{34.1}$$

where

f is the chosen mathematical function,

y_t is the predicted output at time t,

$y_{t-1}, y_{t-2}, \ldots$ are the successive members of the output time series recorded at corresponding time intervals $t-1, t-2, \ldots,$

$x_t, x_{t-1}, x_{t-2} \ldots$ are the successive members of the input time series recorded at time intervals $t, t-1, t-2, \ldots,$

$\theta_1, \theta_2, \ldots$ are the model parameters found by mathematically minimizing the differences between estimated (calculated) and observed y_t values,

ε_t is the model error (residual) given as the difference between the calculated and the recorded value of the output series at time t.

If the chosen mathematical function f ignores the physical laws that govern the transformation of input(s) into output(s), the model is a pure stochastic one. If, in any form, the mathematical function incorporates physical laws, the model is called a stochastic-conceptual model.

Stochastic models can be used to:

- predict the frequency of occurrence of a certain hydrologic variable (for example, the maximum water table elevation);
- issue a short-term forecast with confidence limits (for example, with 90% accuracy we state that the flow at the spring will be 3.5 m^3/s tomorrow);
- extend a short time series using correlation with another time series that has a longer recording period (an example would be the extension of the "recording" period at one gauging station using data from nearby stations that were operating longer);
- generate a synthetic time series to study the characteristics of a hydrologic variable (for example, a stochastic model can generate several thousand time series of the flow in a river which can then be used to find probability of occurrence of a 10,000-year flood).

A very important factor that limits a wider use of stochastic models in hydrogeology is the lack of recorded data. Since these models are based on statistical and probabilistic calculations, very short time series do not allow for meaningful derivation of model parameters. Groundwater levels measured for a couple of years on a quarterly basis are obviously not a good candidate for stochastic modeling. On the other hand, if an appropriate amount of data is available, every attempt should be made to develop one or several stochastic models of the input-output type. This is mainly because the process of building even the simplest stochastic model brings a lot of information on the systems' (aquifer's) structure and connections between various hydrologic variables.

Stochastic models differ based on the mathematical (statistical) functions used to transform input into output, as well as on the number of variables incorporated. One of the techniques often used is multiple (linear) regression. If a hydrologic system of interest is not linear, an attempt is usually made to linearize the corresponding time series and use linear rather than more complicated non-linear regression. The regression can be accompanied with different operators whose function is to attenuate or otherwise mathematically "prepare" input data series for the modeling. For example, the impact of precipitation on groundwater levels is more or less delayed and attenuated by porous medium. Accordingly, the influence of antecedent precipitation may be more important for today's water table elevation than yesterday's rainfall. A simple operator (function) applied to precipitation data, such as the moving average, can significantly improve the stochastic model in such cases.

34.1.1 Autocorrelation

Autocorrelation is the correlation between successive values of the same variable. Figure 34.2 illustrates its principle. If a hydrologic variable is measured on a daily basis (as is the spring discharge in our case), for lag 1 autocorrelation we pair values recorded at days 1 and 2, days 2 and 3, days 3 and 4, and so on. The number of pairs in the autocorrelation is N-1 where N is the number of data. For lag 2 we pair days 1 and 3, days 2 and 4, and so on. Consequently, the number of pairs in correlation decreases again and is N-2. Autocorrelation is measured by the *autocorrelation coefficient*, also called *serial correlation coefficient*, whose estimate for any lag k is:

$$r_k = \frac{\frac{1}{N-k}\sum_{i=1}^{N-k}(x_i-\bar{x})(x_{i+k}-\bar{x})}{\frac{1}{N}\sum_{i=1}^{N}(x_i-\bar{x})^2} \qquad (34.2)$$

where

N is the total number of data in the sample,

x_i is the value of the variable (spring discharge in our case) at time $t=i$,

x_{i+k} is the value of the variable at time $t=i+k$,

$\bar{x}$ is the average value of the sample.

The numerator in equation (34.2) is called the *autocovariance* (or just covariance -- COV), and the denominator is called the *variance* (VAR) of the time series. Variance is also denoted as

S_x^2 or σ_x^2

(Note that the square root of the variance is called *standard deviation.*)

Since the autocorrelation coefficient is in practice estimated from a sample, for small sample sizes the variance will change for different lags and the following equation is a more accurate estimate of the autocorrelation coefficient:

$$r_k = \frac{\sum_{i=1}^{N-k} x_i \cdot x_{i+k} - \frac{1}{N-k}\left(\sum_{i=1}^{N-k} x_i\right)\left(\sum_{i=1}^{N-k} x_{i+k}\right)}{\left[\sum_{i=1}^{N-k} x_i^2 - \frac{1}{N-k}\left(\sum_{i=1}^{N-k} x_i\right)^2\right]^{1/2} \cdot \left[\sum_{i=1}^{N-k} x_{i+k}^2 - \frac{1}{N-k}\left(\sum_{i=1}^{N-k} x_{i+k}\right)^2\right]^{1/2}} \tag{34.3}$$

r_k can have values between 1 and −1. For lag 0 it is always 1 since the correlation between two completely identical samples is "absolute".

Autocorrelation coefficients are calculated for various lags and then plotted on a graph called *autocorrelogram.* The number of lags (autocorrelation coefficients) should be approximately 10% of the total number of data for smaller samples. For large samples, such as daily values over one or several years, the number of lags can be up to 30%.

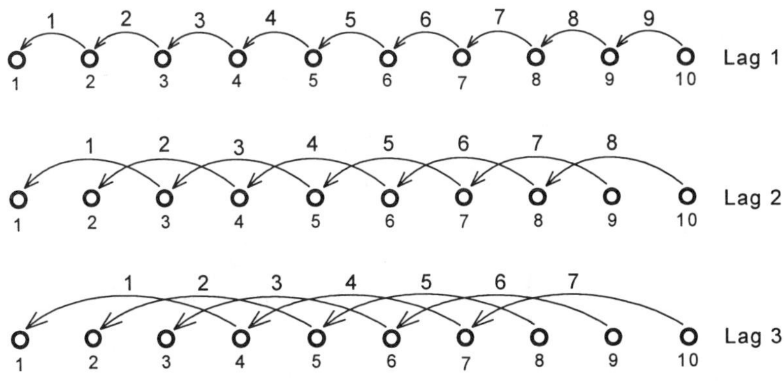

Figure 34.2 Principle of autocorrelation for a sample of a discrete time series. For lag 1 the number of pairs in autocorrelation is N–1 (10–1=9). For lag 2 this number is 8, for lag 3 it is 7 and so on.

If there is some predictability from the past of a series to its present values, the series is *autocorrelated.* Terms that are also often used to describe an autocorrelated series are *persistence* and *memory.* If a series is not autocorrelated it is called independent (*i.e.,* persistence is absent; the series is without memory). The hypothesis that a time series is dependent (autocorrelated) is tested by various statistical tests. One of the simpler tests is proposed by Bartlett (from Gottman, 1981). To be significantly different from zero at the level of confidence 0.05 (*i.e.,* with 95% probability), the autocorrelation coefficient must exceed $2/\sqrt{N}$:

$$r_k > \frac{2}{\sqrt{N}} \tag{34.4}$$

where N is the total number of data in the sample. This test is in hydrologic practice often performed just for the first, or the first two lags which is not recommended. It is more correct to perform a test for the entire correlogram introducing limits of confidence. This may uncover possible delayed or periodic components in the time series that would otherwise be considered as independent if, say, lag 2 was found to be not significantly different from zero.

A test proposed by Anderson gives limits of confidence for the entire correlogram (Prohaska, 1981):

$$LC(r_k) = \frac{1 \pm Z_\alpha \cdot \sqrt{N-k-2}}{N-k-1} \tag{34.5}$$

where

- N is the sample size,
- k is the lag,
- Z_α is the value of the normally distributed standardized variable at the α level of confidence. Values of Z for various levels of confidence can be found in statistical tables and those used most often are given in Table 34.1.

Table 34.1 Values of Z for the most often used levels of significance α (from Spiegel and Meddis, 1980).

Level of significance α	0.10	0.05	0.01	0.005	0.002
Z for one-tail test	± 1.28	± 1.645	± 2.33	± 2.58	± 2.88
Z for two-tail test	± 1.645	± 1.96	± 2.58	± 2.81	± 3.08

In our example (Figure 34.3) the autocorrelogram is calculated for a total of 100 lags. The confidence limits for lag 1, at α = 0.10 (90% probability), are:

$$LC(r_1) = \frac{1 \pm 1.645 \cdot \sqrt{365-1-2}}{365-1-1} = 0.089 \quad and \quad -0.071$$

and for lag 100:

$$LC(r_{100}) = \frac{1 \pm 1.645 \cdot \sqrt{365-100-2}}{365-100-1} = 0.105 \quad and \quad -0.097$$

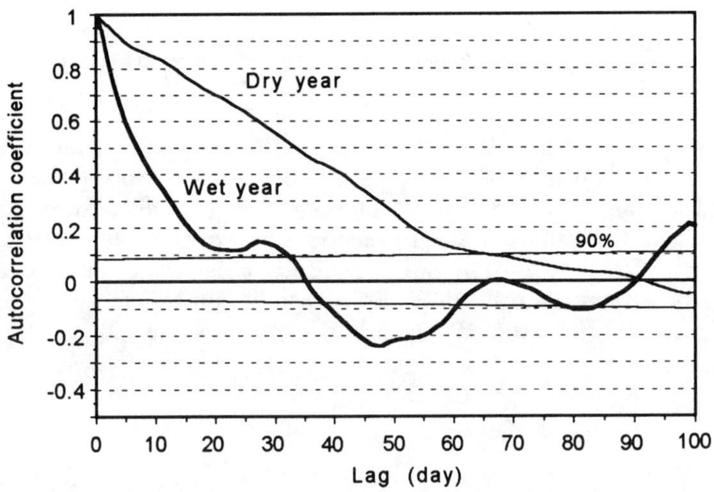

Figure 34.3 Autocorrelograms of the spring flow for a wet and a dry year. (Autocorrelation coefficients are calculated using the Statgraphics software package from Manugistics, Inc., Rockville, Maryland.)

As can be seen in Figure 34.3, the limits of confidence diverge slightly since the number of pairs in correlation (*i.e.*, the accuracy of correlation) decreases with increasing lag. For smaller samples this divergence is more pronounced.

For a dry year, the autocorrelogram exceeds the confidence limits for approximately 60 days indicating that the autocorrelation coefficients for this period are significantly different from zero and that the process of spring discharge is not independent. In other words, the system has a long memory and, generally speaking, flows for up to 60 preceding days are important for today's flow. A physical explanation is that the aquifer storage is significant and that it releases water gradually. This may also indicate that the network of fine fissures and possibly the matrix porosity are the main constituents of effective porosity.

For a wet year the picture is quite different. The system memory is considerably shorter – about 20 days, which could be explained by the fact that the inflow from precipitation is frequent enough to cause a dominant role of groundwater flow through large open fissures/fractures or karst conduits in the case of karst aquifers. These fast-draining paths have high flow amplitudes which can be seen from the spring hydrograph in Figure 34.1.

34.1.2 Cross-correlation

The principle of cross-correlation is illustrated in Figure 34.4 where daily spring flow is represented by the *Y* series, and daily precipitation by the *X* series. In this case Y is the dependent variable (it is influenced by X) and it lags behind X. The time-dependent relationship between the two series is analyzed by computing *coefficients of cross-correlation* for various lags and plotting the corresponding *cross-correlogram*. The cross-correlation coefficient for any lag k is given as:

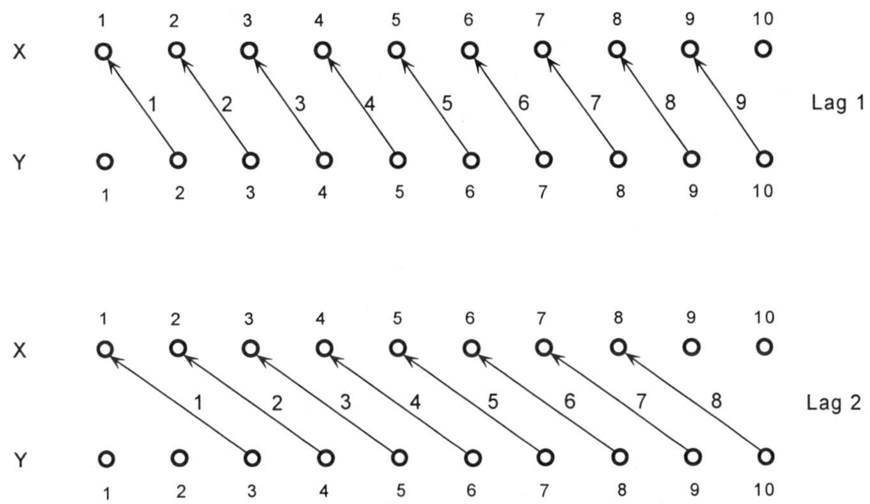

Figure 34.4 Principle of cross-correlation between two discrete time series.

$$r_k = \frac{COV(x_i, y_{i+k})}{(VARx_i \cdot VARy_i)^{1/2}} \tag{34.6}$$

In practice, it is estimated from the sample using the following equation:

$$r_k = \frac{\sum_{i=1}^{N-k} x_i \cdot y_{i+k} - \frac{1}{N-k}(\sum_{i=1}^{N-k} x_i)(\sum_{i=1}^{N-k} y_{i+k})}{[\sum_{i=1}^{N-k} x_i^2 - \frac{1}{N-k}(\sum_{i=1}^{N-k} x_i)^2]^{1/2} \cdot [\sum_{i=1}^{N-k} y_{i+k}^2 - \frac{1}{N-k}(\sum_{i=1}^{N-k} y_{i+k})^2]^{1/2}} \qquad (34.7)$$

Figure 34.5 shows cross-correlograms for the spring flow and the precipitation in its drainage area for typical dry and wet years. In both cases the cross-correlation coefficients are statistically insignificant after 11–12 days which indicates that antecedent precipitation beyond this period has no direct influence on spring discharge. A distinct peak at lag 1 for the wet year shows an important, 1-day-delayed response of large transmitters in the aquifer to significant rainfall events. In a typical dry year this response is much less pronounced but it has the same lag (1 day) which indicates the same discharge mechanism after major rainfall. Generally low values of the cross-correlogram for both years show that the influence of infiltration from precipitation is significantly attenuated by the porous medium.

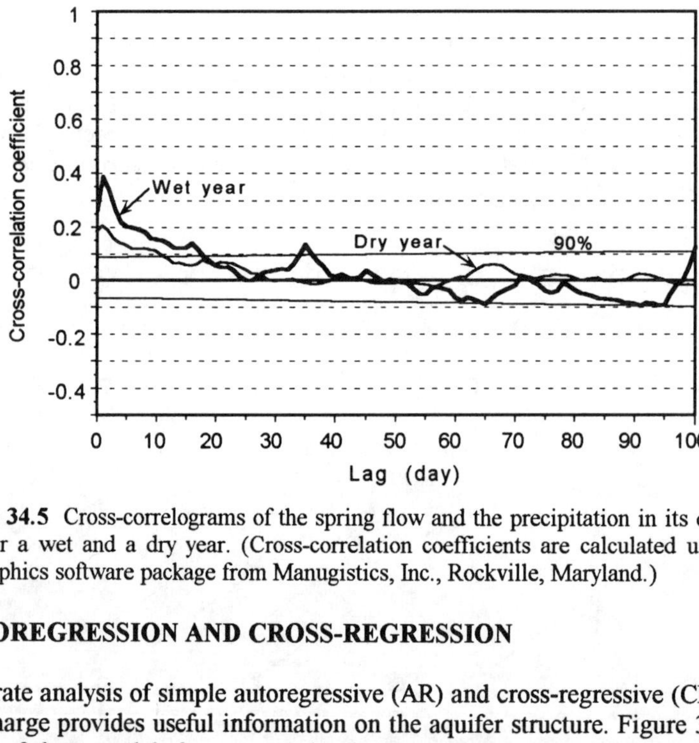

Figure 34.5 Cross-correlograms of the spring flow and the precipitation in its drainage area for a wet and a dry year. (Cross-correlation coefficients are calculated using the Statgraphics software package from Manugistics, Inc., Rockville, Maryland.)

34.2 AUTOREGRESSION AND CROSS-REGRESSION

A separate analysis of simple autoregressive (AR) and cross-regressive (CR) models of spring discharge provides useful information on the aquifer structure. Figure 34.6 shows a comparison of these models for a wet and a dry year. The autoregressive model of order p is given as:

$$Q_t = a + b_1 Q_{t-1} + b_2 Q_{t-2} + \ldots + b_p Q_{t-p} + \varepsilon_t \qquad (34.8)$$

where

Q_t is the predicted discharge at time t,

$Q_{t-1},..,Q_{t-p}$ is the discharge for $1,...,p$ preceding days,

$a, b_1,...,b_p$ are the model parameters,

ε_t is the model error.

The cross-regressive model of order q is:

$$Q_t = a + c_1 P_{t-1} + c_2 P_{t-2} + \ldots + c_q P_{t-q} + \varepsilon_t \tag{34.9}$$

where

Q_t	is the predicted discharge at time t,
$P_{t-1},..,P_{t-q}$	is the precipitation for $1,\ldots,q$ preceding days,
$a, c_1,\ldots,c_q$	are the model parameters,
ε_t	is the model error.

As can be seen in Figure 34.6, autoregressive flow models have much higher coefficients of multiple regression (which is the measure of the model goodness) than cross-regressive models. This is mainly due to the fact that the system has a long internal memory resulting from the gradual release of groundwater from storage. The same can be

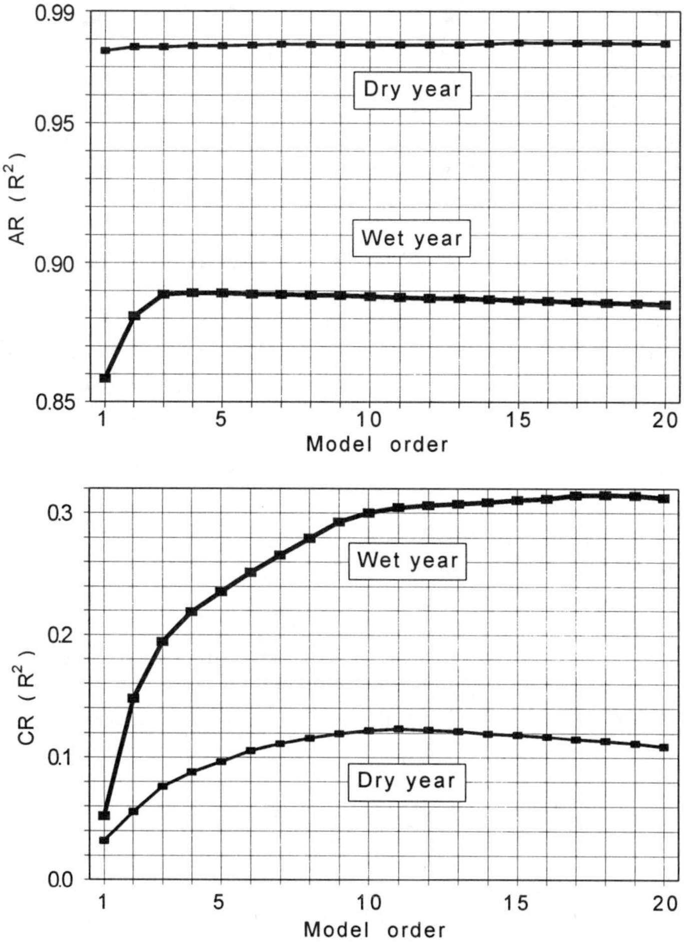

Figure 34.6 Comparison of simple autoregressive (top) and cross-regressive (bottom) models of the spring discharge. The goodness of the models is measured through the squared coefficient of multiple regression (R^2).

concluded from the shape of the autocorrelograms in Figure 34.4. In addition, the cross-regressive model uses only gross precipitation and the time distribution of effective infiltration remains unknown. This introduces a lot of zero values into the multiple cross-regression and significantly decreases its coefficient.

The autoregression model coefficients remain consistently high for both years. For the dry year they are higher than the wet year and change insignificantly for all tested model orders. This again can be explained by the fact that the flow generated by the slow draining of small fissures and filled cavities is always statistically predominant. In the wet year, however, groundwater flow through large fractures (or karst channels) for the preceding three to four days is more important (see maximum on the autoregression graph in Figure 34.6). The cross-regression graph for the wet year changes its slope considerably for the order of 3 and, less abruptly, again for the order of 10, after which it remains steady. This also indicates that, in the wet year, discharge of the spring is statistically influenced by the precipitation of the preceding ten days, with the first three days being the most important.

Because of their statistical insignificance, simple cross-regressive models cannot be used for the prediction of spring discharge: the coefficient of multiple regression for the wet and dry years for all orders is less than 0.32 and 0.13 respectively. Autoregressive models for both years have a very high coefficient of multiple regression which for the dry year approaches almost a functional dependence (R^2 is close to 0.98). Therefore they are good predictive models and can be used for forecasting when the spring discharge is not under the direct influence of precipitation, *i.e.*, in the periods of recession. Autoregressive models, however, cannot be used for forecasting based on expected precipitation, or for generating flow data from the historic precipitation record.

34.3 AUTOREGRESSIVE-CROSS-REGRESSIVE (ARCR) MODELS

ARCR models belong to the category of multivariate time series models and are similar to ARMAX models, *i.e.*, ARMA models with exogenous variables (Salas, 1993). However, they are less complex since the error term is not included in the parameter estimation. The order of the models is generally low -- discharge and precipitation for up to 3 or 4 preceding days are included. Higher orders do increase multiple correlation coefficients (R) as a measure of the models' goodness, but these increases are usually statistically insignificant. However, when a drainage area is large (fifty or more square kilometers) and/or there is a significant delay in the influence of precipitation on the spring discharge, an increase in its cross-regressive part may considerably improve the validity of an ARCR model.

In our case, the combination of autoregressive and cross-regressive models significantly improves validity of the new ARCR model for the wet year. For the dry year this merging increases the coefficient of multiple regression insignificantly since its value for the simple autoregressive model of all tested orders is already close to 0.98. The ARCR model is given as:

$$Q_t = a + b_1 Q_{t-1} + \ldots + b_p Q_{t-p} + c_1 P_{t-1} + \ldots + c_q P_{t-q} + \varepsilon_t \qquad (34.10)$$

where the notification is the same as in equations (34.8) and (34.9).

The model order (p,q) can be chosen from various combinations of its autoregressive (p) and cross-regressive (q) parts as shown in Figure 34.7. For the wet year the model coefficient (R^2) has a maximum for p=3 and q=3. A further increase of the orders of AR and CR parts results only in slight negative changes of R^2 which remains close to 0.925. The model for the dry year does not have a single maximum of multiple regression coefficient. The autoregressive part has the first abrupt positive change for the order of 2 and then for 7 when it also reaches the first maximum. R^2 increases again for p=14 and 15.

However, these changes of the coefficient of multiple regression are with an insignificant magnitude, *i.e.,* less than 0.001. Therefore, introducing more variables into the autoregression by increasing the AR order does not increase the model's validity, but does increase its instability. Generally speaking, when a time series is highly autocorrelated (as in our case), AR models of first and second orders are considered sufficient in hydrologic practice.

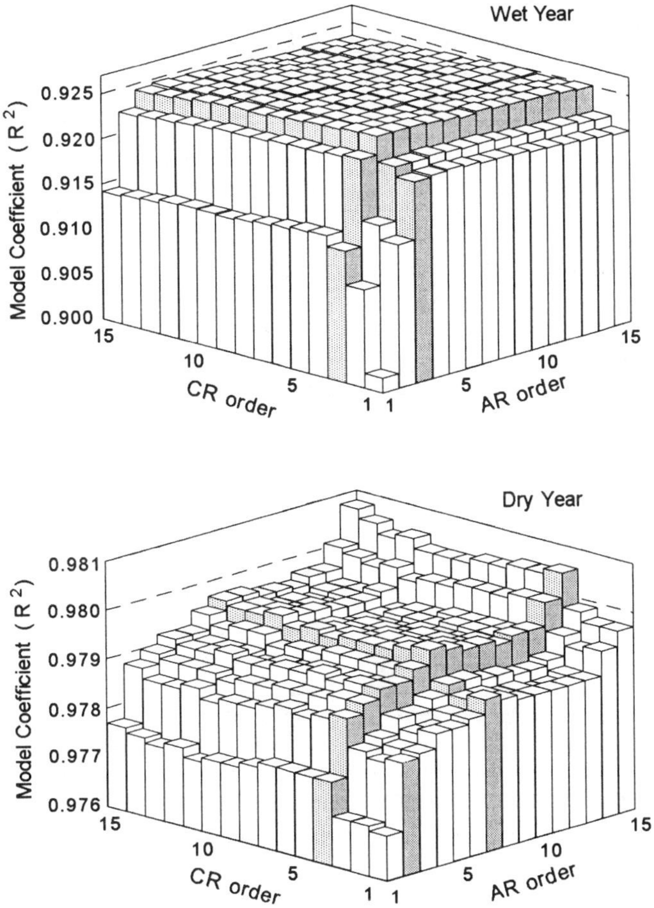

Figure 34.7 Dependence of the ARCR model coefficient (squared coefficient of multiple regression) on the orders of model's cross-regressive and autoregressive parts.

The model's cross-regressive part for the dry year is consistent and its order of 4 has the most significant increase throughout the analyzed domain (see Figure 34.7). The ARCR model order for the dry year can be chosen as p=2, and q=4, *i.e.,* (2,4). Model-predicted versus recorded spring discharge for a part of the wet year is shown in Figure 34.8.

When the model uses actual gross precipitation, the predicted values differ significantly from the measured ones during the recession period (top part in Figure 34.8). This is due to the fact that the model gives the same weight to all precipitation data including isolated summer storms. It does not account for increased evapotranspiration and moisture deficit in the unsaturated zone during the summer months which greatly reduce effective precipitation.

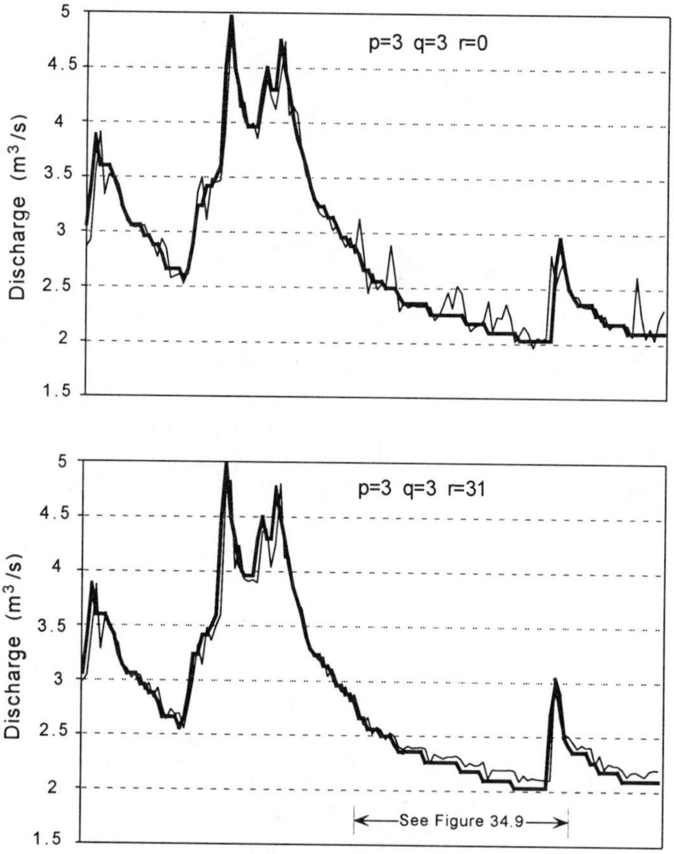

Figure 34.8 Recorded (bold line) versus predicted discharge of the spring for the summer period of a typical wet year. **Top**: Result of the ARCR model without the moving average filter (r=0). Both autoregressive and cross-regressive parts have order of three. **Bottom**: Result of the application of a 31-day moving average linear filter on time series of gross precipitation in the spring drainage area. The outlined section of the hydrograph is also shown in Figure 34.9.

The correct estimation of effective (net) precipitation is a major difficulty in hydrologic modeling of rainfall-runoff relationship. This is particularly true for the groundwater systems where the rainfall infiltration is greatly attenuated by the porous media. The response of aquifers to recharge episodes varies throughout the year depending upon the conditions and thickness of the unsaturated zone, water table levels, position of the discharge areas, and other factors.

A simple way to mathematically describe the attenuation of gross precipitation is to transform data using a moving average linear filter. Figure 34.8 shows the result of one such transformation with the moving average of 31 days. The length of the filter window can be determined visually as shown in Figure 34.9. The ARCR model incorporating modified precipitation input data is given as:

$$Q_t = a + b_1 Q_{t-1} + \ldots + b_p Q_{t-p} + c_1 P_{r,t-1} + \ldots + c_q P_{r,t-q} + \varepsilon_t \tag{34.11}$$

where r is the length of the linear filter window used to transform the precipitation series. The other notification is as in equations (34.9) and (34.10).

The application of the moving average filter actually slightly decreases the model's coefficient of multiple regression: the reason why the visual determination of the filter length is preferred. It can be seen in Figure 34.8 that, for periods of recharge, both gross and transformed precipitation give good predictions. The difference between the two is obvious in the case of recession when the transformed precipitation produces a much better result.

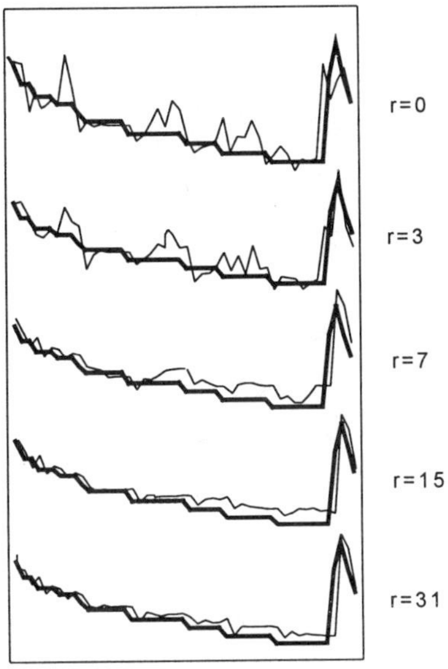

Figure 34.9 Influence of the increasing (top to bottom) moving average window on the predicted discharge of the spring (thin line). The thick line is the recorded spring discharge.

A simple ARCR model can be built by preparing data using any spreadsheet program (Excel, Quattro Pro), and then applying the multiple regression available in most statistical software packages (e.g., Statgraphics). Figure 34.10 shows parts of two time series (spring discharge and precipitation) that are used to illustrate the process of data preparation. For an ARCR(3,3) model, the spring flow is predicted based on precipitation recorded that day, and flows and precipitation for the preceding three days:

$$Q_t = a + b_1 Q_{t-1} + b_2 Q_{t-2} + b_3 Q_{t-3} + c_0 P_t + c_1 P_{t-1} + c_2 P_{t-2} + c_3 P_{t-3} + \varepsilon_t$$

Q	P	Q	P
2.53	*0*	**1**	*1*
2.50	*0*	**2**	*2*
2.50	*2.5*	**3**	*3*
2.47	*0.5*	**4**	*4*
2.45	*0*	**5**	*5*
2.43	*0*	**6**	*6*
2.40	*2.8*	**7**	*7*
2.56	*8.7*	**8**	*8*
2.65	*3.4*	**9**	*9*
2.83	*0*	**10**	*10*
2.92	*0*	**11**	*11*
2.75	*0*	**12**	*12*

Figure 34.10 Parts of two time series (Q -- spring discharge, and P -- precipitation) used to illustrate the principle of preparing data for an ARCR model with spreadsheet programs.

The first column in Figure 34.11-top is the dependent variable in multiple regression (the flow rate to be predicted) and all other columns are "independent" variables which are used to predict the flow. As can be seen, data in columns are shifted one lag at a time so that the above equation of multiple regression is satisfied. The total number of data for

Q	Q-1	Q-2	Q-3	P	P-1	P-2	P-3
2.53				0			
2.5	2.53			0	0		
2.5	2.5	2.53		2.5	0	0	
2.47	2.5	2.5	2.53	0.5	2.5	0	0
2.45	2.47	2.5	2.5	0	0.5	2.5	0
2.43	2.45	2.47	2.5	0	0	0.5	2.5
2.4	2.43	2.45	2.47	2.8	0	0	0.5
2.56	2.4	2.43	2.45	8.7	2.8	0	0
2.65	2.56	2.4	2.43	3.4	8.7	2.8	0
2.83	2.65	2.56	2.4	0	3.4	8.7	2.8
2.92	2.83	2.65	2.56	0	0	3.4	8.7
2.75	2.92	2.83	2.65	0	0	0	3.4
	2.75	2.92	2.83		0	0	0
		2.75	2.92			0	0
			2.75				0

Q	Q-1	Q-2	Q-3	*P*	*P-1*	*P-2*	*P-3*
1				*1*			
2	1			*2*	*1*		
3	2	1		*3*	*2*	*1*	
4	3	2	1	*4*	*3*	*2*	*1*
5	4	3	2	*5*	*4*	*3*	*2*
6	5	4	3	*6*	*5*	*4*	*3*
7	6	5	4	*7*	*6*	*5*	*4*
8	7	6	5	*8*	*7*	*6*	*5*
9	8	7	6	*9*	*8*	*7*	*6*
10	9	8	7	*10*	*9*	*8*	*7*
11	10	9	8	*11*	*10*	*9*	*8*
12	11	10	9	*12*	*11*	*10*	*9*
	12	11	10		*12*	*11*	*10*
		12	11			*12*	*11*
			12				*12*

Figure 34.11 Vertical shifting of original data of the Q and P series in a spreadsheet program. The resulting columns have data arranged for an ARCR(3,3) model. Note that the model should use only data in the outlined boxes since the total number of data in each column is reduced by 3 (this is the largest lag in the model). The "incomplete" rows at the top and the bottom should be deleted before importing the spreadsheet file into a statistical program that has a multiple regression option.

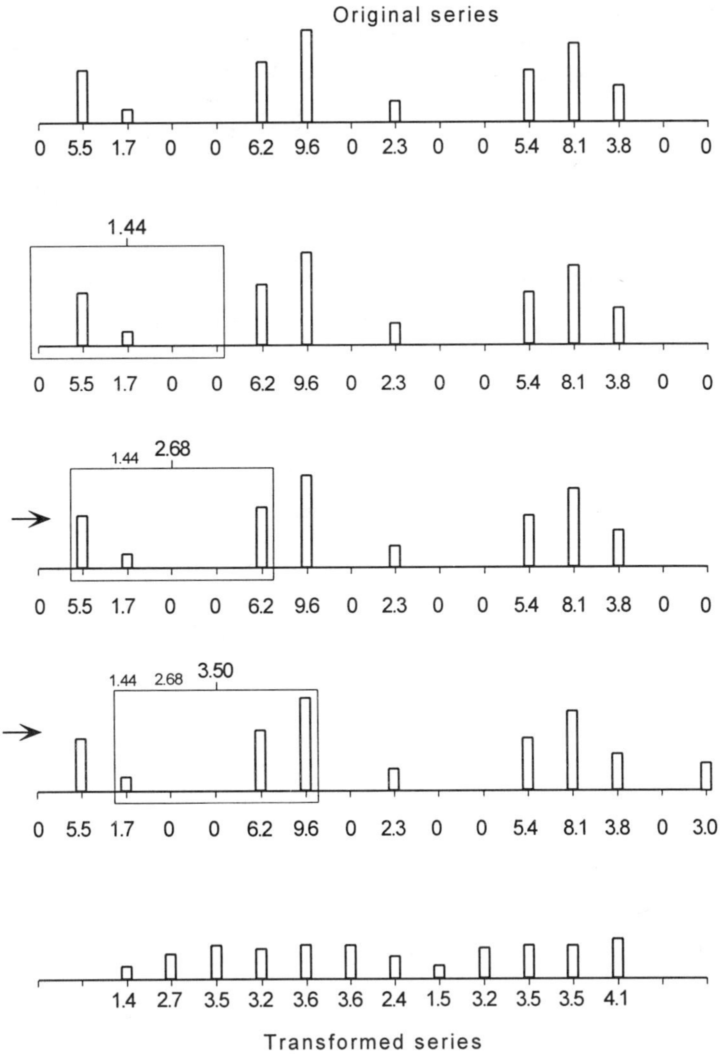

Figure 34.12 Application of a linear moving average filter for the transformation of precipitation series. Note that the total number of data in the transformed series is reduced by $m-1$ where m is the length of the filter window ($m=5$ in our case).

each variable is reduced by 3 which is the highest number of lags in the equation. (For example, if the ARCR model was of (2,6) order, the number of data for each variable in regression would have had to be reduced by 6.) Figure 34.11-bottom shows the same principle illustrated with data order numbers.

As mentioned earlier, cross-regressive models generally have a low coefficient of multiple regression which is one measure of their validity. This is because the perennial daily spring flow is regressed on daily precipitation which is often zero. One way to transform the precipitation series is to apply a simple linear moving average filter as shown in Figure 34.12. In this example, the filter window is 5 which means that five data at a time are summed and their average is assigned to the central point in the window. The window then moves one position and the averaging is repeated. The resulting series is smoother: the high precipitation values decrease and the low values (including zeros) increase. The new transformed series is then included in the ARCR model instead of the raw precipitation data.

More sophisticated stochastic models can include various linear and non-linear transformations of input series, consideration of periodicity (seasons), and filtering (minimizing) of the model residuals -- ε. One widely used linear filter for the residuals is the Kalman filter which usually significantly improves ARMA models. Note, however, that both ARMA and ARCR models give valid predictions for only one step at a time (say, next day) because for the second and following intervals the true values of y (Q) are not known and the error rapidly increases.

For more information about stochastic modeling of time series the reader should consult works by Kalman (1960), Yevjevich (1972), Box and Jenkins (1976), Graupe, Isailovic, and Yevjevich (1976), Adamovski and Hamory (1983), Houston (1983), Weeks and Boughton (1987), and Salas (1993).

Chapter 11

NUMERIC GROUNDWATER FLOW MODELING

PROBLEM 35
WELLS NEAR BOUNDARIES

The possibility of developing a well field for a large-scale water supply is being studied for the unconfined aquifer shown in Figure 35.1. The aquifer is developed in the alluvial sand deposits of a perennial river. The first phase of investigation consists of drilling monitoring wells in and around the aquifer, collecting data on aquifer geometry, water table elevations, and elevations of surface water level along the river. This information is shown in Table 35.1. The average hydraulic conductivity of the aquifer is $2.4x10^{-3}$m/s as determined from slug tests. Using all available data, develop a preliminary groundwater model for assessing the aquifer's potential and planning the second phase of investigation.

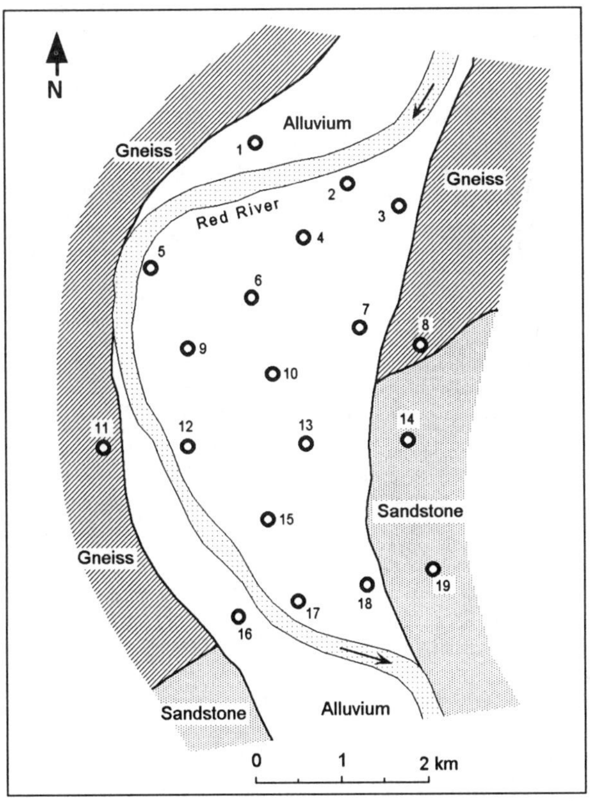

Figure 35.1 Simplified geologic map of the alluvial aquifer considered as a source of a large-scale water supply. Circles with numbers denote monitoring wells.

35.1 INTRODUCTION TO GROUNDWATER MODELING

In recent years groundwater modeling has become a major part of many projects dealing with groundwater exploitation, protection, and remediation. As computer hardware and software continue to be improved and become more affordable, the role of models in highly quantitative earth sciences such as hydrogeology will increase accordingly. It is essential, however, that for any groundwater model to be interpreted and used properly, its limitations should be understood. In addition to strictly "technical" limitations, such as accuracy of computations (hardware/software), the following is true for any model:

- it is based on various assumptions regarding the real natural system being modeled,
- hydrogeologic and hydrologic parameters used by the model are always just an approximation of their actual field distribution which can never be determined with 100% accuracy,
- theoretical differential equations describing groundwater flow are replaced with systems of algebraic equations that are more or less accurate.

It is therefore obvious that a model will be as reliable and "good" as its developer(s), and that it cannot be miss-used as long as all the limitations involved are clearly stated.

Table 35.1 Field data for the project.

Well No.	Altitude (in meters) of the			
	Well bottom	Aquifer base	Water table	Land surface
1	550.0	555.60	570.06	574.81
2	547.0	550.05	570.08	575.05
3	553.0	555.10	570.06	575.26
4			569.95	
5	549.0	552.70	569.79	575.04
6	542.0	546.30	569.82	575.80
7	548.0	551.70	568.83	575.96
8	566.5			587.90
9			569.69	
10			569.70	
11	567.5			587.43
12	548.0	550.20	569.52	575.10
13	547.1	550.70	569.52	575.37
14	555.0	562.20	571.27	592.64
15			569.40	
16	546.5	550.05	564.27	575.32
17	546.0	549.20	569.26	575.16
18	543.0	546.21	569.34	575.30
19	541.0	543.15	571.03	591.26

Altitude of water surface in the river channel

Point	Altitude (m)
a	570.05
b	569.77
c	569.50
d	569.25

35.1.1 Types and Use of Groundwater Models

A model simulates the aerial and temporal properties of a system, or one of its parts, in either *physical* ("real") or *mathematical* ("abstract") way. An example of a physical model in hydrogeology would be a tank filled with sand and saturated with water – equivalent of an unconfined aquifer of limited extent. This miniature "aquifer" can be subject to miniature stresses such as pumping from a perforated tube placed into the sand thus representing a water well. An obvious question when considering similar models is how feasible it is to build a multi-layer "aquifer" exposed to various stresses such as precipitation, surface stream flow, leakage from deep underlying strata, and then change some of its geometric and hydrogeologic properties as needed (if nothing else, more silt and clay brought into the already messy and slippery laboratory will raise an eyebrow or two on a responsible OSHA officer if he ever wanders in!). Consequently, the application of real physical models has been limited to educational and demonstration purposes.

A groundwater system can also be simulated using the analogy between groundwater flow and some other similar physical process such as the flow of electrical currents through conductors. Such models are called *analog* and were often used in hydrogeologic practice before the rapid development of numeric computer modeling (see Rushton and Redshaw, 1979).

Models that use mathematical equations to describe elements of groundwater flow are called *mathematical*. Depending upon the nature of equations involved, these models can be:

- empirical (experimental),
- probabilistic,
- deterministic.

Empirical models are derived from experimental data that are fitted to some mathematical function. A good example is Darcy's law. (Note that Darcy's law was later found to be theoretically grounded and actually became a physical or deterministic law). Although empirical models are limited in scope and are usually site or problem-specific, they can be an important part of a more complex numeric modeling effort. For example, the behavior of a certain pollutant in porous media can be studied in the laboratory or in controlled field experiments, and the derived experimental parameters can then be used for developing numeric models of groundwater transport.

Probabilistic models are based on laws of probability and statistics. They can have various forms and complexity starting with a simple probability distribution of a hydrogeological property of interest, and ending with complicated stochastic, time-dependent models. The main limitations for a wider use of probabilistic (stochastic) models in hydrogeology are that: 1) they require large data sets needed for parameter identification, and 2) they cannot be used to answer (predict) many of the most common questions from hydrogeologic practice such as for example effects of a future pumping.

Deterministic models assume that the stage or future reactions of the system (aquifer) studied are predetermined by physical laws governing groundwater flow. An example is the flow of groundwater toward a fully penetrating well in a confined aquifer as described with the Theis equation. Most problems in traditional hydrogeology are solved using deterministic models which can be as simple as the Theis equation or as complicated as a multiphase flow through a multilayered, heterogeneous, anisotropic aquifer system. There are two large groups of deterministic models depending basically upon the type of mathematical equations involved:

- analytical, and
- numeric.

Simply stated, *analytical models* solve one equation of groundwater flow at a time and the result can be applied to one point or "line of points" in the analyzed flow field (aquifer). For example, if we want to find (*i.e.*, to "model") what the drawdown at 50 m from the pumping well would be after 24 hours of pumping, we would apply one of the equations describing flow toward a well depending upon the aquifer and well characteristics (confined, unconfined, leaky aquifer; fully or partially penetrating well). To find the drawdown at 1000 m from the well, we would have to solve the same equation (say, the Theis equation) for this new distance. If the aquifer is not homogeneous, these solutions would be applicable just for a limited radial distance of 50 or 1000 m within the same distribution of aquifer transmissivity. Obviously, if our aquifer is quite heterogeneous, and we want to know drawdown at "many" points, we might spend a rather long period of time solving the same equation (with slightly changed variables) again and again. If the situation gets really complicated, such as when there are several boundaries, more pumping wells, and several hydraulically connected aquifers, the feasible application of analytical models terminates.

Numeric models describe the entire flow field of interest at the same time, providing solutions for as many data points as specified by the user. The area of interest is subdivided into many small areas (often referred to as *cells* or *elements*) and a basic groundwater flow equation is solved for each cell usually considering its water balance (water inputs and outputs). The solution of a numeric model is the distribution of hydraulic heads at points representing individual cells. These points can be placed in the center of the cell, at intersections between adjacent cells, or elsewhere. The basic differential flow equation for each cell is replaced (approximated) by an algebraic equation so that the entire flow field is represented by x equations with x unknowns where x is the number of cells. This system of algebraic equations is solved numerically, through an iterative process, thus name *numeric models*. Based on various methods of approximating differential flow equations, and methods used for numerically solving the resulting system of new algebraic equations, numeric models are divided into several groups. The two most widely applied groups are:

- finite differences (numeric models), and
- finite elements (numeric models).

Both types of models have their advantages and disadvantages and for certain problems one may be more appropriate than the other. However, because they are easier to design and understand, and require much less mathematical involvement, finite-difference models have prevailed in hydrogeologic practice. In addition, several excellent finite-difference models have been developed by the Unites States Geological Survey and are in the public domain which ensures their widest possible use. One of these is MODFLOW ("A modular three-dimensional finite-difference ground-water flow model") by McDonald and Harbaugh (1988). MODFLOW is probably the most widely used, tested and verified model today and it has become the industry standard thanks to its versatility and an open structure: highly independent subroutines called "modules" are grouped into "packages" which deal with the specific hydrologic features to be simulated. New modules and packages can be easily added to the program without modifying the existing packages or the main code. For these reasons MODFLOW has been chosen as the standard for explaining modeling principles and solving Problems 35 and 36 in this book as well. For a more detailed explanation about finite element and finite difference models in general, and some other computer programs, the reader should consult the excellent work by Anderson and Woessner (1992).

Groundwater models can be used for three general purposes:

- To *predict* or *forecast* expected artificial or natural changes in the system (aquifer) studied. The term *predictive* is more appropriately applied to deterministic (numeric) models since it carries a higher degree of certainty, while *forecasting* is the term used with probabilistic (stochastic) models. Predictive models are by far the largest group of models built in hydrogeologic practice.

- To *describe* the system in order to analyze various assumptions about its nature and dynamics. Descriptive models help to better understand the system and plan future investigations. Although not originally planned as a predictive tool, they often grow to be full predictive models.
- To *generate* a hypothetical system that will be used to study principles of groundwater flow associated with various general or more specific problems. Generic models are used for training and are often created as part of a new computer code development.

35.2 MODEL DESIGN

Every groundwater flow model design consists of the following thirteen stages:

1. Concept development which is the most important part of modeling and the basis for all further activities;
2. Selection of a computer code that can most effectively simulate the concept and meet the purpose of modeling;
3. Definition of the model geometry (lateral and vertical extent of the area to be modeled defined by model boundaries, grid layout, and position and number of layers);
4. Definition of boundary array, *i.e.,* cell types (active, inactive, constant-head cells);
5. Input of hydrogeologic parameters for each cell such as hydraulic conductivity, (horizontal and vertical, including possible anisotropy) transmissivity, storage properties, porosity;
6. Definition of boundary conditions (boundaries with known head, known flux, or head-dependent flux);
7. Definition of initial conditions (distribution of hydraulic head);
8. Definition of stresses acting upon the system such as aerial recharge, evapotranspiration, well pumpage, outflow through springs, drains, inflow of water from other sources (recharge wells, adjacent aquifers);
9. Model run which includes choosing a mathematical method for solving the system of algebraic equations, iteration criterion, and acceptable error criterion for terminating the iteration process;
10. Calibration and sensitivity analysis. This is probably the lengthiest and most demanding part of any modeling process;
11. Verification of model validity. The calibrated model is checked against another set of field data that was not used in model design;
12. Prediction which is in most cases the purpose of model design;
13. Presentation of results which includes both the prediction results and all relevant data documenting stages of model design.

35.2.1 Concept Development

Developing a modeling concept is the initial and the most important part of every modeling effort. It requires a thorough understanding of hydrogeology, hydrology and dynamics of groundwater flow in and around the area of interest. The final result is a computerized data base, and simplified maps and cross-sections that will be used in model design.

Figure 35.2 is a hydrogeologic map of the study area showing water table contours for the alluvial aquifer and general directions of groundwater flow. The Red River is in most part a gaining stream except in the far north where water enters the aquifer from the river. The alluvium is in contact with practically impermeable gneiss in the northeast and does not receive any significant lateral recharge. Gneiss is the aquifer base for the most part except in the south where sandstone is also present (see cross-sections in Figure 35.3).

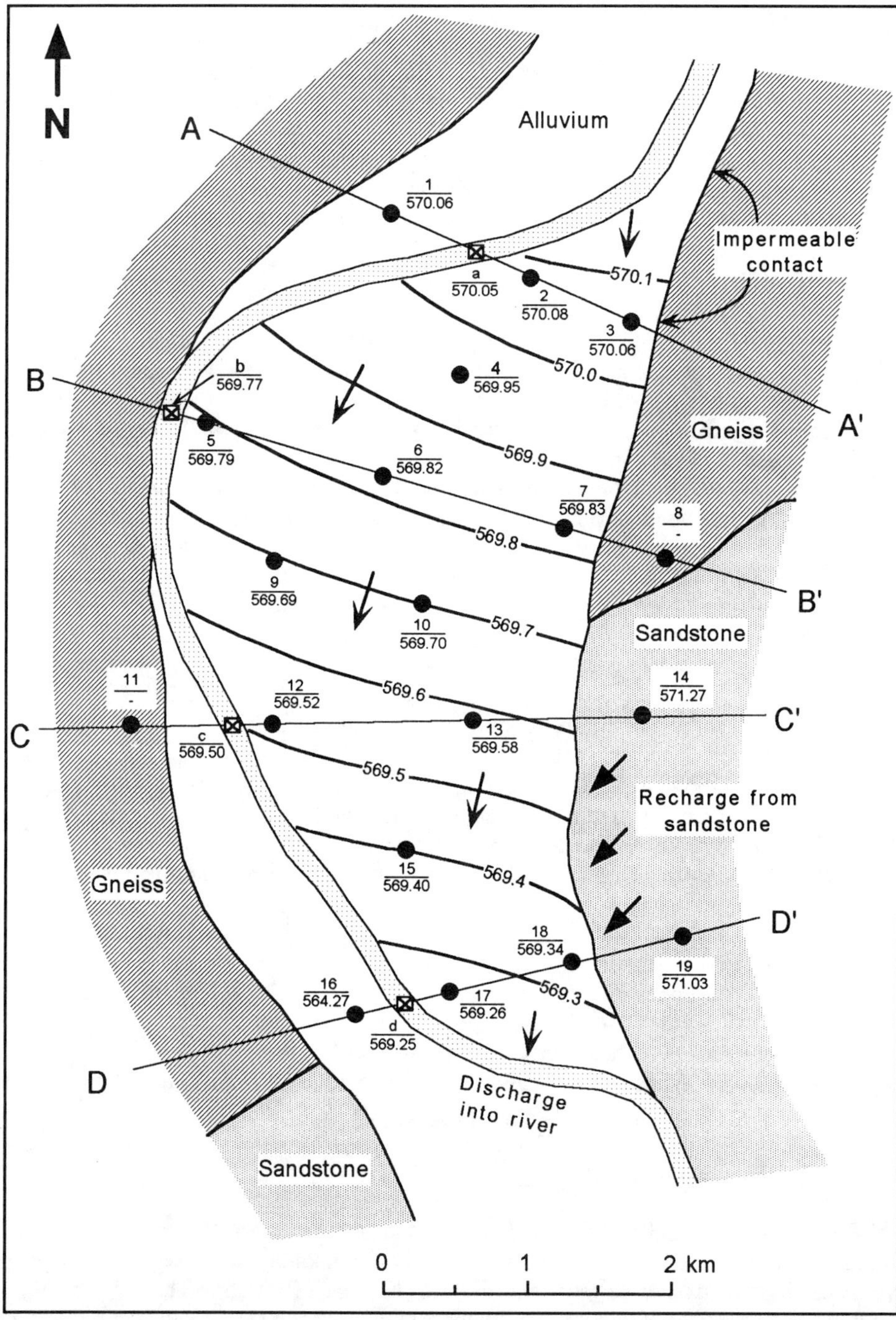

Figure 35.2 Map showing water table contour lines and remarks on the boundaries of the flow field. Arrows indicate general direction of groundwater flow.

Where there is contact with the sandstone, the aquifer receives lateral recharge which increases southward as the area of contact increases (see cross-sections CC′ and DD′ in Figure 35.3).

The hydraulic gradient of the water table is fairly uniform indicating homogeneous transmissivity of the alluvial deposits. The borehole data show that the aquifer has similar thickness throughout its central part which means that the hydraulic conductivity is also homogeneous. The water table is recorded in two wells completed in sandstone aquifer (wells 14 and 19). Based on these two data and the projected water table in the alluvial aquifer, the groundwater flow in sandstone is oriented toward the alluvium aquifer. The hydraulic gradient in the sandstone aquifer is steeper than in the alluvial aquifer indicating lower hydraulic conductivity.

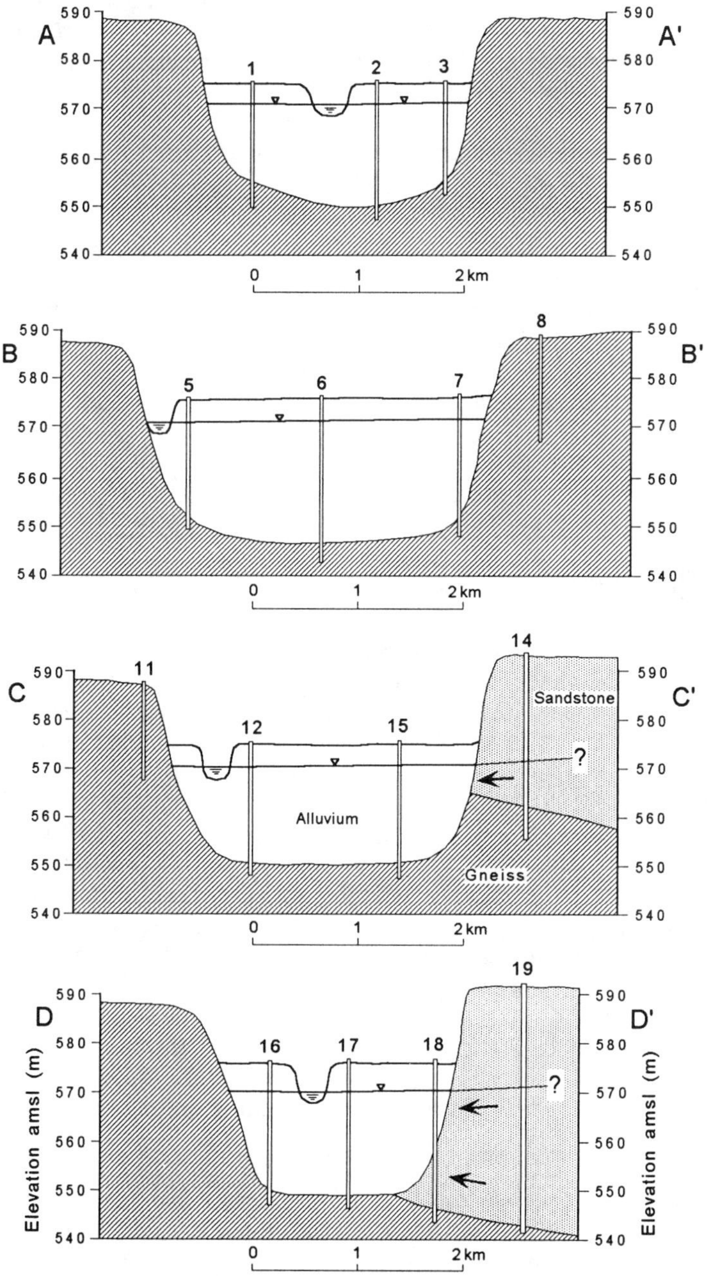

Figure 35.3 Hydrogeologic cross-sections of the alluvial aquifer shown in Figure 35.1 (the vertical scale is exaggerated).

According to the borehole logs, there is no significant silty or clayey cover over the alluvial sands and the aquifer is directly exposed to recharge from precipitation. The exact amount of precipitation in the area is not known. Data for a 23-mile distant gauging station in slightly different orographic conditions show that the average annual precipitation is 32 inches. The site is located in a relatively humid moderate climate in the northern hemisphere.

The water table and river stage measurements were taken only once during the month of May. Although there is no information about the river flow and stages prior to the measurements, it could be concluded, based on the climate characteristics, that the collected data are representative for the season of a relatively high water table and major aquifer recharge.

A relatively simple modeling concept can be developed based on the above facts:

- the groundwater flow system to be modeled consists of a single unconfined aquifer;
- there are five well defined boundaries of the flow field:
  1) equipotential boundary along the contact between the river and the alluvial aquifer,
  2) impermeable lateral boundary between the alluvial aquifer and gneiss,
  3) impermeable base boundary,
  4) water table boundary directly exposed to the surface,
  5) lateral boundary between the alluvial and the sandstone aquifer.
- The alluvial aquifer receives recharge from precipitation and from the adjacent sandstone aquifer. In both cases the recharge will have to be estimated and calibrated.
- Since there is only one set of water table data, the model will be calibrated in steady state (see Section 35.2.8) and it cannot be verified (Section 35.2.10).

35.2.2 Selection of Computer Code

As mentioned earlier, a computer program for groundwater modeling numerically solves a system (matrix) of algebraic equations. This matrix represents an approximation of the mathematical model formulated by partial differential equations of groundwater flow. Finite difference and finite element methods are two common ways in which this approximation is made. (Note that there are some other methods yet to be widely accepted such as integrated finite differences, the boundary integral equation method and analytic elements – see Ligget and Liu, 1983; Ligget, 1987; Strack, 1987, 1988.). Which method will be used depends largely on the type of problem and the knowledge of the model designer. Finite elements more easily describe irregular model boundaries and internal boundaries such as faults. They are also more appropriate in handling point sources and sinks and large variations in water table. However, finite difference models are easier to program, require less data and are more friendly for data input.

"A modular three-dimensional finite-difference ground-water flow model" by McDonald and Harbaugh (1988) is considered by many to be the most reliable, verified and utilized groundwater flow model available. This model, better known as MODFLOW, is developed by The United States Geological Survey and is in the public domain. After its original release, MODFLOW was supplemented by various new packages and its compiled versions for various computer platforms and PC configurations are sold by different vendors. There are several integrated, user-friendly pre- and post-processing software packages for MODFLOW that provide easy data input and visualization of modeling results in forms of contour maps and graphs.

MODFLOW utilizes a numerical solution for the equation governing groundwater flow through porous media:

$$\frac{\partial}{\partial x}\left(K_{xx}\frac{\partial h}{\partial x}\right)+\frac{\partial}{\partial y}\left(K_{yy}\frac{\partial h}{\partial y}\right)+\frac{\partial}{\partial z}\left(K_{zz}\frac{\partial h}{\partial z}\right)-W=S_s\frac{\partial h}{\partial t} \tag{35.1}$$

where K_{xx}, K_{yy} and K_{zz} are values of hydraulic conductivity along the x, y and z coordinate axes, which are assumed to be parallel to the major axes of hydraulic conductivity (Lt^{-1});

h is the potentiometric head (L);

W is a volumetric flux per unit volume and represents sources and/or sinks of water (t^{-1});

S_S is the specific storage of the porous material (L^{-1}); and

t is time.

For a derivation of equation (35.1) see for example Rushton and Redshaw (1979). In general, S_S, K_{XX}, K_{XY}, and K_{ZZ} may be functions of space ($S_S=S_S(x, y, z)$, $K_{XX}=K_{XX}(x, y, z)$, etc.) and W may be a function of space and time $W=(W(x, y, z, t))$. Equation (35.1) describes groundwater flow under non-equilibrium conditions in a heterogeneous and anisotropic medium, provided the principal axes of hydraulic conductivity are aligned with the coordinates directions (McDonald and Harbaugh, 1988). The complete explanation of the numeric approximation of equation (35.1) used in MODFLOW is given in Appendix F.

35.2.3 Model Geometry

Model geometry defines the size and the shape of the model. It consists of model boundaries, both *external* and *internal*, and model *grid.*

Boundaries

There are two types of external model boundaries: physical ("real") and hydraulic ("artificial"):

- *Physical boundaries* are well defined geologic and hydrologic features that permanently influence the pattern of groundwater flow. Examples are impermeable contact between two geologic units, a fault (either as a preferential flow path or a low-permeable barrier), contact between the porous medium and a large body of surface water, or contact between the porous medium and a man-made structure such as a slurry wall. It is preferable to have real physical boundaries as external model boundaries. If that is not possible because of the model scale limitations (*i.e.*, the real boundaries are too far and it is not feasible to include them), the hydraulic boundaries need to be defined.
- *Hydraulic boundaries* are derived from the groundwater flow net and are therefore "artificial" boundaries set by the model designer. They can be no-flow boundaries represented by chosen streamlines (remember that there is no flow across the streamline, see Problem 15), or boundaries with known hydraulic head represented by equipotential lines. It is less desirable to model with hydraulic boundaries since they are not permanent features and can change in time. In addition, their location, although guided by general characteristics of flow nets, is subjective. The main requirement (and difficulty) in setting hydraulic boundaries is that they have to be far enough from the area of interest within the model in order not to influence the future flow pattern created by projected activity (e.g., pumping, injection). The major consequence of this requirement is that the hydraulic boundaries very often have to be placed arbitrarily because of the lack of data beyond the area of an immediate interest. For example, many contaminated sites are relatively small in terms of regional groundwater flow pattern and the related hydrogeologic investigations, for various reasons, focus almost exclusively on the site itself.

As seen from Figures 35.2 and 35.3, most boundaries for the groundwater flow field in our case are physical and are developed in the alluvial aquifer: the large perennial stream (Red River), and the impermeable contact between alluvial deposits and gneiss. The contact between the alluvial aquifer and the sandstone aquifer is well defined and represents the boundary between the two distinct flow fields that are connected. This is also a logical boundary of the model since the main interest is in the alluvial aquifer. The influence of the sandstone aquifer and the role of the boundary between the two aquifers can be modeled with an appropriate boundary condition (Section 35.2.4). Only one hydraulic boundary (equipotential line) must be inferred in the far north where the river runs parallel to the contact between the aquifer and gneiss which leaves the flow field open.

Surface streams should be carefully analyzed before deciding what their role will be in the model design. Fully penetrating perennial streams offer a straightforward answer – if the flow in the river is large enough to sustain an extensive potential groundwater withdrawal in the adjacent aquifer, they are ideal equipotential boundaries. If the river is partially penetrating, and/or the amount of groundwater withdrawal is such that the flow in it might be affected to the point of drying, the river cannot be set as the external boundary. It will still, however, be an important internal boundary and must be described by an appropriate boundary condition.

Grid

Laying out the grid is the starting point in the actual computer model design. In MODFLOW, which is a finite difference model, the grid is formed by two sets of parallel lines that are orthogonal. The blocks formed by these lines are called *cells*. In the center of each cell is the *node* – the point at which the model calculates hydraulic head. It is assumed that hydraulic and hydrogeologic properties are uniform over the extent of a cell so that the cell is represented by its node. This type of grid utilized in MODFLOW is called *block-centered grid*.

Figure 35.4 illustrates the principle and general nomenclature of a block-centered grid for a hypothetical flow system. The grid consists of 5 rows, 9 columns, and 5 layers. The location of each cell in the grid is described with an *i,j,k* indexing system where

- *i* is the row index: i=1,2,..., "nrow";
- *j* is the column index: j=1,2,..., "ncol";
- *k* is the layer index: k=1,2,..., "nlay".

Although application of MODFLOW does not require designation of *x*, *y*, or *z* in terms of Cartesian coordinates (the only requirement is that rows and columns fall along consistent orthogonal coordinates within the layers), the following association is commonly assumed:

- columns (J) correspond to *x* coordinate,
- rows (I) correspond to *y* coordinate,
- layers (K) correspond to *z* coordinate.

There are two important differences between the MODFLOW and the Cartesian coordinate systems: 1) in MODFLOW the origin is in the upper left corner of the grid, and 2) the *k* index corresponds to a decrease in elevation, *i.e.*, the layers are numbered from the top down.

The grid mesh can be *uniform*, when all cells have same dimensions, and *custom*, when cell size varies. Although the uniform grid is preferred from a mathematical standpoint, it will often be necessary to design a custom grid. A uniform grid is a better choice when:

- available data on aquifer (system) characteristics are evenly distributed over the model area;

- there is not more interest in some parts of the model area than others, *i.e.*, the entire flow field is equally important;
- number of cells (size of the model) is not an issue either from computer memory or the modeling time standpoints.

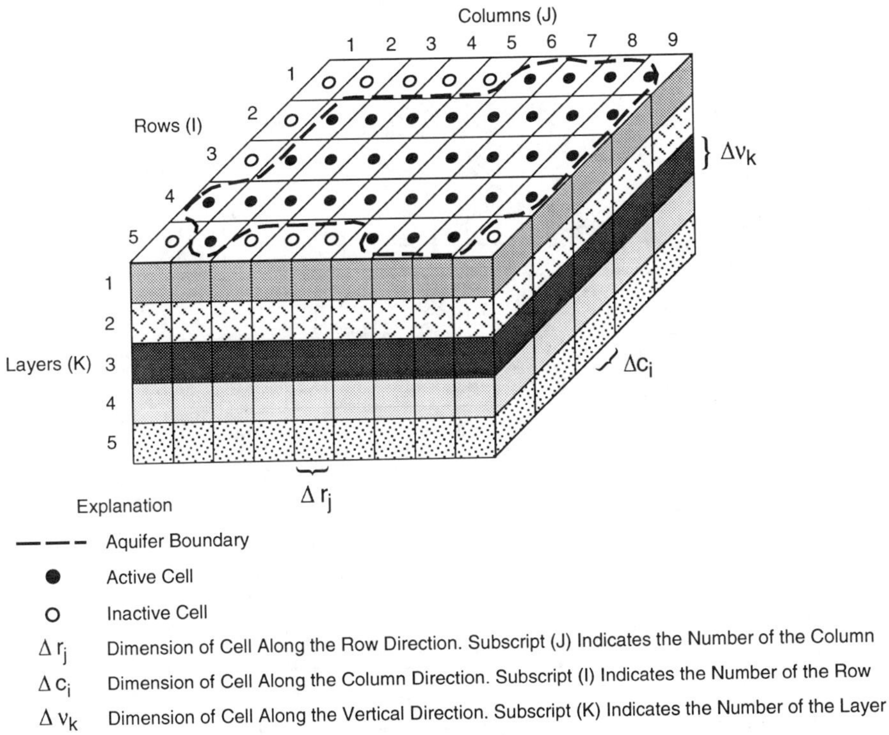

Figure 35.4 A discretized hypothetical aquifer system (McDonald and Harbaugh, 1988).

The custom grid is more appropriate when there is less or no data available for certain parts of the model area and/or there is specific interest in one or more smaller areas within the flow field. For example, a model for design of a groundwater extraction system for containment and cleanup of a plume is a "perfect candidate" for a custom grid because the following is usually true:

- extent of the plume is much smaller than the required model area -- remember that if there are hydraulic (artificial) model boundaries, they must be far enough away to not influence future pumping within the plume;
- data on aquifer (system) characteristics beyond the extent of the plume are limited or not available;
- a small cell size ("fine resolution") within the plume is needed in order to optimize the multiple-well system, *i.e.*, analyze their interaction and interference.

In general, a custom grid can greatly reduce computational time by decreasing the model size (number of cells). The areas of interest are discretized into smaller cells which give them more weight and modeling accuracy, while distant portions of the flow field are described with fewer and larger cells thus keeping the model size reasonable. Custom grids can also more accurately describe model boundaries. This is very important in the case of real physical boundaries (e.g., surface streams, impermeable contacts), especially when they are close to areas of interest.

The grid is laid out keeping in mind the modeling concept and particularly the position and shape of the areas of interest and model boundaries. Figure 35.5 shows a custom grid laid over the alluvial aquifer in our case. The smallest cell size is in the west central part of the aquifer closest to the river. This is based on the assumption that the highest pumping rates can be achieved by placing wells closer to the river and away from impermeable boundaries. The cell size increases somewhat outside this area of interest but not significantly since all boundaries are relatively close. Cell size in the row direction

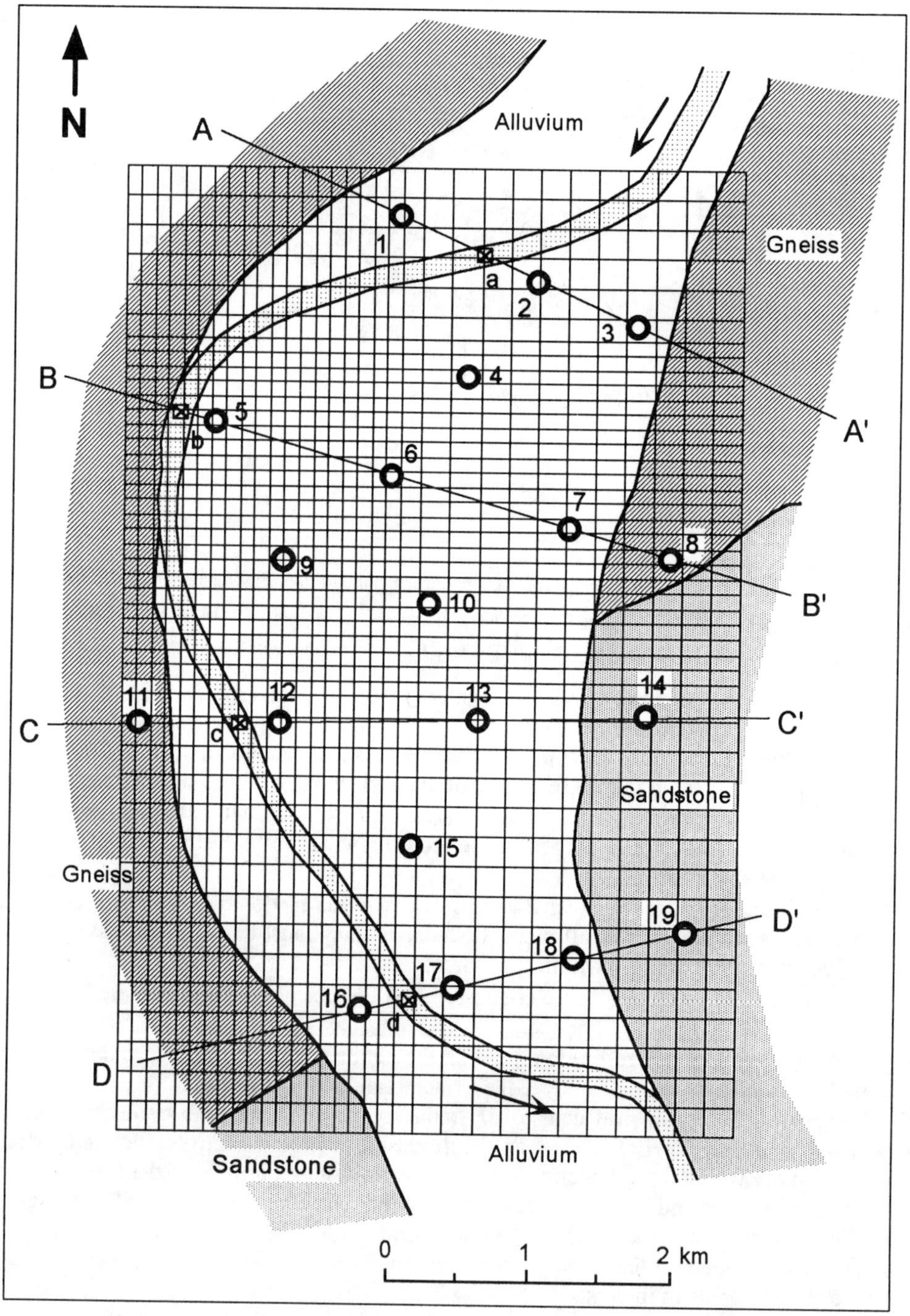

Figure 35.5 Custom finite difference grid laid over the alluvial aquifer.

(y axis direction if viewed in Cartesian coordinate system) could have been larger in the far south if there were no data on aquifer characteristics. However, the grid remains relatively fine since the water table measurements are available in wells 15 through 19 and will be useful during model calibration. In addition, the model has only one layer (unconfined aquifer) and its overall size of 31 columns x 45 rows does not pose any problem for the major versions of MODFLOW and common PC platforms. (Note that most current versions on the market are limited only by available computer memory which means that a standard Pentium PC with 16 to 32 MB of RAM can handle tens of millions of cells.)

A rule of thumb when designing a custom grid is that the size of a cell, in all three directions (row, column, layer), cannot be more than 1.5 times larger (or smaller) than the size of the adjacent cells (see Figure 35.6). This is necessary in order to preserve the mathematical stability of the numerical solution.

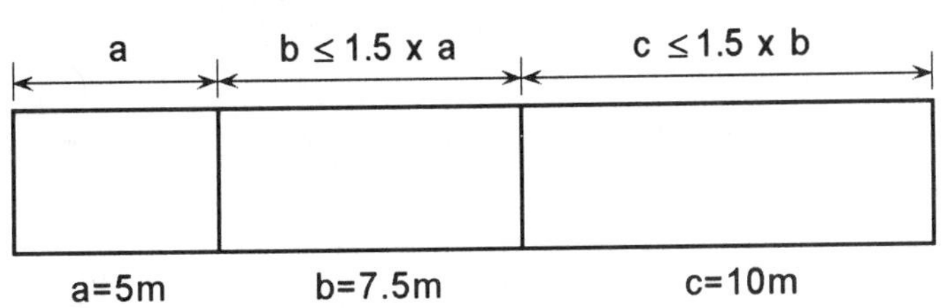

Figure 35.6 Size of adjacent cells in a custom grid cannot increase or decrease by more than 1.5 times in any of the three directions (*i.e.*, along column, row, and layer directions).

It is a good practice to dimension the overall grid size larger than the present estimated model area since first model runs may show that, for example, hydraulic boundaries are placed too close and/or adjacent aquifer(s) should be included in the model. It is important to remember that, once the grid size (*i.e.*, the total lengths in row and column directions) is set, it cannot be changed later without having to re-input all other parameters.

Grid Orientation

When the aquifer is isotropic, *i.e.*, when its hydraulic conductivity is the same in all directions, orientation of the grid is not critical. In such cases it is still recommended to orient the grid so that the number of inactive cells is minimized. Figure 35.7a shows an isotropic system of irregular shape without any dominant direction. Here a common south-north orientation seems most "natural": columns and rows are aligned with *x* and *y* axes respectively. Figure 35.7b shows an isotropic system that is elongated in one direction. In this case it is recommended that one of the model axes (either row or column) be aligned with the predominant direction. This reduces the number of inactive cells (see Section 35.2.4) and the size of the model. Note that inactive cells take space in the computer memory even though the model does not calculate hydraulic heads for them.

When the aquifer is anisotropic, the model coordinate axes must be aligned with the main axes of the hydraulic conductivity tensor. These axes, *i.e.*, directions of maximum and minimum hydraulic conductivities, are always perpendicular to each other. Finding the main axes of anisotropy and assigning the corresponding values of hydraulic conductivity is explained in Section 35.2.5.

Layer Type

There are four basic types of layers in MODFLOW:

- *Type 0.* This layer type is used to simulate *confined conditions* (layers/aquifers) in which transmissivity of each cell remains constant for the entire simulation time. For transient simulations this layer type requires the confined storage coefficient (specific storage x layer thickness) which is used to calculate the rate of change in storage.

- *Type 1.* This layer type is used strictly for *unconfined conditions* and is valid for the first (uppermost) layer only. It requires specific yield for transient conditions. Transmissivity of the layer varies as saturated thickness of the aquifer changes during simulation. The alluvial aquifer in our case is modeled as one layer of type 1.

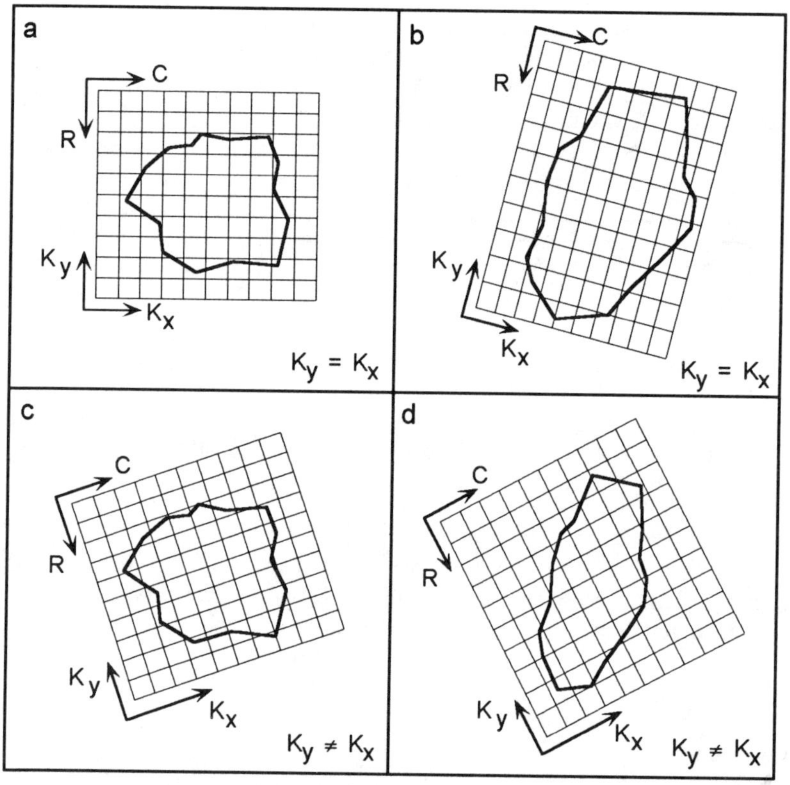

Figure 35.7 Grid orientation is dependent on aquifer's shape and degree of anisotropy. a) Isotropic aquifer with irregular shape. b) Isotropic aquifer elongated in one direction. c) Anisotropic aquifer with irregular shape. d) Anisotropic aquifer elongated in direction which is different from two main directions of hydraulic conductivity anisotropy.

- *Type 2.* This layer type is used when the aquifer alternates between *confined* and *unconfined.* However, it is assumed that the saturated thickness remains everywhere a high fraction of the layer thickness so that recalculation of transmissivity is not necessary (it is constant throughout the period of simulation). In transient simulations it is needed to specify both the storage coefficient for fully saturated confined conditions, and the specific yield for unconfined flow. If the layer completely desaturates, vertical leakage from above ceases.
- *Type 3.* A layer of this type is also used for *confined-unconfined* transitions. It includes varying transmissivity which is recalculated at each iteration using hydraulic conductivity and new saturated thickness. Confined storage coefficient and specific yield are both needed for transient simulations. Vertical leakage from above terminates when the layer is completely dry.

There are four additional types of layers supported by several processing software packages for MODFLOW such as PMWIN (Processing MODFLOW for Windows) by Chiang and Kinzelbach (1996). A semi-confining unit does not have to be simulated as active in MODFLOW. Its influence can be described by a term named vertical leakance between two active model layers. Models which do not directly include semi-confining layers are called *quasi three-dimensional models*. In original MODFLOW and some commercially available processors, the leakance must be calculated directly by the user (see Figure 35.12 and equation 35.4). PMWIN can calculate the leakance term or it can be left to the user if desired.

Although a quasi-3D model may sometimes seem easier to design and more appropriate (particularly when there is no data on the semi-confining layer), it is better to use the full potential of MODFLOW and design a real 3D model. For example, the model may be used later to analyze flow paths between aquifers and the exact role of the separating layer, or it may become part of a groundwater transport model. In both cases it is necessary to fully include the semi-confining layer and assign realistic values of effective porosity and other transport parameters.

Four additional layer types associated with quasi-3D modeling and supported by PMWIN and some other commercial processors are:

- *Type 4* -- Layer of type 0 associated with a semi-confining layer.
- *Type 5* -- Layer of type 1 associated with a semi-confining layer.
- *Type 6* -- Layer of type 2 associated with a semi-confining layer.
- *Type 7* -- Layer of type 3 associated with a semi-confining layer.

It should be noted that, if necessary, one hydrostratigraphic unit (aquifer) can be modeled with more layers of the same hydraulic conductivity. This is typically done when steep hydraulic gradients are present, such as in aquitards, or expected to occur due to an imposed stress such as well pumpage or drain installation. More layers are also recommended when a well is partially penetrating thus allowing for the accurate placement of the well screen.

35.2.4 Boundary Array

The next step after laying out the grid is to describe model boundaries, *i.e.*, to determine the type of each cell in the model.

Cell Type

There are two types of cells in MODFLOW used to describe boundaries: inactive ("no-flow") cells and constant-head cells. *Inactive cells* (designated with number 0 in MODFLOW's boundary array -- IBOUND) are those for which no flow into or out of the cell occurs during the entire time of simulation. The model does not calculate hydraulic head for these cells and they are used to "delete" model array beyond the boundary of the flow field. Figure 35.8 shows the principle of delineating an impermeable boundary with inactive cells. If more than one half of a cell is within the impermeable unit it is designated as an inactive cell. This estimate is quickly done by focusing on the cell side closest to the boundary. As seen in Figure 35.8, all other cells beyond the boundary cells are also designated as inactive (no-flow). The remaining cells are called *active*, or *variable-head cells*. They are designated with number 1 in MODFLOW's boundary array (IBOUND). The hydraulic heads for active cells are calculated by the model and are free to vary in time.

Constant-head cells are used to describe model boundaries with known heads such as aquifer contacts with major surface water features. The hydraulic head at these cells is specified in advance and does not change during a simulation period. Constant-head cells

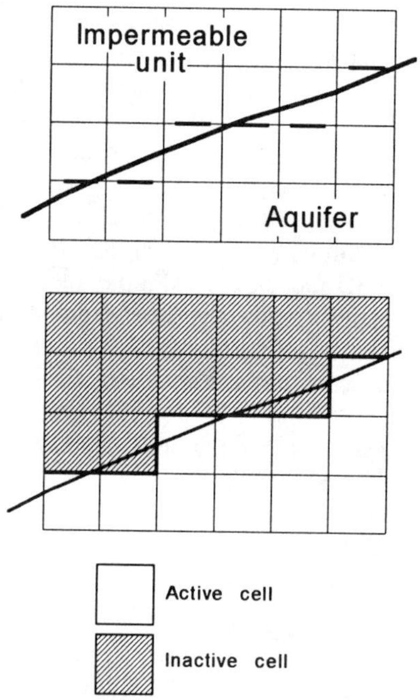

Figure 35.8 Delineation of an impermeable boundary with inactive (no-flow) cells. If more than half of the cell is in the impermeable unit, the cell is designated as inactive.

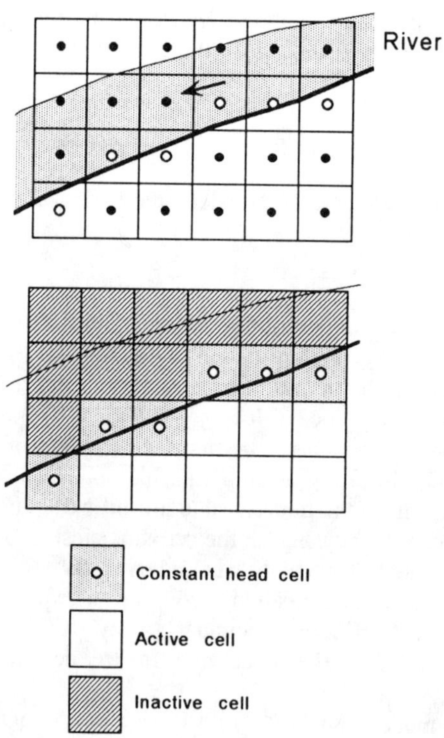

Figure 35.9 Delineation of a surface stream boundary with constant-head cells. The cell whose center is the closest to the contact between the stream and the aquifer is designated as constant head cell.

are designated with number –1 in the IBOUND array. Figure 35.9 shows the principle of delineating the contact between an aquifer and a surface stream which is at the same time the external model boundary (as in our case). This is done simply by noting cell nodes that are closest to the contact. Only the line of the contact is important when designating constant-head cells. In other words, the width of the river is not important and all cells beyond the river bank of interest, including cells in the river, are designated as inactive. When a river is an internal model boundary, the described principle is applied to both river banks and all cells in-between (cells within the river) are designated as constant-head cells. (*Note that it is not always recommended to simulate contact between the aquifer and the surface water body with constant-head cells since they are an inexhaustible source of water. Such boundaries can be simulated with several MODFLOW packages depending upon actual conditions as explained further in this section.*)

Having three cell types allows for accurate description of irregularly shaped flow fields (aquifers). This is important since the model array is always rectangular in outline.

35.2.5 Model Parameters

Model parameters are divided into three groups:

- time,
- space (layer top and bottom),
- hydrogeologic characteristics (hydraulic conductivity, transmissivity, storage parameters, and effective porosity).

Time

Time parameters are specified when modeling transient (time-dependent) conditions. They include time unit, the length and number of time periods, and the number of time steps. During one *time (stress) period* all model parameters associated with boundary conditions and various stresses remain constant. Having more time periods allows these parameters to change in time. For example, a pumping well can change its pumping rate in different time periods, and the river stage or recharge from precipitation can vary according to seasons. A time period is further divided into time steps which are useful for analyzing changes in hydraulic head and drawdown. Time steps do not have to be of same length. This option, controlled by the time step multiplier, is particularly useful when simulating a well pumping test: each time step is longer than the preceding one and the results can be plotted over several logarithmic cycles. More time steps increase the accuracy of iterative computations.

Note that the length of stress periods is not relevant for steady state simulations.

Layer Top and Bottom

The elevation of the layer top and bottom is required if the user wants the program to calculate aquifer (layer) transmissivity, vertical leakance or confined storage coefficient which is recommended. Original MODFLOW reads the top elevation only for layers of type 2 and 3, and the bottom elevation for layers of type 1 or 3. Hydraulic conductivity is needed only when modeling unconfined aquifers, *i.e.*, layers of type 1 and 3, in which case MODFLOW calculates transmissivity by multiplying K with saturated thickness. In all other conditions the transmissivity and storage coefficient (for transient modeling) are specified by the user. This is convenient when the actual layer thickness is unknown or the layer shape is too complicated. Some processors like PMWIN have both options: the user can either specify hydraulic conductivity, and top and bottom for all (or some) layers and then let the program calculate transmissivity, or specify transmissivity directly (Figure 35.10). It is important to remember that the layer top for unconfined aquifers is the water table and not the land (topographic) surface.

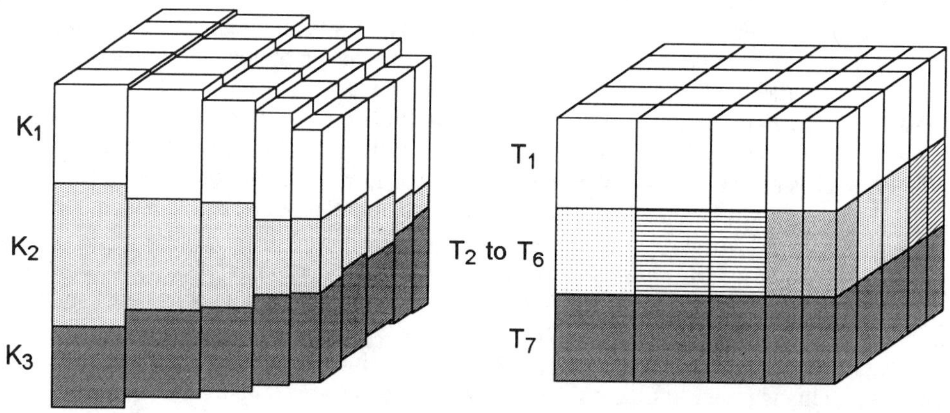

Figure 35.10 Modeling layers of varying thickness. **Left**: With hydraulic conductivities, and actual top and bottom. **Right**: With varying transmissivity, and horizontal top and bottom.

Whenever possible, it is best to input top and bottom for all layers since it is important to relate hydraulic heads in different layers with their respective geometry. This is particularly useful when analyzing possible hydraulic connection and/or migration of contaminants between layers (aquifers).

One common problem when defining layer top and bottom is the question of interpolation. Numerical groundwater models require that these and other areally distributed model parameters be assigned to each cell. Since available data are usually limited to irregularly scattered locations in the model domain, it will be necessary to estimate (interpolate) the parameter values for each cell. A nightmare for every modeler using original MODFLOW was the manual input of model parameters for large multi-layered models of highly heterogeneous aquifers. Today, most major MODFLOW processors, such as PMWIN, offer several convenient options for data input including zones and various interpolators that generate data for each cell. Parameter arrays can also be defined using some external interpolator such as SURFER and then imported in the model. This greatly reduces modeling time and increases accuracy.

As seen from cross-sections in Figure 35.3, elevation of the bottom of the alluvial aquifer in our case is fairly uniform and it increases sharply only very close to the river valley sides. The aquifer base is therefore assumed to be horizontal for the modeling purpose and set at the elevation of 550 m amsl. The aquifer top is equal to water table and is interpolated from the data recorded in monitoring wells.

Horizontal Hydraulic Conductivity and Transmissivity

Hydraulic conductivity is in many cases the most critical and sensitive modeling parameter. Every attempt should be made to design a model with realistic values of K obtained in the field, preferably by pumping tests. By varying the spatial distribution of hydraulic conductivity arbitrarily, an experienced modeler can almost always produce a modeling result that closely resembles the field data set used for calibration. Another question is if that distribution is justified. Only after all other calibration avenues such as adjusting boundaries, boundary conditions, and stresses acting upon the system are exhausted, should the user engage in changing K distribution: there are already enough models designed with oddly shaped "K rectangles" scattered throughout the flow domain without any geological/hydrogeological support, logic or explanation.

In MODFLOW the horizontal hydraulic conductivity is the conductivity along grid rows (K_x). If the aquifer (layer) is isotropic, K along model columns (K_y) will be the same.

Otherwise, the horizontal anisotropy is specified with the anisotropy factor which multiplies K_x to obtain K_y. The original MODFLOW and most processors require input of the anisotropy factor (1 for isotropic conditions) while some programs allow for direct input of K_y.

During field work it is not uncommon to discover two dominant directions of hydraulic conductivity anisotropy that are at a certain angle to each other. This information can be used to define the two axes of K that are perpendicular to each other -- important when orienting the model grid (see Figure 35.11). The following equation relates the principal conductivity components K_x and K_y to the resultant K_r in any angular direction θ (adapted from Freeze and Cherry, 1979, p. 35–36):

$$K_y = \frac{K_r \cdot K_x \cdot \sin^2 \theta}{K_x - K_s \cdot \cos^2 \theta} \tag{35.2}$$

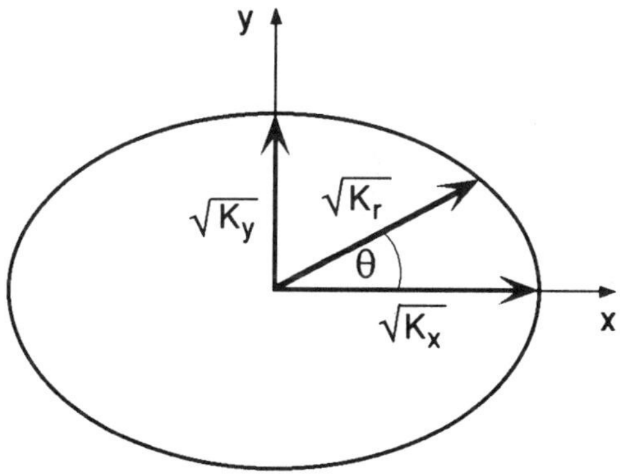

Figure 35.11 Hydraulic conductivity ellipse for determining the second major axis of hydraulic conductivity anisotropy from two distinct directions found in the field that are not perpendicular to each other. (Note that the directions of maximum and minimum hydraulic conductivities are always perpendicular to each other in an anisotropic flow field.)

The hydraulic conductivity is required for layers of type 1 and 3 (unconfined conditions), and the user can specify transmissivity for layers of type 0 and 2 directly (confined conditions). Most MODFLOW processors can calculate transmissivity for a layer of any type by multiplying hydraulic conductivity with the layer thickness derived from the layer top and bottom elevations which is recommended.

Vertical Hydraulic Conductivity and Leakance

For quasi-3D models with more than one layer and for full 3D models, MODFLOW requires the input of the vertical leakance between two layers. The following equation is used when two layers belong to same hydostratigraphic unit, or when two hydrostratigraphic units (layers) are in direct contact (Figure 35.12 top):

$$Vcont_{i,j,k+1/2} = \frac{1}{\dfrac{(\Delta v_k)/2}{K_{z\ i,j,k}} + \dfrac{(\Delta v_{k+1})/2}{K_{z\ i,j,k+1}}} \tag{35.3}$$

where $K_{z\ i,j,k}$ is the vertical hydraulic conductivity of the upper layer in cell i,j,k, and $K_{z\ i,j,k+1}$ is the vertical hydraulic conductivity of the lower layer in cell $i,j,k+1$.

When two layers are separated by a low-permeable unit which is not directly modeled (quasi 3D model), the leakance is calculated using the following equation (Figure 35.12 bottom):

$$Vcont_{i,j,k+1/2} = \frac{1}{\frac{\Delta z_u / 2}{Kzu} + \frac{\Delta z_c}{Kzc} + \frac{\Delta z_L / 2}{KzL}} \qquad (35.4)$$

where K_{zu}, K_{zc}, and K_{zL} are the vertical hydraulic conductivities of the upper aquifer, the semi-confining layer, and the lower aquifer respectively.

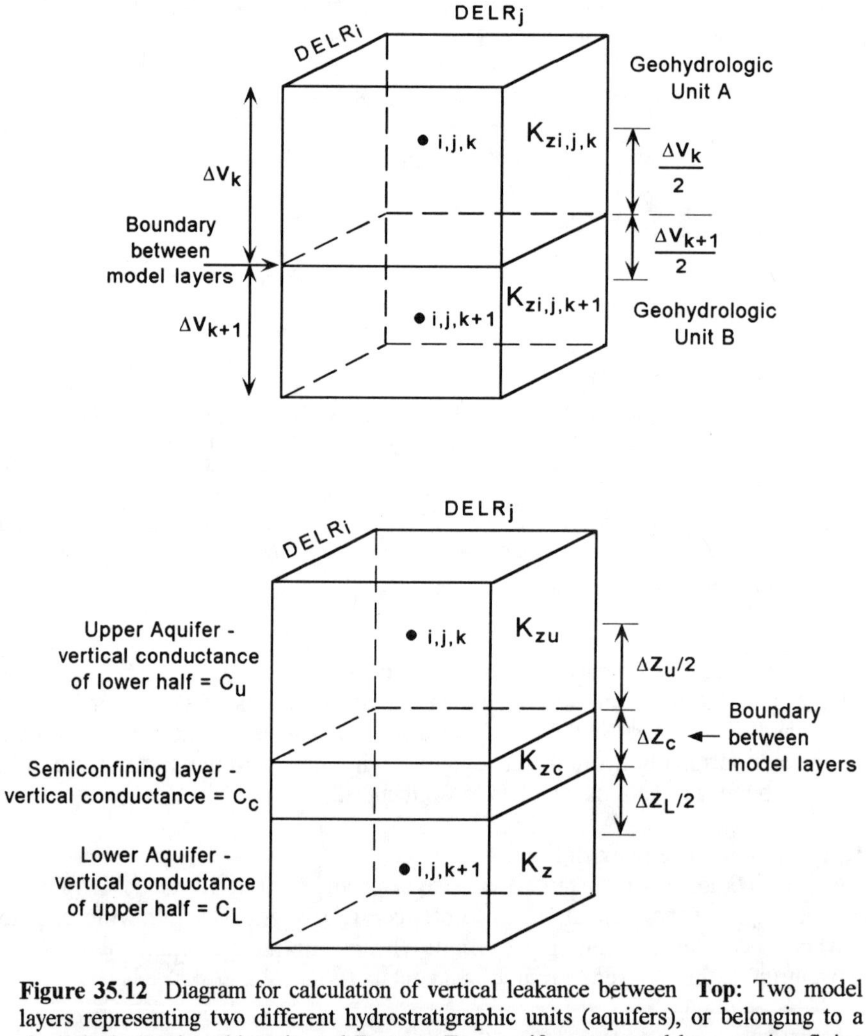

Figure 35.12 Diagram for calculation of vertical leakance between **Top:** Two model layers representing two different hydrostratigraphic units (aquifers), or belonging to a single hydrostratigraphic unit, and **Bottom:** Two aquifers separated by a semiconfining unit which is not directly modeled (after McDonald and Harbaugh, 1988).

If desired, processors such as PMWIN can calculate the vertical leakance for each layer from the information specified by the user: layer top and bottom elevations and the vertical hydraulic conductivity. Unless accurately determined from pumping tests (see for example Problem 29), the vertical hydraulic conductivity is usually assumed and/or calibrated. For example, it is common to assume that the vertical hydraulic conductivity is one magnitude lower than the horizontal conductivity for most stratified sedimentary rocks. If the vertical grid axis is not aligned with the principal axis of vertical hydraulic conductivity (such as in dipping strata), a correction should be made as it was in the case of horizontal anisotropy (see Figure 35.11 and equation 35.2).

Storage Terms

Storage terms – *storage coefficient* for confined layers (layer type 0, 2 and 3) and *specific yield* for unconfined layers (layer type 1,2 and 3) are required only for transient simulations. The storage coefficient (S) is the product of layer thickness (b) and specific storage (S_s):

$$S = b \cdot S_s \qquad \text{(dimensionless)} \tag{35.5}$$

Note that S is the storage term determined by pumping tests in confined aquifers as explained in Chapter 8. Pumping tests in unconfined aquifers give specific yield (also dimensionless) which is several orders of magnitude larger than the storage coefficient. Typical values for the specific yield are between 0.3 and 0.001, and for the storage coefficient between 0.001 and 0.00001.

Effective Porosity

Effective porosity is required only if the model will be used in conjunction with other programs that use results generated by MODFLOW to calculate the average linear velocity of groundwater flow. This velocity is needed to track water particles as they move through the porous medium. Two widely used particle tracking computer programs based on MODFLOW are MODPATH (Pollock, 1989) and PATH3D (Zheng, 1990).

35.2.6 Boundary Conditions

After assigning time parameters, and space and hydrogeologic parameters for each model cell, the next step is to accurately describe boundary conditions using the most appropriate MODFLOW packages. Since the choice is not straightforward, the package that will be used for a particular boundary will often be decided during model calibration.

Boundary with Known Head

A boundary with known head, such as an equipotential line or a contact between the aquifer and a large surface water body, can be simulated by constant-head cells as explained earlier. However, such simulation should be done with great caution since constant-head cells represent an inexhaustible source of water regardless of changes at the boundary and in the aquifer. The head at constant-head cells remains the same throughout the time period of simulation. For example, a small stream may dry out because of the nearby well pumpage and the constant-head cells simulating it will continue to supply water to the aquifer.

PMWIN supports an option developed by Leake and Prudic (1988) called the *Time-Variant Specified-Head Package (CHD1)*. This package is used in transient simulations and allows constant-head cells to take on different head values for each time step during a simulation time period. The user specifies the value of the head in the cell at the start of the stress period (h_s), and at the last time step in the stress period (h_e). The package then linearly interpolates boundary heads (h) for each cell using the following equation:

$$h = h_s + (h_e - h_s) \cdot \frac{PERTIM}{PERLEN} \tag{35.6}$$

where *PERTIM* is the starting time of a time step in a stress period and *PERLEN* is the length of the whole stress period.

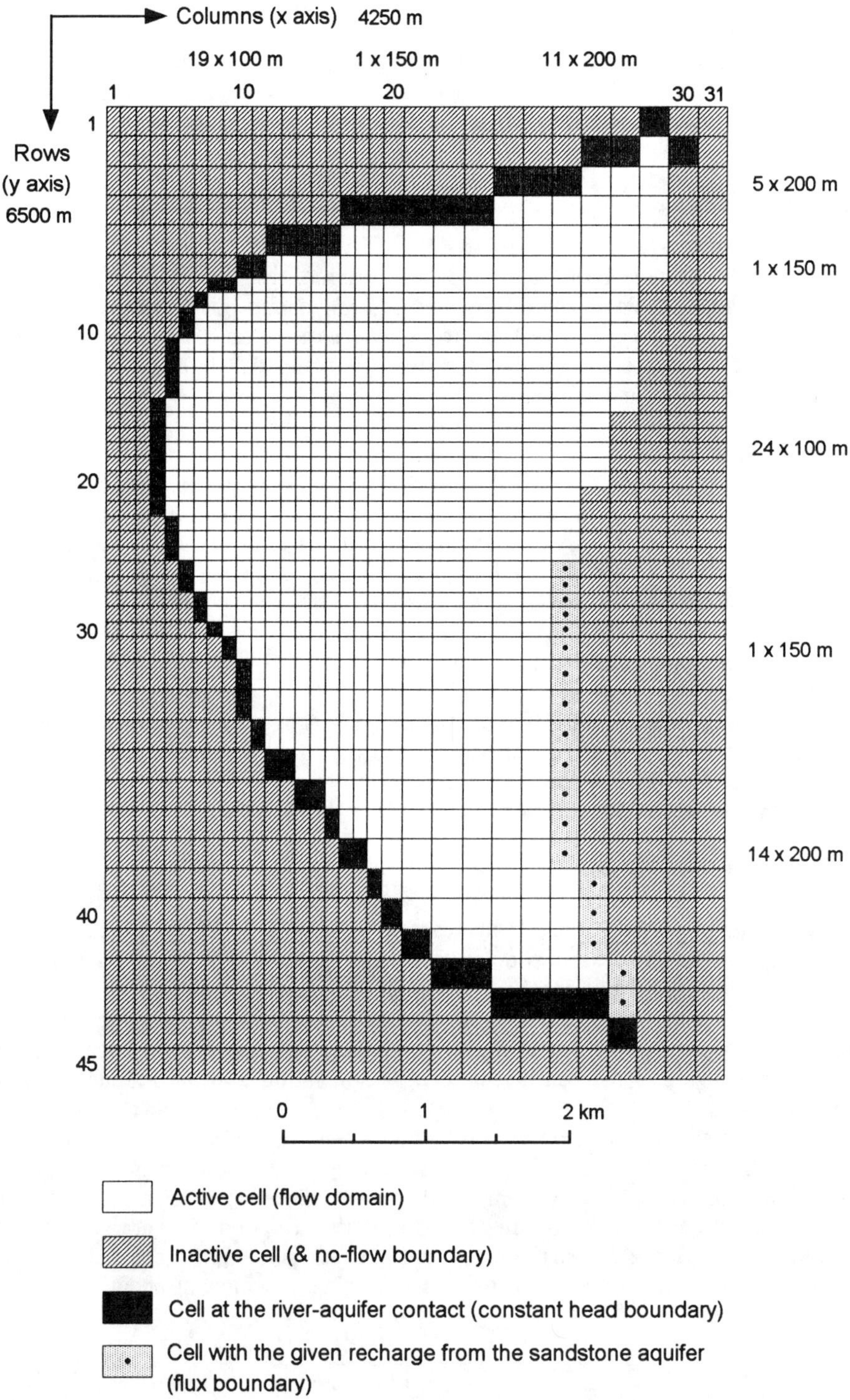

Figure 35.13 Boundaries and boundary conditions for the flow model in our case (alluvial aquifer shown in Figures 35.1, 35.2, and 35.5).

The Time-Variant Specified-Head Package (CHD1) is useful when, for example, data on weekly, monthly, or seasonal variations in river stages are available.

Figure 35.13 shows boundaries and boundary conditions for our example. The Red River is simulated by constant-head cells since there is no data on time variations in river stage, river flows, or river bed characteristics. In addition, it is certain that the river is perennial with a substantial flow during the summer-fall period. If some or all of the above mentioned data were available, it would be recommended to apply the MODFLOW packages describing head-dependent flux boundaries.

Boundary with Head-Dependent Flux

The boundary for which the flow rate is calculated, based on the head difference between the boundary cells and the adjacent aquifer cells, is called the head-dependent flux boundary. Three MODFLOW packages can simulate this condition:

- River Package
- Drain Package
- General-Head Boundary Package

Common to all three packages is that the user must specify a conductance term which can often be completely defined only during the calibration process. Despite this "inconvenience", all three packages offer versatility in describing various external and internal model boundaries.

River Package

The River Package simulates the flow between an aquifer and a surface water body such as a river or a lake. The rate of flow is calculated from the difference in hydraulic heads at the boundary cell and the adjacent aquifer cell using the following equation (McDonald and Harbaugh, 1988):

$$QRIV = \frac{KLW}{M} \cdot (HRIV - h) \tag{35.7}$$

or

$$QRIV = CRIV \cdot (HRIV - h) \tag{35.7a}$$

where

QRIV is the flow between the stream and the aquifer, taken as positive if it is directed into the aquifer,
HRIV is the head in the stream,
h is the head at the node in the cell underlying the stream reach,
CRIV is the hydraulic conductance of the stream-aquifer interconnection:

$$CRIV = \frac{KLW}{M} \tag{35.8}$$

K is the hydraulic conductivity of the streambed material,
L is the length of the river channel in the cell,
W is the width of the river channel in the cell,
M is the thickness of the riverbed material.

When the hydraulic head in the aquifer falls below the river channel bottom, the flow remains constant and is given by:

$$QRIV = CRIV \cdot (HRIV - RBOT) \tag{35.9}$$

where *RBOT* is the elevation of the river channel bottom. As can be seen from equation (35.8), the River Package should be used with care since there is no adjustment for river flow and stage – the supply of water to the aquifer from a losing stream can be more than the flow in the stream. For a more detailed explanation of the River Package see Appendix F.

Drain Package

The Drain Package simulates both closed (buried) and open drains. It is different from the River Package in that the flow is directed only from the aquifer toward the drain and it stops when the head in the aquifer drops below the elevation of the drain. This makes the Drain Package a useful tool for modeling springs. The rate of flow entering the drain (QD) is calculated from the following equations (McDonald and Harbaugh, 1988):

$$QD = CD \cdot (h - d) \quad \text{for } h > d \tag{35.10}$$

$$QD = 0 \quad \text{for } h \leq d \tag{35.11}$$

where

CD is drain hydraulic conductance,

h is the hydraulic head in the aquifer cell,

d is the elevation of the drain in the drain cell.

CD depends on aquifer hydraulic conductivity near the drain, distribution and characteristics of the fill material, number and size of the drain pipe openings, and characteristics of the clogging materials. In practice, many of these data are not available and the drain hydraulic conductivity has to be estimated and adjusted during model calibration. For open drains, CD can be calculated using equation (35.7) for the conductance of riverbed material. For a more detailed explanation of the Drain Package see Appendix F.

General-Head Boundary Package

The head-dependent flow (Q_b) through a boundary simulated by the General-Head Boundary Package is calculated from

$$Q_b = C_b \cdot (h_b - h) \tag{35.12}$$

where

C_b is the conductance of the boundary analogous to the streambed conductance,

h_b is the hydraulic head at the boundary cell,

h is the hydraulic head at the aquifer cell adjacent to the boundary.

The General-Head Boundary package assumes unlimited linear flow between the boundary and the aquifer and should be used with care. It is useful for transient simulations of known head boundaries since it allows changes in C_b and h_b for different time periods. When a large value of conductance is used, the cell simulated by this package is equivalent to a constant-head cell. For a more detailed explanation of the General-Head Boundary package see Appendix F

Boundary with Known Flux

Typical boundaries with known flux are water tables in unconfined aquifers, inflow/outflow through lateral contacts between different aquifers, seepage to or from adjacent layers, spring flow, and recharge to or from surface streams. Water table flux and a lateral inflow are boundary conditions present in our case as shown in Figure 35.13. Whenever possible, the model should have boundary conditions described with known (prescribed) head and/or head-dependent flux rather than with flux only. This is important for the following reasons - the model result is a distribution of hydraulic heads and the iteration process is more accurate if it starts with known heads, and it is easier to measure hydraulic head in the field.

In general, known flux conditions can be simulated by one of the following packages:

Recharge Package

Recharge to a model cell is specified as recharge intensity and must have dimensions of velocity expressed in same length-time units as other parameters. The model calculates the recharge flux by multiplying the intensity with the cell area. The cell within each vertical column to which the recharge is applied is specified through the recharge option menu. The options include:

1. Application of the recharge to the first model layer. If the cell is inactive, it will not receive recharge.
2. Application of the recharge to any cell in the vertical column as specified by layer numbers. If the cell is inactive, it will not receive recharge.
3. Application of the recharge to the uppermost active cell in the vertical column, provided there is no constant cell above it in the column. If there is a constant cell in a vertical column of cells and there is no active cell above, then no recharge is applied to this column because it is assumed that any recharge would be intercepted by the constant head source.

See Appendix F for a more detailed explanation of the Recharge Package.

In our problem the recharge is estimated as 10% of the average annual precipitation of 630 mm. This number is based on the fact that the climate in the area is semi-arid with high temperatures causing a high rate of evapotranspiration most of the year. In general, the exact recharge rate is hard to define accurately since it depends on various factors such as climate, land cover and land use, topography, underlying geology, etc. For more information on determining recharge, the reader should consult a textbook on hydrology (e.g., Maidment, 1993).

Evapotranspiration Package

The Evapotranspiration Package (ET) simulates the effects of plant transpiration and direct evaporation from a shallow water table. The maximum ET occurs at the user-specified water table elevation, usually the elevation of the land surface. ET ceases when the water table drops below the user-specified "extinction depth". Between these two extreme elevations the evapotranspiration rate varies linearly.

ET is drawn from only one cell in the vertical column, usually from the uppermost cell (layer 1). However, there is an option to specify the cell within the vertical column from which ET will withdraw water. In both cases ET is set automatically to zero by the model if the designated cell is either a no-flow (inactive) cell or a constant head cell.

The model calculates the volume of water extracted from a cell by ET using the following information provided by the user:

- maximum ET rate (dimensions of velocity),
- elevation of the ET surface,
- ET extinction depth,
- layer indicator.

Similar to recharge, it is difficult to accurately determine the evapotranspiration rate since it is influenced by many factors. For more information about the nature of evapotranspiration and methods for its determination, the reader should consult some of the textbooks on general hydrology (e.g., Maidment, 1993). A more detailed explanation of the ET package is given in Appendix F.

Well Package

The Well Package simulates pumping or injection wells using rates (Q) specified by the user for each stress (time) period of the simulation. Negative values of Q indicate a withdrawal well, while positive values of Q indicate a recharging well. Since the rate of withdrawal/injection is independent of both the cell area and the head in the cell, the Well Package is the ideal tool for simulating boundaries with known flux. In addition, the inflow rate can be assigned to any cell in any layer (with the exception of no-flow cells and constant head cells).

In our example, the influx from the sandstone aquifer is simulated by recharge wells placed in the cells along the boundary between the alluvium and the sandstone. Figure 35.14 illustrates the principle of assigning injection rate to a well that simulates the boundary with known flux. This rate can vary for each stress period thus simulating seasonal changes in the lateral recharge.

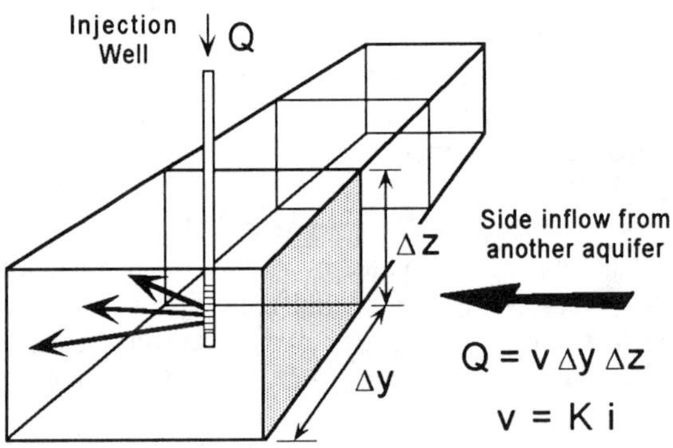

Figure 35.14 Determination of injection (recharge) rate for a well that simulates boundary with known flux. The rate must be calculated externally using (or assuming) the hydraulic conductivity of the adjacent aquifer (K), the hydraulic gradient (i; usually determined from the well data), and the area of the cell side.

When simulating several real wells that fall within a one-cell area, the user must sum the individual pumping rates. If the well penetrates more layers, it should be represented by a group of single-layer wells, each open to one of the layers tapped by the multi-layer well, and each having an individual Q term specified for each stress period. In such cases the discharge rate for individual-layer wells must be determined externally to the program.

For further explanation about the Well Package see Appendix F.

Stream Package

The Stream Package (or the Streamflow-Routing Package) is actually a combination of a known flux and a head-dependent flux boundary. It is similar to the River Package since it allows flow into and from the stream. However, it is more sophisticated than the River Package because it considers the flow rate in the stream and limits the leakage between the aquifer and the stream accordingly. The package increases stream flow in areas of gaining stream segments (reaches) and reduces the flow by taking water out through riverbed seepage in losing reaches. The package can also calculate the stream stage needed for other calculations by using a set of required data instead of a preassigned stage value.

Because of its versatility, the Stream Package requires intensive data preparation and more information than any other MODFLOW package. For detailed information about the Stream Package see Appendix F.

No-Flow Boundaries

Boundaries with the flux set to zero are called no-flow boundaries. Figure 35.8 shows the principle of delineation of an impermeable boundary with inactive (no-flow) cells. MODFLOW automatically assumes a no-flow boundary around the perimeter of the grid. If the user's intention is to align one or more model edges with some other boundary type, this must be done by assigning appropriate boundary conditions to the last cells along the grid edge.

35.2.7 Initial Conditions

Initial conditions are values of the hydraulic head for each active and constant-head cell in the model. They must be higher than the elevation of the cell bottom and are necessary for the start of iterative model calculations. Initial hydraulic heads in transient simulations should closely resemble the head distribution derived from field data. For steady state conditions only hydraulic heads in constant-head cells must be actual (realistic) values. In our example they are linearly extrapolated along Red River from four measurements of water elevation in the river channel (see Table 35.1 and Figure 35.2). The heads in active cells can be set arbitrarily since the model will calculate their distribution based on boundary conditions.

Once the model is calibrated (in either steady state or transient conditions), the resulting distribution of hydraulic heads will become the new set of initial heads for the prediction phase. In other words, these heads will be imported into the initial heads array over the old estimated heads. This can also be done several times during the calibration phase in order to reduce the calculation time.

35.2.8 Stresses (Packages)

In addition to boundary conditions, all MODFLOW packages described earlier are used to simulate various stresses acting upon the aquifer. An aquifer stress is defined as any supply to or withdrawal of water from the system. This broad statement also shows that various packages can be used for various stresses regardless of their name as seen fit by the user. For example, the Drain Package can be used to simulate a river, or a spring, the River Package can simulate a leaky confined aquifer, and so on.

Following is a brief summary of the packages developed for MODFLOW (see Appendix F for descriptions):

- *River Package*. Simulates the flow between an aquifer and a surface water feature in both directions. Includes riverbed conductance for simulating fine sediment clogging of a river channel. Should not be used for intermittent streams since there is no adjustment for the stream flow and stage once it drops below the channel bottom.
- *Recharge Package*. Typically used to simulate areal infiltration from precipitation or irrigation. Can be used to simulate local recharge from ponds. Flexible in assigning vertical flux to different layers along the vertical.
- *Well Package*. Simulates both extraction and recharge wells. Assumes full penetration of the layer. Partial penetration should be modeled with more layers within the same hydrostratigraphic unit. Good tool for modeling lateral inflow or outflow flux in any layer.

- *Drain Package*. Simulates both closed and open drains. Best tool to model springs since the inflow into the drain ceases when the head in the aquifer drops below drain elevation. Requires intensive calibration.
- *Evapotranspiration Package*. Typically used in irrigation for simulating the effects of plant water intake. Varies linearly with the water table elevation accounting for the user specified maximum and minimum (zero) depth of evapotranspiration.
- *General-Head Boundary Package*. Used for transient simulations of constant-head boundaries since it allows the user to change the head from one time period to another. Can be used to simulate permanent surface water features but does not allow for any limit to discharge depending upon actual flow and stage in surface streams.
- *Stream Package* (Prudic, 1989). Simulates interaction between an aquifer and a surface stream accounting for the flow rate in the stream. Flow into or out of the stream stops when the stream dries out. Requires intensive data preparation and more data input than any other package.
- *Horizontal-Flow Barrier Package* (Hsieh and Freckleton, 1993). Simulates thin low-permeable features such as vertical slurry walls and faults.
- *BCF2 Package* (McDonald et al., 1992). Allows for simulations in which a water table rises into unsaturated model layers. A typical application is the simulation of the recovery of over-stressed aquifers, such as after heavy pumpage, either through artificial recharge or the reduction of stress.
- *Transient Leakage Package* (Leake et al., 1994). Simulates transient leakage from confining units.
- *RES1 Package* (Fenske et al., 1996). Simulates leakage from reservoirs.

35.2.9 Model Run

Assuming that all the above explained model parameters are correctly assigned to each cell, and before actually running the model, the user must choose one of the solving packages ("Solvers"), specify several calculation parameters, and choose which model results will be saved. Common for most solvers are the head change criterion for convergence and the maximum allowed number of iterations. When the maximum absolute value of hydraulic head change in any model cell is less than or equal to the head change criterion, the iteration process stops. Three of the most widely used MODFLOW solvers are the Strongly Implicit Procedure (SIP), the Slice-Successive Overrelaxation method (SSOR), and the Preconditioned Conjugate-Gradient 2 method (PCG2). It is a good practice to try all three solvers and see which one gives better results for the particular problem. For a more detailed explanation of these three solvers and their applicability see McDonald and Harbaugh (1988) and Hill (1990).

The first model run is the fear (or joy) of every model designer. When dealing with a "real-life" model, it is almost certain that the first result will not be a satisfactory match between the calculated and measured hydraulic heads. The number of model runs during calibration will depend on the quantity and quality of available data, desirable accuracy of the model results, and often the patience of the user.

35.2.10 Calibration and Sensitivity Analysis

Although often explained separately in modeling reports, calibration and sensitivity analysis are inseparable and are part of the same process. While performing calibration, which is comprised of numerous single and multiple changes of model parameters, every user determines quickly which parameters are more sensitive to changes with regard to the final model result. By carefully recording all of the changes made during calibration and commenting on their results, the model designer is already engaged in the sensitivity analysis and can effortlessly finalize this part of the modeling effort later.

Calibration is the process of finding a set of boundary conditions, stresses, and hydrogeologic parameters which produces the result that most closely matches field measurements of hydraulic heads and flows. Calibration of every model should have the target of an acceptable error set beforehand. Its range will depend mainly on the model purpose. For example, a groundwater flow model for evaluation of a regional aquifer system can sometimes "tolerate" a difference between calculated and measured heads of up to several feet. This, however, would be an unacceptable error in the case of a model for design of containment and cleanup of a contaminant plume spread over, say, fifty acres.

In many instances the quality of calibration will depend on the amount and reliability of available field data. It is therefore crucial to assess these field data, or *calibration data set*, for their consistency, homogeneity and measurement error. Such assessment is the basis for setting the calibration target.

Model calibration can be performed for steady state conditions, transient conditions, or both. Although steady state calibration has prevailed in modeling practice, every attempt should be made to have a transient calibration as well for the following reasons:

- groundwater flow is transient by its nature, and is often subject to artificial (man-made) changes;
- the usual purpose of the model is prediction which is by a definition time-related;
- steady state calibration does not involve aquifer storage properties which are critical for a viable (transient) prediction.

A limited field data set does not leave choice and predetermines the steady state calibration. In such a case an appropriate approach would be to define boundary conditions and stresses that are representative for the period in which the field data are collected.

When a transient field data set of considerable length is available, some meaningful average measure should be derived from it for a steady state calibration. For example, this can be the mean annual water table elevation or the mean water table for the dry season, the average annual groundwater withdrawal, the mean annual precipitation (recharge), the average base flow in a surface stream, and so on.

Transient calibration typically involves water levels recorded in wells during pumping tests or long-term aquifer exploitation. An ideal set that incorporates all common relevant boundary conditions and stresses would be:

- monthly water table (piezometric surface) elevations,
- monthly precipitation (recharge),
- average monthly river stage,
- average monthly groundwater withdrawal.

Transient calibration based on monthly values is preferred over daily or weekly data since groundwater systems usually react with a certain delay to surface stresses. In addition, monthly data enable accurate analysis of seasonal influences which is very important for long-term predictions.

There are two methods of calibration:

- Trial-and-error ("manual") calibration,
- Automated calibration.

Trial-and-error calibration was the first technique applied in groundwater modeling and is still preferred by most users. Although it is heavily influenced by the user's experience, it is always recommended to perform this type of calibration, at least in part. By changing parameter values and analyzing the corresponding effects, the modeler develops a better feeling for the model and the assumptions on which its design is based. During manual calibration boundary conditions, parameter values and stresses are adjusted for each consecutive model run until calculated heads match the pre-set calibration targets.

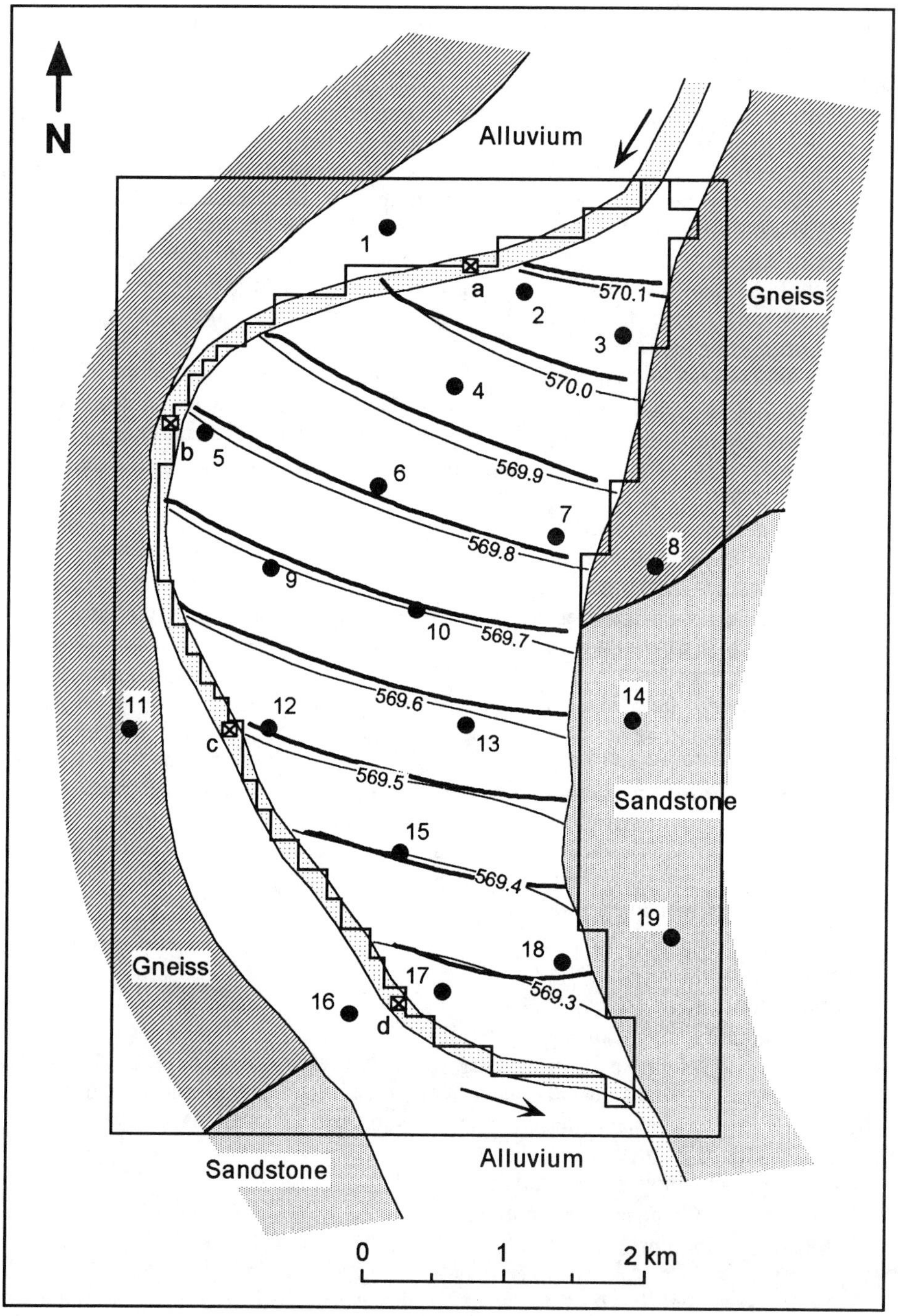

Figure 35.15 Calibrated (bold lines) vs. measured water table contours in the alluvial aquifer. The largest discrepancy is along the contact with sandstone since it was assumed that the inflow from the sandstone aquifer would influence the water table in the alluvial aquifer. However, even for unrealistically high inflow rates this is not the case probably because of the big difference in hydraulic conductivity between the two aquifers.

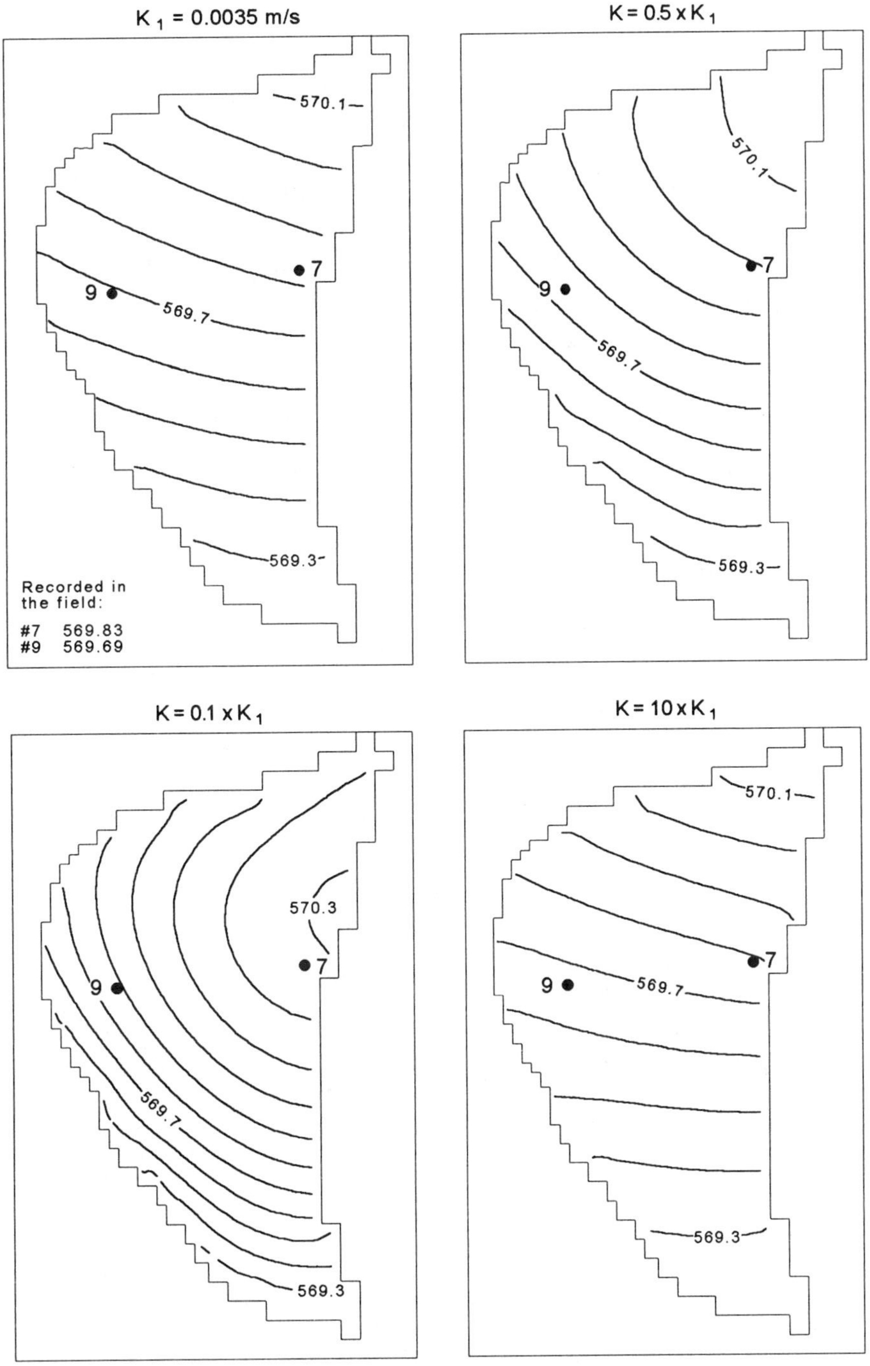

Figure 35.16 Effect of varying uniformly distributed hydraulic conductivity on the water table in the alluvial aquifer.

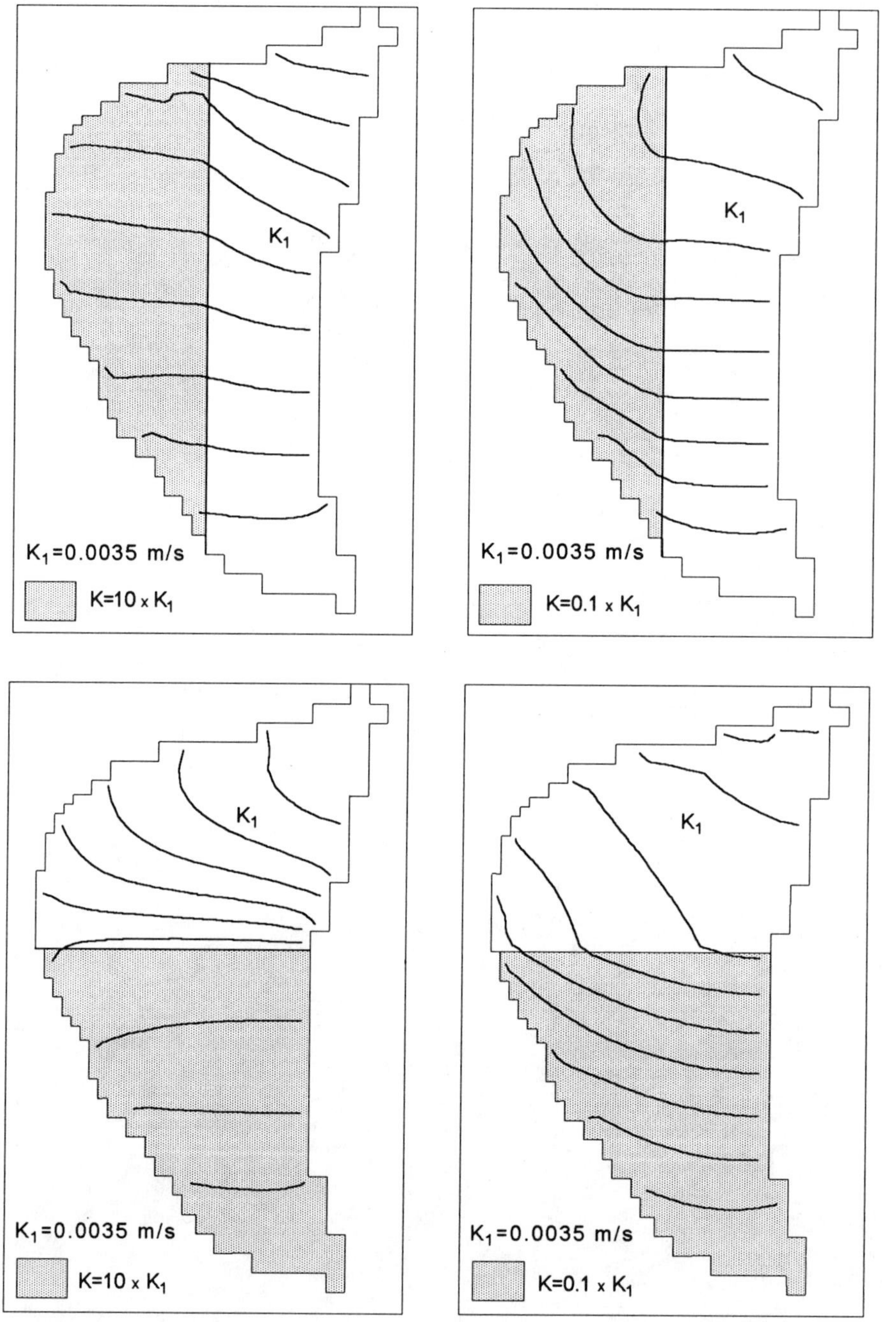

Figure 35.17 Effects of heterogeneous distribution of hydraulic conductivity on the water table contours. Areas of lower hydraulic conductivity create steeper gradients and contour lines change direction at K boundaries. (Note that the straight-line internal boundaries are unnatural and shown here just as an illustration.)

It will sometimes be necessary to change input values and run the model tens of times before reaching the target. The worst case scenario involves a complete redesign of the model with new geometry, boundaries, and boundary conditions.

The first phase of calibration typically ends when there is a good visual match between calculated and measured hydraulic heads as seen on a contour map (Figure 35.15). The next step involves quantification of the model error with various statistical parameters such as standard deviation and distribution of model residuals (differences between calculated and measured values).

During calibration, the user should focus on parameters that are determined with less accuracy or assumed, and change only slightly those parameters that are more certain. For example, hydraulic conductivity determined by several pumping tests should be the last parameter to change "freely" because it is usually the most sensitive. Change of one order of magnitude in hydraulic conductivity can dramatically alter distribution of head as shown in Figure 35.16. Figure 35.17 illustrates how, by changing distribution of hydraulic conductivity, an experienced modeler can produce just about any effect desirable. (Again, another question is if such "regular-geometry" variations are justified.) Most other parameters are less sensitive and can be changed only within certain realistic range: it is obviously not possible to increase precipitation infiltration rate ten times from 10 to 100%. In general, hydraulic conductivity and recharge are two parameters with equivalent quality: an increase in hydraulic conductivity creates the same effect as a decrease in recharge. Since different combinations of parameters can yield similar, or even the same results, trial-and-error calibration is not unique.

During calibration it is recommended to plot calculated vs. recorded heads at several control points (*i.e.*, nodes with monitoring wells) as shown in Figure 35.18. This enables a more accurate determination of the parameter value that produces the best fit.

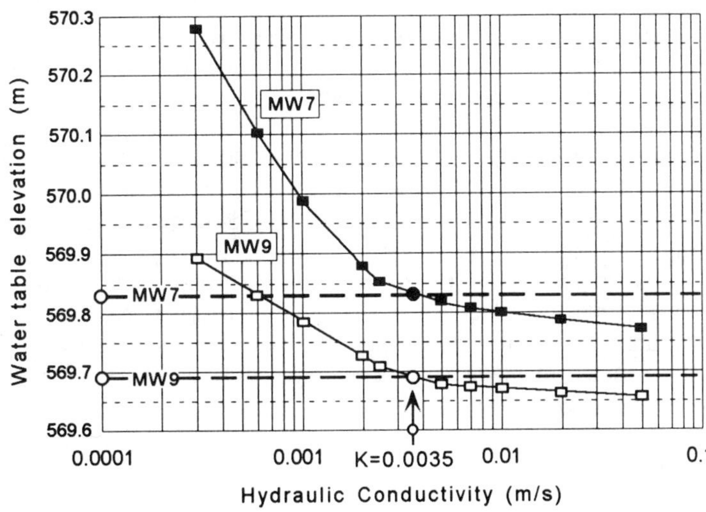

Figure 35.18 Water table elevation at wells No. 7 and No. 9 vs. hydraulic conductivity distribution tested during calibration. The calibrated (chosen) hydraulic conductivity is found as the intersect between calculated curve and the water table value recorded in the field (569.83 for No. 7, 569.69 for No.9).

A very effective measure of calibration is the analysis of water budget calculated by MODFLOW. The model provides flows across boundaries, flows to and from all sources and sinks, and flows generated by the storage. These calculated values should be compared with measurements and/or estimates made during concept development. Unrealistic components of the water balance should be analyzed in order to calibrate the parameter(s) and condition(s) that are causing them.

Automated calibration is a recent technique developed in order to minimize uncertainties associated with the user's subjectivity. Probably because it involves a substantial knowledge of statistics and basically excludes "intuition", this type of calibration gains ground rather slowly. As with any new approach, it has been criticized, particularly because of its instability and non-uniqueness (see Anderson and Woessner, 1992). Most computer codes for automated calibration search an optimal parameter set for which the sum of squared deviations between calculated and measured values is reduced to a minimum. One of the highly regarded codes for parameter estimation developed for MODFLOW is PEST by Doherty, Brebber and Whyte (1994). Detailed explanation and examples of its use are given in PMWIN manual (Chiang and Kinzelbach, 1996). The efficiency of PEST and similar codes, coupled with the trial-and-error input from the user, is arguably the most appropriate calibration method available.

35.2.11 Verification

A model is verified "if its accuracy and predictive capability have been proven to lie within acceptable limits of error by tests independent of the calibration data" (Konikow, 1978, from Anderson and Woessner, 1992). A typical verification is performed for an additional field data set in either steady state or transient simulation. If the calibrated model parameters such as boundary conditions, stresses and distribution of hydrogeologic characteristics (conductivity/transmissivity, effective porosity) are correct, this additional independent field data set should be closely matched by the model for the new boundary conditions and stresses. For example, the calibration data set may be representative of a wet period with high water table, high stage in the nearby river, and high lateral influx from the adjacent aquifer. If the verification data set is collected in the field during a recession period, the corresponding boundary conditions (river stage, lateral influx) and stresses (recharge) must be changed accordingly. If the model run with these changed conditions produces a head distribution that closely matches the new data set, the model is verified. Often, however, it will be necessary to change some of the hydrogeologic parameters to obtain an acceptable match. In such cases the verification actually becomes another calibration and a new independent data set is needed for model verification.

Since models are often calibrated in steady state, it is necessary to estimate storage parameters during transient verification which somewhat decreases its accuracy. Despite this limitation, transient verification is preferred since it offers a better insight into the aquifer responses to time-dependent stresses. Examples of data sets typically used for transient verification are pumping tests, long-term pumpage, and seasonal changes in water table. Unfortunately, in modeling practice often only one data set is available in which case the model cannot be verified.

35.2.12 Prediction

Once the model is calibrated, preferably for both steady state and transient conditions, it can be used for prediction which is the purpose of most modeling efforts. Prediction is the reward for the effort put into the all preceding modeling phases and the stage which model designers, more or less patiently, wait for. It is not uncommon, and it is even recommended, that the user performs a "mock prediction" during early stage of model development. This means that, for example, a number of anticipated wells are placed in the area of interest and pumped heavily in order to test placement of hydraulic (artificial) boundaries. If this pumping is influenced by the boundaries, the model area should be expanded. Doing this at the beginning of the modeling effort, rather than after lengthy

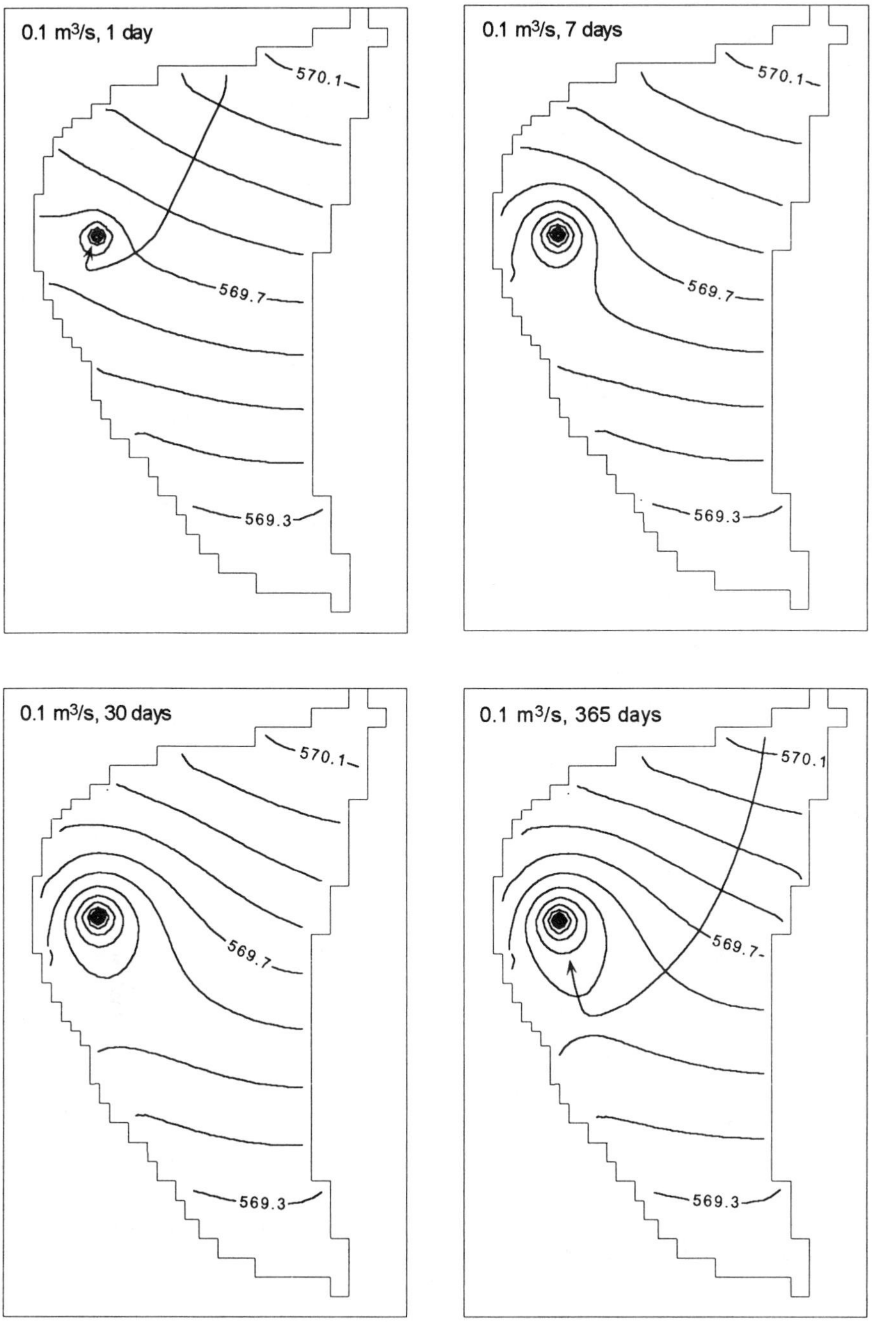

Figure 35.19 Transient effects of pumping 0.1 m^3/s with one well placed close to the river. The capture zone after one day and 365 days of pumping is shown with the boundary streamline. Water table contours are in meters amsl, and the contour interval is 0.1 m.

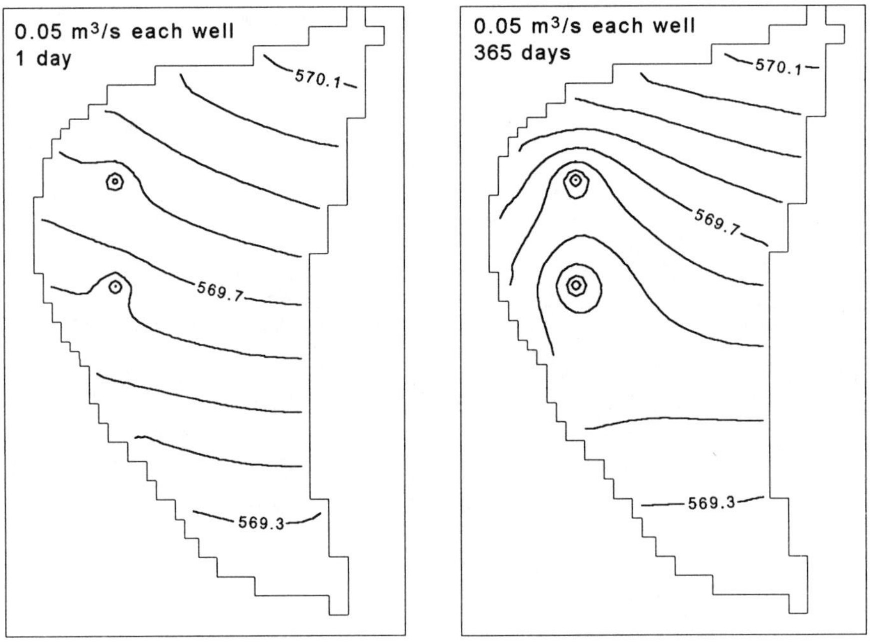

Figure 35.20 Effect on water table of two wells pumping 0.05 m^3/s each.

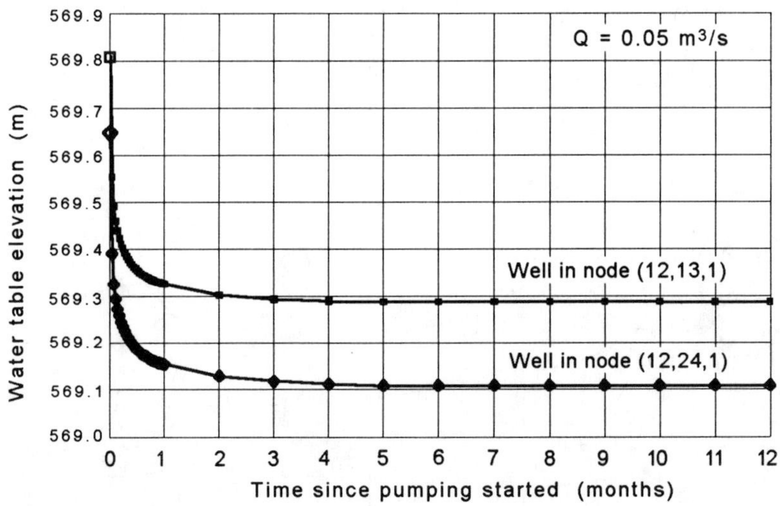

Figure 35.21 Hydraulic head vs. time in two wells shown in Figure 35.20. The main portion of the drawdown develops during the first month of pumping after which it decreases more slowly. A quasi steady state condition is established after approximately 4 months of pumping.

calibration and sensitivity analysis, may save a lot of time, nerves, and spare the colleagues of the user some inappropriate words.

Unless the prediction is made for some very general purposes, in which case the steady state approach may be justified, it is very important to anticipate changes in all boundary conditions and stresses during the prediction period. For example, seasonal variations of recharge, fluctuations in river stage and other equipotential boundaries, or seasonal groundwater withdrawal, should be simulated with an appropriate number of stress periods. Transient prediction is very important for successful water resources management since it allows for the more accurate design of well fields and pumping schedules.

Figures 35.19 through 35.23 show several typical results of transient prediction in the case of the Red River alluvial aquifer (our problem). The calibrated model can give answers to various questions such as "What is the maximum feasible pumping rate of one well placed at this location and what is its capture zone?", or "How much water can be withdrawn with these two wells and will they interfere with each other?" Also, the model

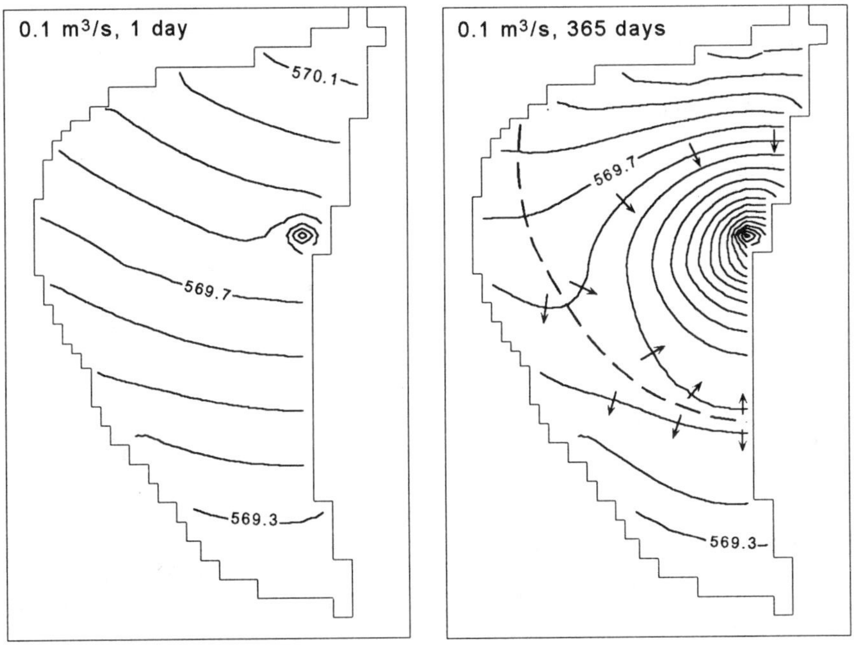

Figure 35.22 Effect on water table of one well pumping 0.1 m^3/s near the impermeable boundary. The capture zone is much larger than in the case of the well pumping near the river (see Figure 35.19).

can be used to test various scenarios that are known to be unrealistic but can give better insight into the system. This takes literally a few minutes with a calibrated model and can provide an invaluable experience to beginners. Figures 35.22 and 35.23 show the effects of transient pumpage from a well placed very close to the impermeable boundary between the alluvial aquifer and gneiss – a situation that is highly unlikely to occur ("who would place a water-supply well right next to an impermeable boundary?").

35.2.13 Presentation of Results

This is the final phase of the modeling effort and arguably the most important from the client's standpoint. A poorly documented and confusing report can ruin days of work and an otherwise excellent model. Every effort should be made to produce an attractive and user-friendly document that will convey clearly all previous phases of the model design. Special attention should be paid to clearly state the model's limitations and uncertainties associated with calibrated parameters.

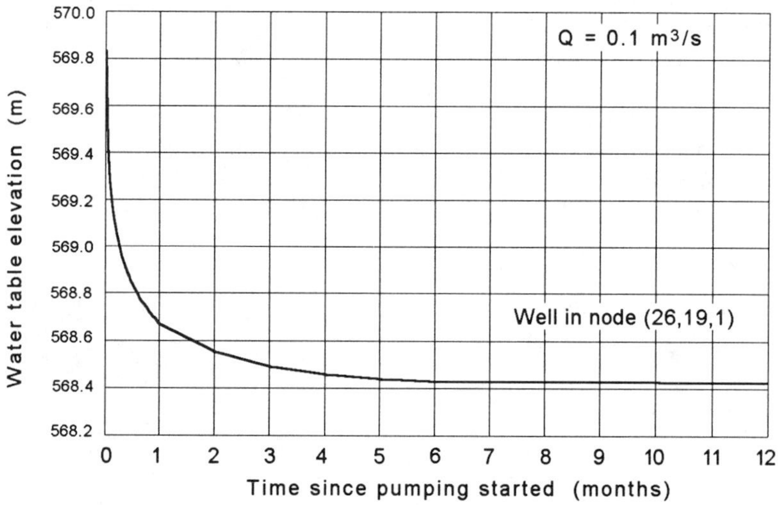

Figure 35.23 Hydraulic head vs. time for the well pumping near the impermeable boundary. A quasi steady state condition is established after approximately 6 months of pumping or 2 months more than at the wells near the river.

PROBLEM 36
THREE-LAYER SYSTEM, PLUME CONTAINMENT

Groundwater contaminant plumes are delineated in the Monahne River alluvial aquifer and the underlying confined aquifer as shown in Figure 36.1. The two aquifers are separated by a low-permeable silty clay layer which is not continuous (see Figure 36.2). Hydrogeologic assessment indicates that the contamination from the upper, unconfined

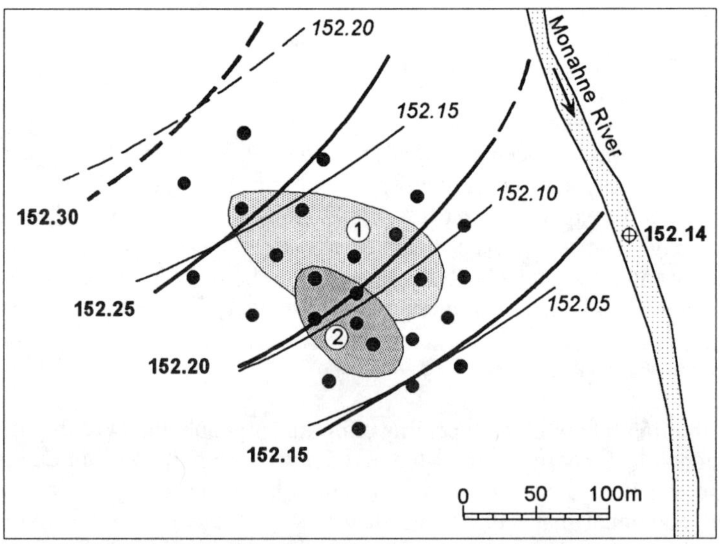

Figure 36.1 Water table contours (thin lines) for the unconfined alluvial aquifer, and the piezometric surface contours (thick lines) for the confined aquifer as determined from the monitoring wells (shown with circles) data. Contour interval is 0.05 m. 1: Contaminant plume in the unconfined aquifer. 2: Contaminant plume in the confined aquifer.

aquifer enters the lower aquifer through the "window" in the silty clay layer due to the difference in hydraulic heads in the two aquifers. Develop a groundwater modeling concept and a model that will be used in the design of a well extraction system for containment of the two plumes.

Hydraulic conductivities of the upper and lower aquifers, as determined from slug tests, are 2.6×10^{-4} and 3.4×10^{-3}m/s, respectively. Hydraulic conductivity of the confining silty clay layer, determined in laboratory from Shelby tube samples, is 6.5×10^{-7} m/s. Average annual recharge rate to the alluvial aquifer is estimated at 2.5×10^{-9} m/s which corresponds to 12% of the gross annual precipitation in the area (660 mm).

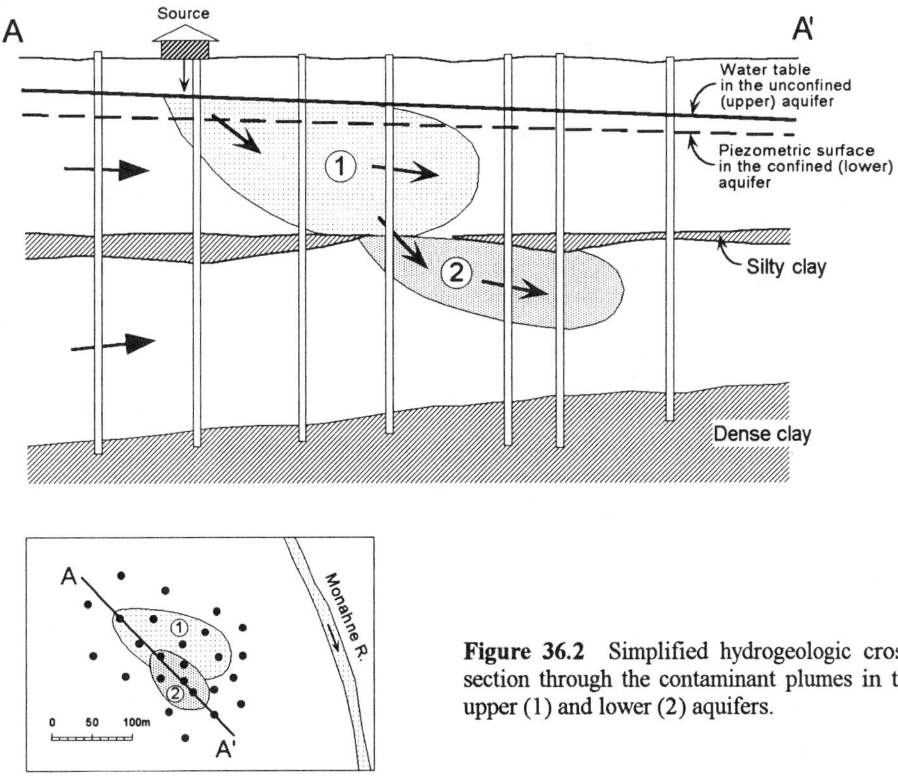

Figure 36.2 Simplified hydrogeologic cross-section through the contaminant plumes in the upper (1) and lower (2) aquifers.

36.1 MODELING CONCEPT

As explained in Problem 35, developing a concept is the most important part of groundwater modeling and requires a thorough understanding of hydrogeology, hydrology and groundwater dynamics. Often, as in our case, investigations are focused on a relatively small area and the collected data have to be supplemented with more regional information in order to develop a valid concept.

As seen from Figure 36.1, the area of the two plumes with the information available from 25 monitoring wells is approximately 250 by 250 m and with only one identified physical boundary of groundwater flow – the Monahne River. Based on water level measurements in the monitoring wells and one measurement of the river stage, this is the flow boundary for the alluvial (upper) aquifer only: the head in the confined (lower) aquifer is below the head in the upper aquifer and the flow directions in the two aquifers are slightly different.

Regional information shows that the Monahne River alluvial plain is approximately 4 miles wide and the stream is effluent, *i.e.*, it gains water. There is no information on water table elevations as far as one mile from the plumes area, and there are four domestic wells within 500-meter radius which were used to measure the piezometric surface in the confined aquifer. The monitoring well logs show a fairly horizontal discontinuous base of the alluvial aquifer (the silty clay layer -- see Figure 36.2). The regional pre-Quaternary base of the confined aquifer (dense clay) is known to be thick and continuous, and it dips toward NW in the area of two plumes.

Water level measurements are collected only once, in late October, which is a month with approximately average precipitation (on yearly basis) for the area.

The following modeling concept is developed based on the above facts:

- The groundwater flow system consists of two aquifers separated by an aquitard, and will be modeled as a three-layer, full 3D system.
- Since the model might grow to be a fate and transport model for predicting the exact role of the aquitard, interactions between the two aquifers, concentrations of pollutants and cleanup times, the layers will be presented with their real thicknesses (top and bottom) and hydraulic conductivities rather than with the transmissivities only.
- There is only one physical (real) boundary for groundwater flow and only in the unconfined aquifer which means that all other boundaries will have to be assumed from the flow nets in two aquifers. These hydraulic (artificial) boundaries will have to be placed far enough in order not to influence the stress created by pumping wells for containment of the two plumes.
- The model will be calibrated in steady state since there is only one set of data available. The predictions will be made for transient conditions taking into account seasonal variations in the river stage, recharge and groundwater levels.

36.2 MODEL GEOMETRY

36.2.1 Boundaries

Model geometry defines the size and the shape of the model. In general, it consists of model boundaries, both external and internal, for all layers, and model grid. Whenever possible, real physical boundaries should be chosen over artificial (hydraulic) boundaries. When the real boundaries are far from the area of interest and are impractical to model, as in our case (except for the Monahne River), it may be necessary to set hydraulic boundaries. Figure 36.3 shows the general principle of setting such boundaries from the flow net. Since the hydraulic boundaries have to be set far from the area of interest, a more regional flow net often has to be assumed because of lack of data beyond the area of site-specific hydrogeologic investigation. This is the case with the alluvial aquifer in our example.

It is not rare in the modeling practice to assume that the rectangular model edges are far enough and there is no need to "bother" with setting some realistic flow boundaries. (Note, however, that the model edges are by definition no-flow boundaries unless otherwise specified by the modeler.) Such an approach is justified only if there is a strong physical internal boundary that will ensure formation of a realistic ("natural") flow net in and around the area of interest.

Figure 36.4 shows contours of the estimated piezometric surface in the confined aquifer. In this case there are no identified physical boundaries around the area of interest (the Monahne River is not influencing the flow in the confined aquifer) and all model boundaries will have to be assumed.

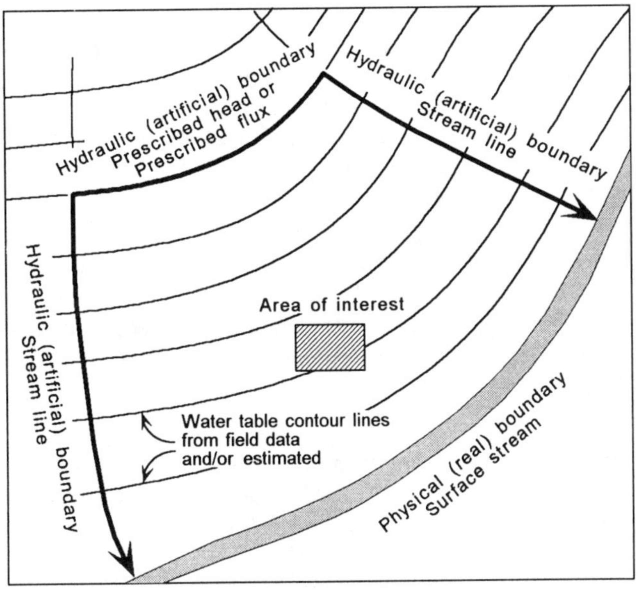

Figure 36.3 In the absence of real (physical) boundaries, the artificial (hydraulic) boundaries should be placed far enough from the area of interest in order not to influence the future flow pattern created by anticipated activity (well pumpage for the plume containment in our case). The artificial no-flow boundary corresponds to the chosen streamline, and the artificial known-head boundary corresponds to the equipotential line.

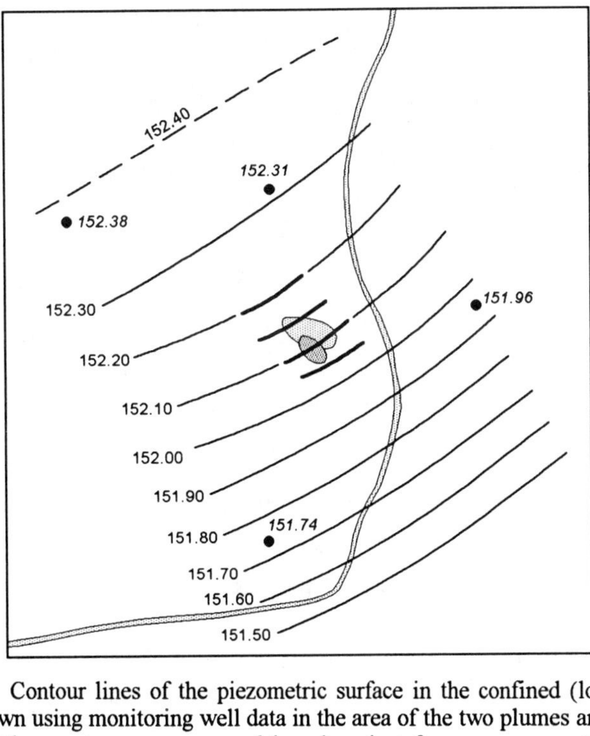

Figure 36.4 Contour lines of the piezometric surface in the confined (lower) aquifer. Contours drawn using monitoring well data in the area of the two plumes are shown with bold lines. Other contours are assumed based on just four measurements in domestic wells (shown with circles) and are therefore less accurate.

36.2.2 Grid and Layers

Once the possible external model boundaries are identified, the next step is to create the model grid. The overall dimensions of the regular rectangular grid should be large enough to encompass all (usually) irregular boundaries of the flow fields in all layers. If the flow field areas in different layers are not of the same size, the grid is set for the largest combined area. The size and number of cells in the row and column directions is then set having in mind the following (see Figure 36.5):

- Area(s) of interest, that will be exposed to various stresses (e.g., well pumpage), and area(s) with more data available, will be covered with a finer mesh.
- Cell size along the physical boundaries should allow for their accurate representation.
- Length of successive cells in the row, column and layer directions cannot increase by more than 1.5 times.

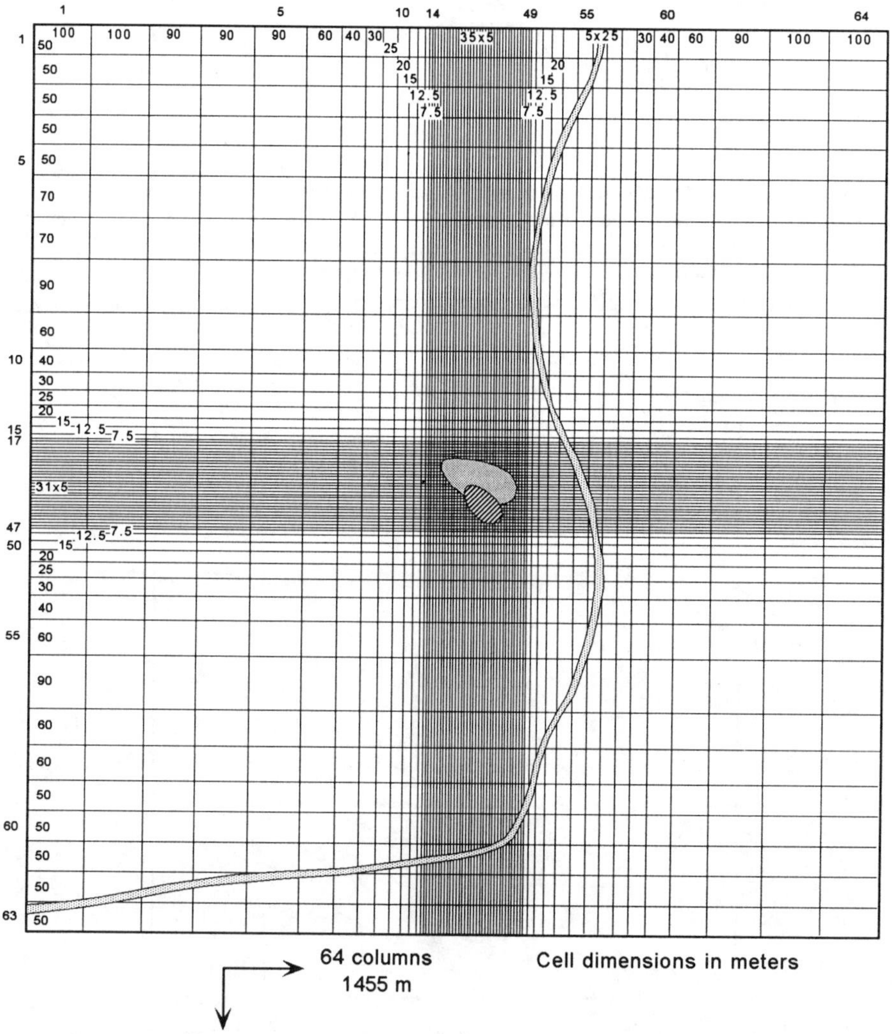

Figure 36.5 Custom grid for the model. The mesh is finer in the two-plume area.

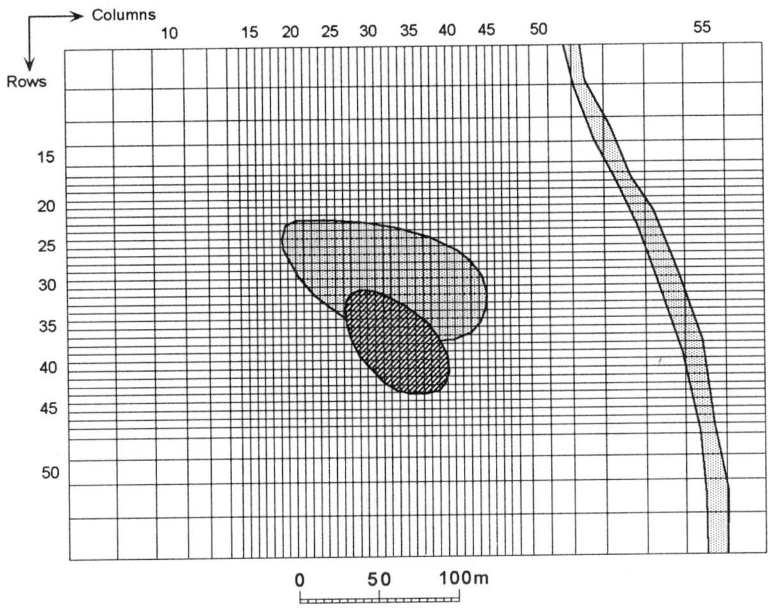

Figure 36.6 Fine grid spacing in the area of the two plumes. Note a gradual increase in cell dimensions beyond the area of interest following the rule of thumb in groundwater modeling – size of adjacent cells cannot increase (or decrease) by more than 1.5 times in row and column directions.

The model geometry along the vertical axis is controlled by the type, number and thickness of layers. In our case the alluvial aquifer is layer 1 of type 1 (strictly unconfined). The silty clay aquitard is layer 2 of type 0 (confined, the transmissivity is not expected to change), as is the confined aquifer (layer 3). This configuration can be changed later if it was concluded that the influence of future well pumpage on water levels should be modeled more precisely. In such cases more layers in each hydrostratigraphic unit can be easily added with most MODFLOW processors such as PMWIN. These new layers set in the same unit will have the same hydrogeologic parameters (hydraulic conductivity, storage coefficient, porosity) as the original layer. Among other things, more layers set in the same aquifer or aquitard are useful when:

- the hydraulic gradients created by pumping are steep,
- the pumping wells are partially penetrating and the well intake area has to be correctly simulated,
- it is important to accurately simulate vertical migration (fate and transport) of pollutants.

Discontinuous aquitard layer in the area where it is not present can be modeled by adjusting its thickness and hydraulic conductivity as shown in Figure 36.7. A realistic hydraulic connection between the two aquifers in the "window" area is simulated by assigning the hydrogeologic parameters (horizontal and vertical hydraulic conductivities, storage coefficient, effective porosity) of the layer above or below the aquitard. Alternatively, this portion of layer 2 (aquitard) can be given the average values of the parameters in layers 1 and 3.

The top of layer 1 is the water table in the unconfined aquifer. It is interpolated from the monitoring well data in the area of the two plumes and then extrapolated to the rest of the model assuming an uniform gradient. (Note that the uniform gradient is assumed since there is no data on water table elevations or aquifer thickness beyond the two-plume area.) Water table contours (*i.e.*, layer top) should be drawn to reflect the influence of the Monahne River which is a gaining stream.

Bottom of layer 1 (which is also top of layer 2) is assumed to be horizontal throughout the model and set at 142 m amsl which is the average value derived from the monitoring well logs. The bottom of layer 2 (silty clay aquitard) is also relatively horizontal and set at 138 m which gives the average thickness of 4 m. This thickness gradually decreases toward the "window" area where the bottom elevation is set at 141 m, *i.e.*, the thickness of layer 2 is reduced to 1 m only.

Elevation of the bottom of layer 3 linearly decreases from 131 m in SE to 128 m in NW. This increase in thickness is assumed from the slope determined from the well logs and the piezometric surface recorded in two domestic wells in N and NE (see Figure 36.4). Note, however, that the wider spacing between contour lines in Figure 36.4 can be a consequence of two things: 1) an increase in thickness, and 2) an increase in hydraulic conductivity.

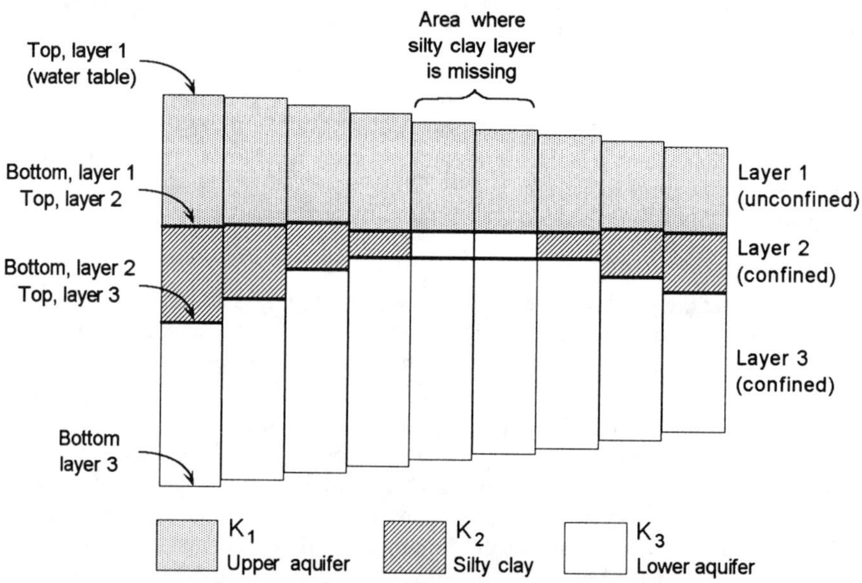

Figure 36.7 Principle of modeling discontinuous layers (the silty clay confining layer in our case; see also cross-section in Figure 2). The area where the actual layer is missing is still modeled as having a certain thickness (usually similar to adjacent cells where it is present) but the hydraulic conductivity (K) is the same as in layer 1 or 2. Note that, following the rule of thumb, the thickness of successive cells cannot increase by more than 1.5 times.

36.2.3 Boundary and Initial Conditions

All boundaries in layers 2 and 3 are artificial and derived from the flow net shown in Figure 36.4. Prescribed head (or constant head) boundaries in northwest and southeast correspond to equipotential lines and provide for a realistic groundwater flow in the confined aquifer from NW toward SE. This flow field can be laterally limited by two boundary streamlines as shown in Figure 36.8 Bottom in an attempt to accurately simulate the flow net in the confined aquifer. A more pragmatic (but less natural) way to model is to assume that the model edges are far enough from the area of interest and let the two equipotential boundaries "take care" of the flow field – Figure 36.8 Top.

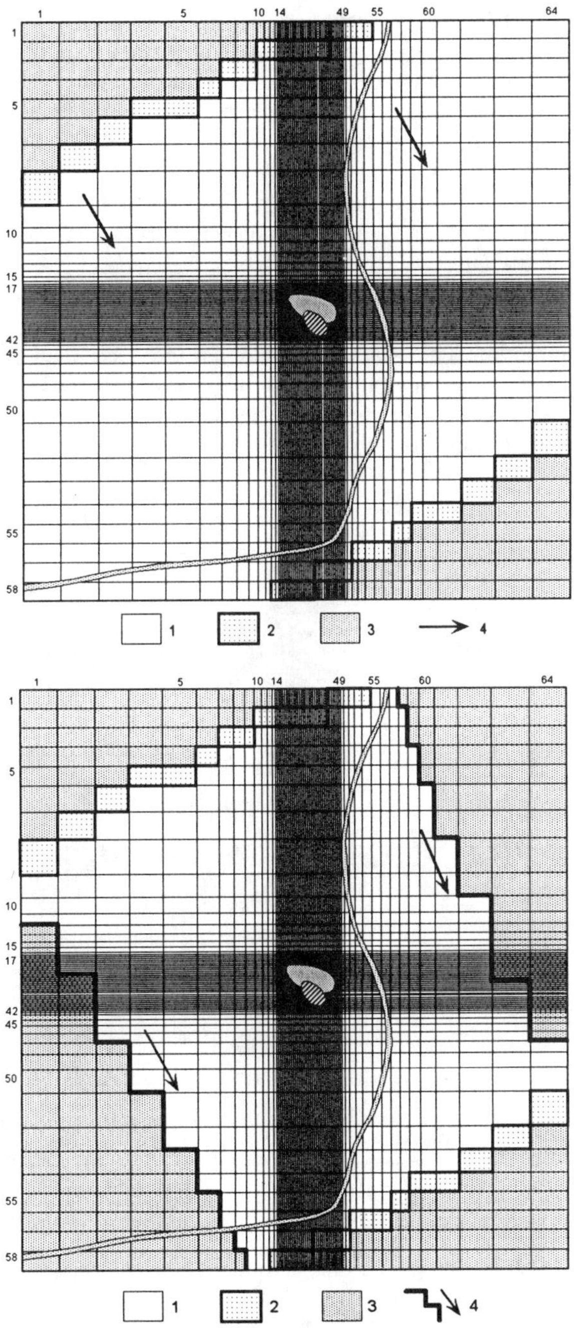

Figure 36.8 Two possible sets of artificial (hydraulic) boundaries for the confined aquifer (layer 3) and the silty clay aquitard (layer 2) based on the assumed contour map of the piezometric surface (see Figure 36.4). **Top**: Prescribed-head boundaries correspond to two contour lines and the model edges (no-flow boundaries) are assumed to be distant enough not to influence the natural flow pattern between the two contours. **Bottom**: No-flow boundaries are placed along two streamlines which is a more natural (and recommended) simulation of the actual flow field. 1: Active cell. 2: Prescribed (known) head cell. 3: Inactive cell. 4: Direction of groundwater flow and streamline (bottom).

Figure 36.9 shows boundaries of the model for layer 1. Although the Monahne River is the only physical boundary, its position and influence as a strong gaining stream are certain to create a flow pattern very similar to the natural ones even if additional hydraulic boundaries were not defined. However, in order to create as realistic flow net as possible, and achieve better computational stability, one constant head boundary simulating an equipotential line is set in the northwestern part of the model. Northern and eastern model edges coincide with the assumed streamlines. Note that all cells east of the river are inactive since the river is a complete physical boundary for the flow in the alluvial aquifer.

The boundary condition along the river can be simulated with constant heads (which is the simplest way), the MODFLOW River Package or the General-Head Boundary Package (see Apendix F). In any case, the hydraulic head in the river cells should closely resemble the hydraulic gradient in the river, which means that the head should decrease in each cell in the downstream direction. This gradient usually has to be assumed and adjusted during model calibration. In our case only one measurement of the river elevation was available near the two-plume area and the gradient was linearly interpolated between this and two other values which were estimated as shown in Figure 36.10.

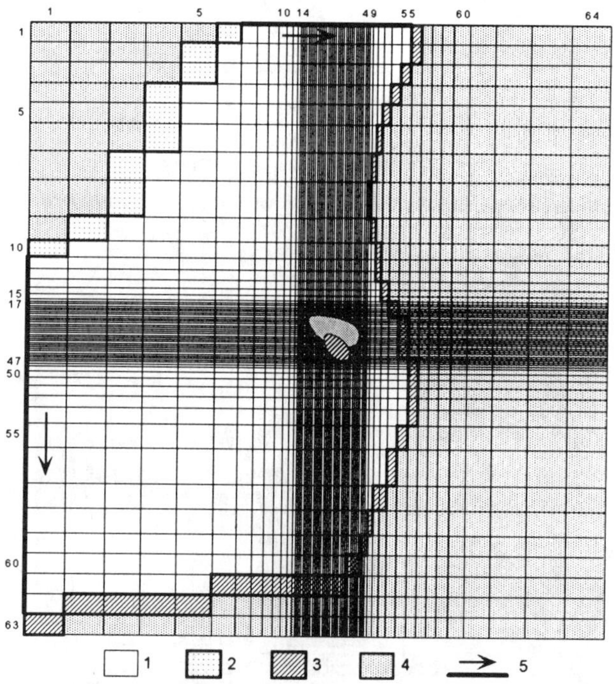

Figure 36.9 Model boundaries for the unconfined alluvial aquifer (layer 1). The Monahne River is the major physical boundary that can be simulated with various MODFLOW packages (see text). The northwestern boundary is along an assumed equipotential line placed far from the area of interest. Northern and western boundaries are along assumed streamlines which coincide with the model edges. 1: Active cell. 2: Constant-head boundary (or head-dependent flux boundary). 3: River cell. 4: Inactive cell. 5: Direction of groundwater flow and streamline (bottom).

Note that simulation of known-head boundaries with constant-head cells should be performed with care since the constant head cells act as an inexhaustible souce of water. Also, the hydraulic head in the constant-head cells cannot change during the model simulation and they are therefore not suitable for transient calibration or prediction. The *Time-Variant Specified-Head Package (CHD1)* by Leake and Prudic (1991) allows constant-head to take on different user-specified head values for each simulation time period and each time step within one period. The head at the boundary can also be specified by the user for different time periods in the General-Head Boundary and River Packages. These three packages are therefore more appropriate for transient modeling of known-head or head-dependent flux boundaries than simple simulation with the constant-head cells.

36.3 MODEL PARAMETERS

As explained in Problem 35, model parameters are divided into three groups: 1) time, 2) space, and 3) hydrogeologic characteristics.

Time parameters are irrelevant for steady state simulations, *i.e.*, calibration in our case. They will be important for the transient prediction allowing for seasonal variations in aquifer recharge and boundary conditions. A common approach is to simulate one hydrologic year with four seasons or modeling time periods in case of humid temperate climates. If more accurate predictions are needed, year can be simulated with twelve time periods provided that there are enough data avilable to represent average monthly hydrologic conditions in the area.

Space parameters, *i.e.*, layer top and bottom, are needed when the user lets the model calculate transmissivity for each layer from its thickness and the hydraulic conductivity which is the approach taken in this example. These parameters are described in Section 36.2.2.

Hydrogeologic parameters are transmissivity, horizontal and vertical hydraulic conductivities, leakance, storage parameters (specific yield for unconfined aquifers and storage coefficient for confined aquifers) and effective porosity. As already mentioned, PMWIN and most other MODFLOW processors can calculate transmissivity for layer of any type from its thickness and the horizontal hydraulic conductivity. Vertical leakance between layers, required by MODFLOW, can also be calculated by the model from the thicknesses of adjacent layers and their vertical hydraulic conductivities (see equations 35.3 and 35.4 in Problem 35). This is recommended over often lengthy manual calculations. The vertical hydraulic conductivity, unless determined from aquifer pumping tests or estimated in some other way, is commonly assumed to be 10 times smaller than the horizontal hydraulic conductivity. Storage properties are important for transient simulations, while effective porosity is required only if the model will be used in conjuction with a particle tracking code, in which case it is needed for calculations of linear flow velocities.

36.4 CALIBRATED MODEL

Figures 36.10 through 36.13 show calibrated vs. measured heads in the alluvial and confined aquifers. Assumptions on relatively homogeneous hydraulic conductivities in all three layers proved to be valid since the initial K values were only slightly changed during the model calibration. The calibrated hydraulic conductivities in layers 1,2 and 3 are 2.3×10^{-4}, 6.5×10^{-7} and 3.7×10^{-4} m/s, respectively. The vertical hydraulic conductivities are 10 times lower than the horizontal ones in all three layers. The recharge rate applied to layer 1 (unconfined aquifer) is calibrated at 2.9×10^{-9} m/s which is a bit more than the assumed value of 2.5×10^{-9} m/s and corresponds to approximately 14% of the average annual gross precipitation.

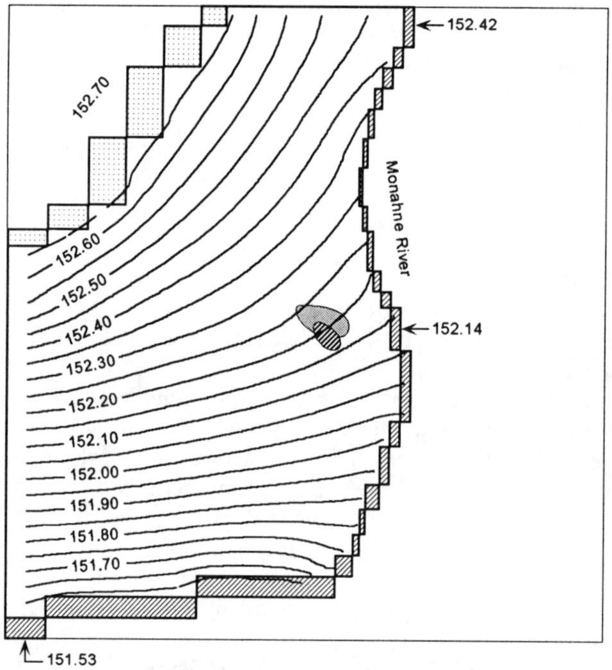

Figure 36.10 Contours of the calibrated water table in the unconfined aquifer. Contour interval is 0.05 m. The model check is done by comparing measured and calculated heads in the area of the two plumes (see Figure 36.11). The calculated heads cannot be checked beyond this area because of lack of data. They are controlled mainly by the assigned hydraulic gradient along the river to follow some logical flow pattern (three head values in the river are shown for illustration).

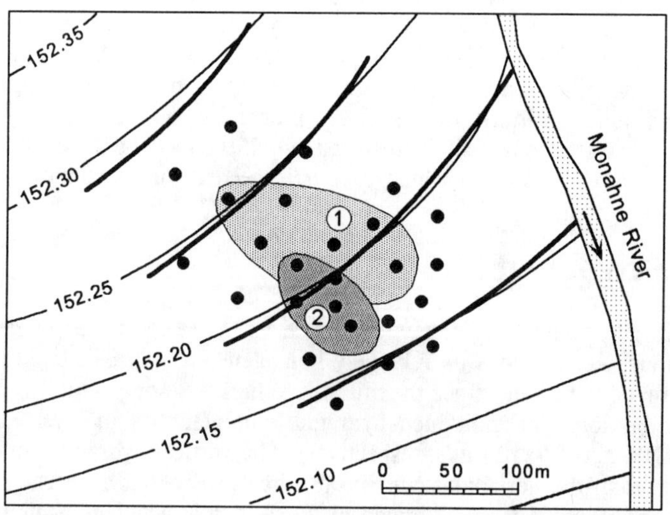

Figure 36.11 Contours of the calibrated (thin lines) and the measured (heavy lines) water table in the unconfined aquifer in the area of the two plumes. Contour interval is 0.05 m. Contours are strongly controlled by the assigned hydraulic head in the cells simulating the river.

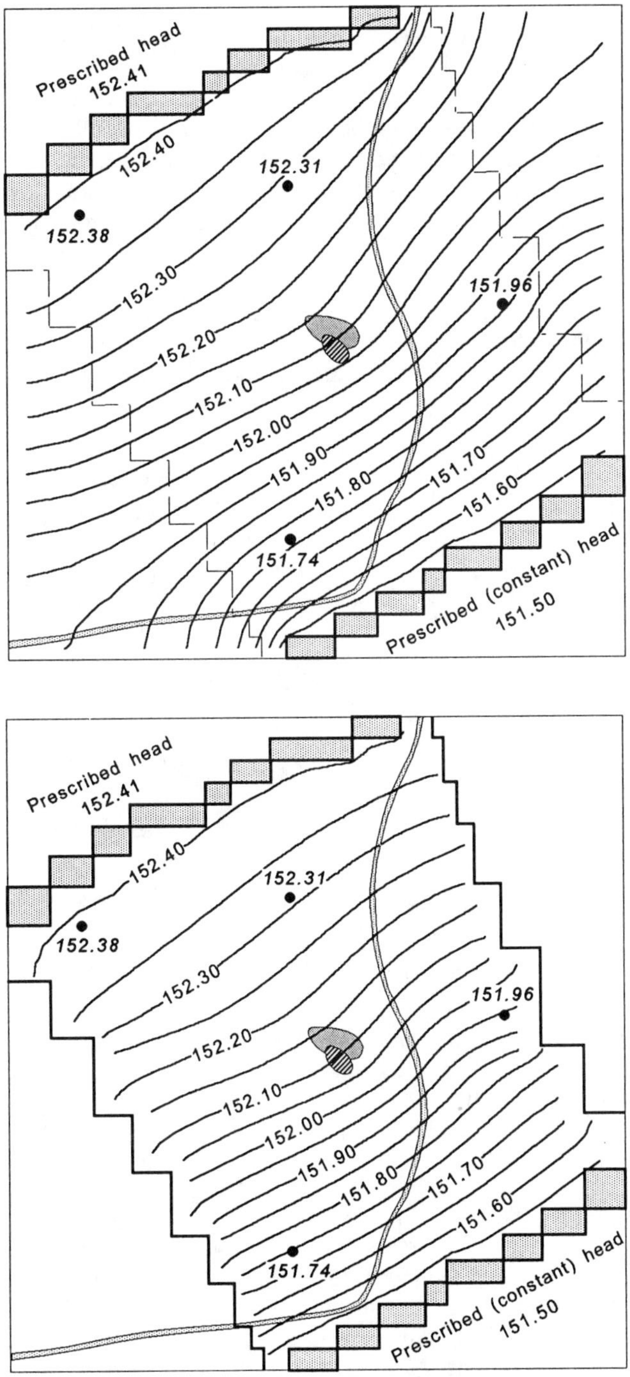

Figure 36.12 Contours of the calibrated piezometric surface in the confined aquifer. Contour interval is 0.05 m. The model check is done by comparing measured and calculated heads in the area of the two plumes (see Figure 36.13) and with field measurements in four domestic wells shown with circles. The contours for the no-flow boundaries with realistic streamlines (**bottom**) show a better match. Note that the contours end at the right ângle along the model edges which are also no-flow boundaries by definition (**top**).

As expected, two major conclusions were made during the flow net calibration in the alluvial (unconfined) aquifer:

- adjusting distribution of the hydraulic head along the river cells is a very efficient way to closely match the calculated and the measured water table, and
- the horizontal hydraulic conductivity and recharge are two parameters with equivalent quality – an increase in hydraulic conductivity creates the same effect as a decrease in recharge.

The most effect on the calculated piezometric surface in the confined aquifer, in addition to changing the horizontal hydraulic conductivity, has adjustments of the hydraulic head at the two constant-head boundaries.

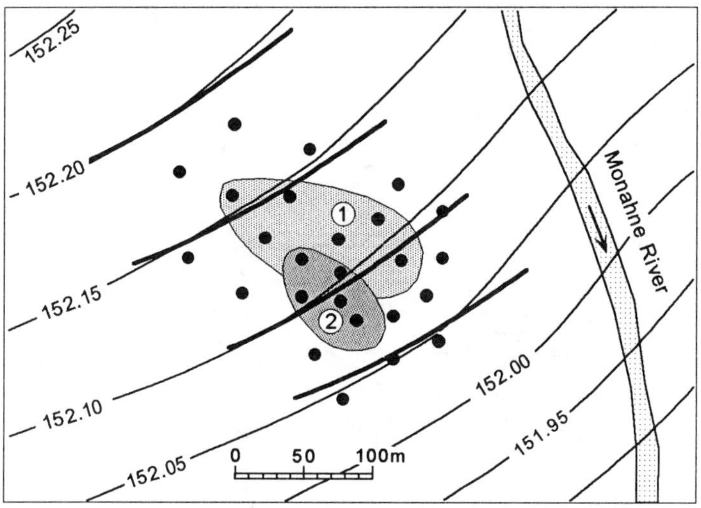

Figure 36.13 Contours of the calibrated (thin lines) and the measured (heavy lines) piezometric surface in the confined aquifer in the area of the two plumes. Contour interval is 0.05 m. Note that the contours are not influenced by the river which is simulated only in layer 1 (the aquifer is separated from the river by the silty clay layer).

36.5 CONTAINMENT OF THE PLUMES

The final test before using the model for various predictions is to check if the artificial boundaries are set correctly. It is recommended that the modeler performs a test prediction during early stages of model development rather than at the end of calibration process. This is done by placing a number of anticipated wells in the area of interest and pumping them heavily in order to check if the hydraulic boundaries are far enough. Figure 36.14 shows that pumping in the upper aquifer is not influenced by the northwestern constant-head boundary. It also does not reach the northern and western model edges representing two streamlines (no-flow boundaries). This check can be done visually, by comparing position of the contours closest to the boundaries with their position before the pumping, *i.e.*, for the steady state calibrated model. Another, more objective and quantitative way is to analyze water budget for the model and see if any portion of the well extraction budget is

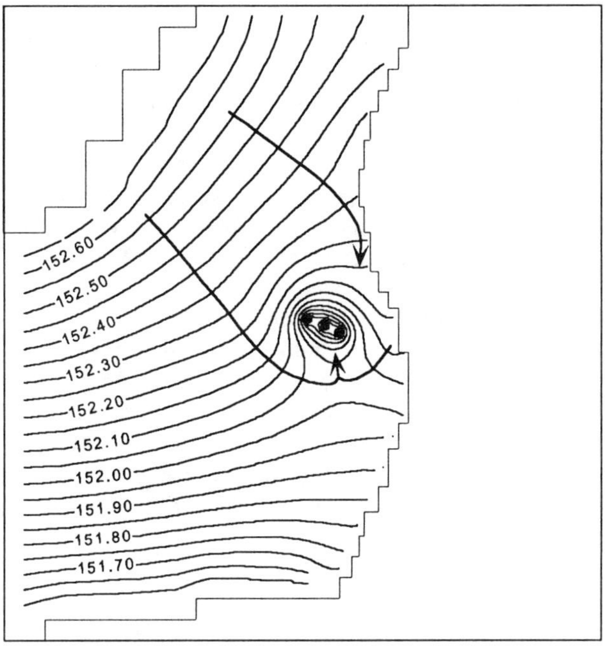

Figure 36.14..Water table contours in the unconfined (upper) aquifer as the result of pumpage with three wells containing the plume. Contour interval is 0.05 m. Note that this pumping stress does not reach the constant-head boundaries which are placed far enough (compare with Figure 36.10). The capture zone of the three wells is shown with streamlines.

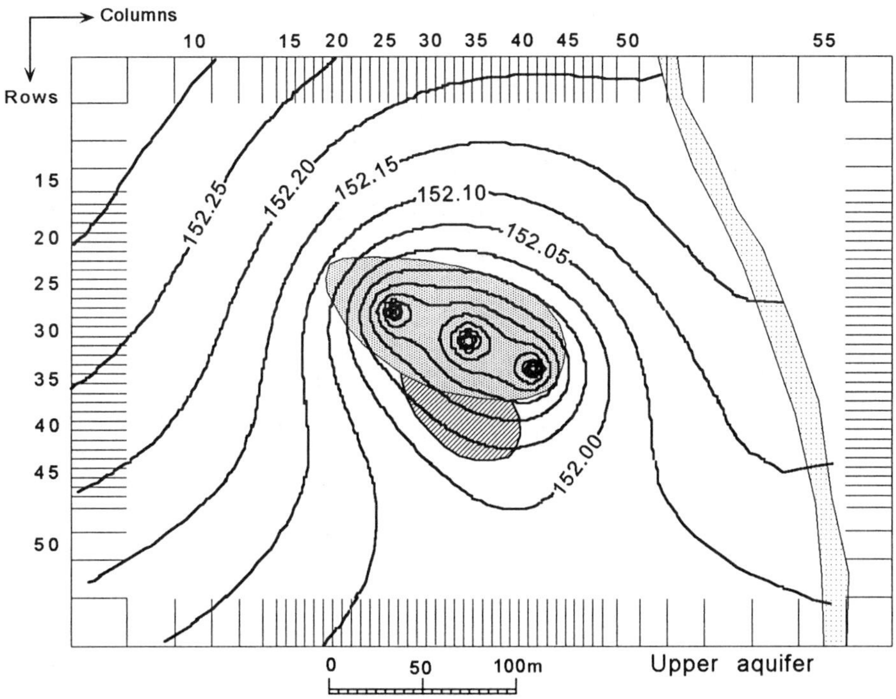

Figure 36.15 Containment of the plume in the upper aquifer with three wells pumping 0.0015 m^3/s each. Water table contours are in meters amsl.

derived from the constant-head boundary. If this is the case, the boundary is set too close and should be moved farther before the model can be used for accurate predictions.

Figures 36.15 and 36.16 show containment of both plumes with three wells pumping the upper aquifer and two wells simultaneously pumping the lower aquifer. The main advantage of the numeric model is that it can be used to quickly analyze different pumping scenarios including varying pumping locations, schedule and rates. The model is irreplaceable when designing the system that would satisfy the following important requirements for an adequate containment of the two plumes in our example:

- First lower the water table in the upper aquifer below the piezometric surface in the lower aquifer to prevent further downward migration of the contaminant.
- Start pumping the lower aquifer and at the same time increase the pumping in the upper aquifer to keep the water table slightly below the piezometric surface (compare Figures 36.15 and 36.16).
- Adjust pumping rates for the seasonal changes in the aquifer recharge, the river stage, and the position of the water levels in the two aquifers (*i.e.*, elevation of the known-head boundaries).

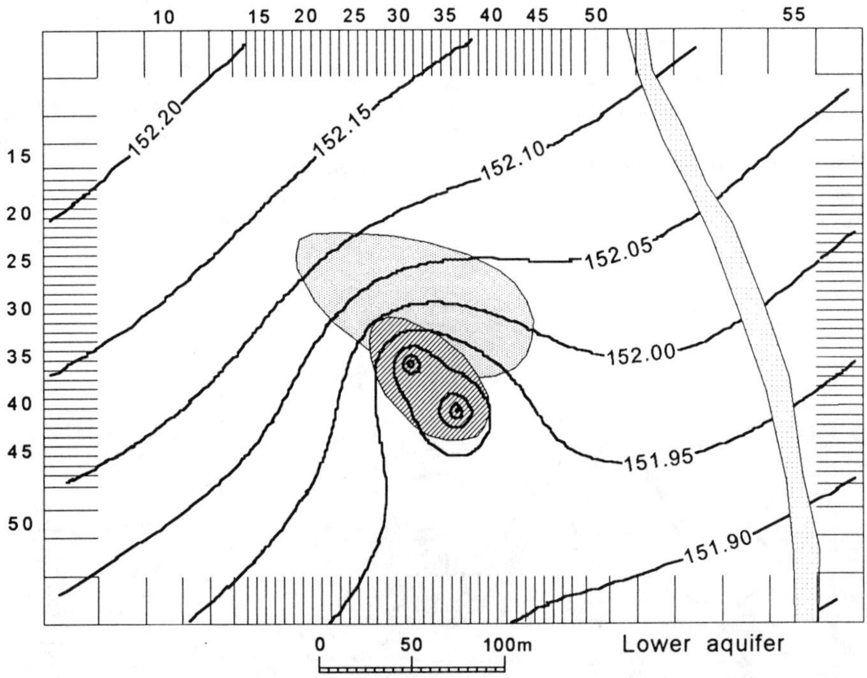

Figure 36.16 The plume in the lower (confined) aquifer contained by two wells pumping 0.001 m^3/s each. Contours of the piezometric surface are in meters amsl.

REFERENCES AND RECOMMENDED READINGS

Adamovski, K., and T. Hamory, 1983. A stochastic systems model of groundwater level fluctuations. *Journal of Hydrology*, 62, p. 129-141.

Anderson, M.P., and W.W. Woessner, 1992. *Applied Ground Water Modeling; Simulation of Flow and Advective Transport*. Academic Press, San Diego, 381 p.

Bentall, R., compiler, 1963. Shortcuts and special problems in aquifer tests. *U.S. Geological Survey Water-Supply Paper* 1545-C.

Bentall, R., compiler, 1963. Methods of determining permeability, transmissibility, and drawdown. *U.S. Geological Survey Water-Supply Paper* 1536-I.

Boreli, M., editor, 1968. *Seminar on ground water balance* (in Serbo-Croatian). Yugoslav Committee for the International Hydrologic Decade. Belgrade, 227 p.

Boreli, M., 1968. Radius of influence of wells. In: Boreli, M., editor: *Seminar on ground water balance* (in Serbo-Croatian). Yugoslav Committee for the International Hydrologic Decade. Belgrade, p. 132-141.

Boulton, N.S., 1954a. Unsteady radial flow to a pumped well allowing for delayed yield from storage. *Internat. Assoc. Sci. Hydrology Pub*. 37, p. 472-477.

Boulton, N.S., 1954b. The drawdown of the water table under nonsteady conditions near a pumped well in an unconfined formation. *Inst. Civil Engineers Proc.* (London), pt.3, p. 564-579.

Boulton, N.S., 1963. Analysis of data from non-equilibrium pumping tests allowing for delayed yield from storage. *Inst. Civil Engineers Proc.* (London), Vol. 26, p.469-482.

Boulton, N.S., 1970. Analysis of data from pumping tests in unconfined anisotropic aquifers. *Journal of Hydrology*, 10, p.369

Boulton, N.S., and J.M.A. Pontin, 1971. An extended theory of delayed yield from storage applied to pumping tests in unconfined anisotropic aquifers. *Journal of Hydrology*, 19, p. 157-169.

Boussinesq, J., 1904. Recherches théoriques sur l'écoulement des nappes d'eau infiltrées dans le sol et sur les débits des sources. *Journal de Mathématiques Pures et Appliquées*, Paris, 10, p. 5-78.

Bouwer, H., 1989. The Bouwer and Rice slug test -- an update. *Ground Water*, Vol. 27, No. 3, p. 304-309.

Bouwer, H., and R.C. Rice, 1976. A slug test for determining hydraulic conductivity of unconfined aquifers with completely or partially penetrating wells. *Water Resources Research*, 12, p.423-428.

Box, G.E.P., and G.M. Jenkins, 1976. *Time Series Analysis. Forecasting and Control.* Holden-Day, San Francisco, 575 p.

Bozic, B., 1977. Analiza i primena stohastickih procesa u hidrologiji (Analysis and application of stochastic processes in hydrology, in Serbo-Croatian). Treci seminar iz hidrologije (Trebinje), racunski primeri, Beograd, IX, p. 1-65.

Brandon, T.W., editor, 1986. *Groundwater: Occurrence, Development and Protection.* Water Practice Manuals, The Institution of Water Engineers and Scientists, London, 615 p.

Bureau of Reclamation, U.S. Department of Interior, 1985. *Ground Water Manual.* U.S. Government Printing Office, Denver, 480 p.

Carlson and Foley, 1991. *Radial Basis Interpolation on Track Data*, Lawrence Livermore National Laboratory, UCRL-JC-1074238.

Chiang, W.H., and W. Kinzelbach, 1996. Procesing Modflow for Windows (PMWIN).(Computer program and manual for the MODFLOW, MODPATH and MT3D processor distributed in the U.S. by Scientific Software, Washington, D.C.)

Chow, V., 1959. *Open-Channel Hydraulics.* McGraw-Hill, New York, 680 p.

Clark, I., 1979. *Practical Geostatistics.* Applied Science Publishers. London, 129 p.

Cooper, H.H., Jr., 1963. Type curves for nonsteady radial flow in an infinite leaky artesian aquifer. In: Bentall, R., Compiler: Shortcuts and special problems in aquifer tests. *U.S. Geological Survey Water-Supply Paper* 1545-C, p. C48-C55.

Cooper, H.H., Jr., and C.E. Jacob, 1946. A generalized graphical method for evaluationg formation constants and summarizing well-field history. *Transactions, American Geophysical Union*, Vol. 27, p. 526-534.

Dawson, K., and J. Istok, 1992. *Aquifer Testing; Design and Analysis.* Lewis Publishers, Boca Raton, Florida, 280 p.

Doherty, J., L. Brebber, and P. Whyte, 1994. PEST - Model-Independent Parameter Estimation. User's Manual, Watermark Computing, Australia.

Domenico, P. A. and F.W. Schwartz, 1990. *Physical and Chemical Hydrogeology.* John Wiley & Sons, New York, 824 p.

Draper, N., and H. Smith, 1966. *Applied Regression Analysis.* John Wiley & Sons, New York, 406 p.

Dreiss, S.J., 1989a. Regional scale transport in a karst aquifer. 1. Component separation of spring flow hydrographs. *Water Resources Research*, Vol. 25, No. 1, p. 117-125.

Dreiss, S.J., 1989b. Regional scale transport in a karst aquifer. 2. Linear systems and time moment analysis. *Water Resources Research*, Vol. 25, No. 1, p. 126-134.

Driscoll, F.G., 1989. *Groundwater and Wells.* (Third Printing). Johnson Filtration Systems Inc, St. Paul, Minnesota, 1089 p.

Drogue, C., 1972. Analyse statistique des hydrogrammes de decrues des sources karstiques. *Journal of Hydrology*, Vol. 15, p. 49-68.

Fenske, J.P., S.A. Leake, and D.E. Prudic, 1996. Documentation of a computer program (RES1) to simulate leakage from reservoirs using the modular finite-difference ground-water flow model (MODFLOW). *U.S. Geological Survey Open-File Report 96-364*, 51 p.

Ferris, J.G., D.B. Knowles, R.H. Brown, and R.W. Stallman, 1962. Theory of aquifer tests. *U.S. Geological Survey Water Supply Paper 1536-E*, Washington, D.C., 173 p.

Fetter, C.W., 1994. *Applied Hydrogeology.* Macmillan, New York, 691 p.

Forchheimer, P., 1914. *Hydraulic.* Leipzig: B.G. Teubner.

Freeze, R.A., and J.A. Cherry, 1979. *Groundwater.* Prentice-Hall, Englewood Cliffs, NJ, 604 p.

Giles, R.V., J.B. Evett, and L. Chiu, 1994. *Fluid Mechanics and Hydraulics* (Third Edition). Schaum's Outlines Series, McGraw Hill, New York, 378 p.

Goode, D.J., and C.A. Appel, 1992. Finite-difference interblock transmissivity for unconfined aquifers and for aquifers having smoothly varying transmissivity. *U.S. Geological Survey Water-Resources Investigations Report 92-4124*, 79 p.

Gottman, J.M., 1981. Time-series analysis. A comprehensive introduction for social scientists. Cambridge University Press, Cambridge, 400 p.

Graupe, D., D. Isailovic, and V. Yevjevich, 1976. Prediction model for runoff from karstified catchments. In: Yevjevich, V., editor, *Karst Hydrology and Water Resources.* Proceedings of the U.S.-Yugoslavian Symposium, Dubrovnik, June 2-7, 1975, Water Resources Publications, Fort Collins, CO, Vol. 1, p. 277-300.

Hantush, M.S., 1956. Analysis of data from pumping tests in leaky aquifers. *Transactions, American Geophysical Union*, Vol. 37, No. 6, p. 702-714.

Hantush, M.S., 1960. Modification of the theory of leaky aquifers. *Journal of Geophysical Research*, Vol. 65, p. 3713-3725.

Hantush, M.S., and C.E. Jacob, 1955. Nonsteady radial flow in an infinite leaky aquifer. *Transactions, American Geophysical Union*, Vol. 36, No. 1, p. 95-100.

Hill, M.C., 1990. Preconditioned conjugate-gradient 2 (PCG2), a computer program for solving ground-water flow equations. *U.S. Geological Survey Water-Resources Investigations Report 90-4048*, U.S. Government Printing Office, Washington, D.C., 43 p.

Houston, J.F.T., 1983. Ground-water systems simulation by time series techniques. *Ground water*, 21(3), p. 301-310.

Hsieh, P.A., and J.R. Freckleton, 1993. Documentation of a computer program to simulate horizontal-flow barriers using the U.S. Geological Survey modular three-dimensional finite-difference ground-water flow model. *U.S. Geological Survey Open-File Report 92-477*, 32 p.

Hurlbut, C.S., Jr., reviser, 1959. *Dana's Manual of Mineralogy*. John Wiley & Sons, New York, 609 p.

Isaaks, E.H., and R.M. Srivastava, 1989. *An Introduction to Applied Geostatistics*. Oxford University Press, New York, 561 p.

Jacob, C.E., 1963. Determining the permeability of water-table aquifers. In: Bentall, R., editor, Methods of determining permeability, transmissibility, and drawdown. *U.S. Geological Survey Water-Supply Paper 1536-I*, p. I245-I271.

Jevdjevic, V., 1956. *Hidrologija, I deo* (Hydrology, Part One, in Serbo-Croatian). Posebna izdanja, Knjiga 4, Hidrotehnicki institut Ing. Jaroslav Cerni, Beograd, 404 p.

Kalman, R.E., 1960. A new approach to linear filtering and prediction problems. *Trans. ASME, J. Basic Engineering*, Vol. 82, p. 35-43.

Kashef, A.A.I., 1987. *Groundwater Engineering*. McGraw Hill, Singapore, 512 p.

Keckler, D., 1994. *SURFER for Windows, Version 5.02* (October 16, 1994). Manual, Golden Software, Inc., Golden, Colorado.

Kresic, N., 1991. *Kvantitativna hidrogeologija karsta sa elementima zastite podzemnih voda (Quantitative Karst Hydrogeology with Elements of Groundwater Protection,* in Serbo-Croatian). Naucna Knjiga, Belgrade, 192 p.

Kresic, N., 1994. Stochastic properties of spring discharge in different karst environments. In: Dutton, A.R., editor, *Toxic Substances and the Hydrologic Sciences*, American Institute of Hydrology, Minneapolis, p. 582-590.

Kruseman, G.P., and N.A. de Ridder, 1990. *Analysis and Evaluation of Pumping Test Data* (2nd edition). International Institute for Land Reclamation and Improvement, Wageningen, The Netherlands, 377 p.

Lambe, T.W., 1951. *Soil Testing for Engineers*. John Wiley & Sons, New York, 165 p.

Lange, O.K., 1958. *Basic Hydrogeology* (in Russian: *Osnovi gidrogeologii*). Moskovskii Gosudarstvenii Univerzitet, Moscow.

Leake, S.A., and D.E. Prudic, 1988. Documentation of a computer program to simulate aquifer-system compaction using the modular finite-difference ground-water flow model. *U.S. Geological Survey Open-File Report 88-482*, 80 p.

Leake, S.A., P.P. Leahy, and A.S. Navoy, 1994. Documentation of a computer program to simulate transient leakage from confining units using the modular finite-difference ground-water flow model. *U.S. Geological Survey Open-File Report 94-59*, 70 p.

Lebedev, A.V., 1968. Determination of hydrogeological parameters by means of piezometric data. In: Boreli, M., editor: *Seminar on ground water balance* (in Serbo-Croatian). Yugoslav Committee for the International Hydrologic Decade. Belgrade, 227 p.

Ligget, J.A., 1987. Advances in the boundary integral equation method in subsurface flow. *Water Resources Bulletin*, 23(4), p. 637-651.

Ligget, J.A., and P.L.F. Liu, 1983. *The Boundary Integral Equation Method for Porous Media Flow*. Allen and Unwin, 255 p.

Linsley, R.K., M.A. Kohler, and J.L.H. Pauhlus, 1975. Hydrology for Engineers. McGraw-Hill, New York, 482 p.

Lohman, S.W., 1972. Ground-water hydraulics. *U.S. Geological Survey Professional Paper 708*, Washigton, D.C., 70 p.

Maidment, D.R., Editor in Chief, 1993. Handbook of Hydrology. McGraw Hill, New York, (various paging).

Maillet, E., editor, 1905. *Essais d'hydraulique souterraine et fluviale*. Herman et Cie, Paris, Vol. 1, 218 p.

Marsily de, G., 1986. *Quantitative Hydrogeology; Groundwater Hydrology for Engineers*. Academic Press, Orlando, 441 p.

McCuen, R.H., and W.M. Snyder, 1986. *Hydrologic modeling. Statistical Methods and Applications*. Prentice-Hall, Englewood Cliffs, NJ, 568 p.

McDonald, M.G., and A.W. Harbaugh, 1988. A Modular Three-Dimensional Finite-Difference Ground-Water Flow Model. *Techniques of Water-Resources Investigations of the United States Geological Survey, Book 6, Chapter A1*, US Government Printing Office, Washington, D.C.

McDonald, M.G., A.W. Harbaugh, B.R. Orr, and D.J. Ackerman, 1992. A method of converting no-flow cells to variable-head cells for the U.S. Geological Survey Modular Finite-Difference Ground-Water Flow Model. *U.S. Geological Survey Open-File Report 91-536*. Reston, VA, 99 p.

Neuman, S.P., 1972. Theory of flow in unconfined aquifers considering delayed response to the water table. *Water Resources Research*, Vol. 8, No. 4, p. 1031-1045.

Neuman, S.P., 1975. Analysis of pumping test data from anisotropic unconfined aquifers considering delayed gravity response. *Water Resources Research*, Vol. 11, No. 2, p. 329-342.

Neuman, S.P., and P.A. Witherspoon, 1969. Applicability of current theories of flow in leaky aquifers. *Water Resources Research*, Vol. 5, p. 817-829.

Ozbilgin, M.M., and D.C. Dickerman, 1984. A modification of the finite-difference model for simulation of two dimensional ground-water flow to include surface-ground water relationships. *U.S. Geological Survey Water-Resources Investigations Report 83-4251*, 98 p.

Pollock, D.W., 1989. Documentation of computer program to compute and display pathlines using results from the U.S. Geological Survey Modular Three-Dimensional Finite-Difference Ground-Water Flow Model. *U.S. Geological Survey Open File Report 89-381*, Reston, Virginia, 188 p.

Prickett, T.A., 1965. Type-curve solution to aquifer tests under water-table conditions. *Ground Water*, Vol. 3, No. 3, p. 5-14.

Prohaska, S., 1981. Stohasticki model za dugorocno prognoziranje recnog oticaja (Stochastic model for long-term prognosis of river flow, in Serbo-Croatian). *Vode Vojvodine*, Special Edition 1981, Novi Sad, 106 p.

Prudic, D.E., 1989. Documentation of a computer program to simulate stream-aquifer relations using a modular, finite-difference, ground-water flow model. *U.S. Geological Survey Open-File Report 88-729*, 113 p.

Rushton, K.R., and S.C. Redshaw, 1979. *Seepage and groundwater flow; numerical analysis by analog and digital methods*. John Wiley & Sons, New York.

Salas, J.D., 1993. Analysis and modeling of hydrologic time series. In: Maidment, D.R., editor, *Handbook of Hydrology*, McGraw-Hill, New York, p. 19.1-19.72.

Sestakov, V.M., 1979. *Dinamika podzemnih vod* (Dynamics of Groundwater, in Russian). Izdatelstvo Moskovskogo Universiteta, Moskva, 368 p.

Spiegel, M.R., and R. Meddis, 1980. *Probability and statistics*. Schaum's Outline Series, McGraw-Hill, New York, 372 p.

Stallman, R.W., 1961a. Boulton's integral for pumping-test analysis. In: Short Papers in the Geologic and Hydrologic Sciences, *U.S. Geological Survey Professional Paper 424-C*, Washington, D.C., p. C24-C29.

Stallman, R.W., 1961b. The significance of vertical flow components in the vicinity of pumping wells in unconfined aquifers. In: Short Papers in the Geologic and Hydrologic Sciences, *U.S. Geological Survey Professional Paper 424-B*, Washington, D.C., p. B41-B43.

Stallman, R.W., 1963. Type curves for the solution of single boundary problems. In: Shortcuts and Special Problems in Aquifer Tests. *U.S. Geological Survey Water Supply Paper 1545-C*, Washington, D.C., p. C45-C47.

Stallman, R.W., 1965. Effects of water-table conditions on water-level changes near pumping wells. *Water Resources Research*, Vol. 1, No. 2, p. 295-312.

Star, J., and J. Estes, 1990. *Geographic Information Systems. An Introduction*. Prentice Hall, Englewood Cliffs, 303 p.

Strack, O.D.L., 1987. The analytic element method for regional groundwater modeling. In: *Solving Ground Water Problems with Models*, National Water Well Association, Columbus, Ohio, p. 929-941.

Strack, O.D.L., 1988. *Groundwater Mechanics*. Prentice-Hall, Englewood Cliffs, N.J., 732 p.

Theis, C.V., 1935. The lowering of the piezometric surface and the rate and discharge of a well using ground-water storage. *Transactions, American Geophysical Union*, Vol. 16, p. 519-524.

Thiem, G., 1906. *Hydrologische methoden*. Leipzig: Gebhardt, 56 p.

Todd, D.K., 1980. *Ground Water Hydrology*. John Wiley & Sons, New York, 530 p.

Vukovic, M., and A. Soro, 1984. *Dinamika Podzemnih Voda (Dynamics of Groundwater*, in Serbo-Croatian). Posebna Izdanja, Knjiga 25, Institut Jaroslav Cerni, Beograd, 500 p.

Vukovic, M., and A. Soro, 1992. *Determination of Hydraulic Conductivity of Porous Media from Grain-Size Composition*. Water Resources Publications, Littleton, CO, 86 p.

Walton, W.C., 1960. Leaky artesian aquifer conditions in Illinois. *Illinois State Water Survey Report of Investigation*, 39.

Walton, W.C., 1962. Selected analytical methods for well and aquifer evaluation. *Illinois State Water Survey Bulletin*, 49, 81 p.

Weeks, W.D., and W.C. Boughton, 1987. Tests of ARMA model forms for rainfall-runoff modelling. *Journal of Hydrology*, 91, p. 29-47.

White, F.M., editor. *Fluid Mechanics*. McGraw-Hill, New York, 701 p.

Yevjevich, V., 1972. *Stochastic processes in hydrology*. Water Resources Publications, Fort Collins, CO.

Zheng, C., 1990. *PATH3D, a groundwater path and travel time simulator*. S.S. Papadopulos & Associates, Inc., Bethesda, Maryland.

Appendix A

ROCK AND FLUID PROPERTIES

Specific Gravity of Common Rock-Forming Minerals

(data from: *Dana's Manual of Mineralogy*, 1959,
Hurlbut C.S.Jr., ed., John Wiley & Sons, New York, 609 p.)

Magmatic Rocks
(granite, rhyolite, granodiorite, dacite, diorite, andesite,
syenite, trachyte, gabro, basalt, diabase, peridotite, dunite)

Mineral	Specific Gravity (G)
Quartz	2.65
K-Feldspar	2.54-2.62
Plagioclase	2.62-2.76
Feldspathoids	2.15-2.65
Micas	2.76-3.20
Amphiboles	3.00-3.40
Pyroxenes	3.15-3.60
Olivine	3.27-4.37

Metamorphic Rocks
(slate, marble, schist, gneiss, quartzite, serpentinite,
amphibolite, granulite, eclogite, skarn, hornfels)

Mineral	Specific Gravity (G)
Quartz	2.65
K-Feldspar	2.54-2.62
Plagioclase	2.62-2.76
Micas	2.76-3.20
Talc	2.70-2.80
Chlorite	2.60-3.30
Serpentine	2.50-2.60
Amphiboles	2.85-3.60
Pyroxenes	3.15-3.60
Epidote	3.35-3.45
Sillimanite	3.23
Kyanite	3.55-3.66
Garnet	3.50-4.30
Calcite	2.71

Sedimentary Rocks
(shale, siltstone, sandstone, conglomerate, limestone, dolomite, chert, gypsum)

Mineral	Specific Gravity (G)
Quartz	2.65
K-Feldspars	2.54-2.62
Plagioclase	2.62-2.76
Micas	2.76-3.20
Amphiboles	2.85-3.60
Pyroxenes	3.15-3.60
Calcite	2.71
Dolomite	2.85
Gypsum	2.32
Kaolinite	2.6
Montmorillonite	2.2-2.74
Illite	2.7-3.1

Porosity of Rocks

Representative ranges in percent

Unconsolidated Sediments

Uniform sand, fine	40-45
Uniform sand, coarse	38-40
Gravelly sand	25-35
Loamy sand	37-50
Non-uniform sand	25-40
Fine to medium gravel	25-33
Coarse gravel	20-25
Glacial till	15-30
Silt	35-50
Silt loam	42-58
Silty clay	42-53
Clay	40-53
Soft glacial clay	50-55
Stiff glacial clay	35-40
Soft organic clay	60-75
Soft bentonite	75-84
Loess	45-50

Consolidated Sedimentary Rocks

Sandstone	5-30
Shale, weathered	10-35
Shale, fresh	<1-10
Siltstone	10-30
Carbonate mud	40-70
Holocene coral limestone	30-50
Tertiary limestone (lms)	20-35

Paleozoic-Mesozoic lms	<1-10
Oolitic limestone	1-25
Karstified limestone	5-50
Dolomite	<1-15
Chalk	15-45
Bedded salt	0.1-0.5
Magmatic Rocks	
Granite, fresh	<1
Granite, fractured	1-10
Granite, weathered	10-40
Basalt, fresh and dense	<1
Vesicular basalt	10-50
Tuff	15-40
Weathered volcanic rocks	15-60
Recent volcanic ash	30-50
Pumice	60-90
Metamorphic Rocks	
Fresh, unfractured	0-0.1
Fractured	<0.1-10
Weathered	10-45

Hydraulic Conductivity of Rocks

(From: *Ground Water Manual*, 1985, U.S. Department of Interior, Bureau of Reclamation, Government Printing Office, Denver, 480 p.)

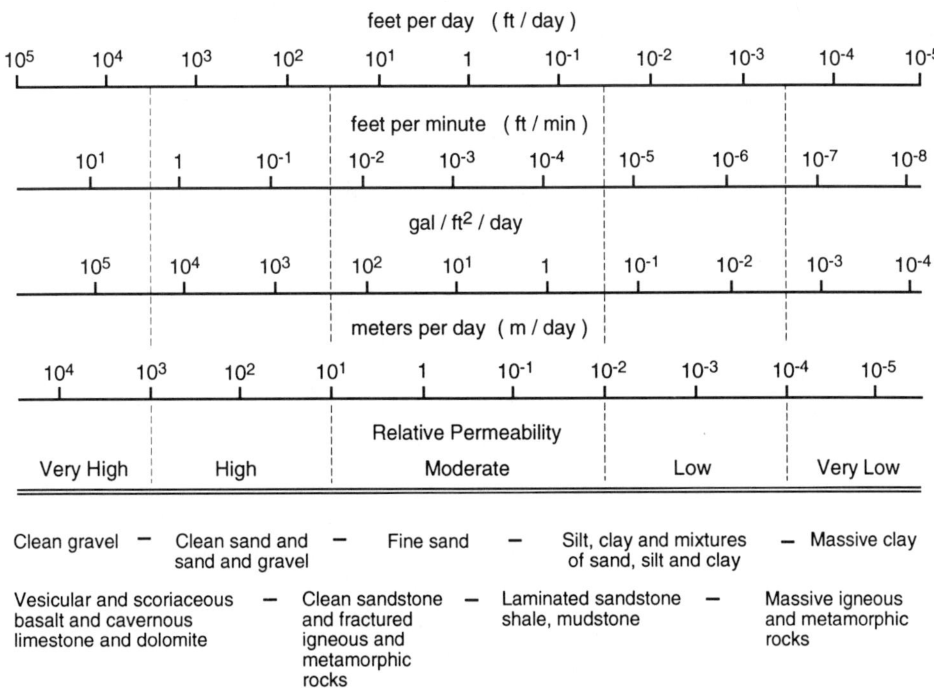

Density of Water vs. Temperature

(Data from: *Handbook of Hydrology*, 1993, David R. Maidment, Editor-in-Chief, McGraw-Hill, New York, p. 11.4)

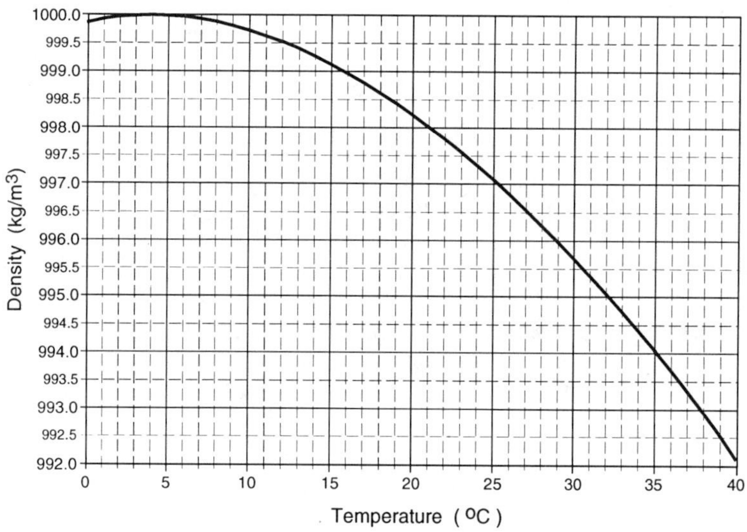

Dynamic Viscosity of Water vs. Temperature

(Data from: Giles et al., 1994. *Fluid Mechanics and Hydraulics*. Schaums' Outline Series, McGraw-Hill, New York, p. 337)

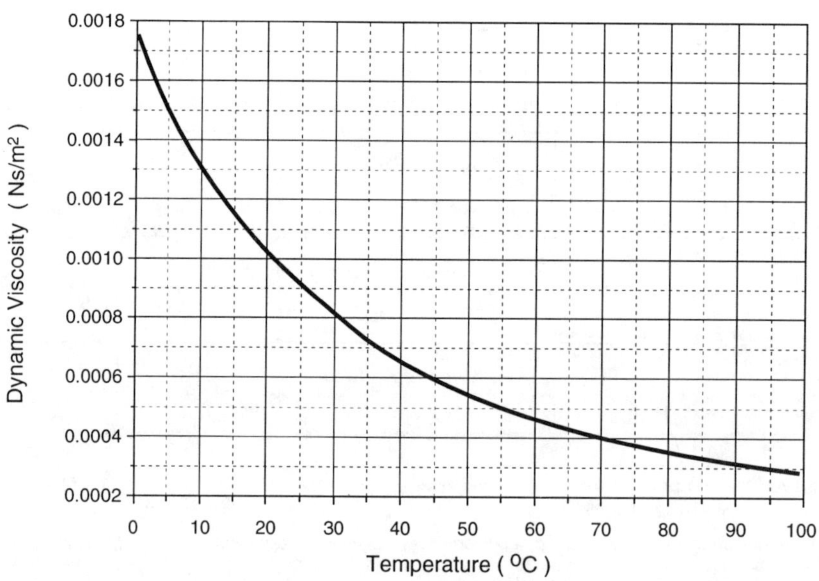

Kinematic Viscosity of Water vs. Temperature

(Data from: *Handbook of Hydrology*, 1993, David R. Maidment, Editor-in-Chief, McGraw-Hill, New York, p. 11.4)

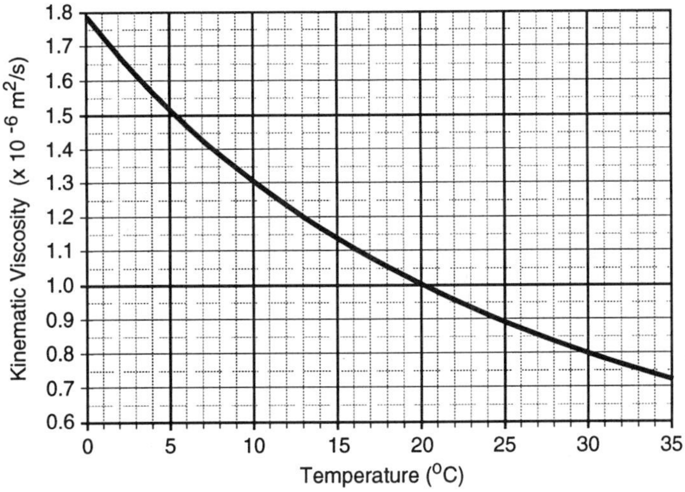

Kinematic Viscosity of Two Common Groundwater Contaminants vs. Temperature

(Data from: Giles et al., 1994. *Fluid Mechanics and Hydraulics*. Schaums' Outline Series, McGraw-Hill, New York, p. 337)

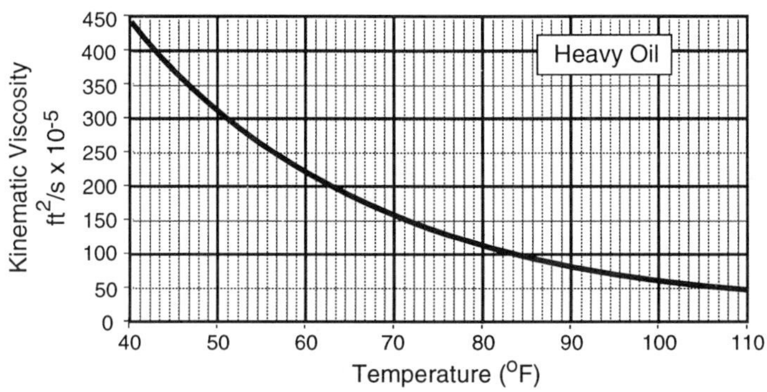

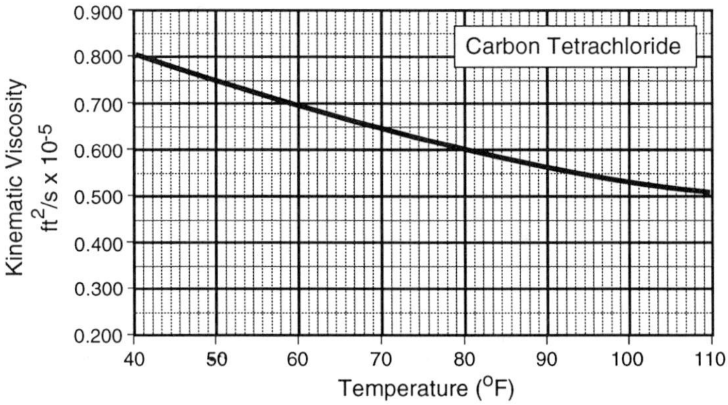

Appendix B

GRAIN SIZE ANALYSIS

Grain Size Distribution Curve

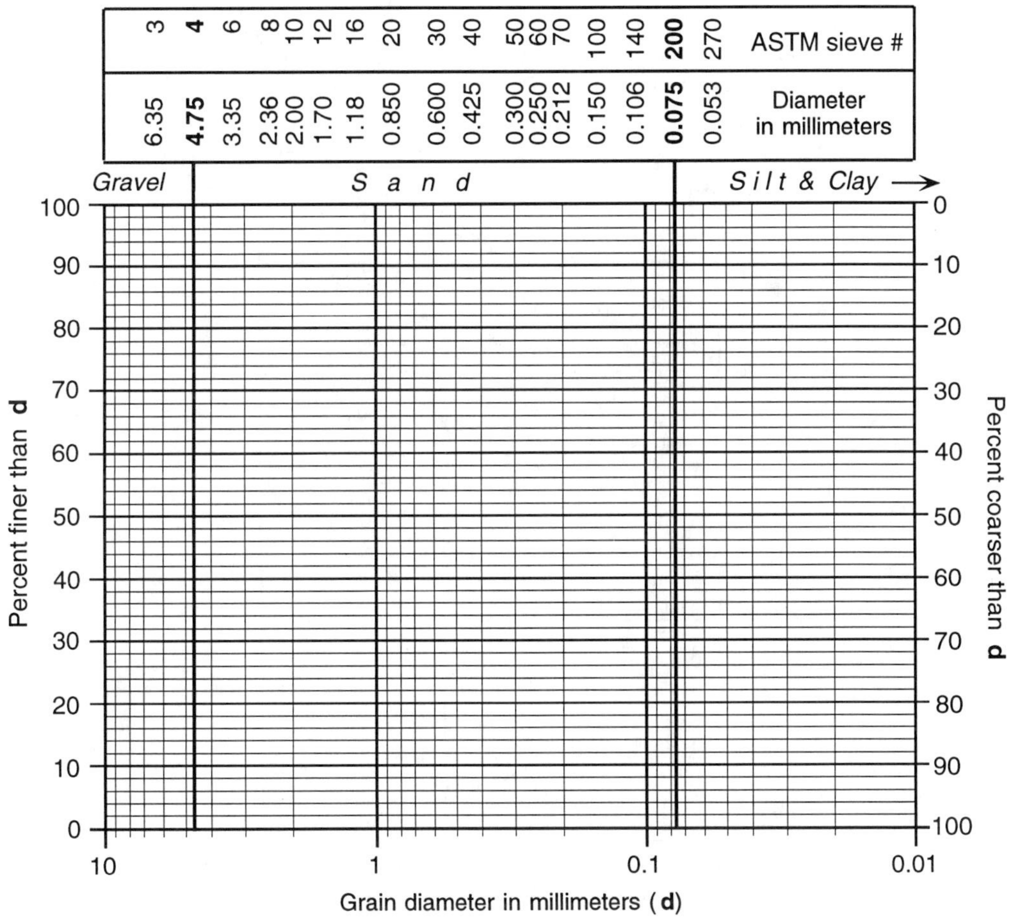

Curve No.	Coefficient of Uniformity $U = d_{60}/d_{10}$	Effective Grain Size $d_e = d_{10}$ (mm)

Dimensions and Designations of Standard ASTM Sieves
(courtesy of Boart Longyear Company)

Standard Designation	Sieve Opening (mm)	Alternative Designation
100 mm	101.6	4 in.
90 mm	90.5	3-1/2 in.
75 mm	76.1	3 in.
63 mm	64	2-1/2 in.
50 mm	50.8	2 in.
45 mm	45.3	1-3/4 in.
38.1mm	38.1	1-1/2 in.
31.5 mm	32	1-1/4 in.
25.0 mm	25.7	1 in.
22.4 mm	22.6	7/8 in.
19.0 mm	19	3/4 in.
16.0 mm	16	5/8 in.
12.5 mm	12.7	1/2 in.
11.2 mm	11.2	7/16 in.
9.5 mm	9.51	3/8 in.
8.0 mm	8	5/16 in.
6.3 mm	6.35	1/4 in.
5.6 mm	5.66	No. 3-1/2
4.75 mm	4.76	No. 4
4.00 mm	4	No. 5
3.35 mm	3.36	No. 6
2.36 mm	2.38	No. 8
2.00 mm	2	No. 10
1.70 mm	1.68	No. 12
1.40 mm	1.41	No. 14
1.18 mm	1.19	No. 16
1.00 mm	1	No. 18
850 μm	0.841	No. 20
710 μm	0.707	No. 25
600 μm	0.595	No. 30
500 μm	0.5	No. 35
425 μm	0.42	No. 40
300 μm	0.297	No. 50
250 μm	0.25	No. 60
212 μm	0.21	No. 70
180 μm	0.177	No. 80
150 μm	0.149	No. 100
125 μm	0.125	No. 120
106 μm	0.105	No. 140
90 μm	0.088	No. 170
75 μm	0.074	No. 200
63 μm	0.063	No. 230
53 μm	0.053	No. 270
45 μm	0.044	No. 325
38 μm	0.037	No. 400

Calculation of Hydraulic Conductivity with Dimensionally Homogeneous Formulas

g	Gravitational acceleration	(m/s^2)
ν	Kinematic viscosity	(m^2/s)
C	Empirical Coefficient	(dimensionless)
U	Coefficient of Uniformity	(dimensionless)
d_e	Effective grain size (d_{10})	(mm)
n	Porosity	(dimensionless)

Breyer's formula:

$$K = \frac{g}{\nu} \cdot C_b \cdot d_e^2$$

$$C_b = 6 \times 10^{-4} \cdot \log \frac{500}{U}$$

Applicable for:
0.06 mm < d10 < 0.6 mm
1 < U < 20

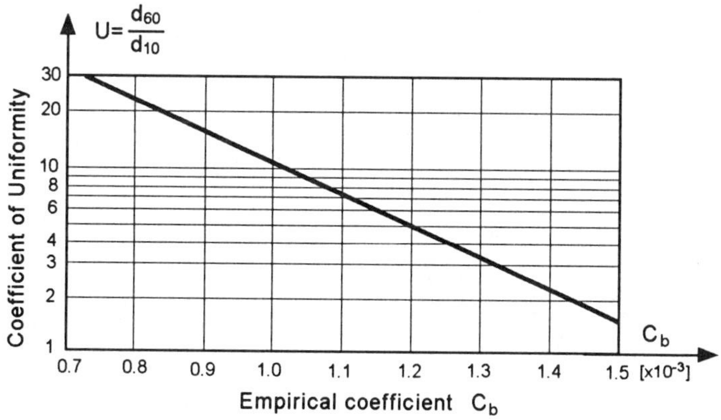

Kozeny's formula:

$$K = \frac{g}{\nu} \cdot C_k \cdot \frac{n^3}{(1-n)^2} \cdot d_e^2$$

$$C_k = 8.3 \times 10^{-3}$$

Applicable for coarse sand

Hazen's formula:

$$K = \frac{g}{\nu} \cdot C_h \cdot f(n) \cdot d_e^2$$

$$C_h = 6 \times 10^{-4}$$

$$f(n) = [1 + 10 \cdot (n - 0.26)]$$

Applicable for:
0.1 mm < d_{10} < 3 mm
U < 5

Appendix C

WELL SCREENS AND CASING DIAMETERS

Dimensions of telescope sizes and pipe sizes standard construction steel screens
(courtesy of Cook Screen Technologies, Inc.)

Nominal Screen Size Diameter of Casing	Screen O.D. (outside diameter) inches	mm	Screen I.D. (inside diameter) inches	mm
3	2-3/4	70	2-1/4	57
4	3-3/4	95	3-1/4	83
5	4-3/4	121	4-1/4	108
6	5-5/8	143	5	127
8	7-1/2	191	6-7/8	175
10	9-1/2	241	8-7/8	225
12	11-1/4	286	10-1/2	267
14	12-3/4	324	12	305
16	14	356	13-1/8	333
18	16	406	15	381
20	18	457	17	432

Cook telescope size screens are commonly lowered through API standard pipe to the bottom of the well. The well casing is then pulled back to expose the screen surface to the aquifer.

Cook pipe size screens are manufactured for direct attachment to API pipe. These screens are so constructed that their dimension is equal to or larger than an API pipe of the same diameter.

Intake areas of telescope sizes and pipe sizes standard construction steel screens
in square inches per lineal foot of screen
(courtesy of Cook Screen Technologies, Inc.)

Pipe size	10 slot	20 slot	30 slot	40 slot	60 slot	80 slot	100 slot	125 slot
2	15	26	35	41	52	59	65	70
3	20	35	47	57	71	81	88	96
4	26	45	60	72	90	102	112	121
5	30	53	71	85	106	121	132	143
6	25	45	62	77	100	118	131	145
8	33	59	65	81	108	130	126	144
10	41	74	81	101	135	162	157	179
12	37	69	77	97	132	162	186	212
14	41	75	84	107	145	177	205	233
16	46	86	96	122	166	203	234	266
18	52	97	108	137	187	228	263	300

Dimensions of telescope sizes and pipe sizes 304 stainless steel screens, sizes 22″ to 46″ (courtesy of Cook Screen Technologies, Inc.)

Screen nominal size inches	Screen O.D. inches	Screen I.D. inches	Screen size inches	Screen O.D. inches	Screen I.D. inches
22	20	19	36	34	33
24	22	21	38	36	35
26	24	23	40	38	37
28	26	25	42	40	39
30	28	27	44	42	41
32	30	29	46	44	43
34	32	31			

Weights and dimensions of PVC pipe and slotted screen (courtesy of Boart Longyear Company)

Schedule 40	Pipe size (inches)	Pipe O.D. (inches)	Pipe I.D. (inches)	Approx ship wt (lb/ft)	Rows of slots
	1	1.315	1.049	0.32	2
	1 1/4	1.660	1.380	0.43	3
	1 1/2	1.90	1.610	0.51	3
	2	2.375	2.067	0.70	4
	3	3.50	3.068	1.45	5
	4	4.500	4.026	2.00	6
	5	5.563	5.047	2.77	6
	6	6.625	6.065	3.45	8
	8	8.625	7.961	5.63	8
	10	10.75	10.02	7.75	10
	12	12.75	11.93	10.30	10

Schedule 80	Pipe size (inches)	Pipe O.D. (inches)	Pipe I.D. (inches)	Approx ship wt (lb/ft)	Rows of slots
	3/4	1.05	0.824	0.21	n/a
	1	1.315	0.975	0.32	2
	1 1/4	1.660	1.279	0.43	3
	2	2.375	1.939	0.93	3
	3	3.50	2.900	1.45	5
	4	4.500	3.826	2.70	6
	5	5.563	4.768	3.60	7
	6	6.625	5.761	3.45	8
	8	8.625	7.625	5.63	8
	10	10.75	9.564	7.75	10
	12	12.75	11.376	10.30	10

Volume of open borehole and annulus between casing and hole
(courtesy of Boart Longyear Company)

Hole Diameter (inches)	Volume/ Lineal ft of Hole		Nominal Casing Diameter (inches)	Volume/ Lineal ft of Annulus	
	gallon	cubic ft		gallon	cubic ft
7.25	2.14	0.29	1.25	2.03	0.27
7.25	2.14	0.29	2	1.91	0.26
7.75	2.45	0.33	2	2.22	0.30
8.25	2.78	0.37	2	2.55	0.34
10.25	4.29	0.57	2	4.06	0.54
8.25	2.78	0.37	3	2.28	0.30
10.25	4.29	0.57	3	3.79	0.51
12.25	6.13	0.82	3	5.62	0.75
8.25	2.78	0.37	4	1.95	0.26
10.25	4.29	0.57	4	3.46	0.46
12.25	6.13	0.82	4	5.30	0.71
12.25	6.13	0.82	6	4.33	0.58

Appendix D

ERROR AND RELATED FUNCTIONS

Values of Error Function
(adapted from Lebedev, 1968)

λ	erf(λ)	λ	erf(λ)	λ	erf(λ)
0.00	0.000	0.42	0.448	0.84	0.765
0.01	0.011	0.43	0.457	0.85	0.771
0.02	0.023	0.44	0.466	0.86	0.776
0.03	0.034	0.45	0.476	0.87	0.781
0.04	0.045	0.46	0.485	0.88	0.787
0.05	0.056	0.47	0.494	0.89	0.792
0.06	0.067	0.48	0.503	0.90	0.797
0.07	0.079	0.49	0.512	0.91	0.802
0.08	0.090	0.50	0.520	0.92	0.807
0.09	0.101	0.51	0.529	0.93	0.812
0.10	0.113	0.52	0.538	0.94	0.816
0.11	0.124	0.53	0.547	0.95	0.821
0.12	0.135	0.54	0.555	0.96	0.825
0.13	0.146	0.55	0.563	0.97	0.830
0.14	0.157	0.56	0.572	0.98	0.834
0.15	0.168	0.57	0.580	0.99	0.839
0.16	0.171	0.58	0.588	1.00	0.843
0.17	0.190	0.59	0.596	1.05	0.862
0.18	0.201	0.60	0.604	1.10	0.880
0.19	0.211	0.61	0.612	1.15	0.897
0.20	0.223	0.62	0.619	1.20	0.910
0.21	0.234	0.63	0.627	1.25	0.928
0.22	0.244	0.64	0.635	1.30	0.934
0.23	0.255	0.65	0.642	1.35	0.944
0.24	0.266	0.66	0.649	1.40	0.952
0.25	0.277	0.67	0.657	1.45	0.960
0.26	0.287	0.68	0.664	1.50	0.966
0.27	0.297	0.69	0.671	1.60	0.976
0.28	0.308	0.70	0.678	1.70	0.984
0.29	0.318	0.71	0.685	1.80	0.989
0.30	0.329	0.72	0.691	1.90	0.993
0.31	0.339	0.73	0.698	2.00	0.9952
0.32	0.349	0.74	0.705	2.10	0.9970
0.33	0.359	0.75	0.711	2.20	0.9981
0.34	0.369	0.76	0.718	2.30	0.99883
0.35	0.379	0.77	0.724	2.40	0.99930
0.36	0.389	0.78	0.730	2.50	0.99958
0.37	0.399	0.79	0.736	2.60	0.99976
0.38	0.409	0.80	0.742	2.70	0.99987
0.39	0.419	0.81	0.748	2.80	0.99992
0.40	0.428	0.82	0.748	2.90	0.99996
0.41	0.438	0.83	0.760	3.00	0.99998

Error function can also be calculated as: $erf(\lambda) = \sqrt{1 - e^{\left(\frac{-4\lambda^2}{\pi}\right)}}$ (within 0.7% error)

Values of Complementary Error Function
(adapted from Lebedev, 1968)

λ	erfc (λ)	λ	erfc (λ)	λ	erfc (λ)
0.00	1.000	0.42	0.552	0.84	0.235
0.01	0.989	0.43	0.543	0.85	0.229
0.02	0.977	0.44	0.534	0.86	0.224
0.03	0.966	0.45	0.524	0.87	0.219
0.04	0.955	0.46	0.515	0.88	0.213
0.05	0.944	0.47	0.506	0.89	0.208
0.06	0.933	0.48	0.497	0.90	0.203
0.07	0.921	0.49	0.488	0.91	0.198
0.08	0.910	0.50	0.480	0.92	0.193
0.09	0.899	0.51	0.471	0.93	0.188
0.10	0.887	0.52	0.462	0.94	0.184
0.11	0.876	0.53	0.453	0.95	0.179
0.12	0.865	0.54	0.445	0.96	0.175
0.13	0.854	0.55	0.437	0.97	0.170
0.14	0.843	0.56	0.428	0.98	0.166
0.15	0.832	0.57	0.420	0.99	0.161
0.16	0.829	0.58	0.412	1.00	0.157
0.17	0.810	0.59	0.404	1.05	0.138
0.18	0.799	0.60	0.396	1.10	0.120
0.19	0.789	0.61	0.388	1.15	0.103
0.20	0.777	0.62	0.381	1.20	0.090
0.21	0.766	0.63	0.373	1.25	0.072
0.22	0.756	0.64	0.365	1.30	0.066
0.23	0.745	0.65	0.358	1.35	0.056
0.24	0.734	0.66	0.351	1.40	0.048
0.25	0.723	0.67	0.343	1.45	0.040
0.26	0.713	0.68	0.336	1.50	0.034
0.27	0.703	0.69	0.329	1.60	0.024
0.28	0.692	0.70	0.322	1.70	0.016
0.29	0.682	0.71	0.315	1.80	0.011
0.30	0.671	0.72	0.309	1.90	0.007
0.31	0.661	0.73	0.302	2.00	0.0048
0.32	0.651	0.74	0.295	2.10	0.0030
0.33	0.641	0.75	0.289	2.20	0.0019
0.34	0.631	0.76	0.282	2.30	0.00117
0.35	0.621	0.77	0.276	2.40	0.00070
0.36	0.611	0.78	0.270	2.50	0.00042
0.37	0.601	0.79	0.264	2.60	0.00024
0.38	0.591	0.80	0.258	2.70	0.00013
0.39	0.581	0.81	0.252	2.80	0.00008
0.40	0.572	0.82	0.252	2.90	0.00004
0.41	0.562	0.83	0.240	3.00	0.00002

$$erfc(\lambda) = 1 - erf(\lambda)$$

Values of Function R(λ) for Various Values of λ

(adapted from Lebedev, 1968)

λ	R(λ)	λ	R(λ)
0.00	1.0000	0.43	0.3408
0.01	0.9776	0.44	0.3312
0.02	0.9752	0.45	0.3224
0.03	0.9340	0.46	0.3132
0.04	0.9128	0.47	0.3048
0.05	0.8920	0.48	0.2960
0.06	0.8716	0.49	0.2880
0.07	0.8516	0.50	0.2800
0.08	0.8120	0.52	0.2644
0.09	0.8124	0.54	0.2492
0.10	0.7736	0.56	0.2332
0.11	0.7748	0.58	0.2220
0.12	0.7568	0.60	0.2092
0.13	0.7388	0.62	0.1968
0.14	0.7212	0.64	0.1852
0.15	0.7040	0.66	0.1744
0.16	0.6872	0.68	0.1640
0.17	0.6704	0.70	0.1528
0.18	0.6540	0.72	0.1448
0.19	0.6384	0.74	0.1360
0.20	0.6228	0.76	0.1276
0.21	0.6072	0.78	0.1196
0.22	0.5924	0.80	0.1120
0.23	0.5776	0.82	0.1048
0.24	0.5632	0.84	0.0984
0.25	0.5492	0.86	0.0920
0.26	0.5352	0.88	0.0860
0.27	0.5216	0.90	0.0804
0.28	0.5080	0.92	0.0748
0.29	0.4956	0.94	0.0700
0.30	0.4828	0.96	0.0652
0.31	0.4704	0.98	0.0608
0.32	0.4580	1.00	0.0568
0.33	0.4464	1.10	0.0396
0.34	0.4348	1.20	0.0272
0.35	0.4232	1.30	0.0184
0.36	0.4120	1.40	0.0120
0.37	0.3992	1.50	0.0080
0.38	0.3904	1.60	0.0052
0.39	0.3800	1.70	0.0032
0.40	0.3700	1.80	0.0020
0.41	0.3600	1.90	0.0012
0.42	0.3490	2.00	0.0008

$$R(\lambda) = 4i^2 erfc(\lambda)$$

Values of Function $i^2erfc(\lambda)$ for Various Values of λ
(adapted from Lebedev, 1968)

λ	$i^2erfc(\lambda)$	λ	$i^2erfc(\lambda)$
0.00	0.2500	0.43	0.0852
0.01	0.2444	0.44	0.0828
0.02	0.2438	0.45	0.0806
0.03	0.2335	0.46	0.0783
0.04	0.2282	0.47	0.0762
0.05	0.2230	0.48	0.0740
0.06	0.2179	0.49	0.0720
0.07	0.2129	0.50	0.0700
0.08	0.2080	0.52	0.0661
0.09	0.2031	0.54	0.0623
0.10	0.1985	0.56	0.0583
0.11	0.1937	0.58	0.0555
0.12	0.1892	0.60	0.0523
0.13	0.1847	0.62	0.0492
0.14	0.1803	0.64	0.0463
0.15	0.1760	0.66	0.0436
0.16	0.1718	0.68	0.0410
0.17	0.1676	0.70	0.0382
0.18	0.1635	0.72	0.0362
0.19	0.1596	0.74	0.0340
0.20	0.1557	0.76	0.0319
0.21	0.1518	0.78	0.0299
0.22	0.1481	0.80	0.0280
0.23	0.1444	0.82	0.0262
0.24	0.1408	0.84	0.0246
0.25	0.1373	0.86	0.0230
0.26	0.1338	0.88	0.0215
0.27	0.1304	0.90	0.0201
0.28	0.1270	0.92	0.0187
0.29	0.1239	0.94	0.0175
0.30	0.1207	0.96	0.0163
0.31	0.1176	0.98	0.0152
0.32	0.1145	1.00	0.0142
0.33	0.1116	1.10	0.0099
0.34	0.1087	1.20	0.0068
0.35	0.1058	1.30	0.0046
0.36	0.1030	1.40	0.0030
0.37	0.0998	1.50	0.0020
0.38	0.0976	1.60	0.0013
0.39	0.0950	1.70	0.0008
0.40	0.0925	1.80	0.0005
0.41	0.0900	1.90	0.0003
0.42	0.0875	2.00	0.0002

Appendix E

WELL FUNCTIONS, GRAPH PAPERS

Values of $W(u)$ for Fully Penetrating Wells in a Confined, Isotropic Aquifer, for Values of u Between 10^{-11} and 9.9

(adapted from Ferris, Knowles, Brown, and Stalman, 1962)

u	W(u)	u	W(u)	u	W(u)	u	W(u)	u	W(u)	u	W(u)
1.0E-11	24.7512	1.0E-09	20.1460	1.0E-07	15.5409	1.0E-05	10.9357	1.0E-03	6.3315	1.0E-01	1.8229
1.5E-11	24.3458	1.5E-09	19.7406	1.5E-07	15.1354	1.5E-05	10.5303	1.5E-03	5.9266	1.5E-01	1.4645
2.0E-11	24.0581	2.0E-09	19.4529	2.0E-07	14.8477	2.0E-05	10.2426	2.0E-03	5.6394	2.0E-01	1.2227
2.5E-11	23.8349	2.5E-09	19.2298	2.5E-07	14.6246	2.5E-05	10.0194	2.5E-03	5.4167	2.5E-01	1.0443
3.0E-11	23.6526	3.0E-09	19.0474	3.0E-07	14.4423	3.0E-05	9.8317	3.0E-03	5.2349	3.0E-01	0.9057
3.5E-11	23.4985	3.5E-09	18.8933	3.5E-07	14.2881	3.5E-05	9.6830	3.5E-03	5.0813	3.5E-01	0.7942
4.0E-11	23.3649	4.0E-09	18.7598	4.0E-07	14.1546	4.0E-05	9.5495	4.0E-03	4.9482	4.0E-01	0.7024
4.5E-11	23.2471	4.5E-09	18.6420	4.5E-07	14.0368	4.5E-05	9.4317	4.5E-03	4.8310	4.5E-01	0.6253
5.0E-11	23.1418	5.0E-09	18.5366	5.0E-07	13.9314	5.0E-05	9.3263	5.0E-03	4.7261	5.0E-01	0.5598
5.5E-11	23.0465	5.5E-09	18.4413	5.5E-07	13.8361	5.5E-05	9.2310	5.5E-03	4.6313	5.5E-01	0.5034
6.0E-11	22.9595	6.0E-09	18.3543	6.0E-07	13.7491	6.0E-05	9.1440	6.0E-03	4.5448	6.0E-01	0.4544
6.5E-11	22.8794	6.5E-09	18.2742	6.5E-07	13.6691	6.5E-05	9.0640	6.5E-03	4.4652	6.5E-01	0.4115
7.0E-11	22.8053	7.0E-09	18.2001	7.0E-07	13.5950	7.0E-05	8.9899	7.0E-03	4.3916	7.0E-01	0.3738
7.5E-11	22.7363	7.5E-09	18.1311	7.5E-07	13.5260	7.5E-05	8.9209	7.5E-03	4.3231	7.5E-01	0.3403
8.0E-11	22.6718	8.0E-09	18.0666	8.0E-07	13.4614	8.0E-05	8.8563	8.0E-03	4.2591	8.0E-01	0.3106
8.5E-11	22.6112	8.5E-09	18.0060	8.5E-07	13.4008	8.5E-05	8.7957	8.5E-03	4.1990	8.5E-01	0.2840
9.0E-11	22.5540	9.0E-09	17.9488	9.0E-07	13.3437	9.0E-05	8.7386	9.0E-03	4.1423	9.0E-01	0.2602
9.5E-11	22.4999	9.5E-09	17.8948	9.5E-07	13.2896	9.5E-05	8.6845	9.5E-03	4.0887	9.5E-01	0.2387
1.0E-10	22.4486	1.0E-08	17.8435	1.0E-06	13.2383	1.0E-04	8.6332	1.0E-02	4.0379	1.0	0.2194
1.5E-10	22.0432	1.5E-08	17.4380	1.5E-06	12.8328	1.5E-04	8.2278	1.5E-02	3.6374	1.5	0.1000
2.0E-10	21.7555	2.0E-08	17.1503	2.0E-06	12.5451	2.0E-04	7.9402	2.0E-02	3.3547	2.0	0.04890
2.5E-10	21.5323	2.5E-08	16.9272	2.5E-06	12.3220	2.5E-04	7.7172	2.5E-02	3.1365	2.5	0.02491
3.0E-10	21.3500	3.0E-08	16.7449	3.0E-06	12.1397	3.0E-04	7.5348	3.0E-02	2.9591	3.0	0.01305
3.5E-10	21.1959	3.5E-08	16.5591	3.5E-06	11.9855	3.5E-04	7.3807	3.5E-02	2.8099	3.5	0.00698
4.0E-10	21.0623	4.0E-08	16.4572	4.0E-06	11.8520	4.0E-04	7.2472	4.0E-02	2.6813	4.0	0.00378
4.5E-10	20.9446	4.5E-08	16.3394	4.5E-06	11.7342	4.5E-04	7.1295	4.5E-02	2.5684	4.5	0.00207
5.0E-10	20.8392	5.0E-08	16.2340	5.0E-06	11.6289	5.0E-04	7.0242	5.0E-02	2.4679	5.0	0.00115
5.5E-10	20.7439	5.5E-08	16.1387	5.5E-06	11.5336	5.5E-04	6.9289	5.5E-02	2.3775	5.5	0.000641
6.0E-10	20.6569	6.0E-08	16.0517	6.0E-06	11.4465	6.0E-04	6.8420	6.0E-02	2.2953	6.0	0.000360
6.5E-10	20.5768	6.5E-08	15.9717	6.5E-06	11.3665	6.5E-04	6.7620	6.5E-02	2.2201	6.5	0.000203
7.0E-10	20.5027	7.0E-08	15.8976	7.0E-06	11.2924	7.0E-04	6.6879	7.0E-02	2.1508	7.0	0.000116
7.5E-10	20.4337	7.5E-08	15.8286	7.5E-06	11.2234	7.5E-04	6.6190	7.5E-02	2.0867	7.5	6.58E-05
8.0E-10	20.3692	8.0E-08	15.7640	8.0E-06	11.1589	8.0E-04	6.5545	8.0E-02	2.0269	8.0	3.77E-05
8.5E-10	20.3086	8.5E-08	15.7034	8.5E-06	11.0982	8.5E-04	6.4939	8.5E-02	1.9711	8.5	2.16E-05
9.0E-10	20.2514	9.0E-08	15.6462	9.0E-06	11.0411	9.0E-04	6.4368	9.0E-02	1.9187	9.0	1.25E-05
9.5E-10	20.1973	9.5E-08	15.5922	9.5E-06	10.9870	9.5E-04	6.3828	9.5E-02	1.8695	9.5	7.19E-06

Values of *W(u, r/B)* for Leaky Confined Aquifers

Source: Hantush, M.S., 1956. Analysis of Data from Pumping Tests in Leaky Aquifers. *Transactions, American Geophysical Union*, Vol. 37, No. 6, p. 702-714.

u \ r/b	0.01	0.015	0.03	0.05	0.075	0.1	0.15	0.2	0.3	0.4
0.000001	9.4413	8.6313	7.2450	6.2282	5.4228	4.8530	4.0595	3.5054	2.7428	2.2290
0.000005	9.4413	8.6313	7.2450	6.2282	5.4228	4.8530	4.0595	3.5054	2.7428	2.2290
0.00001	9.4176	8.6313	7.2450	6.2282	5.4228	4.8530	4.0595	3.5054	2.7428	2.2290
0.00005	8.8827	8.4533	7.2450	6.2282	5.4228	4.8530	4.0595	3.5054	2.7428	2.2290
0.0001	8.3983	8.1414	7.2122	6.2282	5.4228	4.8530	4.0595	3.5054	2.7428	2.2290
0.0005	6.9750	6.9152	6.6219	6.0821	5.4062	4.8530	4.0595	3.5054	2.7428	2.2290
0.001	6.3069	6.2765	6.1202	5.7965	5.3078	4.8292	4.0595	3.5054	2.7428	2.2290
0.005	4.7212	4.7152	4.6829	4.6084	4.4713	4.2960	3.8812	3.4567	2.7428	2.2290
0.01	4.0356	4.0326	4.0167	3.9795	3.9091	3.8150	3.5725	3.2875	2.7104	2.2253
0.05	2.4675	2.4670	2.4642	2.4576	2.4448	2.4271	2.3776	2.3110	1.9283	1.7075
0.1	1.8227	1.8225	1.8213	1.8184	1.8128	1.8050	1.7829	1.7527	1.6704	1.5644
0.5	0.5598	0.5597	0.5596	0.5594	0.5588	0.5581	0.5561	0.5532	0.5453	0.5344
1.0	0.2194	0.2194	0.2193	0.2193	0.2191	0.2190	0.2186	0.2179	0.2161	0.2135
5.0	0.0011	0.0011	0.0011	0.0011	0.0011	0.0011	0.0011	0.0011	0.0011	0.0011

u \ r/b	0.5	0.6	0.7	0.8	0.9	1.0	1.5	2.0	2.5
0.000001	1.8486	1.5550	1.3210	1.1307	0.9700	0.8409	0.4271	0.2278	0.1174
0.000005	1.8486	1.5550	1.3210	1.1307	0.9700	0.8409	0.4271	0.2278	0.1174
0.00001	1.8486	1.5550	1.3210	1.1307	0.9700	0.8409	0.4271	0.2278	0.1174
0.00005	1.8486	1.5550	1.3210	1.1307	0.9700	0.8409	0.4271	0.2278	0.1174
0.0001	1.8486	1.5550	1.3210	1.1307	0.9700	0.8409	0.4271	0.2278	0.1174
0.0005	1.8486	1.5550	1.3210	1.1307	0.9700	0.8409	0.4271	0.2278	0.1174
0.001	1.8486	1.5550	1.3210	1.1307	0.9700	0.8409	0.4271	0.2278	0.1174
0.005	1.8486	1.5550	1.3210	1.1307	0.9700	0.8409	0.4271	0.2278	0.1174
0.01	1.8486	1.5550	1.3210	1.1307	0.9700	0.8409	0.4271	0.2278	0.1174
0.05	1.4927	1.2955	1.2955	1.1210	0.9700	0.8409	0.4271	0.2278	0.1174
0.1	1.4422	1.2115	1.1791	1.0505	0.9297	0.8190	0.4271	0.2278	0.1174
0.5	0.5206	0.5044	0.4860	0.4658	0.4440	0.4210	0.3007	0.1944	0.1174
1.0	0.2103	0.2065	0.2020	0.1970	0.1914	0.1855	0.1509	0.1139	0.0803
5.0	0.0011	0.0011	0.0011	0.0011	0.0011	0.0011	0.0011	0.0011	0.0009

Values of s_D for the Construction of Type A Curves for Fully Penetrating Wells in Unconfined Aquifers with Delayed Response

Source: Neuman, S.P., 1975. Analysis of Pumping Test Data From Anisotropic Aquifers Considering Delayed Gravity Response. *Water Resources Research*, (11)2:329–42. Copyright by the American Geophysical Union.

t_s	β=0.001	β=0.004	β=0.01	β=0.03	β=0.06	β=0.1	β=0.2	β=0.4	β=0.6	β=0.8
1.00E-01	2.48E-02	2.43E-02	2.41E-02	2.35E-02	2.30E-02	2.24E-02	2.14E-02	1.99E-02	1.88E-02	1.70E-02
2.00E-01	1.45E-01	1.42E-01	1.40E-01	1.36E-01	1.31E-01	1.27E-01	1.19E-01	1.08E-01	9.88E-01	8.49E-02
3.50E-01	3.58E-01	3.52E-01	3.45E-01	3.31E-01	3.18E-01	3.04E-01	2.79E-01	2.44E-01	2.17E-01	1.75E-01
6.00E-01	6.62E-01	6.48E-01	6.33E-01	6.01E-01	5.70E-01	5.40E-01	4.83E-01	4.03E-01	3.43E-01	2.96E-01
1.00E+00	1.02E+00	9.92E-01	9.63E-01	9.05E-01	8.49E-01	7.92E-01	6.88E-01	5.42E-01	4.38E-01	3.60E-01
2.00E+00	1.57E+00	1.52E+00	1.46E+00	1.35E+00	1.23E+00	1.12E+00	9.18E-01	6.59E-01	4.97E-01	3.91E-01
3.50E+00	2.05E+00	1.97E+00	1.88E+00	1.70E+00	1.51E+00	1.34E+00	1.03E+00	6.90E-01	5.07E-01	3.94E-01
6.00E+00	2.52E+00	2.41E+00	2.27E+00	1.99E+00	1.73E+00	1.47E+00	1.07E+00	6.96E-01	5.07E-01	3.94E-01
1.00E+01	2.97E+00	2.80E+00	2.61E+00	2.22E+00	1.85E+00	1.53E+00	1.08E+00	6.96E-01	5.07E-01	3.94E-01
2.00E+01	3.56E+00	3.30E+00	3.00E+00	2.41E+00	1.92E+00	1.55E+00	1.08E+00	6.96E-01	5.07E-01	3.94E-01
3.50E+01	4.01E+00	3.65E+00	3.23E+00	2.48E+00	1.93E+00	1.55E+00	1.08E+00	6.96E-01	5.07E-01	3.94E-01
6.00E+01	4.42E+00	3.93E+00	3.37E+00	2.49E+00	1.94E+00	1.55E+00	1.08E+00	6.96E-01	5.07E-01	3.94E-01
1.00E+02	4.77E+00	4.12E+00	3.43E+00	2.50E+00	1.94E+00	1.55E+00	1.08E+00	6.96E-01	5.07E-01	3.94E-01
2.00E+02	5.16E+00	4.26E+00	3.45E+00	2.50E+00	1.94E+00	1.55E+00	1.08E+00	6.96E-01	5.07E-01	3.94E-01
3.50E+02	5.40E+00	4.29E+00	3.46E+00	2.50E+00	1.94E+00	1.55E+00	1.08E+00	6.96E-01	5.07E-01	3.94E-01
6.00E+02	5.54E+00	4.30E+00	3.46E+00	2.50E+00	1.94E+00	1.55E+00	1.08E+00	6.96E-01	5.07E-01	3.94E-01
1.00E+03	5.59E+00	4.30E+00	3.46E+00	2.50E+00	1.94E+00	1.55E+00	1.08E+00	6.96E-01	5.07E-01	3.94E-01
2.00E+03	5.59E+00	4.30E+00	3.46E+00	2.50E+00	1.94E+00	1.55E+00	1.08E+00	6.96E-01	5.07E-01	3.94E-01
3.50E+00	5.59E+00	4.30E+00	3.46E+00	2.50E+00	1.94E+00	1.55E+00	1.08E+00	6.96E-01	5.07E-01	3.94E-01

t_s	β=1.0	β=1.5	β=2.0	β=2.5	β=3.0	β=4.0	β=5.0	β=6.0	β=7.0
1.00E-01	1.70E-02	1.53E-02	1.38E-02	1.25E-02	1.13E-02	9.33E-03	7.72E-03	6.39E-03	5.30E-03
2.00E-01	8.49E-02	7.13E-02	6.03E-02	5.11E-02	4.35E-02	3.17E-02	2.34E-02	1.74E-02	1.31E-02
3.50E-01	1.75E-01	1.36E-01	1.07E-01	8.46E-02	6.78E-02	4.45E-02	3.02E-02	2.10E-02	1.51E-02
6.00E-01	2.56E-01	1.82E-01	1.33E-01	1.01E+00	7.67E-02	4.76E-02	3.13E-02	2.14E-02	1.52E-02
1.00E+00	3.00E-01	1.99E-01	1.40E-01	1.03E-01	7.79E-02	4.78E-02	3.13E-02	2.15E-02	1.52E-02
2.00E+00	3.17E-01	2.03E-01	1.41E-01	1.03E-01	7.79E-02	4.78E-02	3.13E-02	2.15E-02	1.52E-02
3.50E+00	3.17E-01	2.03E-01	1.41E-01	1.03E-01	7.79E-02	4.78E-02	3.13E-02	2.15E-02	1.52E-02
6.00E+00	3.17E-01	2.03E-01	1.41E-01	1.03E-01	7.79E-02	4.78E-02	3.13E-02	2.15E-02	1.52E-02
1.00E+01	3.17E-01	2.03E-01	1.41E-01	1.03E-01	7.79E-02	4.78E-02	3.13E-02	2.15E-02	1.52E-02
2.00E+01	3.17E-01	2.03E-01	1.41E-01	1.03E-01	7.79E-02	4.78E-02	3.13E-02	2.15E-02	1.52E-02
3.50E+01	3.17E-01	2.03E-01	1.41E-01	1.03E-01	7.79E-02	4.78E-02	3.13E-02	2.15E-02	1.52E-02
6.00E+01	3.17E-01	2.03E-01	1.41E-01	1.03E-01	7.79E-02	4.78E-02	3.13E-02	2.15E-02	1.52E-02
1.00E+02	3.17E-01	2.03E-01	1.41E-01	1.03E-01	7.79E-02	4.78E-02	3.13E-02	2.15E-02	1.52E-02
2.00E+02	3.17E-01	2.03E-01	1.41E-01	1.03E-01	7.79E-02	4.78E-02	3.13E-02	2.15E-02	1.52E-02
3.50E+02	3.17E-01	2.03E-01	1.41E-01	1.03E-01	7.79E-02	4.78E-02	3.13E-02	2.15E-02	1.52E-02
6.00E+02	3.17E-01	2.03E-01	1.41E-01	1.03E-01	7.79E-02	4.78E-02	3.13E-02	2.15E-02	1.52E-02
1.00E+03	3.17E-01	2.03E-01	1.41E-01	1.03E-01	7.79E-02	4.78E-02	3.13E-02	2.15E-02	1.52E-02
2.00E+03	3.17E-01	2.03E-01	1.41E-01	1.03E-01	7.79E-02	4.78E-02	3.13E-02	2.15E-02	1.52E-02
3.50E+00	3.17E-01	2.03E-01	1.41E-01	1.03E-01	7.79E-02	4.78E-02	3.13E-02	2.15E-02	1.52E-02

Values of s_D for the Construction of Type B Curves for Fully Penetrating Wells in Unconfined Aquifers with Delayed Response

Source: Neuman, S.P., 1975. Analysis of Pumping Test Data From Anisotropic Aquifers Considering Delayed Gravity Response. *Water Resources Research*, (11)2:329–42. Copyright by the American Geophysical Union.

t_y	β=0.001	β=0.004	β=0.01	β=0.03	β=0.06	β=0.1	β=0.2	β=o.4	β=0.6	β=0.8
1.00E-04	5.62E+00	4.30E+00	3.46E+00	2.50E+00	1.94E+00	1.56E+00	1.09E+00	6.97E-01	5.08E-01	3.95E-01
2.00E-04	5.62E+00	4.30E+00	3.46E+00	2.50E+00	1.94E+00	1.56E+00	1.09E+00	6.97E-01	5.08E-01	3.95E-01
3.50E-04	5.62E+00	4.30E+00	3.46E+00	2.50E+00	1.94E+00	1.56E+00	1.09E+00	6.97E-01	5.08E-01	3.95E-01
6.00E-04	5.62E+00	4.30E+00	3.46E+00	2.50E+00	1.94E+00	1.56E+00	1.09E+00	6.97E-01	5.08E-01	3.95E-01
1.00E-03	5.62E+00	4.30E+00	3.46E+00	2.50E+00	1.94E+00	1.56E+00	1.09E+00	6.97E-01	5.08E-01	3.95E-01
2.00E-03	5.62E+00	4.30E+00	3.46E+00	2.50E+00	1.94E+00	1.56E+00	1.09E+00	6.97E-01	5.09E-01	3.96E-01
3.50E-03	5.62E+00	4.30E+00	3.46E+00	2.50E+00	1.94E+00	1.56E+00	1.09E+00	6.98E-01	5.10E-01	3.97E-01
6.00E-03	5.62E+00	4.30E+00	3.46E+00	2.50E+00	1.94E+00	1.56E+00	1.09E+00	7.00E-01	5.12E-01	3.99E-01
1.00E-02	5.62E+00	4.30E+00	3.46E+00	2.50E+00	1.94E+00	1.56E+00	1.09E+00	7.03E-01	5.16E-01	4.03E-01
2.00E-02	5.62E+00	4.30E+00	3.46E+00	2.50E+00	1.94E+00	1.56E+00	1.09E+00	7.10E-01	5.24E-01	4.12E-01
3.50E-02	5.62E+00	4.30E+00	3.46E+00	2.50E+00	1.94E+00	1.56E+00	1.10E+00	7.20E-01	5.37E-01	4.25E-01
6.00E-02	5.62E+00	4.30E+00	3.46E+00	2.50E+00	1.95E+00	1.57E+00	1.11E+00	7.37E-01	5.57E-01	4.47E-01
1.00E-01	5.62E+00	4.30E+00	3.46E+00	2.51E+00	1.96E+00	1.58E+00	1.13E+00	7.63E-01	5.89E-01	4.83E-01
2.00E-01	5.62E+00	4.30E+00	3.46E+00	2.52E+00	1.98E+00	1.61E+00	1.18E+00	8.29E-01	6.67E-01	5.71E-01
3.50E-01	5.63E+00	4.31E+00	3.47E+00	2.54E+00	2.01E+00	1.66E+00	1.24E+00	9.22E-01	7.80E-01	6.97E-01
6.00E-01	5.63E+00	4.31E+00	3.49E+00	2.57E+00	2.06E+00	1.73E+00	1.35E+00	1.07E+00	9.54E-01	8.89E-01
1.00E+00	5.63E+00	4.32E+00	3.51E+00	2.62E+00	2.13E+00	1.83E+00	1.50E+00	1.29E+00	1.20E+00	1.16E-01
2.00E+00	5.64E+00	4.35E+00	3.56E+00	2.73E+00	2.31E+00	2.07E+00	1.85E+00	1.72E+00	1.68E+00	1.66E-01
3.50E+00	5.65E+00	4.38E+00	3.63E+00	2.88E+00	2.55E+00	2.37E+00	2.23E+00	2.17E+00	2.15E+00	2.15E-01
6.00E+00	5.67E+00	4.44E+00	3.74E+00	3.11E+00	2.86E+00	2.75E+00	2.68E+00	2.66E+00	2.65E+00	2.65E-01
1.00E+01	5.70E+00	4.52E+00	3.90E+00	3.40E+00	3.24E+00	3.18E+00	3.15E+00	3.14E+00	3.14E+00	3.14E-01
2.00E+01	5.76E+00	4.71E+00	4.22E+00	3.92E+00	3.85E+00	3.83E+00	3.82E+00	3.82E+00	3.82E+00	3.82E-01
3.50E+01	5.85E+00	4.94E+00	4.58E+00	4.40E+00	4.38E+00	4.58E+00	4.38E+00	4.58E+00	4.37E+00	4.37E-01
6.00E+01	5.99E+00	5.23E+00	5.00E+00	4.92E+00	4.91E+00	4.91E+00	4.91E+00	4.91E+00	4.91E+00	4.91E-01
1.00E+02	6.16E+00	5.59E+00	5.46E+00	5.42E+00	5.42E+00	5.42E+00	5.42E+00	5.42E+00	5.42E+00	5.42E-01

t_y	β=1.0	β=1.5	β=2.0	β=2.5	β=3.0	β=4.0	β=5.0	β=6.0	β=7.0
1.00E-04	3.18E-01	2.04E-01	1.42E-01	1.03E-01	7.80E-02	4.79E-02	3.14E-02	2.15E-02	1.53E-02
2.00E-04	3.18E-01	2.04E-01	1.42E-01	1.03E-01	7.81E-02	4.80E-02	3.15E-02	2.16E-02	1.53E-02
3.50E-04	3.18E-01	2.04E-01	1.42E-01	1.03E-01	7.83E-02	4.81E-02	3.16E-02	2.17E-02	1.54E-02
6.00E-04	3.18E-01	2.04E-01	1.42E-01	1.04E-01	7.85E-02	4.84E-02	3.18E-02	2.19E-02	1.58E-02
1.00E-03	3.18E-01	2.04E-01	1.42E-01	1.04E-01	7.89E-02	4.78E-02	3.21E-02	2.21E-02	1.58E-02
2.00E-03	3.19E-01	2.05E-01	1.43E-01	1.05E-01	7.99E-02	4.96E-02	3.29E-02	2.28E-01	1.64E-02
3.50E-03	3.21E-01	2.07E-01	1.45E-01	1.07E-01	8.14E-02	5.09E-02	3.41E-02	2.39E-02	1.73E-02
6.00E-03	3.23E-01	2.09E-01	1.47E-01	1.09E-01	8.38E-02	5.32E-02	3.61E-02	2.57E-02	1.89E-02
1.00E-02	3.27E-01	2.13E-01	1.52E-01	1.13E-01	8.79E-02	5.68E-02	3.93E-02	2.86E-02	2.15E-02
2.00E-02	3.37E-01	2.24E-01	1.62E-01	1.24E-01	9.80E-02	6.61E-02	4.78E-02	3.62E-02	2.84E-02
3.50E-02	3.50E-01	2.39E-01	1.78E-01	1.39E-01	1.13E-01	8.06E-02	6.12E-02	4.86E-02	3.98E-02
6.00E-02	3.74E-01	2.65E-01	2.05E-01	1.66E-01	1.40E-01	1.06E-01	8.53E-02	7.14E-02	6.14E-02
1.00E-01	4.12E-01	3.07E-01	2.48E-01	2.10E-01	1.84E-01	1.49E-01	1.28E-01	1.13E-01	1.02E-01
2.00E-01	5.06E-01	4.10E-01	3.57E-01	3.23E-01	2.98E-01	2.66E-01	2.45E-01	2.31E-01	2.20E-01
3.50E-01	6.42E-01	5.62E-01	5.17E-01	4.89E-01	4.70E-01	4.45E-01	4.30E-01	4.19E-01	4.11E-01
6.00E-01	8.50E-01	7.92E-01	7.63E-01	7.45E-01	7.33E-01	7.18E-01	7.09E-01	7.03E-01	6.99E-01
1.00E+00	1.13E+00	1.10E+00	1.08E+00	1.07E+00	1.07E+00	1.06E+00	1.06E+00	1.05E+00	1.05E+00
2.00E+00	1.65E+00	1.63E+00	1.63E+00	1.63E+00	1.63E+00	1.63E+00	1.63E+00	1.63E+00	1.63E+00
3.50E+00	2.14E+00	2.14E+00	2.14E+00	2.14E+00	2.14E+00	2.14E+00	2.14E+00	2.14E+00	2.14E+00
6.00E+00	2.65E+00	2.65E+00	2.64E+00	2.64E+00	2.64E+00	2.64E+00	2.64E+00	2.64E+00	2.64E+00
1.00E+01	3.14E+00	3.14E+00	3.14E+00	3.14E+00	3.14E+00	3.14E+00	3.14E+00	3.14E+00	3.14E+00
2.00E+01	3.82E+00	3.82E+00	3.82E+00	3.82E+00	3.82E+00	3.82E+00	3.82E+00	3.82E+00	3.82E+00
3.50E+01	4.37E+00	4.37E+00	4.37E+00	4.37E+00	4.37E+00	4.37E+00	4.37E+00	4.37E+00	4.37E+00
6.00E+01	4.91E+00	4.91E+00	4.91E+00	4.91E+00	4.91E+00	4.91E+00	4.91E+00	4.91E+00	4.91E+00
1.00E+02	5.42E+00	5.42E+00	5.42E+00	5.42E+00	5.42E+00	5.42E+00	5.42E+00	5.42E+00	5.42E+00

Well Pumping Tests Analysis
Cooper-Jacob Method

Well No.: ______ Location: __________________________ Pumping rate: _____

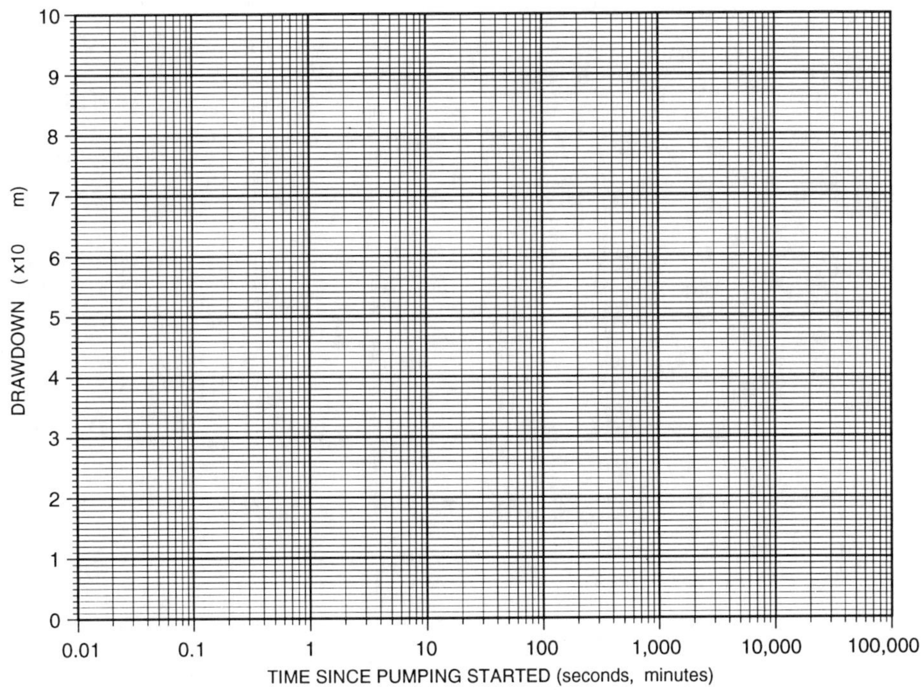

$$T = \frac{0.183Q}{\Delta s}$$

$T = \quad \times 10^{-} \quad m^2/s$

$$S = \frac{2.25Tt_0}{r^2}$$

$S =$

Δs Difference in drawdown between two points that are one log cycle apart and lay on a straight line drawn through recorded data. In metric (SI) units,
Q Pumping rate in metric units,
T Transmissivity,
S Storage coefficient,
t_0 Intercept of the straight line at zero drawdown (time on x axis),
r Distance between the pumping well and the oservation well where the drawdown is recorded.

Well Pumping Tests Analysis
Ratio Between Time and Square Distance

Well No.: ______ Location: ______________________________ Pumping rate: ______

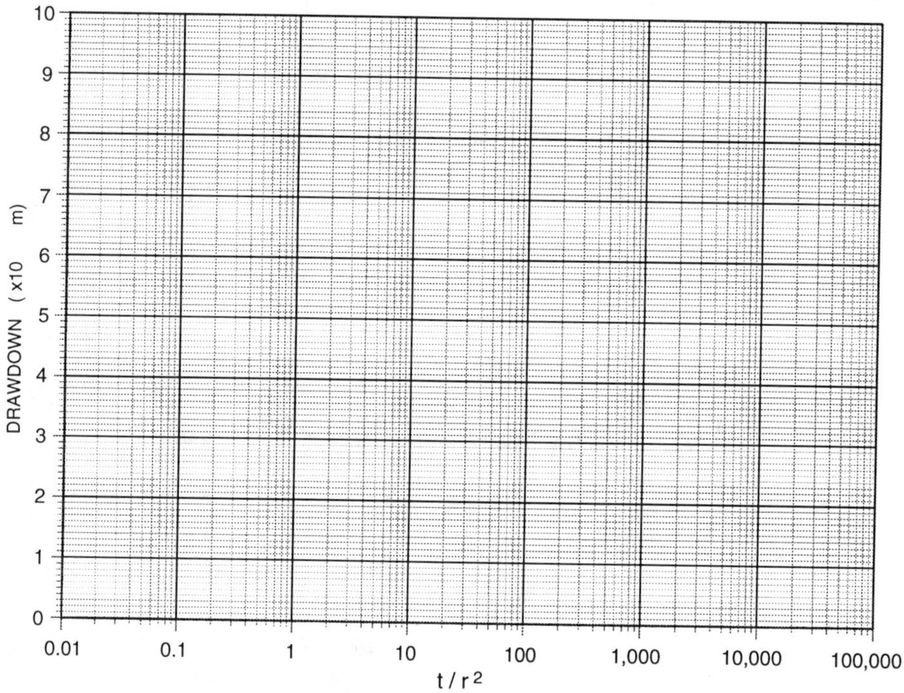

$$T = \frac{0.183Q}{\Delta s}$$

$T = \quad \times 10^{-} \quad m^2/s$

$$S = 2.25T(t/r^2)_0$$

$S =$

Δs	Difference in drawdown between two points that are one log cycle apart and lay on a straight line drawn through recorded data. In metric (SI) units,
Q	Pumping rate in metric units,
T	Transmissivity,
S	Storage coefficient,
t	Time since pumping started,
r	Distance between the pumping well and the oservation well where the drawdown is recorded,
$(t/r^2)_0$	Intercept of the straight line at zero drawdown.

Theis Type Curve

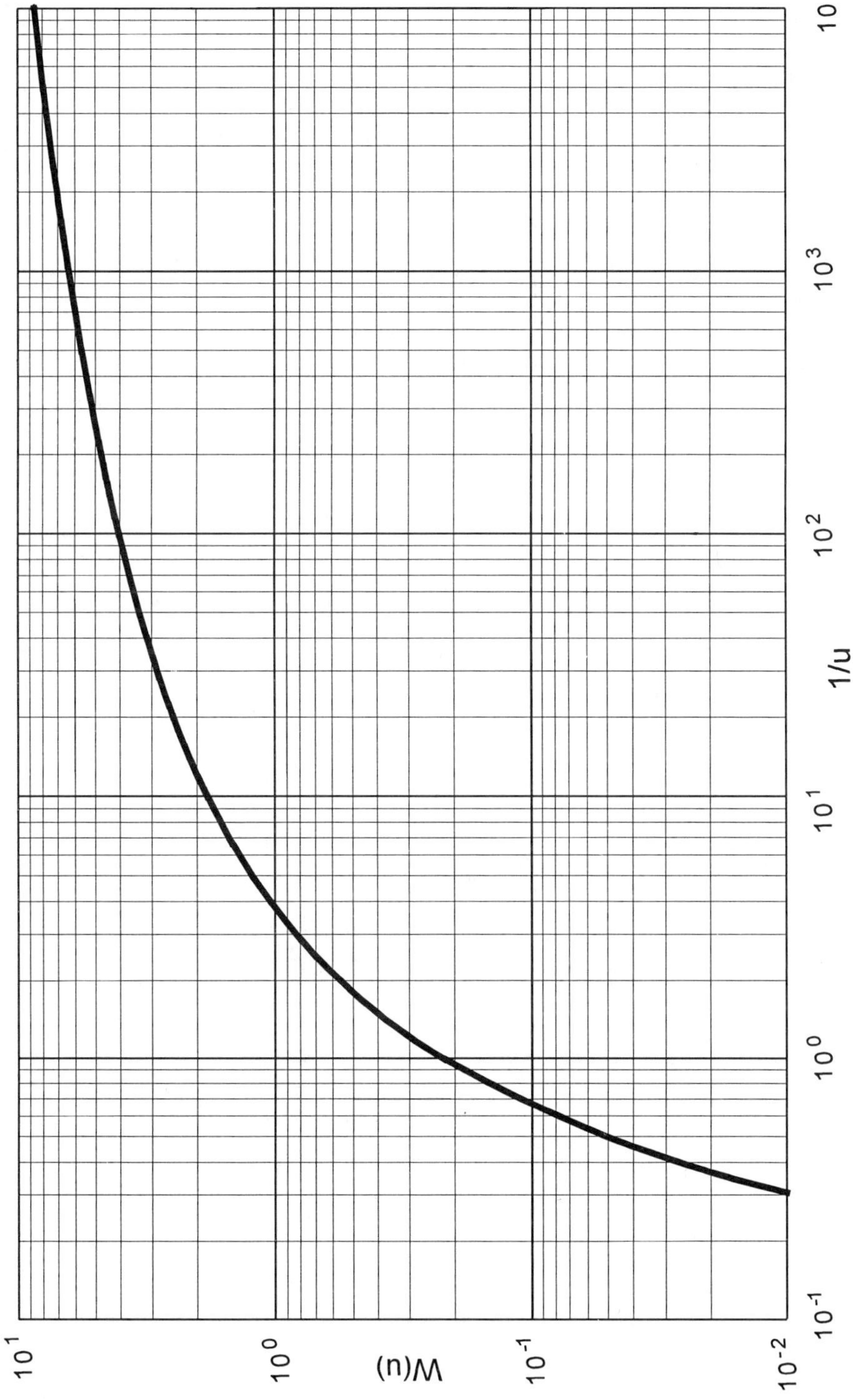

Log-Log Graph for Plotting Field Data

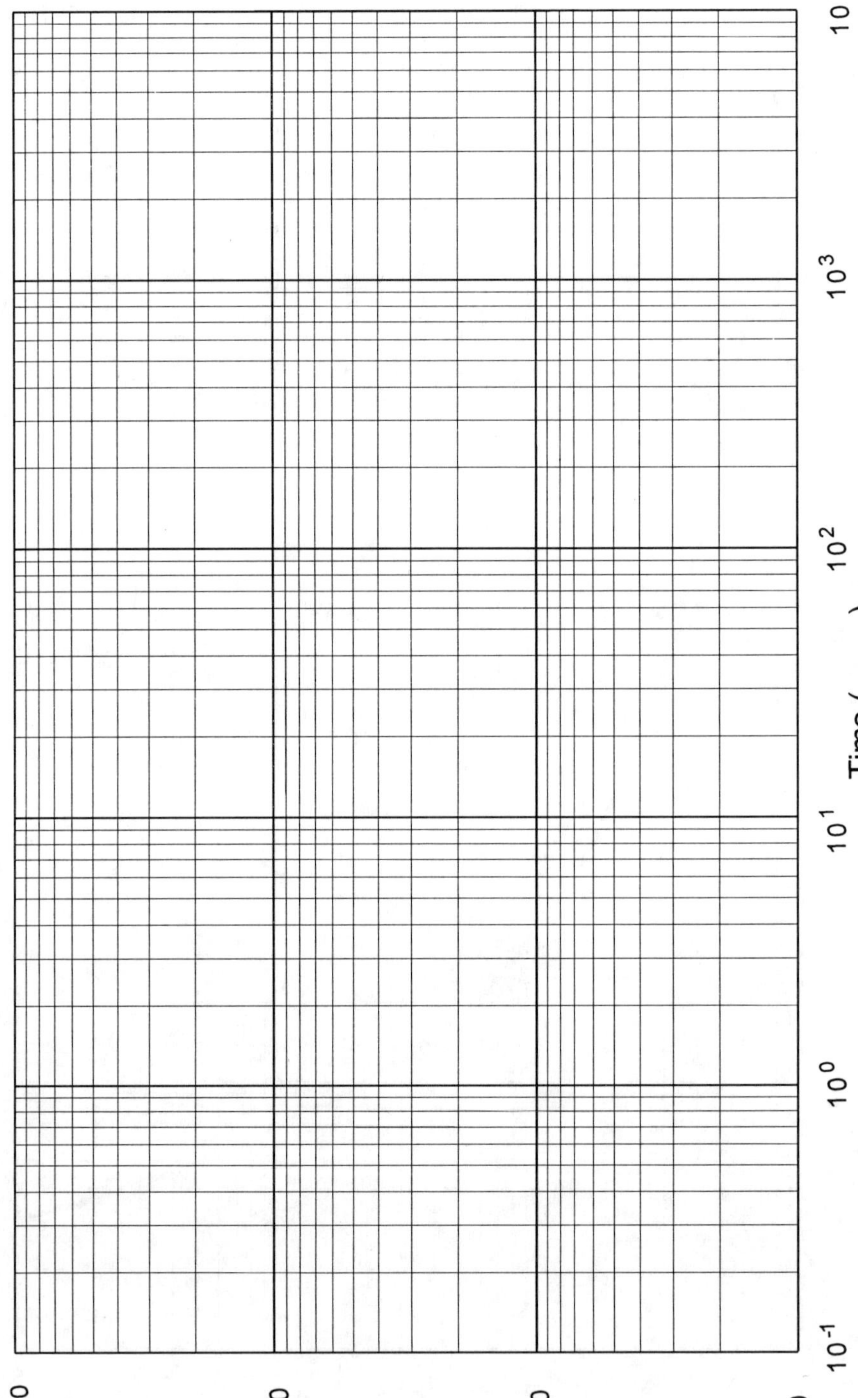

Type Curves for Fully Penetrating Wells in Unconfined Aquifers with Delayed Gravity Response

Source: Neuman, S.P., 1975. Analysis of Pumping Test Data From Anisotropic Aquifers Considering Delayed Gravity Response. *Water Resources Research*, (11)2:329–42. Copyright by the American Geophysical Union.

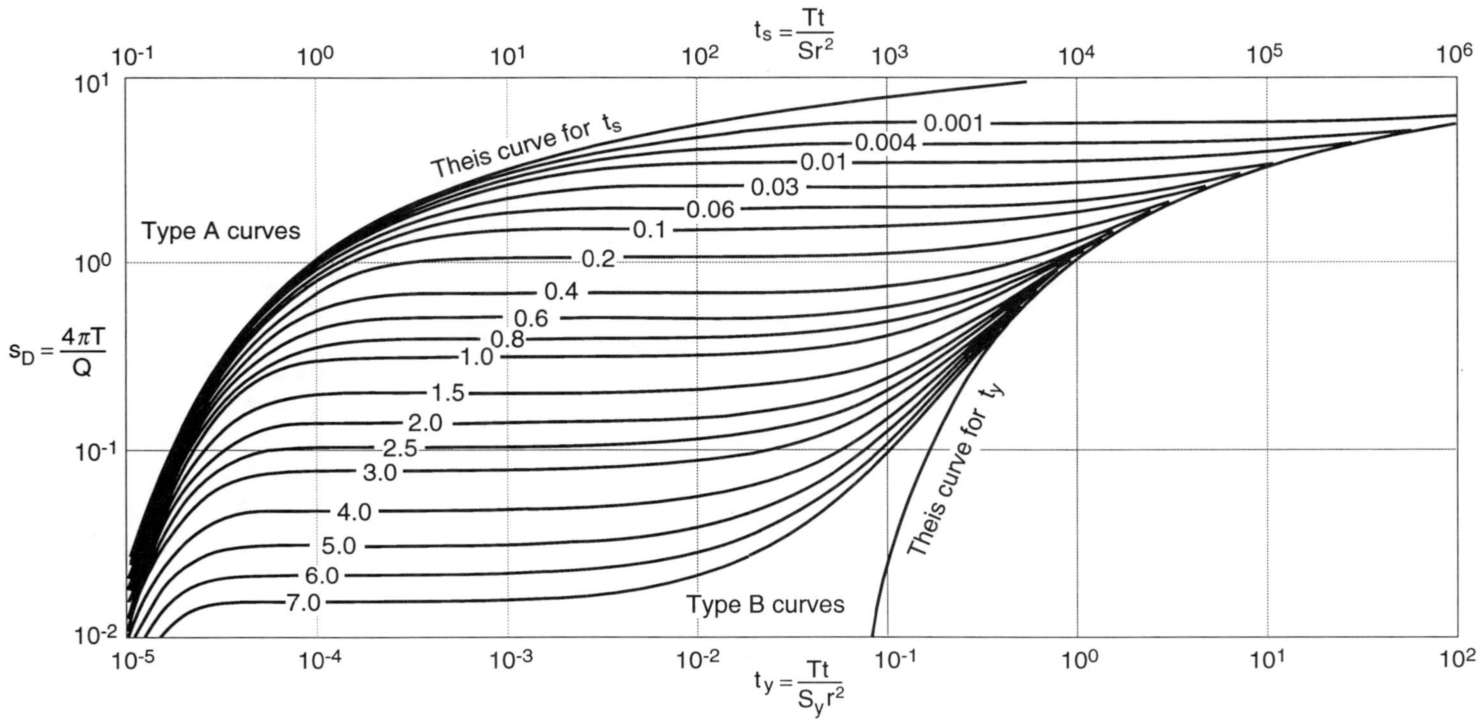

Log-Log Graph for Plotting Field Data

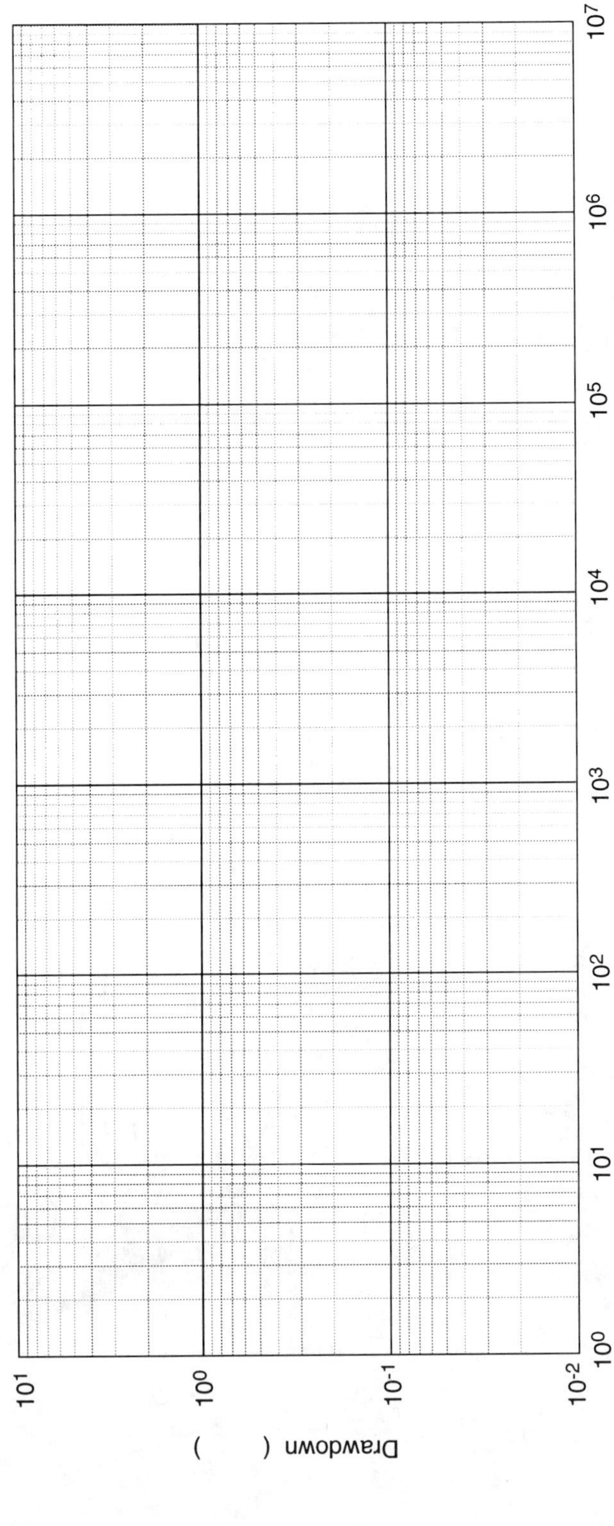

Appendix F

MODFLOW PACKAGES

This appendix contains original text from the publication "A Modular Three-Dimensional Finite-Difference Ground-Water Flow Model" by Michael G. McDonald and Arlen W. Harbaugh. It was published in 1988 in a USGS series entitled *Techniques of Water-Resources Investigations of the United States Geological Survey,* Book 6, Chapter A1 (U.S. Government Printing Office, Washington). Included are important excerpts on derivation of the finite-difference equation, as well as excerpts from the following packages: River Package (Chapter 6), Recharge Package (Chapter 7), Well Package (Chapter 8), Drain Package (Chapter 9), Evapotranspiration Package (Chapter 10), and General-Head Boundary Package (Chapter 11). Also included are the following packages subsequently added to the original MODFLOW by other USGS authors: Stream Package (Prudic, 1989), and Horizontal Flow Barrier Package (Hsieh and Freckleton, 1993).

DERIVATION OF THE FINITE-DIFFERENCE EQUATION

(excerpts from Chapter 2)

MATHEMATICAL MODEL

The three-dimensional movement of ground water of constant density through porous earth material may be described by the partial-differential equation

$$\frac{\partial}{\partial x}\left(K_{xx}\frac{\partial h}{\partial x}\right)+\frac{\partial}{\partial y}\left(K_{yy}\frac{\partial h}{\partial y}\right)+\frac{\partial}{\partial z}\left(K_{zz}\frac{\partial h}{\partial z}\right)-W=S_s\frac{\partial h}{\partial t} \quad (1)$$

where

K_{XX}, K_{yy} and K_{ZZ} are values of hydraulic conductivity along the x, y and z coordinate axes, which are assumed to be parallel to the major axes of hydraulic conductivity (Lt^{-1});
h is the potentiometric head (L);
W is a volumetric flux per unit volume and represents sources and/or sinks of water (t^{-1});
S_S the specific storage of the porous material (L^{-1}); and
t is time (t).

For a derivation of equation (1) see for example Rushton and Redshaw (1979). In general, S_S, K_{XX}, K_{Xy}, and K_{ZZ} may be functions of space (S_S=S_S(x,y,z,), K_{XX}=K_{XX}(x,y,z), etc.) and W may be a function of space and time W=(W(x,y,z,t)); equation (1) describes ground water flow under nonequilibrium conditions in a heterogeneous and anisotropic medium, provided the principal axes of hydraulic conductivity are aligned with the coordinates directions.

Equation (1), together with specification of flow and/or head conditions at the boundaries of an aquifer system and specification of initial head conditions, constitutes a mathematical representation of a ground-water flow system. A solution of equation (1) in an analytical sense, is an algebraic expression giving h(x,y,z,t) such that, when the derivatives of *h* with respect to space and time are substituted in equation (1), the equation and its initial and boundary conditions are satisfied. A time-varying head distribution of this nature characterizes the flow system, in that it measures both the energy of flow and the volume of water in storage, and can be used to calculate directions and rates of movement.

Except for very simple systems, analytical solution of equation (1) are rarely possible, so various numerical methods must be employed to obtain approximate solutions. One such approach is finite-difference method, wherein the continuous system described by equation (1) is replaced by a finite set of discrete points in space and time and the partial derivatives are replaced by terms calculated from the differences in head values at these points. The process leads to systems of simultaneous linear algebraic difference equations; their solution yields values of head at specific points and times. These values constitute an approximation to the time-varying head distribution that would be given by an analytical solution of the partial difference equation of flow.

The finite difference analog of equation (1) may be derived by applying the rules of difference calculus; however, in the discussion presented here, an alternative approach is used with the aim of simplifying the mathematical treatment and explaining the computational procedure in terms of familiar physical concepts regarding the flow system.

DISCRETIZATION CONVENTION

Figure 1 shows a spatial discretization of an aquifer system with a mesh of blocks called cells, the locations of which are described in terms of rows, columns, and layers. An i,j,k indexing system is used. For a system consisting of "nrow" rows and "ncol" columns, and "nlay" layers, i is the row index, i = 1, 2, . . . nrow; j is the column index, j = 1,2,. . . ncol; and k is the layer index, k = 1,2,. . . nlay. For example, figure 1 shows a system with

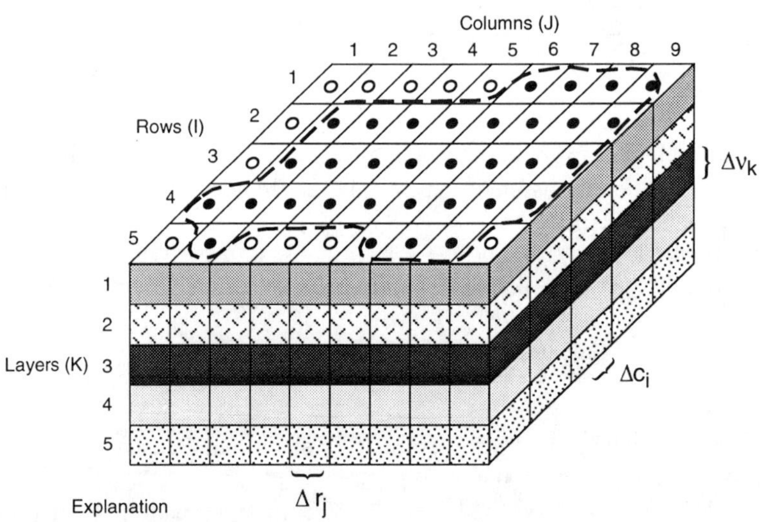

Explanation

— — — Aquifer Boundary

● Active Cell

O Inactive Cell

Δr_j Dimension of Cell Along the Row Direction. Subscript (J) Indicates the Number of the Column

Δc_i Dimension of Cell Along the Column Direction. Subscript (I) Indicates the Number of the Row

Δv_k Dimension of Cell Along the Vertical Direction. Subscript (K) Indicates the Number of the Layer

Figure 1 A discretized hypothetical aquifer system.

nrow=5, ncol=9, and nlay=5. In formulating the equations of the model, an assumption was made that layers would generally correspond to horizontal geohydrologic units or intervals. Thus in terms of Cartesian coordinates, the k index denotes changes along the vertical, z; because the convention followed in this model is to number layers from the top down, an increment in the k index corresponds to a decrease in elevation. Similarly rows would be considered parallel to the x axis, so that increments in the row index, i, would correspond to decreases in y; and columns would be considered parallel to the y axis, so that increments in the column index, j, would correspond to increases in x. These conventions were followed in constructing figure 1; however, application of the model requires only that rows and columns fall along consistent orthogonal directions within the layers, and does not require the designation of x, y, or z coordinate axes.

Following the conventions used in figure 1, the width of cells in the row direction, at a given column, j, is designated Δr_j; the width of cells in the column direction at a given row, i, is designated Δc_i; and the thickness of cells in a given layer, k, is designated Δv_k . Thus a cell with coordinates (i,j,k) = (4,8,3) has a volume of $\Delta r_8\ \Delta c_4\ \Delta v_3$.

Within each cell there is a point called a "node" at which head is to be calculated. Figure 2 illustrates, in two dimensions, two conventions for defining the configuration of cells with respect to the location of the nodes--the block-centered formulation and the point-centered formulation. Both systems start by dividing the aquifer with two sets of

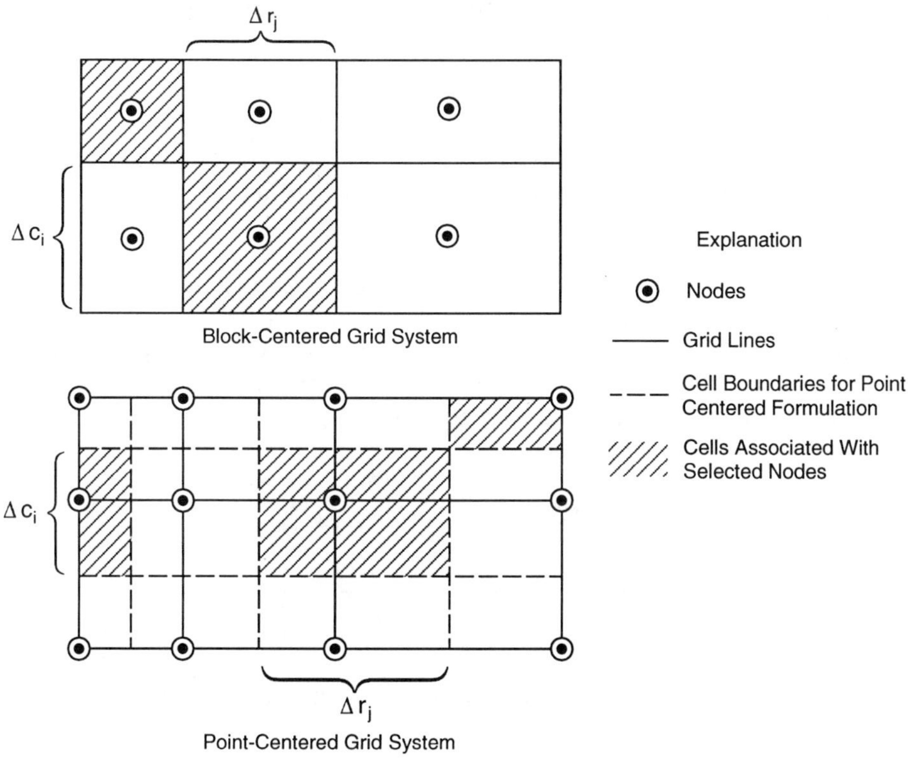

Figure 2 Grids showing the difference between block-centered and point-centered grids.

parallel lines that are orthogonal. In the block-centered formulation, the blocks formed by the sets of parallel lines are the cells; the nodes are at the center of the cells. In the point-centered formulation, the nodes are at the intersection points of the sets of parallel lines, and cells are drawn around the nodes with faces halfway between nodes. In either case, spacing of nodes should be chosen so that the hydraulic properties of the system are, in fact, generally uniform over the extent of a cell. The finite-difference equation developed in the following section holds for either formulation; however, only the block centered formulation is presently used in the model.

In equation (1), the head, h, is a function of time as well as space so that, in the finite-difference formulation, discretization of the continuous time domain is also required.

FINITE-DIFFERENCE EQUATION

Development of the ground-water flow equation in finite-difference form follows from the application of the continuity equation: the sum of all flows into and out of cell must be equal to the rate of change in storage within the cell. Under the assumption that density of ground water is constant, the continuity equation expressing the balance of flow for a cell is

$$\sum Q_i = SS^{\Delta h} \Delta V \tag{2}$$

where

- Q_i is a flow rate into the cell $(L^3 t^{-1})$;
- SS has been introduced as the notation for specific storage in the finite-difference formulation; its definition is equivalent to that of S_S in equation (1), *i.e.*, it is the volume of water which can be injected per unit volume of aquifer material per unit change in head. (L^{-1});
- ΔV is the volume of the cell (L^3); and
- Δh is the change in head over a time interval of length Δt.

The term on the right hand side is equivalent to the volume of water taken into storage over a time interval Δt given a change in head of Δh. Equation (2) is stated in terms of inflow and storage gain. Outflow and loss are represented by defining outflow as negative inflow and loss as negative gain.

Figure 3 depicts a cell i,j,k and six adjacent aquifer cells i–1,j,k; i+1,j,k; i,j–1,k; i,j+1,k; i,j,k–1; and i,j,k+1. To simplify the following development, flows are considered positive if they are entering the cell i,j,k; and the negative sign usually incorporated in Darcy's low has been dropped from all terms. Following these conventions, flow into cell i,j,k, in the row direction from cell i,j–1,k (figure 4), is given by Darcy's law as

$$q_{i,j-1/2,k} = KR_{i,j-1/2,k} \cdot \Delta c_i \Delta v_k \frac{(h_{i,j-1,k} - h_{i,j,k})}{\Delta r_{j-1/2}} \tag{3}$$

where

- $h_{i,j,k}$ is the head at node i, j, k and $h_{i,j-1,k}$ that at node i,j–1,k;
- $q_{i,j-1/2,k}$ is the volumetric fluid discharge through the face between cells i,j,k and i,j–1,k $(L^3 t^{-1})$;
- $KR_{i,j-1/2,k}$ is the hydraulic conductivity along the row between nodes i,j,k and i,j–1,k (Lt^{-1});
- $\Delta c_i \Delta v_k$ is the area of the cell faces normal to the row direction; and
- $\Delta r_{j-1/2}$ is the distance between nodes i,j,k and i,j–1,k (L).

Although the discussion is phrased in terms of flow into the central cell, it can be misleading to associate the subscript j–1/2 of equation (3) with a specific point between the nodes. Rather, the term $KR_{i,j-1/2,k}$ of equation (3) is the effective hydraulic conductivity for the entire region between the nodes, normally calculated as a harmonic mean in the sense described by, for example, Collins (1961). If this is done, equation (3) gives the exact flow, for a one-dimensional steady state case, through a block of aquifer extending from node i,j−1,k to node i,j,k and having a cross sectional area $\Delta c_j \Delta v_k$.

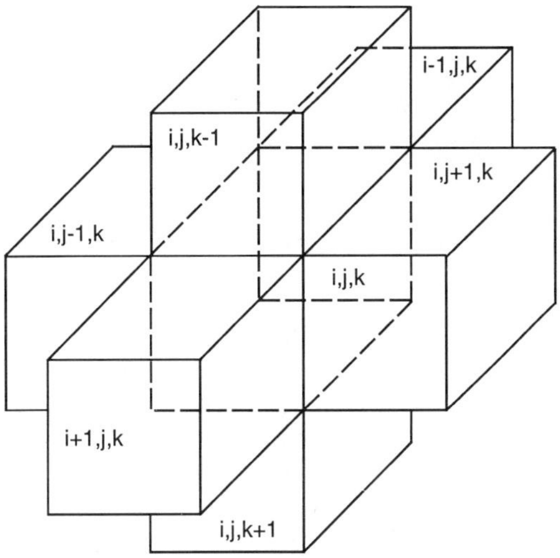

Figure 3 Cell i,j,k and indices for the six adjacent cells.

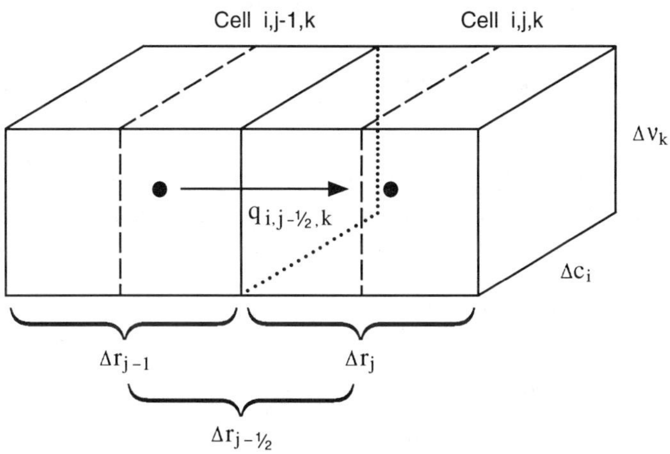

Figure 4 Flow into cell i,j,k from cell i,j-1,k.

Similar expressions can be written approximating the flow into the cell through the remaining five faces, *i.e.*, for flow in the row direction through the face between cells i,j,k and i,j+1,k,

$$q_{i,j+1/2,k} = KR_{i,j+1/2,k} \cdot \Delta c_i \Delta v_k \frac{(h_{i,j+1,k} - h_{i,j,k})}{\Delta r_{j+1/2}} \tag{4}$$

while for the column direction, flow into the block through the forward face is

$$q_{i+1/2,j,k} = KC_{i+1/2,j,k} \cdot \Delta r_j \Delta v_k \frac{(h_{i+1,j,k} - h_{i,j,k})}{\Delta c_{i+1/2}} \tag{5}$$

and flow in to the block through the rear face is

$$q_{i-1/2,j,k} = KC_{i-1/2,j,k} \cdot \Delta r_j \Delta v_k \frac{(h_{i-1,j,k} - h_{i,j,k})}{\Delta c_{i-1/2}} \tag{6}$$

For the vertical direction, inflow through the bottom face is

$$q_{i,j,k+1/2} = KV_{i,j,k+1/2} \cdot \Delta r_j \Delta c_i \frac{(h_{i,j,k+1} - h_{i,j,k})}{\Delta v_{k+1/2}} \tag{7}$$

while inflow through the upper face is given by

$$q_{i,j-1/2,k} = KR_{i,j-1/2,k} \cdot \Delta c_j \Delta v_k \frac{(h_{i,j-1,k} - h_{i,j,k})}{\Delta r_{1-1/2}} \tag{8}$$

Each of these equations expresses the inflow through a face of cell i,j,k in terms of heads, grid dimensions and hydraulic conductivity. The notation can be simplified by combining grid dimensions and hydraulic conductivity into a single constant, the "hydraulic conductance" or, more simply, the "conductance." For example,

$$CR_{i,j-1/2,k} = KR_{i,j-1/2,k} \cdot \Delta c_i \Delta v_k / \Delta r_{j-1/2} \tag{9}$$

where

$CR_{i,j-1/2,k}$ is the conductance in row i and layer k between nodes i,j−1,k and i,j,k ($L^2 t^{-1}$).

Conductance is thus the product of hydraulic conductivity and cross-sectional area of flow divided by the length of the flow path (in this case, the distance between the nodes). Substituting conductance from equation (9) into equation (3) yields

$$q_{i,j-1/2,k} = CR_{i,j-1/2,k} \cdot (h_{i,j-1,k} - h_{i,j,k}) \tag{10}$$

Similarly, equations (4)–(8) can be rewritten to yield

$$q_{i,j+1/2,k} = CR_{i,j+1/2,k} \cdot (h_{i,j+1,k} - h_{i,j,k}) \tag{11}$$

$$q_{i-1/2,j,k} = CC_{i-1/2,j,k} \cdot (h_{i-1,j,k} - h_{i,j,k}) \tag{12}$$

$$q_{i+1/2,j,k} = CC_{i+1/2,j,k} \cdot (h_{i+1,j,k} - h_{i,j,k}) \tag{13}$$

$$q_{i,j,k-1/2} = CV_{i,j,k-1/2} \cdot (h_{i,j,k-1} - h_{i,j,k}) \tag{14}$$

$$q_{i,j,k+1/2} = CV_{i,j,k+1/2} \cdot (h_{i,j,k+1} - h_{i,j,k}) \tag{15}$$

where the conductances are defined analogously to $CR_{i,j-1/2,k}$ in equation (9).

Equations (10)–(15) account for the flow into cell i, j, k from the six adjacent cells. To account for flows into the cell from features or processes external to the aquifer, such as streams, drains, areal recharge, evapotranspiration or wells, additional terms are required. These flows may be dependent on the head in the receiving cell but independent of all other heads in the aquifer, or they may be entirely independent of head in the receiving cell. Flow from outside the aquifer may be represented by the expression

$$a_{i,j,k,n} = p_{i,j,k,n} \cdot h_{i,j,k} + q_{i,j,k,n} \tag{16}$$

where
$a_{i,j,k,n}$ represents flow from the nth external source into cell i, j, k ($L^3\ t^{-1}$), and $p_{i,j,k,n}$ and $q_{i,j,k,n}$ are constants (($L^2\ t^{-1}$) and ($L^3\ t^{-1}$), respectively).

For example, suppose a cell is receiving flow from two sources, recharge from a well and seepage through a riverbed. For the first source (n=1), since the flow from the well is assumed to be independent of head, $p_{i,j,k,1}$ is zero and $q_{i,j,k,1}$ is the recharge rate for the well. In this case,

$$a_{i,j,k,1} = q_{i,j,k,1} \tag{17}$$

For the second source (n=2), the assumption is made that the stream-aquifer interconnection can be treated as a simple conductance, so that the seepage is proportional to the head difference between the river stage and the head in the cell i,j,k (figure 5); thus we have

$$a_{i,j,k,2} = CRIV_{i,j,k,2} \cdot (R_{i,j,k} - h_{i,j,k}) \tag{18}$$

where $R_{i,j,k}$ is the head in the river (L) and $CRIV_{i,j,k,2}$ is a conductance ($L^2\ t^{-1}$) controlling flow from the river into cell i,j,k. For example, in the situation shown schematically in figure 5, CRIV would be given as the product of the vertical hydraulic conductivity of the riverbed material and the area of the streambed as it crosses the cell, divided by the thickness of the streambed material. Equation (18) can be rewritten as

$$a_{i,j,k,2} = -CRIV_{i,j,k,2} \cdot h_{i,j,k} + CRIV_{i,j,k,2} \cdot R_{i,j,k} \tag{19}$$

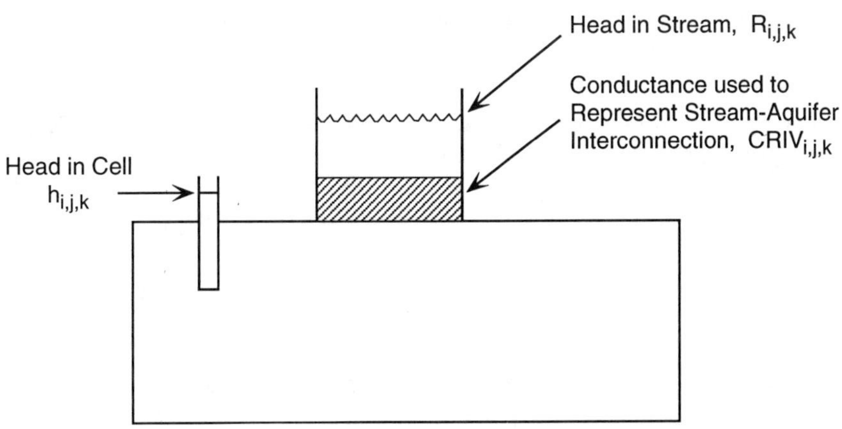

Figure 5 Conceptual representation of leakage through a riverbed into a cell.

The negative conductance term, $-CRIV_{i,j,k,2}$ corresponds to $p_{i,j,k,2}$ of equation (16), while the term $CRIV_{i,j,k,2}$ corresponds to $q_{i,j,k,2}$. Similarly, all other external sources or stresses can be represented by an expression of the form of equation (16). In general, if there are N external sources or stresses affecting the single cell, the combined flow is expressed by

$$QS_{i,j,k} = \sum_{n=1}^{N} a_{i,j,k,n} = \sum_{n=1}^{N} p_{i,j,k,n} \cdot h_{i,j,k} + \sum_{n=1}^{N} q_{i,j,k,n} \tag{20}$$

Defining $p_{i,j,k}$ and $q_{i,j,k}$, by the expressions

$$P_{i,j,k} = \sum_{n=1}^{N} p_{i,j,k,n} \quad \text{and}$$

$$Q_{i,j,k} = \sum_{n=1}^{N} q_{i,j,k,n}$$

the general external flow term for cell i, j, k is

$$QS_{i,j,k} = P_{i,j,k} \cdot h_{i,j,k} + Q_{i,j,k} \tag{21}$$

Applying the continuity equation (2) to cell i, j, k, taking into account the flows from the six adjacent cells, as well as the external flow rate, QS, yields

$$\begin{aligned} &q_{i,j-1/2,k} + q_{i,j+1/2,k} + q_{i-1/2,j,k} + q_{i+1/2,j,k} + q_{i,j,k-1/2} + q_{i,j,k+1/2} + QS_{i,j,k} \\ &= SS_{i,j,k} \cdot \frac{\Delta h_{i,j,k}}{\Delta t} \cdot \Delta r_j \Delta c_i \Delta v_k \end{aligned} \tag{22}$$

where

$\dfrac{\Delta h_{i,j,k}}{\Delta t}$ is a finite-difference approximation for the derivative of head with respect to time,

$SS_{i,j,k}$ represents the specific storage of cell i, j, k (L^{-1}); and $\Delta r_j \Delta c_i \Delta v_k$ is the volume of the cell i, j, k (L^3).

Equations (10) through (15) and (21) may be substituted into equation (22) to give the finite-difference approximation for cell i, j, k as

$$\begin{aligned} &CR_{i,j-1/2,k}\left(h_{i,j-1,k} - h_{i,j,k}\right) + CR_{i,j+1/2,k}\left(h_{i,j+1,k} - h_{i,j,k}\right) \\ &+ CC_{i-1/2,j,k}\left(h_{i-1,j,k} - h_{i,j,k}\right) + CC_{i+1/2,j,k}\left(h_{i+1,j,k} - h_{i,j,k}\right) \\ &+ CV_{i,j,k-1/2}\left(h_{i,j,k-1} - h_{i,j,k}\right) + CV_{i,j,k+1/2}\left(h_{i,j,k+1} - h_{i,j,k}\right) \\ &+ P_{i,j,k} \cdot h_{i,j,k} + Q_{i,j,k} = SS_{i,j,k}(\Delta r_j \Delta c_i \Delta v_k)\Delta h_{i,j,k} / \Delta t \end{aligned} \tag{23}$$

The finite difference approximation for the time derivative of head, $(h_{i,j,k})/\Delta t$ must next be expressed in terms of specific heads and times. Figure 6 shows a hydrograph of head values at node i, j, k. Two values of time are shown on the horizontal axis: t_m which is the time at which the flow terms of equation (23) are evaluated; and t_{m-1}, a time which precedes t_m. The head values at node i, j, k associated with these times are designated by superscript as $h^m_{i,j,k}$ and $h^{m-1}_{i,j,k}$ respectively. An approximation to the time derivative of head at time t_m is obtained by dividing the head difference $h^m_{i,j,k} - h^{m-1}_{i,j,k}$ by the time interval $t_m - t_{m-1}$ that is,

$$\left(\frac{\Delta h_{i,j,k}}{\Delta t}\right)_m = \frac{h^m_{i,j,k} - h^{m-1}_{i,j,k}}{t_m - t_{m-1}}$$

Thus the hydrograph slope, or time derivative, is approximated using the change in the head at the node over a time interval which precedes, and ends with, the time at which flow is evaluated. This is termed a backward-difference approach, in that $\Delta h/\Delta t$ is approximated over a time interval which extends backward in time from t_m, the time at which the flow terms are calculated. There are other ways in which $\Delta h/\Delta t$ could be approximated; for example, we could approximate it over a time interval which begins at the time of flow evaluation and extends to some later time; or over a time interval which is centered at the time of flow evaluation, extending both forward and backward from it. These alternatives, however, may cause numerical instability -- that is, the growth or propagation of error during the calculation of heads at successive times in a simulation. In an unstable situation, errors which enter the calculation for any reason at a particular time will increase at each succeeding time as the calculation progresses, until finally they

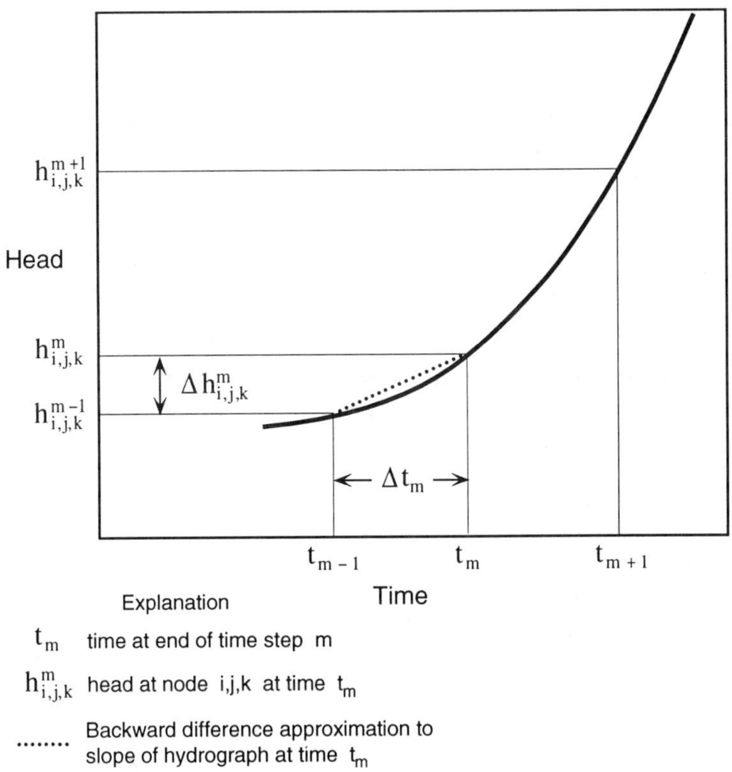

Figure 6 Hydrograph for cell i,j,k.

completely dominate the result. By contrast, the backward-difference approach is always numerically stable -- that is, errors introduced at any time diminish progressively at each succeeding time. For this reason, the backward-difference approach is preferred even though it leads to large systems of equations which must be solved simultaneously for each time at which heads are to be computed .

Equation (23) can be rewritten in backward-difference form by specifying flow terms at t_m, the end of the time interval, and approximating the time derivative of head over the interval t_{m-1} to t_m; that is:

$$\begin{aligned} & CR_{i,j-1/2,k}\left(h^m_{i,j-1,k} - h^m_{i,j,k}\right) + CR_{i,j+1/2,k}\left(h^m_{i,j+1,k} - h^m_{i,j,k}\right) \\ & + CC_{i-1/2,j,k}\left(h^m_{i-1,j,k} - h^m_{i,j,k}\right) + CC_{i+1/2,j,k}\left(h^m_{i+1,j,k} - h^m_{i,j,k}\right) \\ & + CV_{i,j,k-1/2}\left(h^m_{i,j,k-1} - h^m_{i,j,k}\right) + CV_{i,j,k+1/2}\left(h^m_{i,j,k+1} - h^m_{i,j,k}\right) \\ & + P_{i,j,k} \cdot h^m_{i,j,k} + Q_{i,j,k} = SS_{i,j,k} \cdot \frac{(\Delta r_j \Delta c_i \Delta v_k) \cdot (h^m_{i,j,k} - h^{m-1}_{i,j,k}}{t_m - t_{m-1}} \end{aligned} \tag{24}$$

Equation (24) is a backward-difference equation which can be used as the basis for simulation of the partial differential equation of ground water flow, equation (1). Like the term $Q_{i,j,k}$, the coefficients of various head terms in equation (24) are all known, as is the head a the beginning of the time step $h^{m-1}_{i,j,k}$. The seven heads at time t_m, the end of time step, are unknown; that is, they are part of the head distribution to be predicted. Thus equation (24) cannot be solved independently, since it represents a single equation in seven unknowns. However, an equation of this type can be written for each active cell in the mesh; and, since there is only one unknown head for each cell, we are left within a system of "n" equations in "n" unknowns. Such a system can be solved simultaneously.

The objective of transient simulation is generally to predict head distributions at successive times, given the initial head distribution, the boundary conditions, the hydraulic parameters and the external stresses. The initial head distribution provides value of $h^1_{i,j,k}$ at each point in the mesh -- that is, it provides the values of head at the beginning of the first of the discrete time steps into which the time axis is divided in the finite-difference process. The first step in the solution process is to calculate values of $h^2_{i,j,k}$ -- that is, heads at time t_2, which marks the end of the first time step. In equation (25), therefore, the head superscript m is taken as 2, while the superscript m−1, which appears in only one head term, is taken as 1. The equation therefore becomes

$$\begin{aligned} & CR_{i,j-1/2,k}\left(h^2_{i,j-1,k} - h^2_{i,j,k}\right) + CR_{i,j+1/2,k}\left(h^2_{i,j+1,k} - h^2_{i,j,k}\right) \\ & + CC_{i-1/2,j,k}\left(h^2_{i-1,j,k} - h^2_{i,j,k}\right) + CC_{i+1/2,j,k}\left(h^2_{i+1,j,k} - h^2_{i,j,k}\right) \\ & + CV_{i,j,k-1/2}\left(h^2_{i,j,k-1} - h^2_{i,j,k}\right) + CV_{i,j,k+1/2}\left(h^2_{i,j,k+1} - h^2_{i,j,k}\right) \\ & + P_{i,j,k} \cdot h^2_{i,j,k} + Q_{i,j,k} = SS_{i,j,k} \cdot \frac{(\Delta r_j \Delta c_i \Delta v_k) \cdot (h^2_{i,j,k} - h^1_{i,j,k}}{t_2 - t_1} \end{aligned} \tag{25}$$

where again the superscripts 1 and 2 refer to the time at which the heads are taken and should not be interpreted as exponents.

An equation of this form is written for every cell in the mesh in which head is free to vary with time (variable head cells), and the system of equations is solved simultaneously for the heads at time t_2. When these have been obtained, the process is repeated to obtain heads at time t_3, the end of the second time step. To do this, equation (25) is reapplied, now using 2 as time subscript m−1 and 3 as time subscript m. Again, a system of equations is formulated, where the unknowns are now the heads at time t_3; and this set of equations is solved simultaneously to obtain the head distribution at time t_3. This process is continued for as many steps as necessary to cover the time range of interest.

It is important to note that the set of finite-difference equations is reformulated at each time step; that is, at each step there is a new system of simultaneous equations to be solved. The heads at the end of the time step make up the unknowns for which this system must be solved; the heads at the beginning of the step are among the known terms in the equations. The solution process is repeated at each time step yielding a new array of heads for the end of the step.

ITERATION

The model described in this report utilizes iterative methods to obtain the solution to the system of finite-difference equations for each time step. In these methods, the calculation of head values for the end of a given time step is started by arbitrarily assigning a trial value, or estimate, for the head at each node at the end of that step. A procedure of calculation is then initiated which alters these estimated values , producing a new set of head values which are in closer agreement with the system of equations. These new, or interim, head values then take place of the initially assumed heads, and the procedure of calculation is repeated producing a third set of values. This procedure is repeated successively, at each stage producing a new set of interim heads which more nearly satisfies the system of equations. Each repetition of the calculation is termed an "iteration". Ultimately, as the interim heads approach values which would exactly satisfy the set of equations, the changes produced by succeeding stages of calculation become very small. This behavior is utilized in determining when to stop iteration, as discussed in a subsequent paragraph (author's remark: see pages 2–21 to 2–23 of the MODFLOW's manual for further explanation.)

RIVER PACKAGE

(excerpts from Chapter 6)

Conceptualization and Implementation

Rivers and streams contribute water to the ground-water system or drain water from it depending on the head gradient between the stream and the ground-water regime. The purpose of the River Package is to simulate the effects of the flow between surface-water features and ground-water systems. To accomplish this, terms representing seepage to or from the surface features must be added to the ground water flow equation (equation 26) for each cell affected by the seepage.

Figure 32 shows a stream divided into reaches so that each reach is completely contained in a single cell. Stream aquifer seepage is simulated between each reach and the model cell that contains that reach. The cross-section of figure 33-a shows a situation in which the open water of a stream is separated from the ground water system by a layer of low permeability streambed material. Figure 33-b shows an idealization of this system in which the stream-aquifer interconnection is represented as a simple conductance through which one-dimensional flow occurs. The system of figure 33 is helpful in conceptualizing

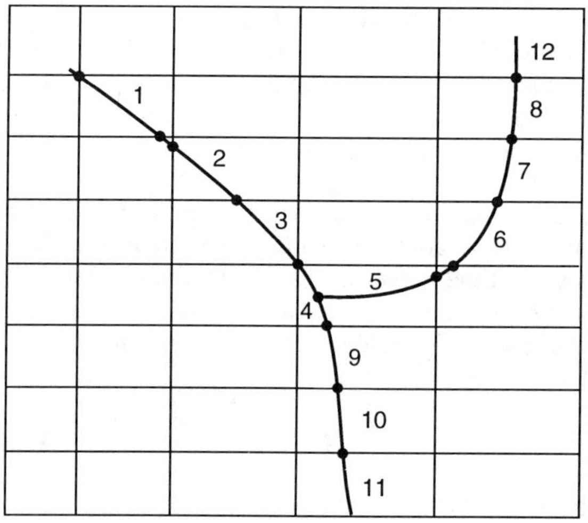

Figure 32 Discretization of a stream into reaches. Some small reaches are ignored.

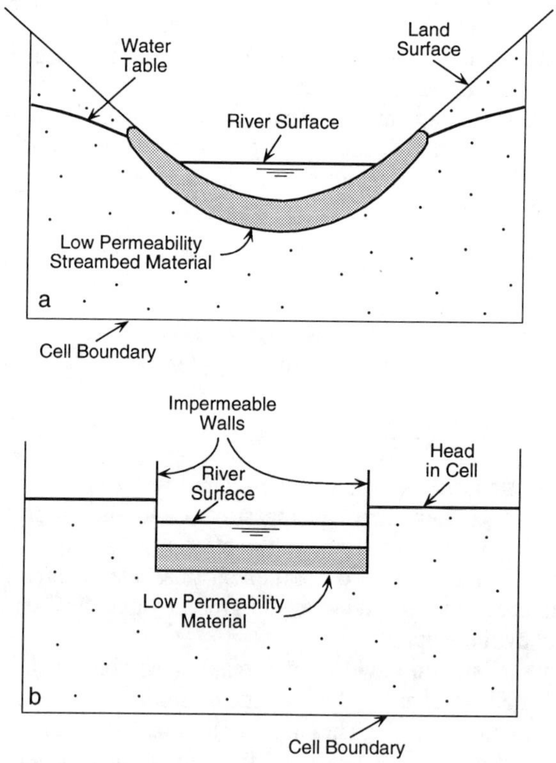

Figure 33 (a) Cross section of an aquifer containing a stream and (b) Conceptual representation of stream-aquifer interconnection in simulation.

and describing the simulation of stream-aquifer interaction; however, it must be recognized that, in many instances, no discrete low-permeability streambed is present. The techniques of simulation developed through the conceptualization of figure 33 can still be applied to represent these situations, provided the proper interpretation is placed on the various terms and parameters that are used.

Figure 34 shows an isolated view of the idealized streambed conductance of figure 33-b, as it crosses an individual cell. The length of the conductance block is taken as the length of the stream, L, as it crosses the node; the width is taken as the stream width, W; the distance of flow is taken as the thickness, M, of the streambed layer; and the hydraulic conductivity of the streambed material is designated K. The assumption is made that measurable head losses between the stream and the aquifer are limited to those across the streambed layer itself--that is, that there is no significant head loss between the bottom of the streambed layer and the point represented by the underlying model node. It is further assumed that the underlying model cell remains fully saturated-- that is, that its water level does not drop below the bottom of the streambed layer. Under these assumptions, flow between the stream and the ground-water system is given by

$$QRIV = \frac{KLW}{M} \cdot (HRIV - h_{i,j,k}) \qquad (63\text{-a})$$

or

$$QRIV = CRIV \cdot (HRIV - h_{i,j,k}) \qquad (63\text{-b})$$

where QRIV is the flow between the stream and the aquifer, taken as positive if it is directed into the aquifer; HRIV is the head in the stream; CRIV is the hydraulic conductance of the stream-aquifer interconnection (KLW/M) and $h_{i,j,k}$ is the head at the node in the cell underlying the stream reach.

If the assumption is satisfied that all significant head loss occurs across a discrete streambed layer, the application of equations (63) is straightforward. More frequently, however, equations (63) must be applied to situations in which no discrete streambed layer can be identified, or in which head losses are not restricted to those across such a layer.

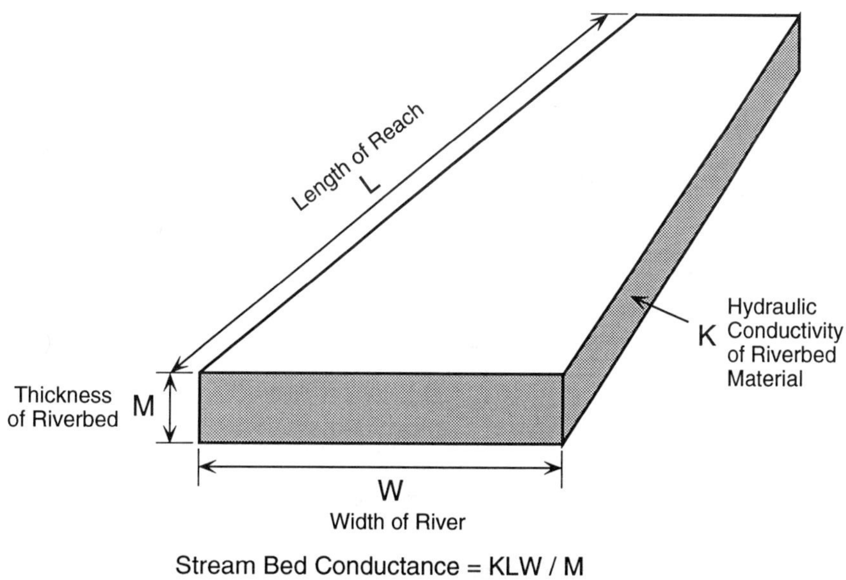

Figure 34 Idealization of streambed conductance in an individual cell.

In these cases, the task is to formulate a single conductance term, CRIV, which can be used in (63-b) to relate flow between the stream and the depth represented by the node i, j, k, to the corresponding head difference. This flow is in general a three-dimensional process, and its representation through single conductance term can never be more than approximate. If reliable field measurements of stream seepage and associated head difference are available, they may be used to calculate an effective conductance. Otherwise, a conductance value must be chosen more or less arbitrarily and adjusted during model calibration. Certain rules can be formulated to guide the initial choice of conductance. For example, the assumed cross-sectional area of flow should normally be of the same order of magnitude as the product of channel width and stream reach length within the cell; the assumed distance of flow should not exceed the vertical interval between the streambed and the node i,j,k; and, if distinct layers can be recognized within this interval, these should be normally treated as conductances in series in formulating an equivalent conductance. In general, however, it should be recognized that formulation of single conductance term to account for a three dimensional flow process is inherently an empirical exercise, and the adjustment during calibration is almost always required.

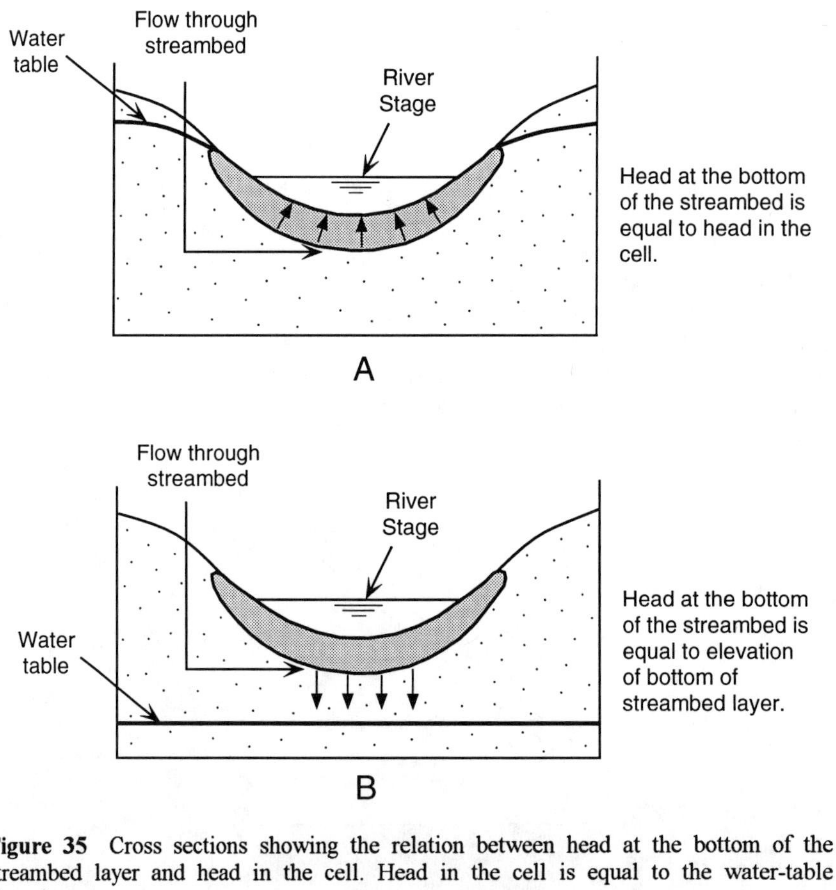

Figure 35 Cross sections showing the relation between head at the bottom of the streambed layer and head in the cell. Head in the cell is equal to the water-table elevation.

Equations (63) normally provide an acceptable approximation of stream-aquifer interaction over a certain range of aquifer head values. In most cases, however, if water levels in the aquifer fall below a certain point, seepage from the stream ceases to depend on the head in the aquifer. This can be visualized by returning to the concept of a discrete streambed layer. Figure 35-a shows the situation described by equations (63); water level in the aquifer is above the bottom of the streambed layer, and flow through that layer is proportional to the head difference between the stream and the aquifer. In figure 35-b, water level in the aquifer has fallen below the bottom of the streambed layer, leaving an unsaturated interval beneath that layer; if it is assumed that the streambed layer itself remains saturated, the head at its base will simply be the elevation at that point. If this elevation is designated RBOT, the flow through the streambed layer is given by

$$QRIV = CRIV \cdot (HRIV - RBOT) \tag{64}$$

where QRIV, CRIV and HRIV are defined for equation (63-b). Obviously, further declines in head below RBOT produce no increase in flow through the streambed layer; the flow simply retains the constant value given by equation (64), as long as head remains below RBOT. The model described herein utilizes these concepts in simulating stream-aquifer interaction--that is, flow beneath a stream and a node i,j,k is simulated according to the equation set

$$\begin{aligned} QRIV &= CRIV \cdot (HRIV - h_{i,j,k}), \quad h_{i,j,k} > RBOT \\ QRIV &= CRIV \cdot (HRIV - RBOT), \quad h_{i,j,k} \le RBOT \end{aligned} \tag{65}$$

Figure 36 shows a graph of flow between the stream and cell i,j,k as a function of the head $h_{i,j,k}$ as calculated using equations (65). Flow is zero when $h_{i,j,k}$ is equal to the water level in the stream, HRIV. For higher values of $h_{i,j,k}$, flow is negative, that is, into the stream; for lower values of $h_{i,j,k}$, flow is positive, that is, into the aquifer. This positive flow increases linearly as $h_{i,j,k}$ decreases, until $h_{i,j,k}$ reaches RBOT; thereafter the flow remains constant.

A relationship similar to that of equations (65) and figure 36 generally prevails in stream-aquifer interaction whether or not a discrete streambed layer is present. For example, once a break in saturation occurs between the stream and the aquifer, seepage

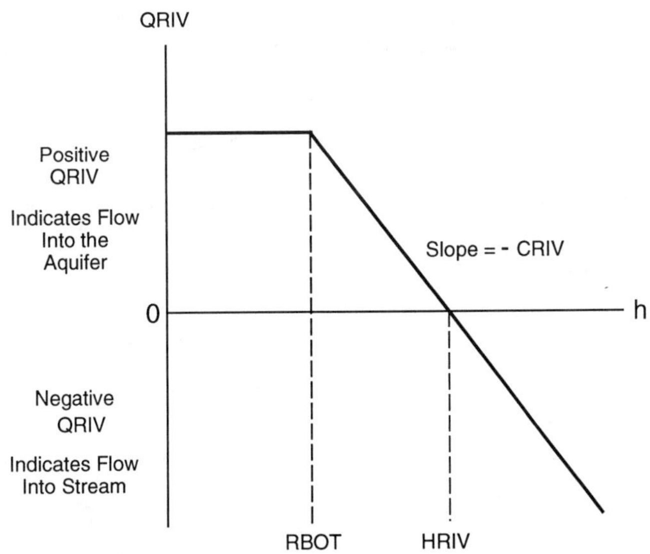

Figure 36 Plot of flow, QRIV, from a stream into a cell as a function of head, h, in the cell where RBOT is the elevation of the bottom of the streambed and HRIV is the head in the stream.

from the stream to the aquifer must become independent of head in the aquifer. In most cases, moreover, this independence is established even before a break in saturation occurs.

Figure 37 shows a situation in which levels in the aquifer have fallen far enough below a stream so that only a narrow saturated connection exists between the streambed and the regional water table. Examination on figure 37 will show that the head gradient in the saturated connection must be approximately unity, and that further lowering of the regional water table will not increase this gradient. Thus, once a condition similar to that in figure 37 is established, seepage from the stream is independent of the further head decline in the aquifer. The situation shown in figure 37 is itself an oversimplification of field conditions, which may often involve complex patterns of saturated and unsaturated material beneath the stream. In all situations, however, seepage from the stream must at same point become independent of head in the aquifer, as that head continues to decline.

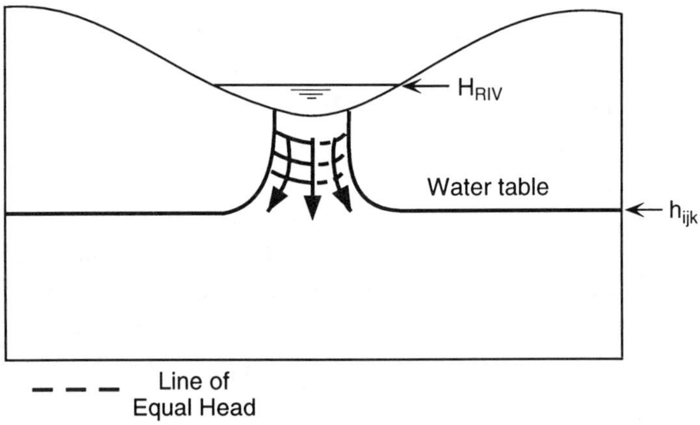

Figure 37 Limiting seepage from a stream at unit hydraulic gradient.

If hydrogeologic conditions indicate that seepage from a stream will increase as the local water table elevation declines, but will reach the limiting condition illustrated in figure 37 when the water table reaches an elevation h_l, RBOT should be taken equal to h_l. Because the vertical hydraulic gradient beneath the stream is approximately one under the conditions of figure 37, seepage from a stream into cell i,j,k is given approximately by the product KLW, where K is the vertical hydraulic conductivity of the material in the saturated column beneath the stream, and again L is the length of the stream as it crosses cell i, j, k, and W is stream width. A value of the stream-aquifer conductance term, CRIV, consistent with this estimated value of limiting seepage and with the selected value of RBOT can be obtained by substituting KLW for QRIV in equation (64) and solving for CRIV. This yields

$$CRIV = \frac{KLW}{HRIV - RBOT} \tag{66}$$

In summary, if the limiting condition of stream seepage is expected to follow the model of figure 37, RBOT should be chosen as the water table elevation at which the transition to this limiting seepage can be expected, and CRIV may be calculated from equation (66). The model simulation technique based on equations (65) should then provide a reasonable approximation to the stream-aquifer interaction.

The simplified model of stream-aquifer interaction utilized here assumes that this interaction is independent of the location of the stream reach within the cell, and that the level of water in the stream is uniform over the reach, and constant over each stress period. The latter assumption implies that conditions of flow in the stream do not vary significantly during the stress period -- for example, that the stream does not go dry or overflow its banks, or that such events are of such short duration as to have no effect on stream-aquifer interaction.

Data describing each river are specified by the user for each stress period. Input consists of six entries for each river reach, specifying the layer, row, and column of the cell containing the reach, and three parameters needed to calculate seepage--stream level or stage (HRIV), the conductance of the stream-aquifer interconnection (CRIV), and the "bottom elevation", or level at which the limiting value of stream seepage is attained (RBOT).

At the start of each iteration, terms representing river seepage are added to the flow equation for each cell containing a river reach .The choice of which river seepage equation to use, equation (63) or (64), is made by comparing the most recent value of head at the cell to the value of RBOT for the reach. Since this process is done at the start of each iteration, the most current value of head (HNEW) is the value from the previous iteration. If equation (63) is selected, the term –CRIV is added to the term HCOF and the term -CRIV*HRIV is added to RHS. If equation (64) is selected, the term –CRIV (HRIV-RBOT) is added to the term RHS.

RECHARGE PACKAGE

(excerpts from Chapter 7)

Conceptualization and Implementation

The Recharge (RCH) Package is designed to simulate areally distributed recharge to the groundwater system. Most commonly, areal recharge occurs as a result of precipitation that percolates to the ground-water system. Recharge applied to the model is defined as

$$Q_{Ri,j} = I_{i,j} * DELR_j * DELC_i \tag{67}$$

where $Q_{Ri,j}$ is the recharge flow rate applied to the model at horizontal cell location (i,j) expressed as a fluid volume per unit time; and $I_{i,j}$ is the recharge flux (in units of length per unit of time) applicable to the map area, $DELR_j$*$DELC_j$ of the cell. The recharge, $Q_{Ri,j}$, is applied to a single cell within the vertical column of cells located at (i,j). There is no need to allow for recharge to occur simultaneously at multiple depths in the same vertical column because natural recharge enters the groundwater system at its top. In the simplest situation, the top of the groundwater system will occur in model layer 1; however, the vertical position of the top of the system may vary with horizontal location and with time as the water-table rises and falls. Three options for specifying the cell in each vertical column of cells that receives the recharge have been implemented as described below. The RCH Package can potentially be used to simulate recharge from sources other than precipitation -- for example, artificial recharge. If the ability to apply recharge to more than one cell in a vertical column of cells is required, then the Well Package, which allows recharge or discharge to be specified at any model cell, can be used.

In the package described herein, values of recharge flux $I_{i,j}$, are read into a two dimensional array, $RECH_{i,j}$, at each stress period (unless an option is exercised to use recharge flux from the previous stress period). These values of recharge flux are immediately multiplied by horizontal cell areas, $DELR_jDELC_i$, to obtain values of $Q_{Ri,j}$, which are maintained in the RECH array. The cell within each vertical column to which the recharge is applied is specified through the recharge option code, NRCHOP, and optional array IRCH. The options include: (1) application of the recharge to model layer 1; (2) application of the recharge to any cell in the vertical column as specified by layer numbers contained in two dimensional array $IRCH_{i,j}$; and (3) application of the recharge to the uppermost active cell in the vertical column, provided there is no constant head cell above it in the column. Under options 1 and 2, if a cell designated to receive recharge is no-flow, then no recharge is added. Under the third option, if there is a constant head cell in a vertical column of cells and there is no active cell above, then no recharge is applied to this column because it is assumed that any recharge would be intercepted by the constant head source. Recharge flux values that are read into the model must be expressed in units that are consistent with the length and time units used to represent all other model parameters.

In the formation of the matrix equations, the recharge flow rate $Q_{R,i,j}$, associated with a given horizontal cell location (i,j) and vertical location, k, that is determined by the recharge option is subtracted from the value of $RHS_{i,j,k}$ (equation (26) or (29)). This is done at each iteration for all cells that receive recharge. Because recharge as defined is independent of aquifer head, nothing is added to the coefficient of head, $HCOF_{i,j,k}$.

Careful consideration should be given to the problem under study and to the other options employed in its simulation before deciding which of three recharge options listed above to utilize in a given situation. For example, figure 38 shows a situation in which a cross sectional model has been used to simulate a hypothetical problem involving recharge, seepage from a stream, and seepage into a stream (figure 38-a). Using the provision described in Chapter 5 for horizontal conductance formulation under water table conditions, the model mesh has been progressively truncated during simulation so that the uppermost active cells in each vertical column fall approximately at the water table. This process yields the final distribution of active, constant head and no flow cells shown in figure 38-b.

Figure 38-c illustrates the recharge distribution to the model if option 1 above is utilized. Under this option recharge is permitted only to the top layer of the model. Thus once, the water table shape has been simulated by the use of no flow cells in the top layer, recharge to the vertical columns beneath those cells is shut off. This clearly fails to simulate the given system.

Figure 38-d illustrates the recharge distribution if option 2 is utilized, assuming that the user specifies recharge cells prior to the simulation on the basis of an estimated water table position, which differs slightly from that finally obtained in the simulation process. Four of the cells which the user had designated as recharge cells have converted to an inactive condition and thus receive no recharge.

Figure 38-e illustrates the simulation under the third option, which turns out to be the one best suited for this particular situation. Under this option, recharge enters the uppermost active cell in each vertical column, except where constant head cells have been used to represent the streams. Thus a continuous distribution of recharge to the water table is simulated.

For the typical situation of recharge from precipitation, option 3 is the easiest to use. The model user does not have to be concerned about determining which is the highest active cell in a vertical column because the program automatically determines this throughout the simulation. Option 1, however, can be useful in situations where recharge should not pass through the no-flow cells in layer 1. For example, some cells may be designated no-flow because they are impermeable. Any recharge specified for those cells should not pass into layer 2. Of course option 3 could still be used in this situation by specifying that the recharge rate is zero at the impermeable cells. The user should select the option that will result in the least effort for specification of input data. Similarly, option 2 may be useful when layers other than layer 1 have outcrop areas and when recharge to the specified areas should not penetrate through no-flow cells to a lower layer. Other factors to consider when choosing the best option are that option 2 uses more computer memory than options 1 and 3, and option 3 uses slightly more computer time than options 1 and 2.

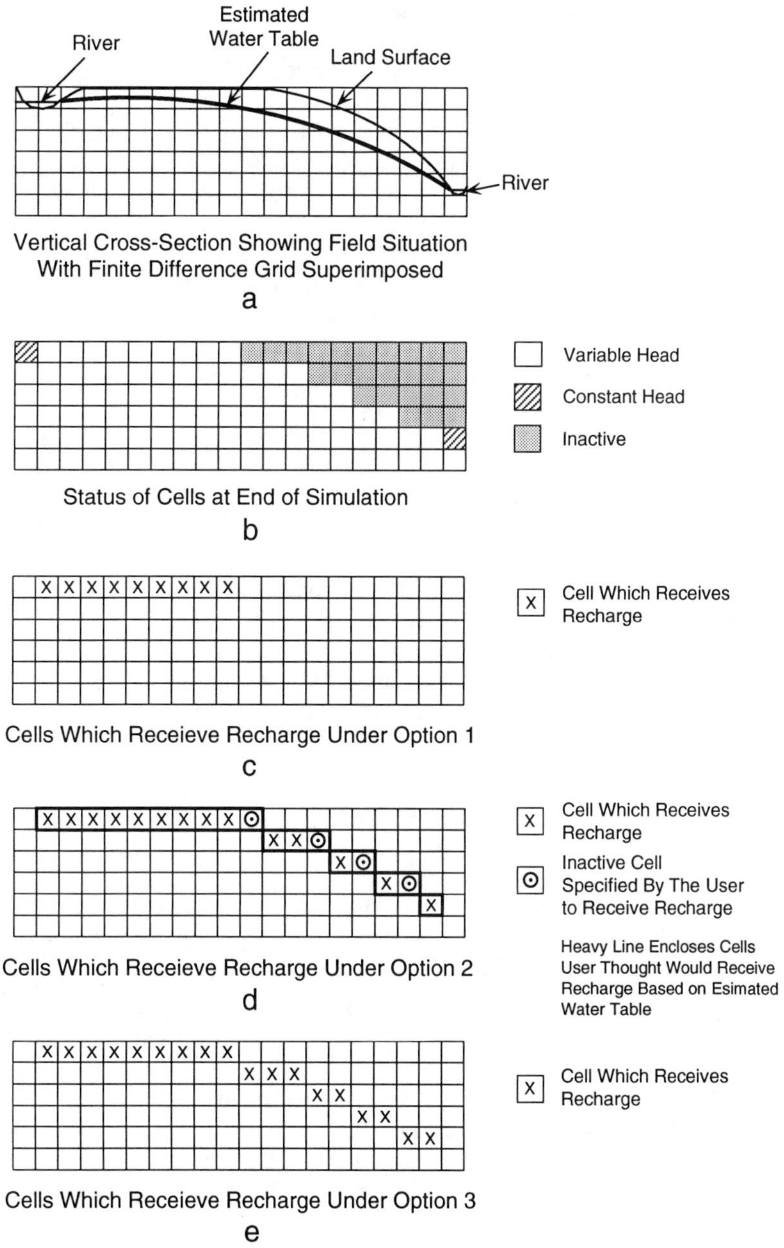

Figure 38 Hypothetical problem showing which cells receive recharge under the three options available in the Recharge Package.

WELL PACKAGE
(excerpts from Chapter 8)

Conceptualization and Implementation

The Well Package is designed to simulate features such as wells which withdraw water from the aquifer (or add water to it) at a specified rate during a given stress period, where the rate is independent of both the cell area and the head in the cell. The discussion in this section is developed on the assumption that the features to be simulated are actually wells, either discharging or recharging.

Well discharge is handled in the Well Package by specifying the rate, Q, at which each individual well adds water to the aquifer or removes water from it, during each stress period of the simulation. Negative values of Q are used to indicate well discharge, while positive values of Q indicate a recharging well.

At the beginning of each stress period, the WEL1RP module reads four values for each well--the row, column and layer number of the cell in which the well is located, and the discharge or recharge rate, Q, of the well during that stress period. At each iteration, as the matrix equations are formulated, the value of Q for each well is subtracted from the RHS value (equation (26) or (29)) for the cell containing that well. Where more than one well falls within a single cell, the calculation is repeated for each well as the RHS term for that cell is assembled. Thus the user specifies the discharge associated with each individual well, and these are in effect summed within the program to obtain the total discharge from the cell.

The Well Package, as it is presently formulated, does not accommodate wells which are open to more than one layer of the model. However, a well of this type can be represented as a group of single-layer wells, each open to one of the layers tapped by the multi-layer well, and each having an individual Q term specified for each stress period. If this approach is used, the discharge of the multi-layer well must be divided or apportioned in some way among the individual layers, externally to the model program. A common method of doing this is to divide the well discharge in proportion to the layer transmissivities *i.e.*,

$$\frac{Q_l}{Q_w} = \frac{T_l}{\Sigma T} \tag{68}$$

where Q_l is the discharge from layer l to a particular well in a given stress period, Q_w is the well discharge in that stress period, T_l is the transmissivity of layer l and ΣT represents the sum of the transmissivities of all layers penetrated by the well. Again, it is important to note that equation (68), or some other method of apportioning the discharge, must be implemented by the user externally to the program for each multi-layer well, and for each stress period.

This approach, in which a multi-layer well is represented as a group of single layer wells, fails to take into account the interconnection between various layers provided by the well itself, and is thus an incomplete representation of the problem. A package which will provide an improved approximation of multi-layer well effects is under development.

DRAIN PACKAGE

(excerpts from Chapter 9)

Conceptualization and Implementation

The Drain Package is designed to simulate the effects of features such as agricultural drains, which remove water from the aquifer at a rate proportional to the difference between the head in the aquifer and some fixed head or elevation, so long as the head in the aquifer is above that elevation, but which have no effect if head falls below that level. The discussion in this section is phrased on the assumption that the features to be simulated are actually agricultural drains.

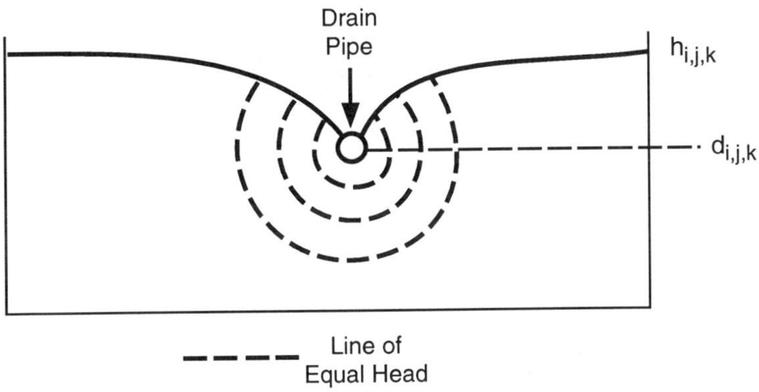

Figure 39 Cross section through cell i,j,k illustrating head loss in convergent flow into drain.

Figure 39 shows a cross section through a cell, illustrating concepts underlying the simulation of drains in the model. The drain is assumed to run only partially full, so that head within the drain is approximately equal to the median drain elevation, $d_{i,,j,k}$. The head computed by the model for cell i,j,k ($h_{i,j,k}$) is actually an average value for the cell, and is normally assumed to prevail at some distance from the drain itself. The drain head, $d_{i,j,k}$ prevails only locally, within the drain -- it does not characterize the cell as a whole. Between the drain and the area in which head $h_{i,j,k}$ prevails there exists a radial or semiradial flow pattern in the vertical plane, normally characterized by progressively steeper head gradients as the drain is approached. The head loss within this converging flow pattern forms one part of the head difference $h_{i,j,k}$-$d_{i,j,k}$. An additional component of head loss may occur in the immediate vicinity of the drain if the hydraulic conductivity in that region differs from the average value used for cell i,j,k -- because of the presence of foreign material around the drain pipe, or channel-bed material in the case of an open drain (figure 40). Finally, head losses occur through the wall of a drain pipe, depending upon the number and size of the openings in the pipe, and the degree to which those openings may be blocked by chemical precipitates, plant roots, etc.

The three processes discussed above -- convergent flow toward the drain, flow through material of different conductivity immediately around the drain, and flow through the wall of the drain -- each generate head losses which may be assumed proportional to the discharge, QD, through the system -- that is, the discharge from cell i,j,k into the drain. Because these head losses occur in series, the total head loss $h_{i,j,k}$ - $d_{i,j,k}$ may also be taken as proportional to QD. This has been done in the method of simulation embodied in the Drain Package. That is, it has been assumed that the drain function is described by the equation pair

$$QD_{i,j,k} = CD_{i,j,k} \cdot (h_{i,j,k} - d_{i,j,k}) \quad \text{for} \quad h_{i,j,k} > d_{i,j,k} \tag{69}$$

$$QD_{i,j,k} = 0 \quad \text{for} \quad h_{i,j,k} \leq d_{i,j,k} \tag{70}$$

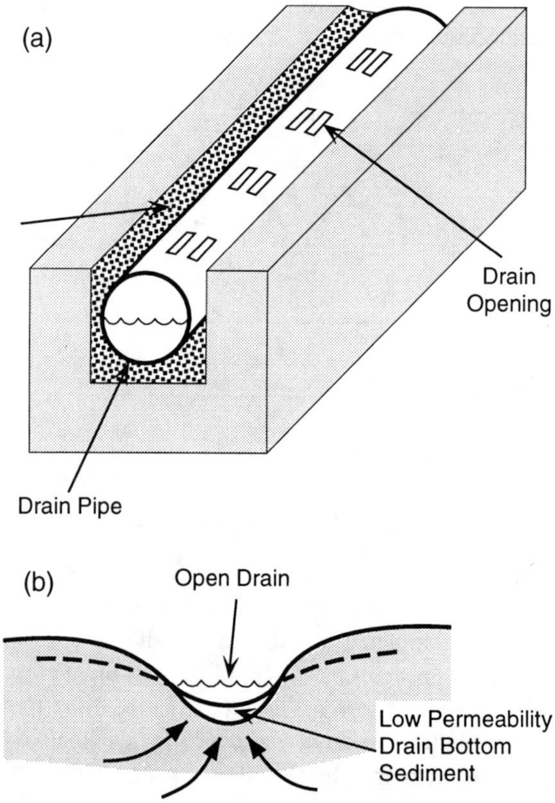

Figure 40 Factors affecting head loss immediately around a drain: (a) buried drain pipe in backfilled ditch and (b) open drain.

The coefficient $CD_{i,j,k}$ of equation (69) is a lumped (or equivalent) conductance describing all of the head loss between the drain and the region of cell i,j,k in which the head $h_{i,j,k}$ can be the assumed to prevail. It depends on the characteristics of the convergent flow pattern toward the drain, as well as on the characteristics of the drain itself and its immediate environment.

One could attempt to calculate values for CD by developing approximate equations for conductance for the three flow processes, and then calculate the equivalent series conductance. The conductance for each process would be based on the formulation of a one-dimensional flow equation. The formulations vary significantly depending on the specific drain system being simulated, so no general formulation for calculating CD is presented here. Also, in most situations a specific formulation would require detailed information that is not usually available, such as detailed head distribution around the drain, aquifer hydraulic conductivity near the drain, distribution of the fill material, hydraulic conductivity of fill material, number and size of the drain pipe openings, the amount of clogging materials, and the hydraulic conductivity of the clogging materials. In practice, it is more common to calculate CD from measured values of QD and h-d using equation (69). If h-d is not accurately known, CD is usually adjusted during model calibration in order to match measured values of QD to model calculated values.

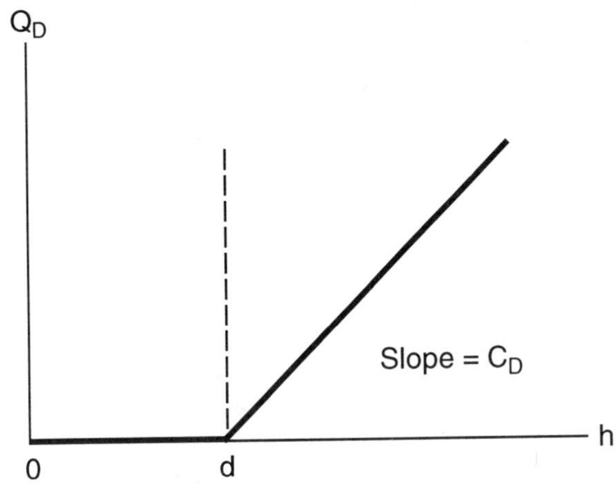

Figure 41 Plot of flow, QD, into a drain as a function of head, h, in a cell where the elevation of the drain is d and the conductance is CD.

Figure 41 shows a graph of QD vs. $h_{i,j,k}$ as defined by equations (69) and (70); the function is similar to that for flow between a surface stream and the aquifer (figure 36) except that flow into the aquifer is excluded, and positive values of QD have been taken as corresponding to flow into the drain. With proper selection of coefficients, the River Package could in fact be utilized to perform the functions of the Drain Package.

Because $QD_{i,j,k}$ in equation (69) has been taken as a flow out of cell i,j,k, it must be subtracted from the left side of equation (24) for each cell affected by a drain, provided the head $h_{i,j,k}$ is above the drain elevation. This is accomplished in the Drain Package by testing to determine whether head exceeds drain elevation, and if so, by adding the term $-CD_{i,j,k}$ to $HCOF_{i,j,k}$ (equation (26)) and adding the term $-CD_{i,j,k}d_{i,j,k}$ to $RHS_{i,j,k}$, as the matrix equations are assembled.

EVAPOTRANSPIRATION PACKAGE
(excerpts from Chapter 10)

Conceptualization and Implementation

The Evapotranspiration (ET) Package simulates the effects of plant transpiration and direct evaporation in removing water from the saturated ground water regime. The approach is based on the following assumptions: (1) when the water table is at or above a specified elevation, termed the "ET surface" in this report, evapotranspiration loss from the water table occurs at a maximum rate specified by the user; (2) when the depth of the water table below the ET surface elevation exceeds a specified interval, termed the "extinction depth" or "cutoff depth" in this report, evapotranspiration from the water table ceases; and (3) between these limits, evapotranspiration from the water table varies linearly with water table elevation.

This can be expressed in equation form as

$$R_{ETi,j} = R_{ETMi,j} \qquad h_{i,j,k} > h_{si,j} \tag{71}$$

$$R_{ETi,j} = 0 \qquad h_{i,j,k} < h_{si,j} - d_{i,j} \tag{72}$$

$$R_{ETi,j} = R_{ETMi,j} \cdot \left(\frac{h_{i,j,k} - (h_{si,j} - d_{i,j})}{d_{i,j}}\right) \qquad (h_{si,j} - d_{i,j}) \le h_{i,j,k} \le h_{si,j} \tag{73}$$

where $R_{ETi,j}$ is the rate of loss per unit surface area of water table due to evapotranspiration, in volume of water per unit area per unit time, within the map area $DELR_jDELC_i$; $h_{i,j,k}$ is the head, or water table elevation in the cell from which the evapotranspiration occurs; $R_{ETMi,j}$ is the maximum possible value of $R_{ETi,j}$; $h_{si,j}$ is the ET surface elevation, or the water table elevation at which this maximum value of evapotranspiration loss occurs; and $d_{i,j}$ is the cutoff or extinction depth, such that when the distance between $h_{si,j}$ and $h_{i,j,k}$ exceeds $d_{i,j}$ evapotranspiration ceases.

In implementing the finite difference approach the volumetric rate of evapotranspiration loss from a given cell is required. This is given as the product of the loss rate per unit area, and the horizontal surface area, $DELR_jDELC_i$ of the cell from which the loss occurs, *i.e.*,

$$Q_{ETi,j} = R_{ETi,j} * DELR_j * DELC_i \tag{74}$$

where $Q_{ETi,j}$ is the evapotranspiration, in volume of water per unit time, through the area $DELR_jDELC_i$. If the maximum value of $Q_{ETi,j}$ (corresponding to $R_{ETi\,,j}$) is designated $Q_{ETi,j}$, equations (71) – (73) can be expressed in terms of volumetric discharge as

$$Q_{ETi,j} = Q_{ETMi,j} \qquad h_{i,j,k} > h_{si,j} \tag{75}$$

$$Q_{ETi,j} = 0 \qquad h_{i,j,k} < h_{si,j} - d_{i,j} \tag{76}$$

$$Q_{ETi,j} = Q_{ETMi,j} \cdot \left\{\frac{h_{i,j,k} - (h_{si,j} - d_{i,j})}{d_{i,j}}\right\} \qquad (h_{si,j} - d_{i,j}) \le h_{i,j,k} \le h_{si,j} \tag{77}$$

Figure 42 shows a graph of evapotranspiration loss, $Q_{ETi,j}$, vs. head in cell i,j,k based on equations (75) – (77). Comparison of the ET function with the river or drain functions shows that the three are mathematically similar, except that the linear portion of the ET function is bounded at both ends by constant values, rather than only at the lower end.

Evapotranspiration is drawn from only one cell in the vertical column beneath the map area $DELR_j*DELC_i$; the user designates the cell (*i.e.*, the layer, k) using one of two options. Under the first option, evapotranspiration is always drawn from the uppermost layer of the model; under the second, the user specifies the cell, within the vertical column at i,j from which the evapotranspiration is to be taken. In either case the computed evapotranspiration has no influence on the simulation if the designated cell is either a no-flow cell or a constant head cell.

For each location, (i,j), in the horizontal plane, and for each stress period (unless an option is exercised to use prior values) the ET package reads values of R_{ETM} (maximum evapotranspiration loss per unit time) into an array labeled EVTR. These rates are immediately multiplied by cell areas, $DELR_j*DELC_i$, to obtain the maximum volumetric

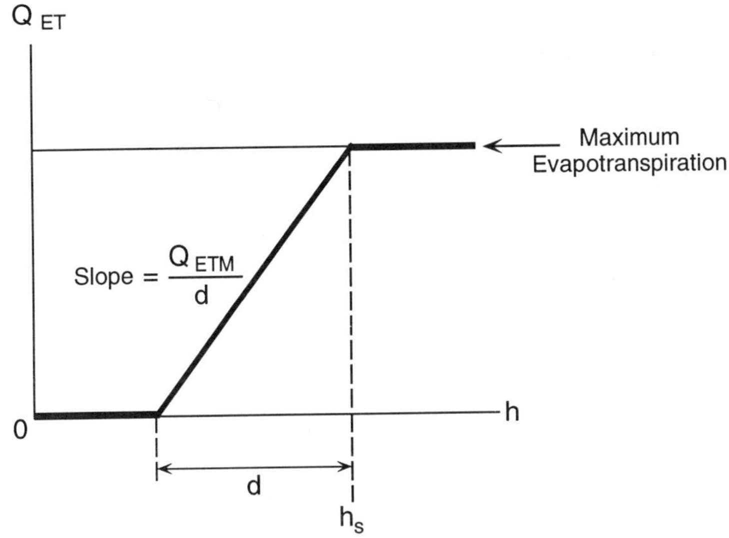

Figure 42 Plot of volumetric evapotranspiration, Q_{ET}, as a function of head, h, in a cell where d is the cutoff depth and h_s is the ET surface elevation.

rate of evapotranspiration from each cell, Q_{ETM}; these maximum volumetric rates then replace the values of $R_{ETMi,j}$ in the EVTR array. Thus, the input to the EVTR array consists of maximum evapotranspiration rates per unit area, and as such must have dimensions Lt^{-1}. In the calculation carried out within the model, however, the entries in the EVTR array appear as maximum volumetric rates, having dimensions L^3t^{-1}.

Values of $h_{si,j}$, the ET surface elevation (or water table elevation at which evapotranspiration is maximum), are read into the two-dimensional array SURF by the ET package; values of the cutoff depth or extinction depth are read into the two-dimensional array EXDP. Because the term $Q_{ETi,j}$ of equations (75) – (77) has been defined as an outflow from the aquifer it must be subtracted from the left side of equation (24). In terms of the expressions HCOF and RHS of equation (26), this is accomplished in the ET package as follows:

1) f $h_{i,j,k} < (h_{si,j,k} - d_{i,j})$ no changes are made in the terms HCOF or RHS for cell i,j,k;

2) if $h_{i,j,k} > h_{si,j}$, $Q_{ETMi,j}$ is added to $RHS_{i,j}$; and

3) if $(h_{si,j,k} - d_{i,j}) \leq h_{i,j,k} \leq h_{si,j,k}$, $- Q_{ETMi,j}/d_{i,j}$ is added to $HCOF_{i,j,k}$ and $-Q_{ETMi,j}\{(h_{si,j} - d_{i,j})/d_{i,j}\}$ is added to $RHS_{i,j,k}$.

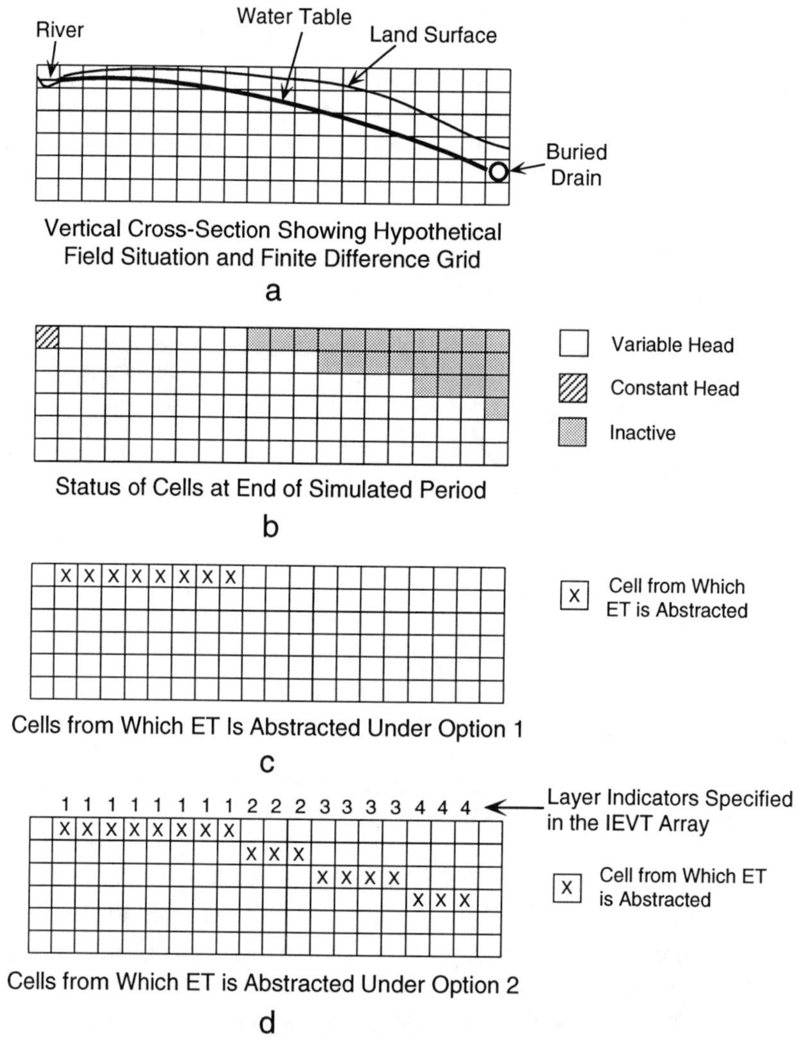

Figure 43 Hypothetical problem showing cells from which ET will be abstracted under the two options available in the ET package.

The value of $h_{s,i,j}$, the water table elevation at which evapotranspiration is maximum, should normally be taken as the average land surface elevation in the map area $DELR_j$ $DELC_i$; the cutoff or extinction depth, $d_{i,j}$,is then frequently assumed to be on the order of six to eight feet below land surface (although considerable variation can be introduced by climatic factors, the presence of deep-rooted phreatophytes, or so on). Where the distance from land surface to water table varies extensively within the area of a cell, care must be exercised in implementing the ET package and in choosing the various parameters of equation (70), or misleading results may be obtained.

The options for selection of the layer from which ET is to be drawn provide some flexibility in adapting the package to special situations, but also require some care in implementation. Figure 43 shows a situation similar to that discussed for the recharge package, in which a cross sectional model has been progressively truncated to follow the water table, using the provision for horizontal conductance formulation under water table conditions (Chapter 5). Figure 43-a shows the hydrologic situation under study, and figure 43-b the final distribution of variable head and inactive (no flow) cells obtained in the simulation.

Under option 1 (figure 43-c), evapotranspiration is drawn only from the uppermost layer of the model; in the problem shown, the presence of no flow cells in this layer deletes evapotranspiration from the right half of the model, so that the simulation fails to represent field conditions.

Figure 43-d shows the situation which could be achieved through the use of option 2, assuming that simulation was carried out in stages and that the user interacted with the simulation process, designating the cells from which evapotranspiration was to be drawn as the truncation of the mesh developed.

GENERAL-HEAD BOUNDARY PACKAGE
(excerpts from Chapter 11)

Conceptualization and Implementation

The function of the General-Head Boundary (GHB) Package is mathematically similar to that of the River, Drain and ET Packages, in that flow into or out of a cell, i,j,k, from an external source is provided in proportion to the difference between the head in the cell, $h_{i,j,k}$, and the head assigned to the external source, $h_{bi,j,k}$. Thus a linear relationship between flow into the cell and head in the cell is established, *i.e.*,

$$Q_{bi,j,k} = C_{bi,j,k} \cdot (h_{bi,j,k} - h_{i,j,k}) \tag{78}$$

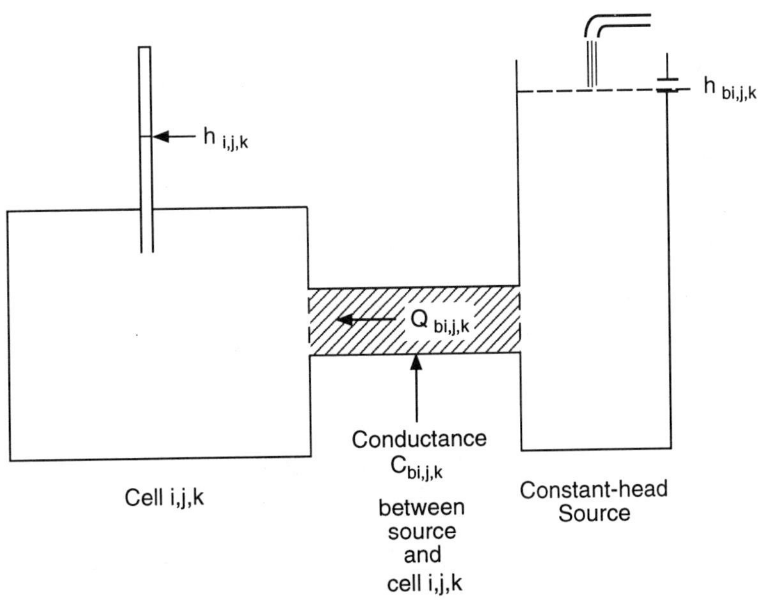

Figure 44 Schematic diagram illustrating principle of general-head boundary package.

where $Q_{bi,j,k}$ is the flow into cell i,j,k from the source; $C_{bi,j,k}$ is the conductance between the external source and cell i,j,k; $h_{bi,j,k}$ is the head assigned to the external source; and $h_{i,j,k}$ is the head in cell i,j,k. The relationship between cell i,j,k and the external source is shown schematically in figure 44. The constant-head source is represented by the apparatus on the right in figure 44, which holds the source head at the level h_b regardless of other factors; the link between the source and cell i,j,k is represented by the block of porous material $C_{bi,j,k}$. Note that figure 44 shows no mechanism to limit flow in either direction as $h_{i,j,k}$ rises or falls.

A graph of $Q_{bi,j,k}$ versus $h_{i,j,k}$ as given by equation (78) is shown in figure 45. In contrast to the River, Drain and ET Packages, the GHB Package provides no limiting value of flow to bound the linear function in either direction; and as the head difference between cell i,j,k and the source increases, flow into or out of the cell continues to increase without limit. Care must accordingly be used in utilizing the GHB Package to insure that unrealistic flows into or out of the system do not develop during the course of simulation. Because $Q_{bi,j,k}$ of equation (78) is defined as an inflow to the aquifer it must be added to the left side of equation (24). In terms of the expressions HCOF and RHS of equation (26), this is accomplished in the model by subtracting the term $C_{bi,j,k}$ from $HCOF_{i,j,k}$ and subtracting the term $C_{bi,j,k}h_{bi,j,k}$ from $RHS_{i,j,k}$ as the matrix equations are assembled.

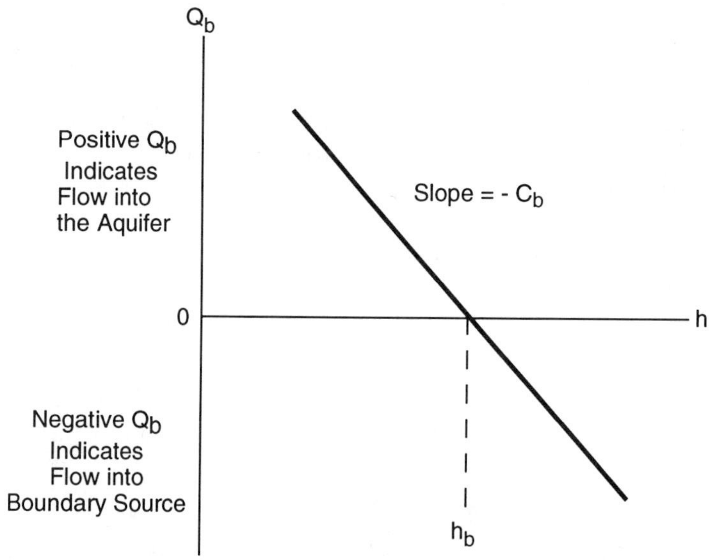

Figure 45 Plot of flow, Q_b, from a general-head boundary source into a cell as a function of head, h, in the cell where h_b is the source head.

STREAM PACKAGE
(excerpts from Prudic, 1989)

CONCEPTUALIZATION OF ROUTING STREAMFLOWS

The Streamflow-Routing Package is a modification of the River Package described by McDonald and Harbaugh (1988, p. 6-1 through 6-30) and is designed to be incorporated into their program. The new package is designed to route flow through one or more rivers, streams, canals, or ditches (all of which are hereafter referred to as streams, for convenience) in addition to computing leakage between the streams and the aquifer system.

Numbering and Ordering of Streams

A brief description of the model developed by McDonald and Harbaugh (1988) is necessary prior to describing the numbering and ordering of ground-water flow equation using finite-difference approximations (p. 2-1 through 2-35). The finite-difference procedure requires that the aquifer be divided into blocks called cells, which have dimensions x, y, and z. The aquifer properties in each cell are assumed uniform. The unknown head in each cell is calculated at a point or node at the center of the cell. The head is calculated by iterating through the finite-difference equations for all nodes until the maximum head change in any cell between the previous iteration and the current iteration is less than a value specified by the user. Once this criterion is met, the program advances to a new time step and the process of computing heads at each node is repeated.

Streams superimposed on the aquifer are divided into reaches and segments. A segment is a stream or diversion in which streamflow from surface sources are added at the beginning of the segment or subtracted (in the case of a diversion) at the end of a segment. A reach is the part of a segment that corresponds to an individual cell in the finite-difference grid used to simulate ground-water flow in an aquifer. A segment may consist of one or more reaches. Segments are numbered sequentially from the farthest segment. Both must also be read into the program in sequential order. The use of segments in the conceptualization of routing streamflows makes the accounting of flows where streams merge to form one, or where diversions subtract flow from a stream, easier to program. Reaches, however, form the basis for the calculation of leakage between surface water and ground water.

An example of a finite-difference grid of a hypothetical aquifer connected to streams is shown in figure 1a. In this example, the locations of cells representing the aquifer are described in terms of rows and columns. Thus, cell (1,3) is the cell located in row 1, column 3. The streams are divided into segments and reaches. Segment 1 has two reaches and begins in cell (1,3) and ends in cell (2,3) where some flow is diverted into segment 2. The remaining flow from segment 1 goes to segment 3, which has 4 reaches. Segments 2 and 4 join in cell (5,3) to form segment 5. Segments 3, 5, and 6 join in cell (5,4) to form segment 7. Two small parts of streams in segments 3 and 7 are not included in the numbering scheme because they only cross the corners of cells (2,4) and (3,5).

A diversion from one segment can later become tributary to another segment, which is the case for segment 2 in figure 1a. Also, more than one reach in different segments can be assigned to the same model cell. For example, reach 2 in segment 5; reach 4 in segment 3; reach 5 in segment 6, and reach 1 in segment 7 are all located in cell (5,4). Leakage to or from the aquifer from all these reaches is the accumulated sum of all the leakage, however, the leakage between each reach and the cell is calculated at the cell node (located in the center of the cell). Thus, only the head at the cell node is used to calculate leakage for each reach.

An alternative numbering scheme is shown in figure 1b whereby only one reach and segment is assigned to a cell. In the example, segment 2 (the diversion canal) begins in cell (3,3) and segment 3 begins in cell (3,4). Segment 5 has only one reach, located in cell (5,3), but it consists of parts of three streams. This scheme results in assigning fewer reaches but causes the combining of flows prior to the confluence of streams and the subtraction of flows after the diversion.

Streamflow Accounting

Streamflow is accounted for by specifying flow for the first reach in each segment that enters the modeled area, and then computing streamflow to adjacent downstream reaches

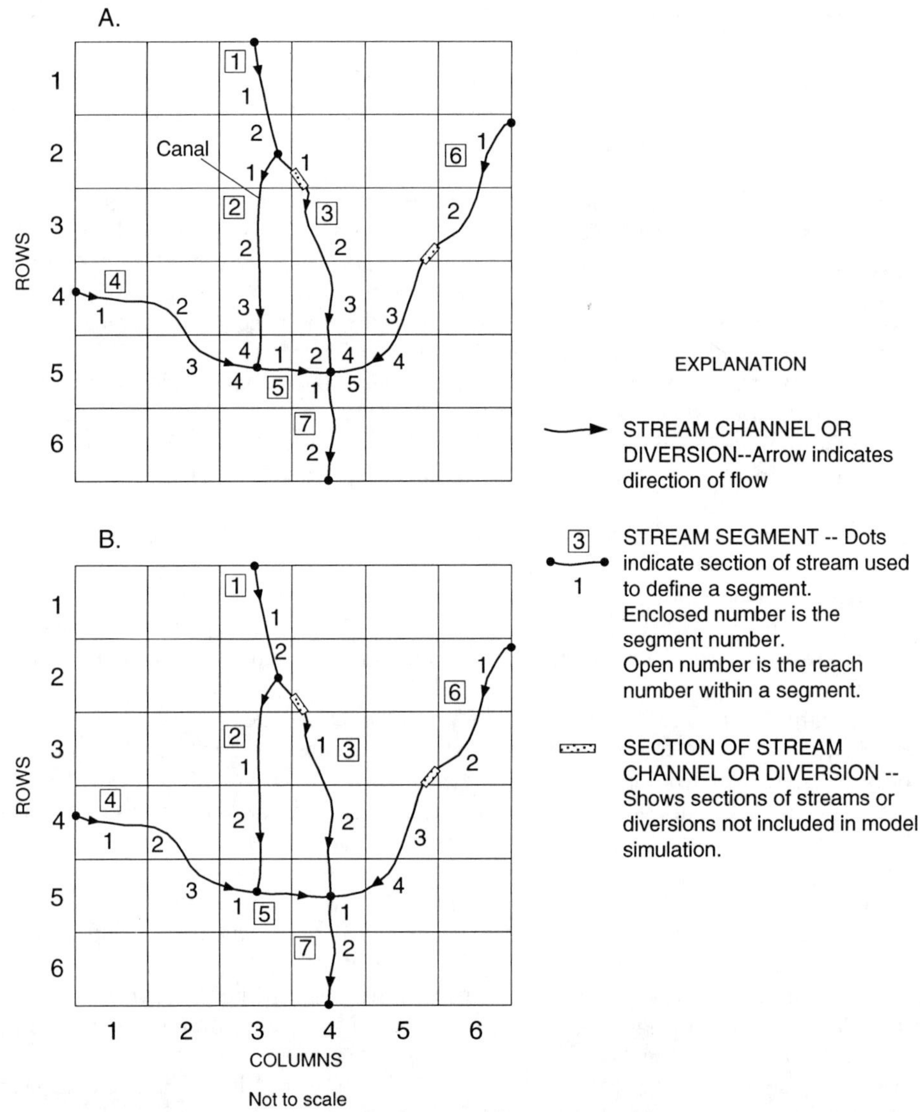

Figure 1 Example numbering systems of streams and diversions for the simulation of stream-aquifer relations with the finite-difference ground-water flow model by McDonald and Harbaugh (1988). A, one method of numbering streams and diversions. B, an alternative method of numbering streams and diversions.

in each segment as equal to inflow in the upstream reach plus or minus leakage from or to the aquifer in the upstream reach. The accounting scheme used assumes that streamflow entering the modeled layer is instantly available to downstream reaches. This assumption is generally reasonable because of the relatively slow rates of ground-water flow.

Streamflow into a segment that is formed from tributary streams is computed by adding the outflows from the last reach in each of the specified tributary segments. If a segment is a diversion, then the specified flow into the first reach of the segment is subtracted form flow in the main stream. However, if the specified flow of the diversion is greater than the flow out of the segment from which flow is to be diverted, then no flow is diverted from the segment.

Reaches in a segment are permitted to go dry whenever downward leakage to the aquifer exceeds stream inflow into the reach. The downstream reaches also go dry (streamflow into the downstream reaches is zero) and leakage into the aquifer is not permitted. Upward leakage (from the aquifer) is permitted even when there is no streamflow in a reach. This happens when the head in the aquifer exceeds the top of the streambed in a dry reach. Flow in the stream resumes and downward leakage back into the aquifer can be simulated in downstream reaches.

Computing Flow Between Stream and Aquifer

Leakage to or from a stream reach is computed by Darcy's law as follows:

$$Q_l = CSTR \cdot (H_s - H_a) \tag{1}$$

where

Q_l leakage to or from the aquifer through the streambed, (L^3/T);

H_s head in the stream, (L);

H_a head in aquifer side of streambed, (L); and

CSTR conductance of the streambed (L^2/T), which is the hydraulic conductivity of the streambed times the product of the width of the stream and its length divided by the thickness of the streambed.

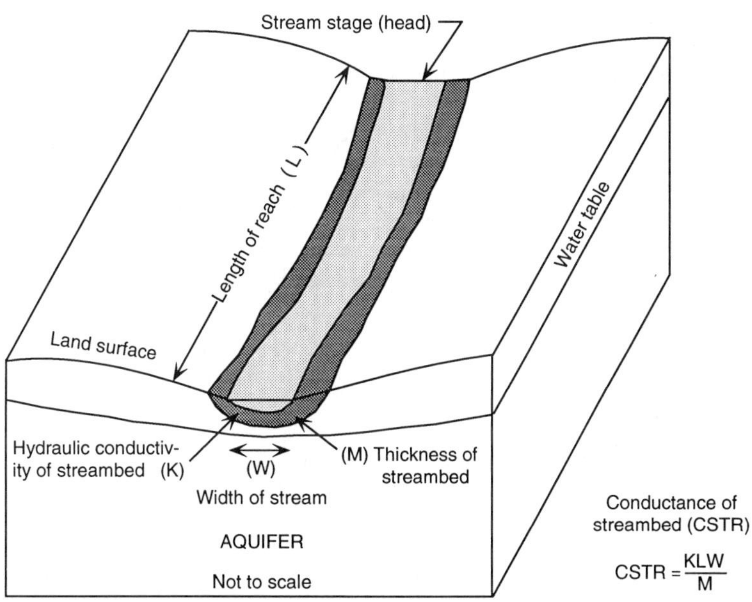

Figure 2 Diagram showing part of an aquifer with a stream over it depicting how conductance of the streambed is calculated for a reach.

A sketch of the properties used in the calculation of stream leakage to or from the aquifer is shown in figure 2. The stage in the stream is for the center of the reach. The head in the aquifer side of the streambed depends upon whether the earth materials beneath the stream are saturated or not. If the materials are saturated, then the head in the aquifer side of the streambed is equal to the head in the model cell beneath the stream reach. If materials are unsaturated, then the head in the aquifer beneath the streambed is equal to the elevation of the bottom of the streambed. The relation between stream leakage and head in the model cell is shown in figure 3. For more detailed information regarding this relation, the reader is referred to McDonald and Harbaugh (1988, p. 6-1 through 6-30). If leakage through the streambed to the aquifer in a stream reach is greater than the amount of streamflow that enters the reach, then the leakage is set equal to the streamflow that enters the reach and flow out of the reach is zero. Inflow into the adjacent downstream reach is also zero.

The terms that represent leakage to or from the stream reaches are added to the finite-difference equations at the start of each iteration in the same manner as in the River Package described by McDonald and Harbaugh (1988, p. 6–12). If streamflow is zero in a reach and the head in the model cell is less than the top of the streambed, then no leakage is computed for that reach. The choice as to which terms to add to the finite-difference equations is made by comparing the most recent value of the head in the model cell to the elevation of the bottom of the streambed; or if flow in the stream reach is zero, to the top of the streambed. These comparisons are made at the start of each iteration, meaning that the most current value of head in the model cell used to calculate leakage between stream reaches and the aquifer is from the previous iteration.

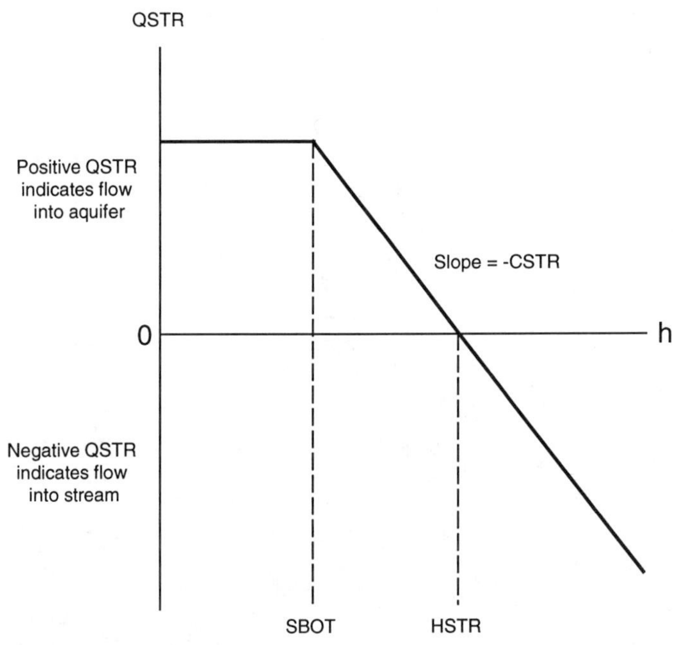

Figure 3 Leakage, QTSR, through a streambed into an aquifer as a function of head, h, in a model cell where SBOT is the streambed bottom and HSTR is the stream stage. Slope is dependent on streambed conductance, CSTR. Modified from McDonald and Harbaugh (1988, p. 6–9).

Computing Stream Stage in Reaches

An option in the package is available that computes the stream stage in each reach. If the option is used, the stream stage is computed assuming incompressible steady flow in the stream at constant depth and using the Manning formula as described by Ozbilgin and Dickerman (1984, p. 3, equation 1) as:

$$Q = \frac{c}{n} \cdot (AR^{2/3} S^{1/2}) \tag{2}$$

where

Q stream discharge, (L^3/T);
n Manning's roughness coefficient, dimensionless;
A cross-sectional area of the stream, (L^2);
R hydraulic radius, (L);
S slope of the stream channel, (L/L); and
C a constant, $(L^{1/3}/T)$, which is 1.486 for units of cubic feet per second or 1.0 for cubic meters per second.

The Manning formula, in particular the units of (C/n), is discussed by Chow (1959, p. 98 – 101). The cross-sectional area and the hydraulic radius for a rectangular channel are:

$$A = wd \tag{3}$$

and

$$R = \frac{wd}{w + 2d} \tag{4}$$

where

d depth of the water in the stream, (L); and
w width of the channel, (L).

Substituting these equations into equation 2 and assuming that the depths of the streams are much less than the width, Ozbilgin and Dickerman (1984, p. 4) derived the following equation that computes stream depth:

$$d = \left[\frac{Qn}{CwS^{1/2}} \right]^{3/5} \tag{5}$$

This is the equation used to approximate the stream stage in each reach in the Streamflow-Routing Package.

A brief discussion of Manning's roughness coefficient is presented by White (1979, p. 602–607). White (p. 605) presented estimates of Manning's roughness coefficient for a variety of stream channels. The estimates are listed in table 1.

Streamflow estimates from the Manning formula are accurate in an intermediate range of roughness coefficients (White, 1979, p. 602–607), but for deep, smooth channels and shallow, rough channels the estimates of flow may be unrealistic. Roughness coefficients can be estimated from actual measurements of stream depth and discharge. However, if field data are not available, values can be estimated from table 1.

The stream stage in each reach is computed prior to calculating leakage to or from the aquifer. For the first iteration, the stage for each reach is calculated using the specified inflow into a stream segment. If no streamflow is specified into a segment, the stage for all reaches in the segment will equal the top of the streambed. Leakage terms are then calculated on the basis of the computed stream stage, streambed conductance, and the initial (starting) head in each corresponding model cell. These terms are added to the finite-difference equations used to calculate a new head for each model cell. The process continues, except in the following iterations, leakage calculated in the previous iteration is added or subtracted from streamflow prior to calculating a new stream stage, and new leakage terms are computed on the basis of the head in the corresponding model cell calculated from the previous iteration.

Table 1 Experimental values of Manning's roughness coefficient (White, 1979, p. 605).

Description of channel	Manning's roughness coefficient (n)
Excavated earth channels:	
Clean	0.022 ± 0.004
Gravelly	0.025 ± 0.005
Weedy	0.030 ± 0.005
Stony, cobbles	0.035 ± 0.010
Natural channels:	
Clean and straight	0.030 ± 0.005
Sluggish, deep pools	0.040 ± 0.010
Major rivers	0.035 ± 0.010
Floodplains:	
Pasture, farmland	0.035 ± 0.010
Light brush	0.050 ± 0.020
Heavy brush	0.075 ± 0.025
Trees	0.150 ± 0.050

Assumptions and Limitations

The streamflow-Routing Package does not include a time function for routing flows specified for the first reach in each segment that enters the modeled area to downstream reaches. The accounting scheme used assumes that streamflow entering the modeled area is instantly available in downstream reaches for each specified period in the model simulation. (The specified period is referred to as a stress period by McDonald and Harbaugh, 1988, p. 1–5.) This assumption is generally valid because of the relatively slow rate of ground-water flow and because stress periods used in model simulations are commonly months or years, whereas changes in streamflow along a course of a stream are usually in days.

Stream stage for a reach can be calculated using the Manning formula and assuming a rectangular channel with a stream width much greater than depth. The assumption of a rectangular channel with width much greater than depth is generally valid but may not be appropriate for all streams. Thus, using this method to calculate a stream stage may not be valid for all cases.

The numerical scheme where information from the previous iteration is used to determine stream stage, streamflow, and the leakage terms in the finite-difference equations can, in some cases, cause instability in the iterative solution process. Such instability may be more prominent when stream leakage is a dominant part of the simulation.

Streambed conductance values used to calculate leakage between a reach and a model cell are specified a constant for each stress period. Widths of a stream channel usually vary depending on the flow and stage in the stream which may result in varying streambed conductance values. In the package presented herein, changes in the streambed conductances due to changing streamflow must be computed separately from the model and then added at the beginning of each stress period.

Another limitation of the model is that leakage from the stream to the aquifer is assumed to be instantaneous, which is reasonable where the thickness of the unsaturated zone between a stream and an aquifer is not great. The assumption may be unreasonable when the thickness of the unsaturated zone is large, because of the time lag between stream infiltration and recharge to the aquifer. In addition, the stream infiltration rate may be greater than the amount of recharge to the aquifer because some of the infiltrated water may be used to replace soil moisture and (or) be used by plants adjacent to the stream.

The Streamflow-Routing Package can be used to simulate the leakage between streams and aquifers. But the user must consider the limitations associated with the package when using it as the package may not be appropriate for all cases. The accuracy of the model results will, to a large extent, be dependent on the size of the model cells, the time intervals used in the simulations, and the closure criteria for completing the iteration cycle in the modular model (McDonald and Harbaugh, 1988, p. 2-20 through 2-25).

HORIZONTAL-FLOW-BARRIER PACKAGE

(excerpts from Hsieh and Freckleton, 1993)

Conceptualization and Implementation

A plan view of a model layer containing a thin, vertical low-permeability geologic feature is shown in figure 2A. This geologic feature is approximated, in figure 2B, as a series of horizontal-flow barriers conceptually situated on the boundary between pairs of adjacent cells in the finite-difference grid. Note that each segment of the geologic feature lying between two adjacent cells is defined as a horizontal-flow barrier. Thus, geologically speaking, figure 2A shows a single low-permeability feature in the model layer. However, in the finite-difference grid (figure 2B), this feature is treated as six horizontal-flow barriers in the model layer.

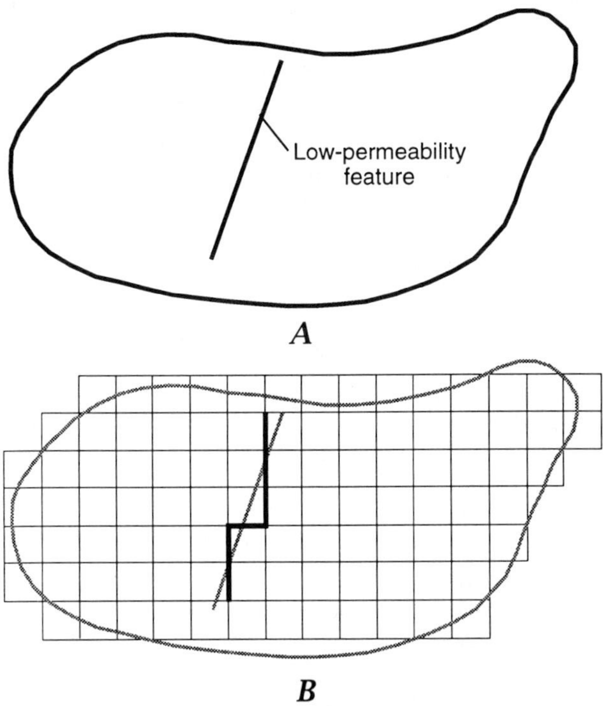

Figure 2 Schematic representation of model layer (A) and grid (B) with a low-permeability feature and its representation in the grid as a series of six horizontal-flow barriers.

The key assumption underlying the Horizontal-Flow-Barrier Package is that the width of the barrier (that is, the shortest distance through the barrier) is negligibly small in comparison with the horizontal dimensions of the cells in the finite-difference grid. Thus, the barrier width is not explicitly considered in the Package, but it is included implicitly in a hydraulic characteristic defined as either (1) barrier transmissivity divided by barrier width, if it is in a constant-transmissivity layer, or (2) barrier hydraulic conductivity divided by barrier width, if it is in a variable-transmissivity layer. Furthermore, the barrier is assumed to have zero storage capacity. Its sole function is to lower the horizontal branch conductance between the two cells that it separates.

Shown in figure 3A is a horizontal-flow barrier situated between cell i,j,k, and cell i,j+1,k -- that is, two adjacent cells in the same row in a model layer. The branch conductance in the row direction between the two cells can be determined as shown by McDonald and Harbaugh (1988, chap. 5, p. 6, eq. 38, and p. 7, fig. 25):

$$\frac{1}{CR_{i,j+1/2,k}} = \frac{1}{\dfrac{TR_{i,j,k}DELC_i}{\dfrac{DELR_j}{2}}} + \frac{1}{TDW_{i,j+1/2,k}DELC_i} + \frac{1}{\dfrac{TR_{i,j+1,k}DELC_i}{\dfrac{DELR_{j+1}}{2}}} \tag{1}$$

where

$CR_{i,j+1/2,k}$	is the branch conductance in the row direction between nodes i,j,k and i,j+1,k $[L^2T^{-1}]$;
$TR_{i,j,k}$	is the transmissivity in the row direction of cell i,j,k $[L^2T^{-1}]$;
$TR_{i,j+1,k}$	is the transmissivity in the row direction of cell i,j+1,k $[L^2T^{-1}]$;
$DELR_j$	is the grid width in the row direction of column j [L];
$DELR_{j+1}$	is the grid width in the row direction of column j+1 [L];
$DELC_i$	is the grid width in the column direction of row i [L]; and
$TDW_{i,j+1/2,k}$	is the barrier transmissivity divided by the width of the barrier between cell i,j,k and cell i,j+1,k $[LT^{-1}]$.

Combining the first and third terms on the right side of equation 1 gives:

$$\frac{1}{CR_{i,j+1/2,k}} = \frac{1}{CR^*_{i,j+1/2,k}} + \frac{1}{TDW_{i,j+1/2,k}DELC_i} \tag{2}$$

where $CR^*_{i,j+1/2,k}$ is the branch conductance if the barrier did not exist, and is given by:

$$CR^*_{i,j+1/2,k} = 2 \cdot DELC_i \frac{TR_{i,j,k}TR_{i,j+1,k}}{TR_{i,j,k}DELR_{j+1} + TR_{i,j+1,k}DELR_j} \tag{3}$$

Solving for $CR_{i,j+1/2,k}$ in equation (2) yields:

$$CR_{i,j+1/2,k} = \frac{CR^*_{i,j+1/2,k}TDW_{i,j+1/2,k}DELC_i}{CR^*_{i,j+1/2,k} + TDW_{i,j+1/2,k}DELC_i} \tag{4}$$

The same process can be applied when the horizontal-flow barrier is situated between two cells on the same column in a layer (fig, 3B). In this case, the branch conductance in the column direction between nodes i,j,k and nodes i+1,j,k is:

$$CC_{i+1/2,j,k} = \frac{CC^{*}_{i+1/2,j,k} TDW_{i+1/2,j,k} DELR_{i}}{CC^{*}_{i+1/2,j,k} + TDW_{i+1/2,j,k} DELR_{i}} \tag{5}$$

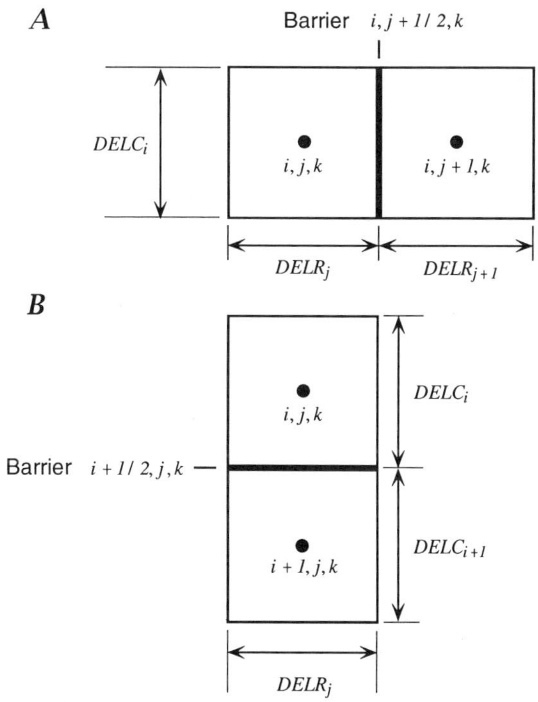

Figure 3. Schematic representation of a horizontal-flow barrier separating two adjacent model cells in the same row (A) and two adjacent model cells in the same column (B).

where $CC^{*}_{i+1/2,j,k}$ is the branch conductance if the barrier did not exist, and is given by:

$$CC^{*}_{i+1/2,j,k} = 2 \cdot DELR_{j} \frac{TC_{i,j,k} TC_{i+1,j,k}}{TC_{i,j,k} DELC_{i+1} + TC_{i+1,j,k} DELC_{i}} \tag{6}$$

and

$TC_{i,j,k}$	is the transmissivity in the column direction of cell i,j,k $[L^2T^{-1}]$; and
$TC_{i+1,j,k}$	is the transmissivity in the column direction of cell i+1,j,k $[L^2T^{-1}]$.

In the computer program, the horizontal branch conductances are computed in two steps. First, the horizontal-flow barriers are neglected and the branch conductances CC* and CR* are computed by the standard method (for example, eqs. 3 and 6; see also McDonald and Harbaugh, 1988, chap. 5, p. 6-10). Next, the effects of the barriers are taken into account by means of equations 4 and 5.

In a model layer that is confined, TDW (in this case, barrier transmissivity divided by barrier width) of a barrier will be constant for the simulation. If a layer is potentially unconfined, new values of TDW must be calculated as the head fluctuates. This is done at the start of each iteration. In this case, TDW is obtained by multiplying the average saturated thickness of the two cells on either side of the barrier by the barrier hydraulic conductivity divided by the width of the barrier.

Appendix G

BASIC MATHEMATICAL EXPRESSIONS

ALGEBRA

Exponents and Roots

If $a = x^b$ then $x = \sqrt[b]{a}$

$$x^a \cdot x^b = x^{a+b}$$

$$\frac{x^a}{x^b} = x^{a-b}$$

$$(x^a)^b = x^{ab}$$

$$x^{-a} = \frac{1}{x^a}$$

$$x^{\frac{1}{a}} = \sqrt[a]{x}$$

$$x^a \cdot y^a = (xy)^a$$

$$\frac{x^a}{y^a} = \left(\frac{x}{y}\right)^a$$

$$x^{\frac{a}{b}} = \sqrt[b]{x^a}$$

$$x^{-\frac{1}{a}} = \sqrt[a]{1/x}$$

$$(\sqrt[a]{x})^a = x$$

$$\sqrt[a]{xy} = \sqrt[a]{x} \cdot \sqrt[a]{y}$$

Logarithms

If $a^x = b$
then x is the logarithm to the base a of the number b, or
$x = log_a b$

For example:

$2^3 = 8$
$\log_2 8 = 3$

Logarithm to the base e, or *natural logarithm*, or *Napierian logarithm* is written as

$\log_e x$ or $\ln x$

where $e = 1 + \frac{1}{1!} + \frac{1}{2!} + \frac{1}{3!} + \frac{1}{4!} + \ldots\ldots \cong 2.71828$

Logarithm to the base 10, or *common logarithm*, or *Briggsian logarithm* is written as

$\log_{10} x$ or $\log x$

<u>*Connections between natural and common logarithm*</u>

$\log_{10} x = 0.4343 \cdot \log_e x$ $(\log x = 0.4343 \cdot \ln x)$

$\log_e 10 = 2.302585$ $(\ln 10 = 2.3026)$

$\log_e x = 2.3026 \cdot \log_{10} x$ $(\ln x = 2.3026 \cdot \log x)$

$\log_{10} e = 0.434294$ $(\log e = 0.4343)$

<u>*Relations valid for any logarithmic base*</u>

$$\log ab = \log a + \log b$$

$$\log \frac{a}{b} = \log a - \log b$$

$$\log \frac{1}{a} = -\log a$$

$$\log 1 = 0$$

$$\log a^b = b \log a$$

$$\log_a a = 1$$

$$b^{\log_a b} = a$$

$$\log \sqrt[b]{a} = \frac{1}{b} \log a$$

$$\log \sqrt[b]{a^c} = \frac{c}{b} \log a$$

If $a^x = b$

then $\log a^x = \log b$ or $x \log a = \log b$

and $x = \dfrac{\log b}{\log a}$

The Quadratic Equation

$ax^2 + bx + c = 0$ $(a, b, c \neq 0)$

Roots of the equation: $x_{1,2} = \dfrac{-b \pm \sqrt{b^2 - 4ac}}{2a}$

If:

$b^2 - 4ac > 0$ the roots are real and not equal $x_1 \neq x_2$

$b^2 - 4ac = 0$ the roots are real and equal $x_1 = x_2$

$b^2 - 4ac < 0$ the roots are imaginary numbers called conjugate-complex numbers with the same real part and the imaginary part having the opposite sign.

$ax^2 + bx + c = a(x - x_1)(x - x_2)$ where x_1 and x_2 are the solutions

Expansions and Splitting into Factors (Factoring)

$$(x+y)^2 = x^2 + 2xy + y^2$$
$$(x-y)^2 = x^2 - 2xy + y^2$$
$$(x+y)^3 = x^3 + 3x^2y + 3xy^2 + y^3$$
$$(x-y)^3 = x^3 - 3x^2y + 3xy^2 - y^3$$
$$x^2 - y^2 = (x-y)(x+y)$$
$$a_1x + b_1y = c_1$$
$$x^2 + y^2 = (x+y-\sqrt{2xy})(x+y+\sqrt{2xy})$$
$$x^3 - y^3 = (x-y)(x^2 + xy + y^2)$$
$$x^3 + y^3 = (x+y)(x^2 - xy + y^2)$$

System of Two Linear Equations with Two Unknowns

$$a_1x + b_1y = c_1 \qquad (1)$$
$$a_2x + b_2y = c_2 \qquad (2)$$

where

a_1,b_1,c_1,a_2,b_2,c_2 are the coefficients, and
x,y are the unknowns.

Solution by method of comparison

From (1) it follows that $$y = \frac{c_1 - a_1x}{b_1} \qquad (3)$$

From (2) it follows that $$y = \frac{c_2 - a_2x}{b_2} \qquad (4)$$

Since the left sides of equations (3) and (4) are equal, the right sides must be equal too:

$$\frac{c_1 - a_1x}{b_1} = \frac{c_2 - a_2x}{b_2}$$

It then follows that

$$x = \frac{b_2c_1 - b_1c_2}{a_1b_2 - a_2b_1}$$

The other unknown (y) is found by substituting the value of x into either equation (3) or (4):

$$y = \frac{a_1c_2 - a_2c_1}{a_1b_2 - a_2b_1}$$

Solution by method of substitution

From (1) it follows that $$y = \frac{c_1 - a_1x}{b_1} \qquad (5)$$

Substituting equation (5) into (2) gives equation with only one unknown (x):

$$a_2 x + b_2 \cdot \frac{c_1 - a_1 x}{b_1} = c_2$$

which, solved for x, gives:

$$x = \frac{b_2 c_1 - b_1 c_2}{a_1 b_2 - a_2 b_1}$$

Substituting the value of x into equation (5) gives the solution for y:

$$y = \frac{a_1 c_2 - a_2 c_1}{a_1 b_2 - a_2 b_1}$$

Solution by method of determinant

Matrix of the system is

$$\left(\begin{array}{cc|c} a_1 & b_1 & c_1 \\ a_2 & b_2 & c_2 \end{array} \right)$$

and the solution is

$$x = \frac{D_1}{D} \qquad y = \frac{D_2}{D}$$

where the determinants are:

$$D = \begin{vmatrix} a_1 & b_1 \\ a_2 & b_2 \end{vmatrix} = a_1 b_2 - a_2 b_1$$

$$D_1 = \begin{vmatrix} c_1 & b_1 \\ c_2 & b_2 \end{vmatrix} = b_2 c_1 - b_1 c_2$$

$$D_2 = \begin{vmatrix} a_1 & c_1 \\ a_2 & c_2 \end{vmatrix} = a_1 c_2 - a_2 c_1$$

Note that the elements in the determinant are multiplied diagonally, starting from the top of each column. The sign is positive for the multiplication from upper left to lower right, and it is negative for the multiplication from upper right to lower left.

System of Three Linear Equations with Three Unknowns

$$a_1 x + b_1 x + c_1 x = d_1 \qquad (1)$$

$$a_2 x + b_2 x + c_2 x = d_2 \qquad (2)$$

$$a_3 x + b_3 x + c_3 x = d_3 \qquad (3)$$

Solution by method of determinant

Matrix of the system is:

$$\left(\begin{array}{ccc|c} a_1 & b_1 & c_1 & d_1 \\ a_2 & b_2 & c_2 & d_2 \\ a_3 & b_3 & c_3 & d_3 \end{array}\right)$$

and the solution is

$$x = \frac{D_1}{D} \qquad y = \frac{D_2}{D} \qquad z = \frac{D_3}{D}$$

where the determinants are given as

$$D = \begin{vmatrix} a_1 & b_1 & c_1 \\ a_2 & b_2 & c_2 \\ a_3 & b_3 & c_3 \end{vmatrix}$$

$$D_1 = \begin{vmatrix} d_1 & b_1 & c_1 \\ d_2 & b_2 & c_2 \\ d_3 & b_3 & c_3 \end{vmatrix}$$

$$D_2 = \begin{vmatrix} a_1 & d_1 & c_1 \\ a_2 & d_2 & c_2 \\ a_3 & d_3 & c_3 \end{vmatrix}$$

$$D_3 = \begin{vmatrix} a_1 & b_1 & d_1 \\ a_2 & b_2 & d_2 \\ a_3 & b_3 & d_3 \end{vmatrix}$$

Note that determinant D_1 (D_2, D_3) is obtained from the basic determinant (D) in which the first (second, third) column is replaced by the free coefficients (d_1, d_2, d_3).

The determinant of the third order is solved by applying Saruss's rule:

1) add the first two columns to the right of the determinant,
2) multiply elements in the "new" determinant diagonally, starting at the top of each column,
3) sum the products following the rule of signs: multiplication from upper right to lower left is positive, and multiplication from upper right to lower left is negative. For example:

$$\begin{vmatrix} a_1 & b_1 & c_1 \\ a_2 & b_2 & c_2 \\ a_3 & b_3 & c_3 \end{vmatrix} \begin{matrix} a_1 & b_1 \\ a_2 & b_2 \\ a_3 & b_3 \end{matrix}$$

$$D = a_1b_2c_3 + b_1c_2a_3 + c_1a_2b_3 - c_1b_2a_3 - a_1c_2b_3 - b_1a_2c_3$$

ANALYTICAL GEOMETRY

Orthogonal (Cartesian) Coordinate System

Orthogonal (Cartesian) coordinate system in a plane is determined by the following:

1) horizontal X axis,
2) vertical Y axis,
3) origin (intersection of X and Y axes).

Position of point P is defined by its coordinates (see Figure G.1):

1) x coordinate (or abscissa), the orthogonal distance of the point from the vertical (Y) axis, and
2) y coordinate (or ordinate), the orthogonal distance of the point from the horizontal (X) axis.

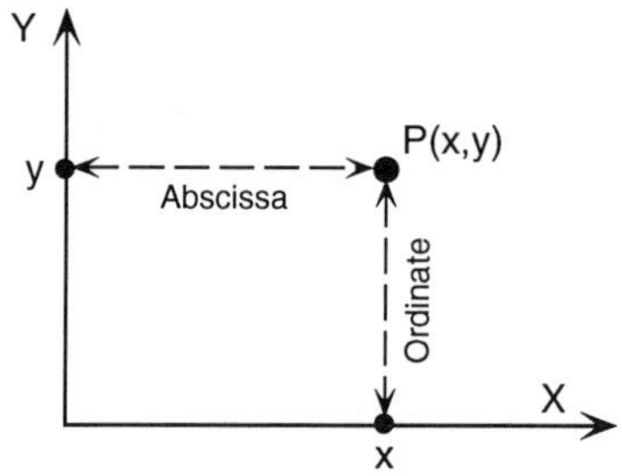

Figure G.1

<u>*Distance between two points (Figure G.2)*</u>

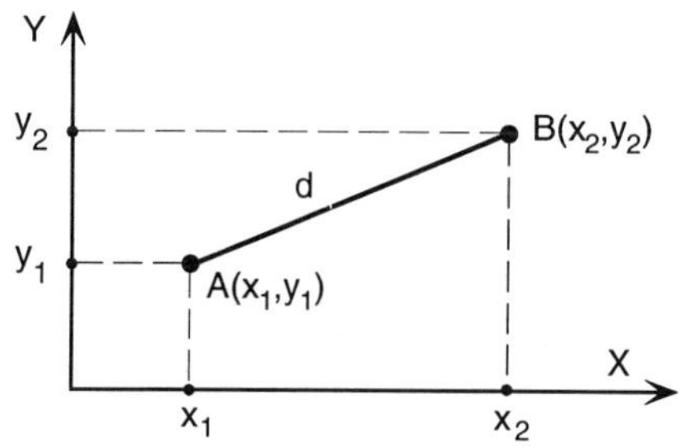

Figure G.2

$$d = \sqrt{(x_2 - x_1)^2 + (y_2 - y_1)^2}$$

Polar Coordinate System

Polar coordinate system is determined by the following (see Figure G.3):

1) pole O,
2) polar axis OX.

Position of point P is defined by its polar coordinates:

1) radius vector r ($\overline{OP}$),
2) angle φ (angle between axis OX and r).

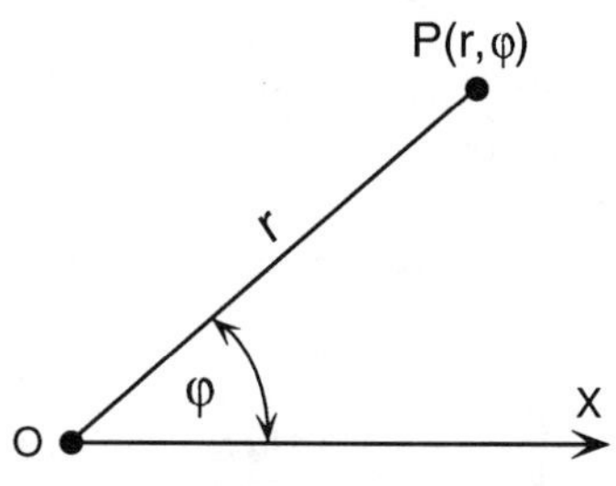

Figure G.3

<u>*Distance between two points (Figure G.4)*</u>

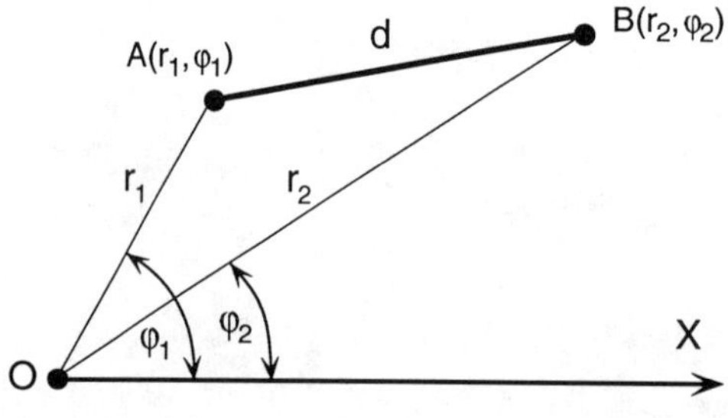

Figure G.4

$$d = \sqrt{r_1^2 + r_2^2 - 2r_1 r_2 \cos(\varphi_2 - \varphi_1)}$$

Connection between Cartesian and Polar Coordinates

If Cartesian coordinates are given (x,y), the polar coordinates are:

$$r = \sqrt{x^2 + y^2} \qquad \text{and} \qquad \varphi = \arctan\frac{y}{x}$$

If polar coordinates are given (r, φ), the Cartesian coordinates are:

$$x = r \cdot \cos\varphi \qquad \text{and} \qquad y = r \cdot \sin\varphi$$

TRIGONOMETRY

Definitions of Trigonometric Functions

$$\sin\alpha = \frac{a}{c}$$

$$\cos\alpha = \frac{b}{c}$$

$$\tan\alpha = \frac{a}{b}$$

$$\cot\alpha = \frac{b}{a}$$

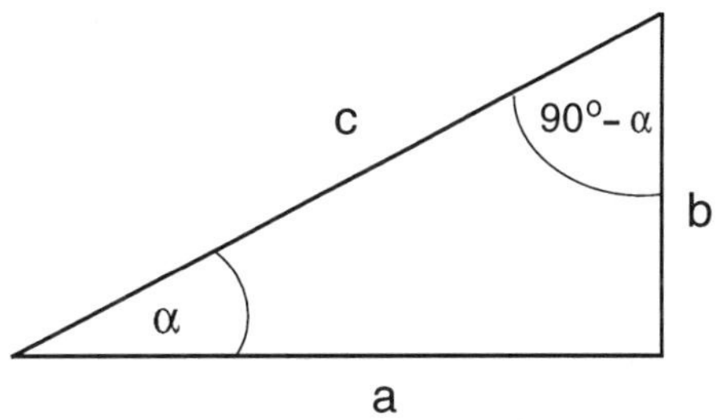

Figure G.5

Trigonometric Functions of Complete Angles

$$\sin(90^o - \alpha) = \cos\alpha$$

$$\cos(90^o - \alpha) = \sin\alpha$$

$$\tan(90^o - \alpha) = \cot\alpha$$

$$\cot(90^o - \alpha) = \tan\alpha$$

$$\sin^2\alpha + \cos^2\alpha = 1$$

Basic Relations

$$\tan^2\alpha + 1 = \frac{1}{\cos^2\alpha}$$

$$1 + \cot^2\alpha = \frac{1}{\sin^2\alpha}$$

$$\tan\alpha = \frac{\sin\alpha}{\cos\alpha}$$

$$\cot\alpha = \frac{\cos\alpha}{\sin\alpha}$$

$$\tan\alpha \cdot \cot\alpha = 1$$

Table G.1 Relationships between trigonometric functions

Function	In terms of, equals			
	sin α	cos α	tan α	cot α
sin α	—	$\sqrt{1-\cos^2\alpha}$	$\sqrt{\dfrac{\tan\alpha}{1+\tan^2\alpha}}$	$\dfrac{1}{\sqrt{1+\cot^2\alpha}}$
cos α	$\sqrt{1-\sin^2\alpha}$	—	$\dfrac{1}{\sqrt{1+\tan^2\alpha}}$	$\dfrac{\cot\alpha}{\sqrt{1+\cot^2\alpha}}$
tan α	$\dfrac{\sin\alpha}{\sqrt{1-\sin^2\alpha}}$	$\dfrac{\sqrt{1-\cos^2\alpha}}{\cos\alpha}$	—	$\dfrac{1}{\cot\alpha}$
ctan α	$\dfrac{\sqrt{1-\sin^2\alpha}}{\sin\alpha}$	$\dfrac{\cos\alpha}{\sqrt{1-\cos^2\alpha}}$	$\dfrac{1}{\tan\alpha}$	—

Table G.1 is derived from the following basic relationships:

$$\sin^2\alpha+\cos^2\alpha=1$$
$$\tan\alpha=\frac{\sin\alpha}{\cos\alpha}$$
$$\tan\alpha\cdot\cot\alpha=1$$

Adding and Subtracting Angles

$$\sin(\alpha+\beta)=\sin\alpha\cdot\cos\beta+\cos\alpha\cdot\sin\beta$$
$$\sin(\alpha-\beta)=\sin\alpha\cdot\cos\beta-\cos\alpha\cdot\sin\beta$$
$$\cos(\alpha+\beta)=\cos\alpha\cdot\cos\beta+\sin\alpha\cdot\sin\beta$$
$$\cos(\alpha-\beta)=\cos\alpha\cdot\cos\beta-\sin\alpha\cdot\sin\beta$$
$$\tan(\alpha+\beta)=\frac{\tan\alpha+\tan\beta}{1-\tan\alpha\tan\beta}$$
$$\tan(\alpha-\beta)=\frac{\tan\alpha-\tan\beta}{1+\tan\alpha\cdot\tan\beta}$$
$$\sin\alpha+\sin\beta=2\sin\frac{\alpha+\beta}{2}\cdot\cos\frac{\alpha-\beta}{2}$$
$$\sin\alpha-\sin\beta=2\sin\frac{\alpha+\beta}{2}\cdot\sin\frac{\alpha-\beta}{2}$$
$$\cos\alpha+\cos\beta=2\cos\frac{\alpha+\beta}{2}\cdot\cos\frac{\alpha-\beta}{2}$$
$$\cos\alpha+\cos\beta=-2\sin\frac{\alpha+\beta}{2}\cdot\sin\frac{\alpha-\beta}{2}$$

$$\tan\alpha + \tan\beta = \frac{\sin(\alpha+\beta)}{\cos\alpha \cdot \cos\beta}$$

$$\tan\alpha - \tan\beta = \frac{\sin(\alpha-\beta)}{\cos\alpha \cdot \cos\beta}$$

Transformation of Products of Trigonometric Functions into Sums

$$\sin\alpha \cdot \cos\beta = \frac{1}{2}\left[\sin(\alpha+\beta) + \sin(\alpha-\beta)\right]$$

$$\cos\alpha \cdot \cos\beta = \frac{1}{2}\left[\cos(\alpha+\beta) + \cos(\alpha-\beta)\right]$$

$$\sin\alpha \cdot \sin\beta = \frac{1}{2}\left[\cos(\alpha+\beta) - \cos(\alpha-\beta)\right]$$

Functions of Multiple Angles and Powers

$$\sin 2\alpha = 2\sin\alpha \cdot \cos\alpha$$

$$\cos 2\alpha = \cos^2\alpha - \sin^2\alpha$$

$$\tan 2\alpha = \frac{2\tan\alpha}{1-\tan^2\alpha}$$

$$\cot 2\alpha = \frac{\cot^2\alpha - 1}{2\cot\alpha}$$

$$\sin 3\alpha = 3\sin\alpha - 4\sin^3\alpha$$

$$\cos 3\alpha = 4\cos^3\alpha - 3\cos\alpha$$

$$\tan 3\alpha = \frac{3\tan\alpha - \tan^3\alpha}{1-3\tan^2\alpha}$$

$$\cot 3\alpha = \frac{\cot^3\alpha - 3\cot\alpha}{3\cot^2\alpha - 1}$$

$$\sin\frac{\alpha}{2} = \pm\sqrt{\frac{1-\cos\alpha}{2}}$$

$$\cos\frac{\alpha}{2} = \pm\sqrt{\frac{1+\cos\alpha}{2}}$$

$$\tan\frac{\alpha}{2} = \pm\sqrt{\frac{1-\cos\alpha}{1+\cos\alpha}}$$

$$\cot\frac{\alpha}{2} = \pm\sqrt{\frac{1+\cos\alpha}{1-\cos\alpha}}$$

$$\sin^2\alpha = \frac{1}{2}(1-\cos 2\alpha)$$

$$\cos^2\alpha = \frac{1}{2}(1+\cos 2\alpha)$$

$$\sin^3\alpha = \frac{1}{4}(3\sin\alpha - \sin 3\alpha)$$

$$\cos^3 \alpha = \frac{1}{4}(3\cos\alpha + \cos 3\alpha)$$

$$\sin^4 \alpha = \frac{1}{8}(\cos 4\alpha - 4\cos 2\alpha + 3)$$

$$\cos^4 \alpha = \frac{1}{8}(\cos 4\alpha + 4\cos 2\alpha + 3)$$

DERIVATIVES

Definition

Derivative of function f(x) at point x=a is called the final limit value at point *a* of the ratio between the increment of the function and the increment of the argument, when the increment of the argument approaches zero (when $x \to a$):

$$f'(a) = \lim_{x \to a} \frac{f(x) - f(a)}{x - a}$$

where $f'(a)$ is the first derivative of the function at point a.

Derivative of function f(x) at point x is:

$$f'(x) = \lim_{\Delta x \to 0} \frac{f(x + \Delta x) - f(x)}{\Delta x}$$

which is also written as:

$$f'(x) = \lim_{\Delta x \to 0} \frac{\Delta y}{\Delta x} \qquad \text{or} \qquad f'(x) = \frac{dy}{dx}$$

Geometric Interpretation of the Derivative

The first derivative is the tangent of the angle between the horizontal (X axis) and the line that touches the curve at a given point as shown in Figure G.6 (note that this line is also called tangent):

$$\frac{f(x) - f(a)}{x - a} = \tan\alpha'$$

Note that:

$$\lim_{x \to a} \frac{f(x) - f(a)}{x - a} = f'(a) = \tan\alpha$$

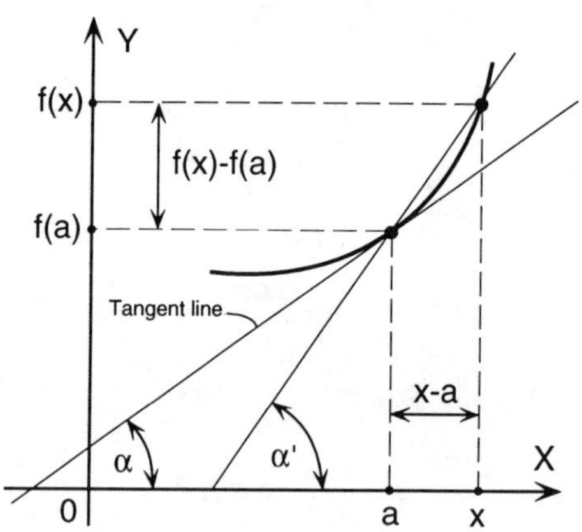

Figure G.6

Elementary Derivatives

Function	First Derivative	
$y = c$	$y' = 0$	
$y = x$	$y' = 1$	
$y = ax + b$	$y' = a$	
$y = x^n$	$y' = nx^{n-1}$	
$y = a^x$	$y' = a^x \ln a$	
$y = e^x$	$y' = e^x$	
$y = \log_a x$	$y' = \frac{1}{x} \log_a e$	
$y = \ln x$	$y' = \frac{1}{x}$	
$y = \log x$	$y' = \frac{1}{x} \cdot 0.4343$	
$y = \sin x$	$y' = \cos x$	
$y = \cos x$	$y' = -\sin x$	
$y = \tan x$	$y' = \frac{1}{\cos^2 x}$	
$y = \text{ctan}x$	$y' = -\frac{1}{\sin^2 x}$	
$y = \arcsin x$	$y' = \frac{1}{\sqrt{1 - x^2}}$	for $(\lvert x \rvert < 1)$
$y = \arccos x$	$y' = -\frac{1}{\sqrt{1 - x^2}}$	for $(\lvert x \rvert < 1)$
$y = \arctan x$	$y' = \frac{1}{1 + x^2}$	
$y = \text{arcctan}x$	$y' = -\frac{1}{1 + x^2}$	

Basic Rules of Derivation

If f(x) and g(x) are functions of an independent variable x, and if they have first derivatives f ′(x) and g ′(x), then it follows:

$$[f(x) \pm g(x)]' = f'(x) \pm g'(x)$$

$$[f(x) \cdot g(x)]' = f'(x) \cdot g(x) \pm f(x) \cdot g'(x)$$

$$\left[\frac{f(x)}{g(x)}\right]' = \frac{f'(x) \cdot g(x) \pm f(x) \cdot g'(x)}{[g(x)]^2} \quad \text{for } g(x) \neq 0$$

$$[c \cdot f(x)]' = c \cdot f'(x)$$

Derivatives of Complex Functions

Function $y=y(u)$, where $u=u(x)$, has the first derivative with respect to x as:

$$y'(x) = y'(u) \cdot u'(x)$$

Examples

Function	First Derivative	
$y=[f(x)]^n$	$y'=n\cdot[f(x)]^{n-1}\cdot f'(x)$	
$y=a^{f(x)}$	$y'=a^{f(x)}\cdot\ln a\cdot f'(x)$	for $a>0$
$y=e^{f(x)}$	$y=e^{f(x)}\cdot f'(x)$	
$y=\ln[f(x)]$	$y'=\dfrac{f'(x)}{f(x)}$	
$y=\sin[f(x)]$	$y'=f'(x)\cos[f(x)]$	
$y=\cos[f(x)]$	$y'=-f'(x)\sin[f(x)]$	
$y=\tan[f(x)]$	$y'=\dfrac{f'(x)}{\cos^2[f(x)]}$	
$y=\text{ctan}[f(x)]$	$y'=-\dfrac{f'(x)}{\sin^2[f(x)]}$	

Derivatives of Higher Order

Derivative of the second order (second derivative) of function $f(x)$ is the derivative of its first derivative:

$$y''(x)=[y'(x)]'$$

Derivative of the n-th order is:

$$y^{(n)}(x)=\left[y^{(n-1)}(x)\right]'$$

INDEFINITE INTEGRAL

Definition

A primitive function of function $f(x)$ is the function $F(x)$ such that its first derivative equals $f(x)$:

$$F'(x)=f(x)$$

Since

$$[F(x)+c]'=F'(x)=f(x) \qquad \text{(because } c'=0\text{)}$$

it follows that if the function $f(x)$ has a primitive function, then there exist an indefinite number of primitive functions of $f(x)$ that differ for the constant c.

The array of all possible primitive functions of the continuous function $f(x)$ is called the indefinite integral of that function:

$$\int f(x)dx = F(x)+c$$

where F(x) is the primitive function of the function f(x), and c is an arbitrary constant of integration.

Elementary Indefinite Integrals

$$\int dx = x+c$$

$$\int xdx = \frac{x^2}{2}+c$$

$$\int x^n dx = \frac{x^{n+1}}{n+1}+c \qquad n \neq -1$$

$$\int \frac{1}{x}dx = \ln x+c$$

$$\int e^x dx = e^x +c$$

$$\int a^x dx = \frac{a^x}{\ln a}+c$$

$$\int \cos xdx = \sin x+c$$

$$\int \sin xdx = -\cos x+c$$

$$\int \frac{1}{\cos^2 x}dx = \tan x+c$$

$$\int \frac{1}{\sin^2 x}dx = -\operatorname{ctan} x+c$$

$$\int \frac{1}{\sqrt{1-x^2}}dx = -\arcsin x+c$$

$$\int \frac{1}{1+x^2}dx = -\arctan x+c$$

$$\int \frac{dx}{\sqrt{x^2+1}} = \ln(x+\sqrt{x^2+1})+c$$

$$\int \frac{dx}{\sqrt{x^2-1}} = \ln(x+\sqrt{x^2-1})+c \qquad (\textit{for } |x|>1)$$

$$\int \frac{dx}{1-x^2}dx = \frac{1}{2}\ln\frac{1+x}{1-x}+c \qquad (\textit{for } |x|<1)$$

$$\int \frac{dx}{x^2-1}dx = \frac{1}{2}\ln\frac{x-1}{x+1}+c \qquad (for\ |x|>1)$$

Integration Rules

$$\int [f(x) \pm \varphi(x)]\, dx = \int f(x)\, dx \pm \int \varphi(x)\, dx$$

$$\int c \cdot f(x)\, dx = c\int f(x)\, dx$$

$$\int f(x)\, dx = \int f[\varphi(u)]\, \varphi'(u)\, du$$

where u is a new variable (method of substitution).

$$\int u\, dv = uv - \int v\, du$$

where u and v are the functions of the independent variable x (partial integration).

DEFINITE INTEGRAL

Definition

The value of the definite integral is

$$\int_a^b f(x)\, dx = [F(x)]_a^b = F(b) - F(a)$$

where F(x) is the primitive function of the function f(x), *i.e.,*:

$$F'(x) = f(x)$$

For example:

$$\int_4^9 5x^2 dx = 5 \cdot \int_4^9 x^2 dx = 5 \cdot \left[\frac{x^3}{3}\right]_4^9 = 5 \cdot \left[\frac{9^3}{3} - \frac{4^3}{3}\right] = 1108.3$$

Geometric Interpretation of the Definite Integral

$$Area = \int_a^b f(x)\, dx$$

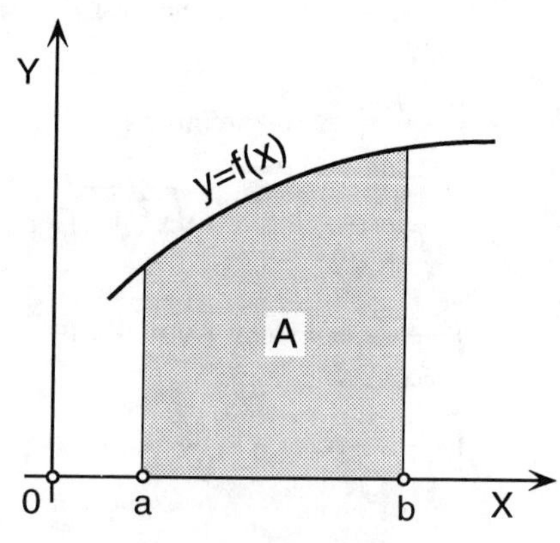

Figure G.7

Properties of the Definite Integral
(Rules for Calculations with Definite Integrals)

$$\int_a^b f(x)\,dx = -\int_b^a f(x)\,dx$$

$$\int_a^a f(x)\,dx = 0$$

$$\int_a^b [f(x)+g(x)-h(x)]\,dx = \int_a^b f(x)\,dx + \int_a^b g(x)\,dx - \int_a^b h(x)\,dx$$

$$\int_a^b f(x)dx + \int_b^c f(x)dx = \int_a^c f(x)dx$$

$$\int_a^b c\,f(x)\,dx = c\int_a^b f(x)\,dx$$

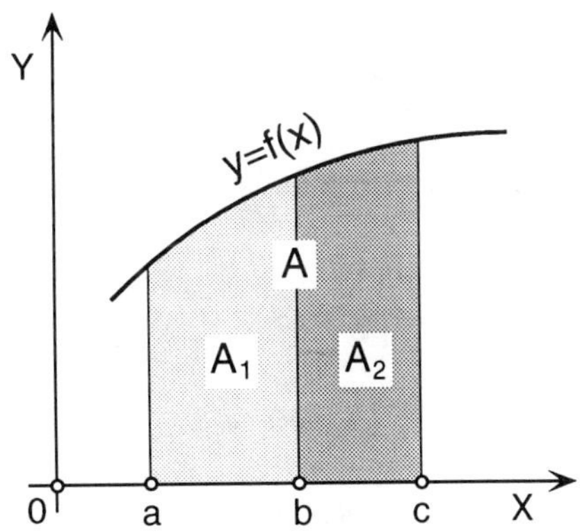

Figure G.8

Rule of the Average Value in the Integral Calculations

F(x) is a primitive function (indefinite integral) of the function *f(x)*.

$$\int_a^b f(x)\,dx = [F(x)]_a^b = F(b) - F(a)$$

and

$$\int_a^b f(x) = [F(x)]_b^a = F(b) - F(a)$$

give

$$\int_a^b f(x)\,dx = -\int_a^b f(x)\,dx$$

Examples of the Definite Integral Calculations

1. Calculate the area bound by curve y = log x, axis 0X, and ordinate of the point x = 4

$$A = \int_1^4 \ln x dx = [x \log x - x]_1^4 =$$
$$= (4\log 4 - 4) - (\log 1 - 1) = 8\log 2 - 3$$

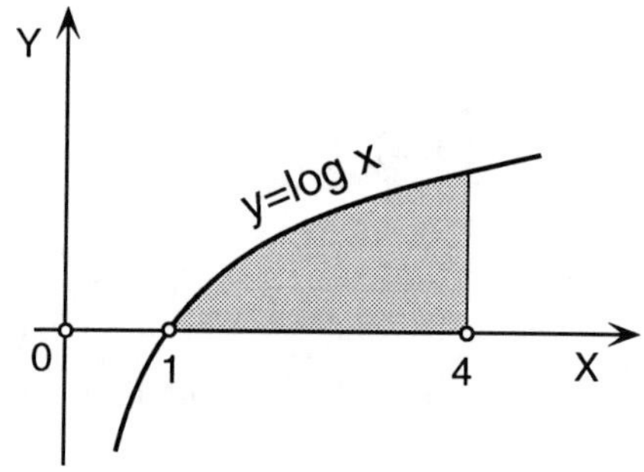

Figure G.9

2. Find the area bound by curves y = f(x), y = φ(x), and lines x = a, and x = b.

The area A is found as the difference between the area under curve y = f(x):

$$A_1 = \int_a^b f(x)\, dx$$

and the area under curve y = φ(x):

$$A_2 = \int_a^b \varphi(x)\, dx$$

which gives:

$$A = \int_a^b f(x)\, dx - \int_a^b \varphi(x)\, dx$$

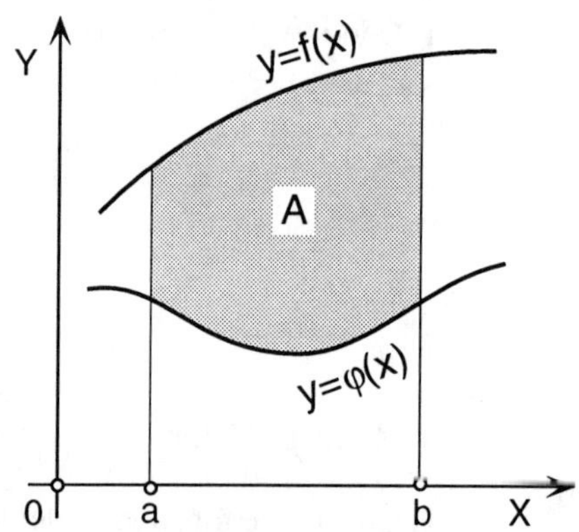

Figure G.10

This equation can also be written as:

$$A = \int_a^b [f(x) - \varphi(x)]\, dx$$

Basic Rule of the Integral Calculus
(Connection Between Definite and Indefinite Integrals)

The area abAB in Figure G.11 is given as

$$A = \int_a^b f(x)\, dx$$

If point x is an arbitrarily chosen point in the interval (a,b), then the area axMA is a function of x:

$$A(x) = \int_a^x f(x)\,dx$$

If x is increased by Δx, then

$$A(x+\Delta x) - A(x) = \int_x^{x+\Delta x} f(x)\,dx$$

$$[F(x)+c]_b^a = [F(x)]_a^b$$

(c is a constant that drops out during the calculation of definite integrals).

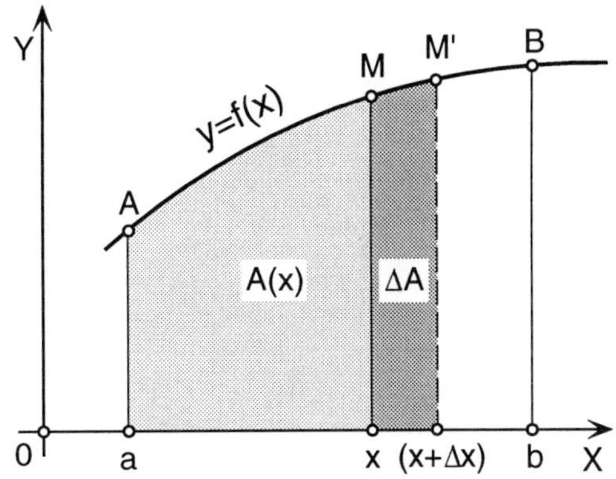

Figure G.11

Appendix H

UNIT CONVERSION TABLES

International System (SI metric) to U.S. Units

Modified from: *Ground Water Manual*, U.S. Department of Interior, Bureau of Reclamation, Government Printing Office, Denver, 1985 (p. 463–471)

Length

To convert from	*To*	*Multiply by*
angstrom units	nanometers (nm)	0.1
	micrometers (μm)	1.0×10^{-4}
	millimeters (mm)	1.0×10^{-7}
	meters (m)	1.0×10^{-10}
	mils	3.937×10^{-6}
	inches (in)	3.937×10^{-9}
micrometers	millimeters	1.0×10^{-3}
	meters	1.0×10^{-6}
	angstrom units	1.0×10^{4}
	mils	0.039
	inches	3.937×10^{-5}
millimeters	micrometers	1.0×10^{3}
	centimeters (cm)	0.1
	meters	1.0×10^{-3}
	mils	39.370
	inches	0.039
	feet (ft)	3.281×10^{-3}
centimeters	millimeters	10.0
	meters	0.01
	mils	0.3937×10^{3}
	inches	0.3937
	feet	3.281×10^{-2}
inches	millimeters	25.40
	meters	0.0254
	mils	1.0×10^{3}
	feet	0.083

To convert from	*To*	*Multiply by*
feet	millimeters	304.8
	meters	0.3048
	inches	12.0
	yards (yd)	0.333
yards	meters	0.9144
	inches	36.0
	feet	3.0
meters	millimeters	1.0×10^{3}
	kilometers (km)	1.0×10^{-3}
	inches	39.370
	feet	3.281
	yards	1.094
	miles (mi)	6.214×10^{-4}
kilometers	meters	1.0×10^{3}
	feet	3.281×10^{3}
	miles	0.621
miles	meters	1.609×10^{3}
	kilometers	1.609
	feet	5280.0
	yards	1760.0
nautical miles (nmi)	kilometers	1.8520
	miles	1.1508

Area

To convert from	*To*	*Multiply by*
square millimeters	square centimeters (cm^2)	0.01
	square inches (in^2)	1.550×10^{-3}
square centimeters	square millimeters (mm^2)	100.0
	square meters (m^2)	1.0×10^{-4}
	square inches	0.1550
	square feet (ft^2)	1.076×10^{-3}
square inches	square millimeters	645.16
	square centimeters	6.4516
	square meters	6.4516×10^{-4}
	square feet	69.444×10^{-4}

To convert from	*To*	*Multiply by*
square yards	square meters	0.0929
	hectares	8.3613×10^{-5}
	square inches	144.0
	acres	2.296×10^{-5}
square feet	square meters	0.0929
	hectares (ha)	9.2903×10^{-6}
	square inches	144.0
	acres	2.066×10^{-4}
square meters	hectares	1.0×10^{-4}
	square feet	10.764
	acres	2.471×10^{-4}
	square yards (yd^2)	1.196
acres	square meters	4046.8564
	hectares	0.405
	square feet	4.356×10^{4}
hectares	square meters	1.0×10^{4}
	acres	2.471
square kilometers	square meters	1.0×10^{6}
	hectares	100
	square feet	107.6391×10^{5}
	acres	247.105
	square miles (mi^2)	0.3861
square miles	square meters	258.999×10^{4}
	hectares	258.999
	square kilometers (km^2)	2.59
	square feet	2.788×10^{7}
	acres	640.0

Volume-Capacity

To convert from	*To*	*Multiply by*
cubic millimeters	cubic centimeters (cm^3)	1.0×10^{-3}
	liters (l)	1.0×10^{-6}
	cubic inches (in^3)	61.024×10^{-6}
cubic centimeters	liters	1.0×10^{-3}
	milliliters (ml)	1.0
	cubic inches	61.024×10^{-3}
	fluid ounces (fl. oz)	33.814×10^{-3}

To convert from	*To*	*Multiply by*
milliliters	liters	1.0×10^{-3}
	cubic centimeters	1.0
cubic inches	milliliters	16.387
	cubic feet (ft^3)	57.870×10^{-5}
liters	cubic meters	1.0×10^{-3}
	cubic feet	0.035
	gallons	0.264
	fluid ounces	33.814
gallons	liters	3.785
	cubic meters	3.785×10^{-3}
	fluid ounces	128.0
	cubic feet	0.134
cubic feet	liters	28.317
	cubic meters (m^3)	28.317×10^{-3}
	cubic dekameters (dam^3)	28.317×10^{-6}
	cubic inches	1728.0
	cubic yards (yd^3)	37.037×10^{-3}
	gallons (gal)	7.480
	acre feet (acre-ft)	22.957×10^{-6}
cubic miles	cubic dekameters	4.168×10^{6}
	cubic kilometers (km^3)	4.168
	acre-feet	3.3792×10^{6}
cubic yards	cubic meters	0.76455
	cubic feet	27.0
cubic meters	liters	1.0×10^{3}
	cubic dekameters	1.0×10^{-3}
	gallons	264.1721
	cubic feet	35.315
	cubic yards	1.308
	acre feet	8.107×10^{-4}
acre feet	cubic meters	1233.482
	cubic dekameters	1.2335
	cubic feet	43.560×10^{3}
	gallons	325.8514×10^{3}
cubic dekameters	cubic meters	1.0×10^{3}
	cubic feet	35.315×10^{3}
	acre-feet	0.811
	gallons	$26.417\ 21 \times 10^{4}$

To convert from	*To*	*Multiply by*
cubic kilometers	cubic dekameters	1.0×10^6
	acre-feet	0.811×10^6
	cubic miles (mi^3)	0.24

Temperature

degrees Celsius (^oC)	t_c
Kelvin (K)	t_k
degrees Fahrenheit (^oF)	t_f
degrees Rankine (R)	t_r

$t_c = (t_f - 32)/1.8$
$= t_k - 273.15$

$t_k = t_c + 273.15$
$= (t_f + 459.67)/1.8$
$= t_r/1.8$
$= t_c/1.8+32$
$= 1.8\ t_k$
$= 1.8\ t_c + 491.68$

Acceleration

To convert from	*To*	*Multiply by*
feet per second squared	meters per second squared (m/s^2)	0.3048
	G's	0.031
meters per second squared	feet per second squared (ft/s^2)	3.281
	G's	0.102
G's (standard gravitational acceleration)	meters per second squared	9.807
	feet per second squared	32.174

Velocity

To convert from	*To*	*Multiply by*
feet per second	meters per second (m/s)	0.3048
	kilometers per hour (km/h)	1.097
	miles per hour (mi/h)	0.682
meters per second	kilometers per hour	3.60
	feet per second (ft/s)	3.281
	miles per hour	2.237

To convert from	*To*	*Multiply by*
kilometers per hour	meters per second	0.278
	feet per second	0.911
	miles per hour	0.6215
miles per hour	kilometers per hour	1.609
	meters per second	0.447
	feet per second	1.467
feet per year (ft/yr)	millimeters per second (mm/s)	9.665 14x10^{-6}

Force

To convert from	*To*	*Multiply by*
pounds	newtons (N)	4.4482
kilograms	newtons	9.801
	pounds	0.225
newtons	pounds	0.225
dynes	newtons	1.0x10^{-5}

Mass

To convert from	*To*	*Multiply by*
grams	kilograms (kg)	1.0x10^{-3}
	ounces (avdp)	0.0353
ounces (avdp)	grams (g)	28.3495
	kilograms	0.028
	pounds (avdp)	0.0625

To convert from	*To*	*Multiply by*
pounds (avdp)	kilograms	0.4536
	ounces (avdp)	16.00
kilograms	kilograms (force)-second squared per meter (kgf·s^2/m)	0.102
	pounds (avdp)	2.205
	slugs	0.0685
slugs	kilograms	14.5939

To convert from	*To*	*Multiply by*
short tons	kilograms	907.1847
	metric tons (t)	0.907
	pounds (avdp)	2000.0
metric tons (tonne or megagram)	kilograms	1.0×10^3
	pounds (avdp)	$2.204\ 62 \times 10^3$
	short tons	1.102
long tons	kilograms	1016.047
	metric tons	1.016
	pounds (avdp)	2240.0
	short tons	1.120

Flow
(Volume Per Unit Time)

To convert from	*To*	*Multiply by*
cubic feet per second	liters per second (l/s)	28.317
	cubic meters per second (m^3/sec)	0.028
	cubic dekameters per day (dam^3/d)	2.4466
	gallons per minute (gal/min)	448.831
	acre-feet per day (acre-ft/d)	1.9835
	cubic feet per minute (ft^3/min)	60.0
gallons per minute	cubic meters per second	6.31×10^{-5}
	liters per second	0.0631
	cubic dekameters per day	5.451×10^{-3}
	cubic feet per second (ft^3/s)	2.288×10^{-3}
	acre-feet per day	4.4192×10^{-3}
acre-feet per day	cubic meters per second	0.014
	cubic dekameters per day	1.2335
	cubic feet per second	0.504
cubic dekameters per day	cubic meters per second	0.0116
	cubic feet per second	0.409
	acre-feet per day	0.811

Viscosity

To convert from	*To*	*Multiply by*
centipoise	pascal-second (Pa·s)	1.0×10^{-3}
	poise	0.01
	pound per foot-hour (lb/ft·h)	2.419
	pound per foot-second (lb/ft·s)	6.7197×10^{-4}
	slug per foot-second (slug/ft·s)	2.0885×10^{-5}

To convert from	*To*	*Multiply by*
pascal-second	centipoise	1000.0
	pound per foot-hour	2.419×10^{3}
	pound per foot-second (lb/ft·s)	0.672
	slug per foot-second	20.8854×10^{-3}
pound per foot hour	pascal-second	4.134×10^{-4}
	pound per foot-second	2.778×10^{-4}
	centipoise	0.4134
pound per foot-second	pascal-second	1.488
	slug per foot-second	31.0809×10^{-3}
	centipoise	1.488×10^{3}
centistokes	square meters per second (m^2/s)	1.0×10^{-6}
	square feet per second (ft^2/s)	10.764×10^{-6}
	stokes	0.01
square feet per second	square meters per second (m^2/s)	9.2903×10^{-2}
	centistokes	9.2903×10^{4}
stokes	square meters per second	1.0×10^{-4}
rhe	1 per pascal-second (1/Pa·s)	10.0

Pressure-Stress
(Force Per Unit Area)

To convert from	*To*	*Multiply by*
pounds per square inch	kilopascals (kPa)	6.895
	[1]meters-head	0.703
	[2]mm of Hg	51.7151
	[1]feet of water	2.3067
	pounds per square foot (lb/ft^2)	144.0
	std. atmospheres	68.046×10^{-3}
pounds per square foot	kilopascals	0.048
	[1]meters-head	4.8826×10^{-3}
	[2]mm of Hg	0.359
	[1]feet of water	16.0189×10^{-3}
	pounds per square inch	6.9444×10^{-3}
	standard atmospheres	0.4725×10^{-3}

To convert from	*To*	*Multiply by*
short tons per square foot	kilopascals	95.7605
	pounds per square inch (lb/in^2)	13.889
[1]meters head	kilopascals	9.806
	[2]mm of Hg	73.554
	[1]feet of water	3.281
	pounds per square inch	1.422
	pounds per square foot	204.81
[1]feet of water	kilopascals	2.989
	[1]meters-head	0.3048
	[2]mm of Hg	22.4193
	[2]inches of Hg	0.883
	pounds per square inch	0.4335
	pounds per square foot	62.4261
kilopascals	newtons per square meter (N/m^2)	$1.0x10^3$
	[2]mm of Hg	7.500
	[1]meters-head	0.102
	[2]inches of Hg	0.2953
	pounds per square foot	20.8854
	pounds per square inch	0.145
	std. atmospheres	$9.8692x10^{-3}$
kilograms (f) per square meter	kilopascals	$9.807x10^{-3}$
	[2]mm of Hg	$73.556x10^{-3}$
	pounds per square inch	$1.4223x10^{-3}$
millibars (mbar)	kilopascals	0.10
bars	kilopascals	100.0
std. atmospheres	kilopascals	101.325
	[2]mm of Hg	760.0
	pounds per square inch	14.70
	[1]feet of water	33.90

[1] Column of water (H_2O) measured at 4^oC.

[2] Column of mercury (Hg) measured at 0^oC.

Density and Mass Capacity
(Mass Per Unit Volume)

To convert from	*To*	*Multiply* by
pounds per cubic foot	kilogram per cubic meter (kg/m^2)	16.0185
	slugs per cubic foot ($slug/ft^3$)	0.031
	pounds per gallon (lb/gal)	0.1337
pounds per gallon	kilograms per cubic meter (kg/m^3)	119.8264
	slugs per cubic foot ($slug/ft^3$)	0.2325
pounds per cubic yard	kilograms per cubic meter (kg/m^3)	0.593
	pounds per cubic foot (lb/ft^3)	0.037
grams per cubic centimeter	kilograms per cubic meter	1.0×10^3
	pounds per cubic yard	1.6856×10^3
ounces per gallon (oz/gal)	grams per liter (g/l)	7.489
	kilograms per cubic meter	7.489
kilograms per cubic meter	grams per cubic centimeter (g/cm^3)	1.0×10^{-3}
	metric tons per cubic meter (t/m^3)	1.0×10^{-3}
	pounds per cubic foot (lb/ft^3)	62.4297×10^{-3}
	pounds per gallon	8.3454×10^{-3}
	pounds per cubic yard	1.6856
long tons per cubic yard	kilograms per cubic meter	1328.939
ounces per cubic inch (oz/in^3)	kilograms per cubic meter	1729.994
slugs per cubic foot	kilograms per cubic meter	515.3788

[1]Hydraulic Conductivity
(Volume Per Cross Sectional Area Per Unit Time, or Length per Unit Time)

To convert from	*To*	*Multiply by*
cubic feet per square foot per day	cubic meters per square meter per day [$m^3/(m^2 \cdot d)$]	0.3048
	cubic feet per square foot per minute [$ft^3/(ft^2 \cdot min)$]	0.6944×10^{-3}
	liters per square meter per day ($l/m^2 \cdot d$)	304.8

To convert from	*To*	*Multiply by*
cubic feet per square foot per day	gallons per square foot per day [gal/(ft^2·d)]	7.4805
	cubic millimeters per square millimeter per day [mm^3/(mm^2·d)]	304.8
	cubic millimeters per square millimeter per hour [mm^3/(mm^2·h)]	25.4
	cubic inches per square inch per hour (in^3/(in^2·h))	0.5
gallons per square foot per day	cubic meters per square meter per day [m^3/(m^2·d)]	40.7458x10^{-3}
	liters per square meter per day (l/m^2·d)	40.7458
	cubic feet per square foot per day [ft^3/(ft^2·d)]	0.1337
meters per second (m/s)	centimeters per second (cm/s)	1.0x10^2
	meters per day (m/d)	8.64x10^4
	feet per second	3.281
	feet per day	2.834x10^5
centimeters per second (cm/s)	meters per second	1.0x10^{-2}
	meters per day	8.64x10^2
	centimeters per day	8.64x10^4
	feet per second	0.0328
	feet per day	2.834x10^3
feet per second	meters per second	0.3048
	centimeters per second	30.48
	meters per day	2.6335x10^4
	feet per day	8.64x10^4
feet per day	meters per second	3.528x10^{-6}
	feet per second	1.157x10^{-5}

[1]Many of these units can be dimensionally simplified. For example, m^3/(m^2·d) can also be written m/d.

[1]Transmissivity
(Volume Per Unit Area Per Unit Time, or Area per Unit Time)

To convert from	*To*	*Multiply by*
cubic feet per foot per day [$ft^3/(ft \cdot d)$]	cubic meters per meter per day ($m^3/(m \cdot d)$)	0.0929
	gallons per foot per day (gal/ft·d)	7.4805
	liters per meter per day (l/(m·d))	92.903
gallons per foot per day	cubic meters per meter per day [$m^3/(m \cdot d)$]	0.0124
	cubic feet per foot per day ($ft^3/ft*d$)	0.1337
square meters per second (m^2/s)	square centimeters per second	$1.0x10^4$
	square meters per day	$8.64x10^4$
	square feet per second	10.764
	square feet per day	$9.301x10^5$
square centimeters per second (cm^2/s)	square meters per second	$1.0x10^{-4}$
	square meters per day	8.64
	square centimeters per day	$8.64x10^4$
	square feet per second	$1.0764x10^{-3}$
	square feet per day	$9.30x10^1$
square feet per second (ft^2/s)	square meters per second	0.0929
	square centimeters per second	$9.29x10^2$
	square meters per day	$8.0266x10^3$
	square feet per day	$8.64x10^4$

[1]Many of these units can be dimensionally simplified. For example, $m^3/(m \cdot d)$ can also be written m^2/d.

Permeability

To convert from	*To*	*Multiply by*
darcys	square meters	$9.87x10^{-13}$
	square centimeters	$9.87x10^{-9}$
	square millimeters	$9.87x10^{-7}$
square meters	darcys	$1.013x10^{12}$
square centimeters	darcys	$1.013x10^8$

INDEX